Discrete-Time Dynamic Models

TOPICS IN CHEMICAL ENGINEERING
A Series of Textbooks and Monographs

SERIES EDITOR
Keith E. Gubbins, Cornell University

ASSOCIATE EDITORS
Mark A. Barteau, University of Delaware
Edward L. Cussler, University of Minnesota
Klavs F. Jensen, MIT
Douglas A. Lauffenburger, MIT

Manfred Morari, ETH
W. Harmon Ray, University of Wisconsin
William B. Russel, Princeton University

Receptors: Models for Binding, Trafficking, and Signalling   D. Lauffenburger and J. Linderman

Process Dynamics, Modeling, and Control   B. Ogunnaike and W. H. Ray

Microstructures in Elastic Media   N. Phan-Thien and S. Kim

Optical Rheometry of Complex Fluids   G. Fuller

Nonlinear and Mixed Integer Optimization: Fundamentals and Applications   C. A. Floudas

An Introduction to Theoretical and Computational Fluid Dynamics   C. Pozrikidis

Mathematical Methods in Chemical Engineering   A. Varma and M. Morbidelli

The Engineering of Chemical Reactions   L. D. Schmidt

Analysis of Transport Phenomena   W. M. Deen

The Structure and Rheology of Complex Fluids   R. Larson

Discrete-Time Dynamic Models   R. Pearson

Discrete-Time Dynamic Models

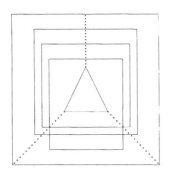

RONALD K. PEARSON

New York   Oxford
Oxford University Press
1999

Oxford University Press

Oxford New York
Athens Auckland Bangkok Bogotá Buenos Aires Calcutta
Cape Town Chennai Dar es Salaam Delhi Florence Hong Kong Istanbul
Karachi Kuala Lumpur Madrid Melbourne Mexico City Mumbai
Nairobi Paris São Paulo Singapore Taipei Tokyo Toronto Warsaw

and associated companies in
Berlin Ibadan

Copyright © 1999 by Oxford University Press, Inc.

Published by Oxford University Press, Inc.,
198 Madison Avenue, New York, New York 10016
http://www.oup-usa.org

Oxford is a registered trademark of Oxford University Press.

All rights reserved. No part of this publication may be reproduced,
stored in a retrieval system, or transmitted, in any form or by any means,
electronic, mechanical, photocopying, recording, or otherwise,
without the prior permission of Oxford University Press.

Library of Congress Cataloging-in-Publication Data
Pearson, Ronald K., 1952–
    Discrete-time dynamic models / Ronald K. Pearson.
        p. cm.—(Topics in chemical engineering)
    Includes bibliographical references and index.
    ISBN 0-19-512198-8
    1. Chemical process control—Mathematical models. 2. Chemical process control—
Computer simulation. I. Title. II. Series: Topics in chemical engineering (Oxford
University Press)
    TP155.75.P43 1999
    660'.2815—dc21   98-53008

9 8 7 6 5 4 3 2 1

Printed in the United States of America
on acid-free paper

Preface

This book is intended as a reference for anyone interested in the development of discrete-time dynamic models to approximate or predict the evolution of complex physical systems. Because I spent 17 years with the DuPont company, most of the examples discussed in this book have a strong process control flavor, but the ideas presented here certainly have much wider applicability, particularly in areas like nonlinear digital signal processing, model-based fault detection, or the newly emerging field of mechatronics, involving integrated microcomputer-based control of electromechanical systems. The general subject of model development is a large one, and this book makes no attempt to cover all of it. Instead, the principal focus here is on the first step in the model development process: that of model structure selection. In fact, the original motivation for this book was the observation that many papers appearing in the literature proceeded roughly as follows: "We investigate the utility of nonlinear model structure X in developing controllers for physical system P ... ." The question of why this *particular* model structure was chosen—or indeed, what other model structures might be reasonable alternatives—was too seldom addressed, yet these questions seemed in many ways the most difficult. The intent of this book is to provide some useful guidance in dealing with these questions.

More specifically, this book has two primary objectives: first, to describe some of the useful model structures that have been proposed in the process control, time-series analysis and digital signal processing literature and second, to describe some of the important qualitative behavior of these model structures to provide a rational basis for matching classes of potentially useful mathematical models with specific applications. I hope this book will be useful to researchers in various fields dealing with the problems of computer-based control of complicated physical systems that exhibit nonlinearities too significant to be ignored. Although the book is primarily intended as a reference for researchers, it should also be of use to industrial professionals like my former colleagues at DuPont, and it can be used either as a supplementary text for an advanced undergraduate course on modeling or as a text for a graduate special topics course. In fact, Frank Doyle has used some preliminary portions of this book in just such a setting at Purdue University and I have used the material from Chapters 1 through 6 of this book as the basis for a course at ETH, Zürich entitled *Discrete-Time Dynamic Models*; I have also used the material from Chapter 8 as the starting point for a subsequent course entitled *Nonlinear Model Identification*.

To the extent possible, I have attempted to make each of the eight chapters in this book reasonably self-contained with frequent cross-references to related topics in other chapters. Chapter 1 offers a broad introduction to the problems of modeling complex physical systems, with particular emphasis on the issues that must be addressed in developing *nonlinear dynamic models of moderate complexity*. The emphasis here is important since detailed mechanistic models can accurately describe process dynamics, but they are very often too complex for direct use in computer-based control strategies; this point is discussed in some detail in Chapter 1. In addition, this chapter also considers six specific forms of observable nonlinear dynamic behavior to provide some practical guidance in deciding whether a nonlinear dynamic model is necessary or not, and these ideas are illustrated with a number of specific examples from the literature. Chapter 2 presents a brief summary of some important results concerning linear dynamic models, with two objectives. First, since the term *nonlinear* defines an enormous class of systems by its lack of linearity, some understanding of what is meant by "linear behavior" is necessary to understand what is meant by "nonlinear behavior." Second, many ideas from the theory of linear dynamic models provide starting points for defining and understanding different classes of nonlinear model structures. Chapter 3 attempts to address the general question of what nonlinear dynamic models are, defining four different classes of these models. The first of these definitions is *structural*, meaning that explicit equations are given describing the input/output behavior of the model class. This approach to nonlinear systems is certainly the most popular one and it is adopted in most of the discussions in subsequent chapters of this book. However, it is also possible to define nonlinear models on the basis of their *qualitative behavior*, and the other three model classes discussed in Chapter 3 are behaviorally defined, both to illustrate the nature and mechanics of this approach and because it leads to some results that are useful in subsequent discussions.

Chapters 4, 5, and 6 consider the broad classes of NARMAX models, Volterra models, and linear multimodels, respectively. These model classes are all defined structurally and each one is the subject of significant research in the development of empirical model identification procedures and the design of model-based control strategies for complex physical systems. The NARMAX class discussed in Chapter 4 is enormously broad, including as special cases most of the other model classes considered in this book. To help in dealing with this breadth, Chapter 4 concentrates on some important model subclasses, including nonlinear moving average (NMAX) models, nonlinear autoregressive (NARX) models, structurally additive (NAARX) models, polynomial NARMAX models and rational NARMAX models. Distinctions between nonlinear autoregressive and nonlinear moving average model classes is particularly important since the differences in their qualitative behavior are enormous, with the NMAX class being both better-behaved and easier to analyze than NARMAX models that include nonlinear autoregressive terms. Conversely, it follows as a corollary that NMAX models are incapable of exhibiting certain types of qualitative behavior that may be important in some applications (e.g., output multiplicity or subharmonic generation in response to periodic inputs). The Volterra model class considered in Chapter 5 represents a highly structured subset of the NMAX class that is, because of this structure, particularly amenable to analysis. Conversely, the number of parameters required to specify a general Volterra model grows rapidly with increasing model order, and this observation motivates interest in further structural restrictions within the Volterra class (e.g., Hammerstein and Wiener models), a topic considered in some detail in Chapter 5. The class of linear multimodels is one of increasing popularity and is motivated by the observation that linear models often provide adequate approximations of a system's dynamics *over a sufficiently narrow operating range*; hence, the linear multimodel approach attempts to "piece together" a number of *local, linear models* to obtain a single *global, nonlinear model*. The details of how these local models are connected together appear to be more important than the local models themselves, a point that is illustrated in Chapter 6 (as Dan Rivera likes to say, "the devil is in the details"); in addition, the advantages of viewing these models as members of the NARMAX class are demonstrated in these discussions.

Chapter 7 explores the relationship between the different model classes discussed in earlier chapters, starting with simple (but not always obvious) relations based on exclusion and inclusion (e.g., Hammerstein and Lur'e models both belong to the larger class of structurally additive NARMAX models, but Wiener models do not). To provide a basis for more systematic comparisons between model classes—and the construction of new nonlinear model classes—Chapter 7 also introduces the basic notions of *category theory*. Essentially, category theory provides a very general framework for discussing relations between different collections of mathematical objects and the motivation for introducing this topic here is to provide a useful basis for extending some of our intuitions about the beahvior of linear dynamic models to the nonlinear case. Because it is very general, category theory is extremely powerful and permits the consideration of a very wide range of system-theoretic topics, including structure-behavior relations, linearization, discrete- vs. continuous-time model differences, equivalent realizations, and input sequence design for model identification, to name only a few. Conversely, because it is somewhat abstract, category theory is not to everyone's taste and for this reason, it is introduced near the end of the book and these ideas are not used in the preceeding discussions. Finally, Chapter 8 represents an extended summary, returning to the original motivating problem: the development of nonlinear, discrete-time dynamic models of moderate complexity to provide adequate approximations of the dynamics of complex physical systems. Topics treated in this chapter include model parameter estimation, data pretreatment (an extremely important issue in practice), input sequence design, and model validation; length restrictions do not permit complete treatments of these topics, but useful references are given for more detailed discussions.

Finally, it is important to acknowledge my debt to three influential institutions and more colleagues than I can reasonably name here. This book began as an annotated table of model structures that I developed during an extremely productive month in Chapel Hill, North Carolina at the UNC Statistics Department's Center for Stochastic Processes, early in 1995; the Center served as an excellent incubator for this book and I benefitted from my interactions with a number of people there, particularly Ross Leadbetter, Andre Russek, and Harry Hurd. Sadly, that month was also one of the last few in the life of Stamatis Cambanis, my sponsor there and one of the most enthusiastic people I have ever met. Following that visit, I returned to my regular job with the DuPont company, whose support and encouragement was invaluable, particularly that from Dave Smith and Jim Trainham. Many of the views on dynamic modeling presented in this book are consequences of my years with DuPont. One of the many advantages of being in Dave Smith's organization was the constant stream of visitors, of varying tenure from various places. Between visitors and DuPont colleagues, I benefitted substantially from discussions and collaborations with many people; a partial list in alphabetical order includes Yaman Arkun, Frank Doyle, Christos Georgakis, Tunde ("Eyeballs") Ogunnaike, Mirek Pawlak, Martin Pottmann, Jim Rawlings, Harmon Ray, Dan Rivera, and Dale Seborg. Ultimately, one of the most influential of these visitors proved to be Frank Allgöwer who remained with our group for a year before leaving to accept a position with the Institut für Automatik at ETH, Zürich; I subsequently joined Frank's group at ETH, where this book has finally come to completion. As at DuPont, I am indebted to many people here, but particularly to the *Heinzelmännchen*: Eric Bullinger, Rolf Findeisen, and Patrick Menold. In addition, special thanks to Bjarne Foss for his useful comments on Chapter 6 and to Oxford University Press for their patience, especially Bob Rogers and Cynthia Garver. In the end, until I can find a sufficiently creative way of blaming someone else for them, I must claim full credit for whatever errors may have crept, unbidden into the manuscript.

Contents

1 Motivations and Perspectives   **3**
   1.1 Modeling complex systems 4
       1.1.1 Fundamental models 7
       1.1.2 Assumptions, approximations, and simplifications 10
       1.1.3 Continuous- vs. discrete-time models 12
       1.1.4 Empirical models 15
       1.1.5 Gray-box models 18
       1.1.6 Indirect empirical modeling 20
   1.2 Inherently nonlinear behavior 21
       1.2.1 Harmonic generation 22
       1.2.2 Subharmonic generation 26
       1.2.3 Chaotic response to simple inputs 29
       1.2.4 Input-dependent stability 30
       1.2.5 Asymmetric responses to symmetric inputs 32
       1.2.6 Steady-state multiplicity 33
   1.3 Example 1: distillation columns 35
       1.3.1 Hammerstein models 36
       1.3.2 Wiener models 38
       1.3.3 A bilinear model 40
       1.3.4 A polynomial NARMAX model 41
   1.4 Example 2: chemical reactors 42
       1.4.1 Hammerstein and Wiener models 45
       1.4.2 Bilinear models 47
       1.4.3 Polynomial NARMAX models 48
       1.4.4 Linear multimodels 49
       1.4.5 The Uryson model 50
   1.5 Organization of this book 53

2 Linear Dynamic Models   **57**
   2.1 Four second-order linear models 58
   2.2 Realizations of linear models 60
       2.2.1 Autoregressive moving average models 61
       2.2.2 Moving average and autoregressive models 64
       2.2.3 State-space models 66

		2.2.4	Exponential and BIBO stability of linear models	66
	2.3	Characterization of linear models		68
	2.4	Infinite-dimensional linear models		72
		2.4.1	Continuous-time examples	72
		2.4.2	Discrete-time slow decay models	74
		2.4.3	Fractional Brownian motion	77
	2.5	Time-varying linear models		80
		2.5.1	First-order systems	81
		2.5.2	Periodically time-varying systems	85
	2.6	Summary: the nature of linearity		89

3 Four Views of Nonlinearity — 93

	3.1	Bilinear models		94
		3.1.1	Four examples and a general result	97
		3.1.2	Completely bilinear models	104
		3.1.3	Stability and steady-state behavior	108
	3.2	Homogeneous models		111
		3.2.1	Homogeneous functions and homogeneous models	111
		3.2.2	Homomorphic systems	115
		3.2.3	Homogeneous ARMAX models of order zero	117
	3.3	Positive homogeneous models		121
		3.3.1	Positive-homogeneous functions and models	121
		3.3.2	PH^0-ARMAX models	125
		3.3.3	TARMAX models	127
	3.4	Static-linear models		129
		3.4.1	Definition of the model class	129
		3.4.2	A static-linear Uryson model	132
		3.4.3	Bilinear models	133
		3.4.4	Mallows' nonlinear data smoothers	135
	3.5	Summary: the nature of nonlinearity		137

4 NARMAX Models — 140

	4.1	Classes of NARMAX models		140
	4.2	Nonlinear moving average models		143
		4.2.1	Two NMAX model examples	144
		4.2.2	Qualitative behavior of NMAX models	148
	4.3	Nonlinear autoregressive models		152
		4.3.1	A simple example	153
		4.3.2	Responses to periodic inputs	156
		4.3.3	NARX model stability	158
		4.3.4	Steady-state behavior of NARX models	161
		4.3.5	Differences between NARX and NARX* models	163
	4.4	Additive NARMAX Models		168
		4.4.1	Wiener vs. Hammerstein models	170
		4.4.2	EXPAR vs. modified EXPAR models	174
		4.4.3	Stability of additive models	177

		4.4.4 Steady-state behavior of NAARX models 179

- 4.5 Polynomial NARMAX models 180
  - 4.5.1 Robinson's AR-Volterra model 182
  - 4.5.2 A more general example 184
- 4.6 Rational NARMAX models 186
  - 4.6.1 Zhu and Billings model 189
  - 4.6.2 Another rational NARMAX example 192
- 4.7 More Complex NARMAX Models 195
  - 4.7.1 Projection-pursuit models 196
  - 4.7.2 Neural network models 198
  - 4.7.3 Cybenko's approximation result 201
  - 4.7.4 Radial basis functions 202
- 4.8 Summary: the nature of NARMAX models 203

5 Volterra Models 209

- 5.1 Definitions and basic results 210
  - 5.1.1 The class $V_{(N,M)}$ and related model classes 210
  - 5.1.2 Four simple examples 213
  - 5.1.3 Stochastic characterizations 216
- 5.2 Four important subsets of $V_{(N,M)}$ 218
  - 5.2.1 The class $H_{(N,M)}$ 219
  - 5.2.2 The class $U^r_{(N,M)}$ 219
  - 5.2.3 The class $W_{(N,M)}$ 221
  - 5.2.4 The class $P^r_{(N,M)}$ 222
- 5.3 Block-oriented nonlinear models 224
  - 5.3.1 Block-oriented model structures 224
  - 5.3.2 Equivalence with the class $V_{(N,M)}$ 226
- 5.4 Pruned Volterra models 227
  - 5.4.1 The class of pruned Volterra models 227
  - 5.4.2 An example: prunings of $V_{(2,2)}$ 228
  - 5.4.3 The PPOD model structure 230
- 5.5 Infinite-dimensional Volterra models 232
  - 5.5.1 Infinite-dimensional Hammerstein models 233
  - 5.5.2 Robinson's AR-Volterra model 234
- 5.6 Bilinear models 235
  - 5.6.1 Matching conditions 236
  - 5.6.2 The completely bilinear case 238
  - 5.6.3 A superdiagonal example 238
- 5.7 Summary: the nature of Volterra models 240

6 Linear Multimodels 243

- 6.1 A motivating example 244
- 6.2 Three classes of multimodels 248
  - 6.2.1 Johansen-Foss discrete-time models 248
  - 6.2.2 A modifed Johansen-Foss model 250
  - 6.2.3 Tong's TARSO model class 251

	6.3	Two important details 253

- 6.3 Two important details 253
  - 6.3.1 Local model selection criteria 253
  - 6.3.2 Affine vs. linear models 254
- 6.4 Input-selected multimodels 255
  - 6.4.1 Input-selected moving average models 255
  - 6.4.2 Steady states of input-selected models 256
  - 6.4.3 J-F vs. modified J-F models 256
  - 6.4.4 Two input-selected examples 259
- 6.5 Output-selected multimodels 263
  - 6.5.1 Output-selected autoregressive models 263
  - 6.5.2 Steady-state behavior 264
  - 6.5.3 Two output-selected examples 264
- 6.6 More general selection schemes 267
  - 6.6.1 Johanssen-Foss and modified models 268
  - 6.6.2 The isola model 271
  - 6.6.3 Wiener and Hammerstein models 274
- 6.7 TARMAX models 276
  - 6.7.1 The first-order model 277
  - 6.7.2 The multimodel representation 280
  - 6.7.3 Steady-state behavior 281
  - 6.7.4 Some representation results 284
  - 6.7.5 Positive systems 287
  - 6.7.6 PHADD models 287
- 6.8 Summary: the nature of multimodels 291

7 Relations between Model Classes 295
- 7.1 Inclusions and exclusions 296
  - 7.1.1 The basic inclusions 297
  - 7.1.2 Some important exclusions 300
- 7.2 Basic notions of category theory 302
  - 7.2.1 Definition of a category 303
  - 7.2.2 Classes vs. sets 304
  - 7.2.3 Some illuminating examples 306
  - 7.2.4 Some simple "non-examples" 309
- 7.3 The discrete-time dynamic model category 310
  - 7.3.1 Composition of morphisms 310
  - 7.3.2 The category **DTDM** and its objects 314
  - 7.3.3 The morphism sets in **DTDM** 316
- 7.4 Restricted model categories 318
  - 7.4.1 Subcategories 319
  - 7.4.2 Linear model categories 320
  - 7.4.3 Structural model categories 323
  - 7.4.4 Behavioral model categories 326
- 7.5 Empirical modeling and IO subcategories 329
  - 7.5.1 IO subcategories 329
  - 7.5.2 Example 1: the category $\mathbf{Aff}^{ss}$ 333

Contents xiii

		7.5.3	Example 2: the category **Gauss**	334
		7.5.4	Example 3: the category **Median**	335
	7.6	Structure-behavior relations		337
		7.6.1	Joint subcategories	337
		7.6.2	Linear model characterizations	338
		7.6.3	Volterra model characterizations	342
		7.6.4	Homomorphic system characterizations	344
	7.7	Functors, linearization and inversion		347
		7.7.1	Basic notion of a functor	347
		7.7.2	Linearization functors	349
		7.7.3	Inverse NARX models	351
	7.8	Isomorphic model categories		353
		7.8.1	Autoregressive vs. moving average models	353
		7.8.2	Discrete- vs. continuous-time	356
	7.9	Homomorphic systems		358
		7.9.1	Relation to linear models	358
		7.9.2	Relation to homogeneous models	359
		7.9.3	Constructing new model categories	360
	7.10	Summary: the utility of category theory		364

8 The Art of Model Development 369

	8.1	The model development process		370
		8.1.1	Parameter estimation	371
		8.1.2	Outliers, disturbances and data pretreatment	376
		8.1.3	Goodness-of-fit is not enough	385
	8.2	Case study 1—bilinear model identification		390
	8.3	Model structure selction		395
		8.3.1	Structural implications of behavior	396
		8.3.2	Behavioral implications of structure	399
	8.4	Input sequence design		404
		8.4.1	Effectiveness criteria and practical constraints	405
		8.4.2	Design parameters	409
		8.4.3	Random step sequences	414
		8.4.4	The sine-power sequences	417
	8.5	Case study 2—structure and inputs		423
		8.5.1	The Eaton-Rawlings reactor model	424
		8.5.2	Exact discretization	425
		8.5.3	Linear approximations	428
		8.5.4	Nonlinear approximations	432
	8.6	Summary: the nature of empirical modeling		445

Bibliography 451

Index 461

Discrete-Time Dynamic Models

Chapter 1

Motivations and Perspectives

This book deals with the relationship between the *qualitative behavior* and the *mathematical structure* of nonlinear, discrete-time dynamic models. The motivation for this treatment is the need for such models in computerized, model-based control of complex systems like industrial manufacturing processes or internal combustion engines. Historically, linear models have provided a solid foundation for control system design, but as control requirements become more stringent and operating ranges become wider, linear models eventually become inadequate. In such cases, nonlinear models are required, and the development of these models raises a number of important new issues. One of these issues is that of *model structure selection*, which manifests itself in different ways, depending on the approach taken to model development (this point is examined in some detail in Sec. 1.1). This choice is critically important since it implicitly defines the range of qualitative behavior the final model can exhibit, for better or worse. The primary objective of this book is to provide insights that will be helpful in making this model structure choice wisely.

One fundamental difficulty in making this choice is the notion of nonlinearity itself: the class of "nonlinear models" is defined precisely by the crucial quality they lack. Further, since much of our intuition comes from the study of linear dynamic models (heavily exploiting this crucial quality), it is not clear how to proceed in attempting to understand nonlinear dynamic phenomena. Because these phenomena are often counterintuitive, one possible approach is to follow the lead taken in mathematics books like *Counterexamples in Topology* (Steen and Seebach, 1978). These books present detailed discussions of counterintuitive examples, focusing on the existence and role of certain critical working assumptions that are required for the "expected results" to hold, but that are not satisfied in the example under consideration. As a specific illustration, the Central Limit Theorem in probability theory states, roughly, that "sums of N independent random variables tend toward Gaussian limits as N grows large."

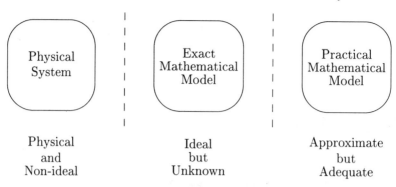

Figure 1.1: The system modeling problem

The book *Counterexamples in Probability* (Stoyanov, 1987) has an entire chapter (67 pages) entitled "Limit Theorems" devoted to achieving a more precise understanding of the Central Limit Theorem and closely related theorems, and to clarifying what these theorems do and *do not* say.

In a similar spirit, this book may be approached as "Counterexamples in Linear Intuition," illustrating some of the ways nonlinear dynamic models can violate this intuition. This book considers both the qualitative behavior of *linear* models, particularly "near the boundary," where even these models may behave somewhat unexpectedly, and the qualitative behavior of *different classes* of nonlinear models. The emphasis on different nonlinear model classes is important since the term *nonlinear* suggests a "homogeneity" of both structure and behavior that is seen in linear systems but absent from nonlinear ones. That is, it is important to realize that different classes of nonlinear models (e.g., the Hammerstein and Wiener models introduced in Sec. 1.2.1) may be as different in behavior from each other as they are from the class of linear models.

1.1 Modeling complex systems

The general modeling problem that motivated this book is illustrated graphically in Fig. 1.1. This figure shows three boxes, each representing a different description of a given physical system in the real world (e.g., a chemical manufacturing process, an ultracentrifuge, or sea-ice dynamics in the Sea of Okhotsk). The left-most of these boxes represents the physical system itself, subject to both controllable and uncontrollable influences from its surrounding environment. It is important to emphasize that this system and these environmental influences are both *physical* and *nonideal*. The middle box in Fig. 1.1 represents an *exact mathematical description* of this physical system. In practice, such a perfect mathematical model is not attainable for either the system or its environment, because all mathematical models of physical systems involve some degree of approximation. That is, in developing mathematical models, it is necessary to make assumptions and proceed, either because certain effects are believed to be negligible or because detailed descriptions are simply not available (both of these points are discussed in detail in Sec. 1.1.2). Thus, *practical mathematical models*, represented in the right-most box in Fig. 1.1, are necessarily approximate since they must be of manageable complexity and they can only be based on the information available to the model developer. Conversely, to be useful, these models must be *adequate* approximations, capturing the *important* dynamic phenomena exhibited by the physical system. Further, the concepts of "importance" and "reasonable complexity" are highly application-dependent. For example, the ultimate complexity limitations on dynamic models developed for improved process understanding are primarily computational; consequently, these models can be extremely complex, a point illustrated in Sec. 1.1.1.

The application area assumed here is computer control, for which practical complexity restrictions are much more severe. More specifically, with the rapid advances in computer technology over the past two or three decades, there is increasing interest in *model-based* control strategies that can take fuller advantage of these computational resources than do traditional control strategies. The complexity of these model-based control approaches depends directly on the structure and complexity of the process models on which they are based. In particular, practical application of these control approaches dictates that an "acceptable" process model must:

1. Describe process evolution from one discrete sample time to the next
2. Be simple enough to serve as a basis for controller design
3. Be sufficiently accurate to yield satisfactory controller performance.

The first of these conditions—the need for a *discrete-time* dynamic model of the process—arises directly from the nature of computer control, which is based on measurements made at discrete time instants t_k. Control actions are then computed from these measurements and translated into manipulated variable changes to be made at the next discrete time instant t_{k+1}. Since even single-loop "local" controllers (e.g., flow controllers) are often implemented digitally, this description is valid at all levels: single loops, more complex cascade and ratio configurations, and "plant-wide" control strategies that coordinate the overall operation of a complex process. Differences between these types of control problems become more apparent when the associated control strategies are considered: single-loop controllers are typically digital implementations of classical structures like the proportional-integral-derivative (PID) controller, whereas the question of how best to approach "plant-wide" control problems is the subject of much investigation, both within industry and academia. The greatest need for nonlinear, discrete-time dynamic models arises in intermediate-level control problems where control approaches like Model Predictive Control (MPC) are both applicable and potentially advantageous (Morari and Zafiriou, 1989; Ogunnaike and Ray, 1994; Meadows and Rawlings, 1997).

The second condition—a bound on model complexity—follows from the fact that the complexity of the resulting controller generally grows with the complexity of the model on which it is based. This conclusion is most easily seen in the case of the optimal linear regulator (Kwakernaak and Sivan, 1972), based on the linear state-space process models defined in Sec. 1.1.1. The solution of this problem is a linear state feedback map, whose dimension grows with the dimension of the underlying state space. Further, if the state variables are not all directly measurable—a common situation in practice—an observer must be designed to reconstruct the unmeasurable states, and the complexity of this observer is normally comparable to that of the original process model. Similarly, if a Kalman filter is required to provide state estimates from noisy measurement data, a linearized process model is incorporated explicitly in the filter equations. Finally, model predictive control applications lead to optimization problems based on predictions of the process response to possible future control inputs. In all of these cases, extremely complex process models lead to intractable control problems.

The third requirement listed above—that the model be sufficiently accurate to yield satisfactory controller performance—may be approached in several ways. One obvious and widely used measure of model fidelity is "goodness-of-fit," typically expressed as a mean square prediction error. A critical issue here, and one that is revisited many times throughout this book, is the choice of datasets on which *any* model assessment is based. In particular, note that different input/output datasets acquired from the same physical process can yield radically different assessments of the goodness-of-fit of any given process model. Another, more subtle, issue is that different model classes generally exhibit different types of *inherent* dynamic behavior. As a specific example, some physical systems can exhibit oscillatory responses to step changes, while others cannot. In the case of *linear*, continuous-time dynamic models, it is well known that at least a second-order model is required to exhibit oscillatory behavior. In the case of nonlinear models, the range of possible behavior is much wider and the relations between model structure and qualitative behavior are not nearly as well understood. Consequently, another important measure of "model fidelity" is that of *qualitative agreement* between the physical process and the approximate model, an idea that is also discussed further throughout the rest of this book. That is, if the process of interest does (or does not) exhibit some particular dynamic behavior (e.g., oscillatory or chaotic step responses), it may be reasonable to insist that a good model also does (or does not) exhibit this behavior. This point is important because these general qualitative tendencies may not be at all apparent from a particular set of observed input/output responses.

The development of models satisfying these three adequacy criteria is a difficult task: a detailed summary of nonlinear model predictive control (NMPC) applications in industry concludes that nonlinear model development is "one of the three most significant obstacles to the practical application of NMPC" (Qin and Badgwell, 1998). Similar conclusions were also presented in a recent survey of model requirements for NMPC (Lee, 1998) and in an earlier survey covering a wider range of nonlinear process control strategies (Bequette, 1991). For complex physical systems, three basic approaches to model development are *fundamental modeling*, *empirical modeling*, and *gray-box modeling*. Fundamental models are developed from knowledge of fundamental physical principles and are sometimes called *white-box* or *glass-box* models because the internal details of these models have clear, direct physical interpretations. In contrast, empirical models are developed from the observed input/output behavior of the system, and they are often called *black-box* models because they generally lack any direct physical interpretation: the model is viewed as a black box that maps input sequences into output sequences. Intermediate between these two extremes are gray-box models, based on some combination of empirical data and fundamental knowledge; the term gray-box refers to the fact that this model is partially interpretable in physical terms, lying somewhere between the extremes of black-box and white-box models.

1.1.1 Fundamental models

Fundamental models describe the balances of force, mass, momentum, energy, electric charge, or other important constituents of a real physical system. Generally speaking, the equations that describe these phenomena are ordinary differential equations, partial differential equations, integral equations, or integro-differential equations, coupled through various constitutive relations and completed with the appropriate initial and boundary conditions (Bird et al., 1960; Melcher, 1981). Consequently, a reasonably complete first-principles description of a real-world system of realistic complexity (e.g., a manufacturing process, a diesel engine, or a plasma) is generally quite complex. In particular, note that the complexity of these models depends on the inherent complexity of the physical process (e.g., the number of physical model constituents, the mathematical complexity of the relations describing these constituents, the manner in which they interact, etc.). Usually, these models are simplified by exploiting geometric symmetries, neglecting certain spatial inhomogeneities (e.g., imperfect mixing in chemical reactors, fringing fields in electrical capacitors, etc.), or assuming simplified idealizations (e.g., gaseous components obey the ideal gas law).

The final result of this development is often a collection of coupled nonlinear ordinary differential equations that may be represented in *state-space* form

$$\dot{\mathbf{x}}(t) = \mathbf{f}(\mathbf{x}(t), \mathbf{u}(t)) \qquad \mathbf{y}(t) = \mathbf{g}(\mathbf{x}(t), \mathbf{u}(t)). \qquad (1.1)$$

Here, $\mathbf{u}(t)$ is a vector of m manipulated or input variables, $\mathbf{y}(t)$ is a vector of p response or output variables, and $\mathbf{x}(t)$ is a vector of n state variables. Given an initial state vector $\mathbf{x}(0)$, the essential idea is that Eq. (1.1) gives a complete description of the system's future evolution in response to any external input vector $\mathbf{u}(t)$. This interpretation motivates the terminology: $\mathbf{x}(t)$ completely represents the *state* of the system at time t. The vector functions $\mathbf{f}(\cdot, \cdot)$ and $\mathbf{g}(\cdot, \cdot)$ may be very general, although some conditions must be imposed to guarantee the existence and uniqueness of a solution $\mathbf{x}(t)$; as a specific example, one common assumption is Lipschitz continuity (Kreyszig, 1978). In the special case of *linear state-space models*, these functions are of the form

$$\mathbf{f}(\mathbf{x}(t), \mathbf{u}(t)) = \mathbf{A}\mathbf{x}(t) + \mathbf{B}\mathbf{u}(t) \qquad \mathbf{g}(\mathbf{x}(t), \mathbf{u}(t)) = \mathbf{C}\mathbf{x}(t) + \mathbf{D}\mathbf{u}(t). \qquad (1.2)$$

In practice, fundamental state-space models are complex because each physical component or phenomenon generally requires one or more state variables to describe its dynamics. For example, Eskinat et al. (1991) present a simple dynamic model for a 25-tray binary distillation column consisting of one mass balance equation for each tray, one for the bottom and reboiler, and another for the condenser and accumulator. The resulting model consists of 27 coupled nonlinear ordinary differential equations, representable as a nonlinear state-space model of dimension 27. More realistic models are more complex, including energy balances for each tray and more detailed descriptions of other effects. As a specific example, Lundstrom and Skogestad (1995) describe a 39-tray distillation column model consisting of 123 coupled nonlinear ordinary differential equations.

The situation is even more complex for *distributed parameter systems* whose fundamental descriptions include partial differential equations. To see this point, first consider the simple linear diffusion model (Arfkin, 1970; Bird et al., 1960; Morse and Feschbach, 1953)

$$\partial \psi / \partial t = \frac{\kappa}{\rho C_p} \nabla^2 \psi. \tag{1.3}$$

One physical situation in which this equation arises is heat conduction in a material slab of constant thermal conductivity κ, density ρ, and specific heat C_p. The solution $\psi(x,t)$ of this equation represents the temperature of the slab at location x and time t, expressible as the infinite series

$$\psi(x,t) = \sum_{n=1}^{\infty} \phi_n(x) \exp\left[-\left(\frac{\pi n a}{\ell}\right)^2 t\right], \tag{1.4}$$

where ℓ is the thickness of the slab, $a = \sqrt{\kappa/\rho C_p}$ and the functions $\phi_n(x)$ are the *normal modes* of this system, determined by the boundary conditions. Using this representation, it is possible to convert the partial differential equation (1.3) and its boundary conditions into an equivalent *infinite-dimensional* system of ordinary differential equations, and to exploit this representation in the design of controllers (Curtain and Pritchard, 1978).

Alternatively, it is also possible to approximate the exact solution $\psi(x,t)$ given in Eq. (1.4), truncating the infinite sum to a finite number of terms. It is important to note that there are situations in which this approximation is inherently inadequate, as the non-exponential decay examples discussed in Chapter 2 illustrate. Conversely, when it is adequate, this approximation reduces the original partial differential equation model to a finite collection of ordinary differential equations. Further, this basic idea can be extended from linear partial differential equations to nonlinear partial differential equations, typical of those encountered in fundamental models of complex physical systems. As a specific example, Michelsen and Foss (1996) apply this approach in their development of a fundamental model for a continuous Kamyr digester, used in the pulp and paper industry to remove lignin from wood chips. This unit operation is physically large (42 meters high) and the resulting model is a collection of coupled nonlinear partial differential equations and algebraic relations. Adopting a spatial discretization procedure, the authors ultimately obtain a system of 226 coupled nonlinear ordinary differential equations and a comparable collection of algebraic relations. Essentially, this approximation corresponds to the replacement of a distributed parameter tubular reactor with a finite collection of *continuous stirred tank reactors* (CSTR's), each of which may be described by a collection of ordinary differential equations. Physically, the CSTR model represents an idealized reaction vessel that is perfectly mixed so that there are no concentration gradients, temperature gradients, or other spatial inhomogeneities within the reaction mixture. The complexity of each CSTR model depends on its level of detail and the number of CSTR models depends on the spatial discretization required to achieve the desired predictive accuracy. Kayihan (1998) also describes CSTR approximations to continuous digester dynamics, partitioning the digester into three functionally distinct components (denoted the cook zone, the mcc zone, and the emcc zone) and develops separate spatial discretizations for each component. Each CSTR is described by 10 state variables (five solid densities, four liquor densities and one temperature), resulting in a model with $N = 10(N_{cook} + N_{mcc} + N_{emcc})$ state variables overall; this approach easily leads to models involving thousands or tens of thousands of state variables. A similar approach is described by Eek (1995) in approximating the population balance partial differential equation that appears in the continuous crystallizer model discussed in Sec. 1.1.2.

The examples described so far have all been concerned with the dynamics of industrial process equipment, but it is important to emphasize that the general character of fundamental models obtained for *any* complex physical system is similar in terms of both development and complexity. The following three examples illustrate this point. First, mass and energy balance equations for diesel engine dynamics are described by Guzzella and Amstutz (1998), who also present a fairly detailed discussion of some of the constitutive relations and working assumptions used in developing detailed models. Second, Basser and Grodzinsky (1994) describe a fundamental model for the transient behavior of an ultracentrifuge for biological applications, based on mass and momentum balances, together with Maxwell's equations to describe the electrostatic phenomena that are central to the dynamic behavior of the ultracentrifuge. Specifically, interactions between charged molecules significantly influence the behavior of the polyelectrolyte solutions, gels, colloids, and tissues present in the ultracentrifuge sample. Third, Yang and Honjo (1996) describe the dynamics of sea-ice in the Sea of Okhotsk in the northwestern Pacific Ocean, lying between Japan and Siberia. The Sea of Okhotsk is particularly interesting because it exhibits a subsurface layer between 100 and 150 meters thick that remains near the freezing point of sea water (-2^oC) all year, despite summertime sea surface temperatures as high as 15^oC. This model is discussed further in Sec. 1.2.2.

Detailed models consisting of hundreds or thousands of differential-algebraic equations (DAE's) are becoming increasingly common, a trend that can be expected to continue as commercial software support for the development, verification, and simulation of such models becomes more widely available. As a specific example, Gross et al. (1998) describe a fundamental model developed for an industrial heat-integrated distillation column, noting that the final result is a differential-algebraic system consisting of approximately 350 differential equations and 3500 algebraic relations. Further, they argue that such complexity is necessary in realistic fundamental models, noting that "sometimes even construction details of a particular equipment item can have a profound influence on process dynamics."

1.1.2 Assumptions, approximations, and simplifications

This last observation emphasizes the importance of working assumptions in the development of fundamental models. For example, Eek (1995) describes the development of a fundamental model of an industrial crystallizer. This model consists of a partial differential equation for the time evolution of the crystal size distribution, together with ordinary differential equations describing heat and mass balances and algebraic relations between various physical properties. In addition to the model equations, 17 working assumptions are also listed (Eek, 1995, Appendix C), including such details as constant (and specified) particle shape, the existence of a single solid phase, no aggregation of crystals or breakage of particles, and constant impeller speed. These working assumptions are imposed on the general crystallizer model formulation; six additional assumptions are imposed on the operation of the specific crystallizer under consideration (e.g., the crystallizer temperature is assumed constant). The key point here is that some such set of working assumptions is necessary in the development of *any* fundamental model. Further, these assumptions generally influence both the mathematical structure and the qualitative behavior of the final model, sometimes significantly.

As a specific illustration of this last point, the first of Eek's 17 assumptions is that the crystallizer volume is perfectly mixed. This assumption is a very standard one, inherent in the definition of the CSTR model used extensively in chemical reaction engineering. Zhang and Ray (1997) explicitly consider the influence of this assumption on the steady-state and dynamic behavior of three industrially important polymerizations reactors. To treat nonuniform mixing, a three-compartment CSTR model is developed, with compartment volumes and intercompartment flows chosen to describe a particular form of nonuniform mixing seen in practice. Their results demonstrate that for these specific systems, nonuniform mixing can result in changes in the steady-state reactor operating conditions, in the stability of those steady-states, and in the physical properties of the final product. Similar consequences can be expected when other commonly invoked working assumptions fail to hold. Conversely, it is important to emphasize that *some* simplifying assumptions are necessary in any fundamental model development since without them, no model development is possible.

In addition to these basic simplifying assumptions, other approximations are generally necessary in the development of fundamental models. That is, detailed fundamental descriptions of certain model components may not be available, either because these components are too complex in themselves to admit a useful first-principles description, or because they are not sufficiently well understood. In such cases, *phenomenological models* may be used to describe these components, typically obtained by analyzing empirical data. For example, one of the components of Eek's crystallizer model is the following relationship between the dynamic solution viscosity η, the crystallizer temperature T and the material solubility w

$$\log_{10} \eta = A_1 + A_2 w + A_3 w^2 + A_4 w^3 + A_5 T + A_6 T^2 + A_7 T^3, \qquad (1.5)$$

where the coefficients A_1 through A_7 are constants, given for a specified range of temperature and solubility. As another example, Guzzella and Amstutz (1998) describe many such component relationships that have been found to be useful in the development of practical models of diesel engine dynamics. One of these equations is

$$\tilde{p}_{me}(t) = e(\omega_e) p_{mf}(t) - p_{mr}(\omega_e), \qquad (1.6)$$

relating the engine's speed ω_e to its mean effective pressure $\tilde{p}_{me}(t)$. A simple expression is given for $p_{mf}(t)$, together with both a range of typical values for $e(\omega_e)$ and a simple quadratic approximation for it derived from passenger car data. A typical range of values for $p_{mr}(t)$ is also given, together with a more complicated phenomenological expression relating it to other internal variables in the engine model. It is noted that "for coarse simulations the two parameters $[e(\omega_e)$ and $p_{mr}(\omega_e)]$ can be kept constant." The key point here is that approximations like these are almost always required in the development of fundamental models; further, the mathematical complexity of these approximations generally grows as the required accuracy increases.

The general complexity of fundamental models has motivated research into a variety of model simplification approaches. One such approach is called *lumping* or *compartmentalization* in which a detailed model is replaced by a coarser approximation model that attempts to describe the behavior of *pseudo-components* on the basis of approximate fundamental principles. This approach is illustrated nicely for the catalytic cracking of gas oil (Coxson and Bischoff, 1986), where it is noted that even if a linearized approximation is reasonable, individual component balances for all of the reactive species involved would typically lead to a coupled system of several thousand equations. Instead, *lumped* component balances are developed in which each "lump" consists of many different but chemically similar species. An initial lumping may be made on the basis of a chemical classification (e.g., heavy paraffins, heavy napthenes, heavy aromatics, etc.). The paper by Coxson and Bischoff starts with the result of such a preliminary lumping of 10 such pseudo species and describes a systematic procedure for further reducing the complexity of this model, while still achieving reasonable predictive accuracy.

This basic approach can be extended to the development of compartmental models for nonlinear systems as well (Benallou et al., 1986), although its effectiveness can depend strongly on the details of its implementation. For example, Horton et al. (1991) apply this approach to a 30-tray binary distillation column to obtain a nonlinear state space model of dimension $N = 5$. The underlying idea behind this simplification is to replace mass balances for each individual tray with overall mass balances for *compartments*, each composed of adjacent groups of trays. Two advantages of this approach are first, that it forces the reduced-order model to exhibit the same steady-state behavior as the original full-order model and second, that the state variables in the reduced-order model have direct physical interpretations. To implement this idea, it is necessary to define the effective composition within each of these compartments, and two different definitions are possible. The authors show that one of these choices leads to a reduced-order model that approximates the dynamics of the column reasonably well, while the other definition does not. In particular, the unacceptable definition leads to *inverse response* in the reduced model that is not present in the full-order model. The nature of inverse response models and their control implications are discussed briefly in Chapter 2; here, it is enough to note that this particular mismatch in qualitative behavior between the full-order model and the reduced model is extremely undesirable for control applications.

1.1.3 Continuous- vs. discrete-time models

In the linear case, there is a close relationship between continuous-time models and discrete-time models of the same order. This point is discussed further in Pearson and Ogunnaike (1997), and specific techniques for converting from continuous-time to the equivalent discrete-time model are discussed in chapter 25 of Ogunnaike and Ray (1994). Briefly, it is possible to express the evolution of the state vector $\mathbf{x}(t)$ of the continuous-time linear model (1.2) as

$$\mathbf{x}(t) = e^{\mathbf{A}(t-s)}\mathbf{x}(s) + \int_s^t e^{\mathbf{A}(\tau-s)}\mathbf{Bu}(\tau)d\tau, \tag{1.7}$$

for any $t > s$ (Curtain and Pritchard, 1977). Defining $x_k = x(t_k)$ where $t_k = t_0 + kT$ for some fixed constants t_0 and T, it follows that the state vector evolves from time t_k to time t_{k+1} according to

$$\mathbf{x}_{k+1} = e^{\mathbf{A}T}\mathbf{x}_k + \int_0^T e^{\mathbf{A}\tau}\mathbf{Bu}(t_k+\tau)d\tau. \tag{1.8}$$

This result further simplifies if $u(t) = u_k$ for $t \in [t_k, t_{k+1})$, since Eq. (1.8) may then be written in the discrete-time linear state-space form

$$\mathbf{x}_{k+1} = \mathbf{F}\mathbf{x}_k + \mathbf{G}u_k. \tag{1.9}$$

Here, the matrices $\mathbf{F}$ and $\mathbf{G}$ are given by

$$\mathbf{F} = e^{\mathbf{A}T} \qquad \mathbf{G} = \int_0^T e^{\mathbf{A}\tau}\mathbf{B}d\tau. \tag{1.10}$$

Note that for a fixed sampling interval T and a piecewise constant input $\mathbf{u}(t)$, Eq. (1.9) is an exact representation of the original continuous-time linear state-space model (1.2).

For the general nonlinear state-space model defined by Eq. (1.1), an analogous discretization is possible *in principle*, although it generally does not lead to useful results in practice. Specifically, if the nonlinear functions appearing in this equation are sufficiently well-behaved for a well-defined, unique solution to exist for all t, this solution may be expressed as

$$\mathbf{x}(t) = \mathbf{s}(\mathbf{x}(0), \mathbf{u}[0, t)), \tag{1.11}$$

where $\mathbf{u}[0, t)$ represents the behavior of the input vector $\mathbf{u}(\tau)$ over the entire interval $\tau \in [0, t)$. Defining $t_k = t_0 + kT$ as before, if $\mathbf{u}(t) = \mathbf{u}_k$ for all $t \in [t_k, t_{k+1})$, the evolution of the state vector from $\mathbf{x}_k = \mathbf{x}(t_k)$ to $\mathbf{x}_{k+1} = \mathbf{x}(t_{k+1})$ is given by

$$\mathbf{x}_{k+1} = \mathbf{s}(\mathbf{x}_k, \mathbf{u}_k), \tag{1.12}$$

and the output vector $\mathbf{y}_k = \mathbf{y}(t_k)$ becomes

$$\mathbf{y}_k = \mathbf{g}(\mathbf{x}_k, \mathbf{u}_k). \tag{1.13}$$

This result establishes the existence of a discrete-time equivalent of the nonlinear state-space model (1.1) at the sample times $\{t_k\}$, provided that the input vector is constant between these times. As is often true of existence proofs, however, this result does not provide a practical means of constructing this solution. Specifically, note that the discrete-time representation given in Eq. (1.12) depends on the exact solution map $\mathbf{s}(\cdot, \cdot)$ of the continuous-time model. Such solution maps are almost never known in practice, especially for models of realistic complexity.

This point is illustrated clearly by the following example, which is discussed in detail in Chapter 8. Eaton and Rawlings (1990) consider a simple continuous-time reactor model which may be expressed in the following form:

$$\frac{dy}{dt} = -h[y^2 + 2\mu y - 2d\mu]. \tag{1.14}$$

Here, $y(t)$ represents the reactor concentration at time t, $\mu(t)$ is a scaled inlet flow rate, and h and d are constants. Owing to its simplicity, this equation can be integrated exactly and the result can be used to implement the discretization scheme described in Eqs. (1.11) through (1.13). The details of this procedure are described in Chapter 8, and the final result is the discrete-time model

$$y(k) = \frac{[1 - \tau(k-1)\mu(k-1)]y(k-1) + 2d\tau(k-1)\mu(k-1)}{1 + \tau(k-1)[y(k-1) + \mu(k-1)]}, \tag{1.15}$$

where the auxiliary variable $\tau(k-1)$ is defined as

$$\tau(k-1) = \frac{\tanh[hT\sqrt{\mu^2(k-1) + 2d\mu(k-1)}]}{\sqrt{\mu^2(k-1) + 2d\mu(k-1)}}. \tag{1.16}$$

This result clearly illustrates that the mathematical structure of the original continuous-time equation (1.14) is not even approximately preserved in the equivalent discrete-time model (1.15). In particular, note the significant increase in the structural complexity of Eqs. (1.15) and (1.16) relative to the original model (1.14); one of the key motivations of empirical modeling is the fact that *much simpler* discrete-time models can often exhibit the basic qualitative behavior of primary interest. This point is illustrated in the discussion of chaotic dynamics in Sec. 1.2.3.

Approximate discretization is also possible for continuous-time nonlinear dynamic models, although this task must be approached carefully. An extremely useful discussion of this topic is given by Kazantzis and Kravaris (1997) who develop approximate discretizations by truncated Taylor-Lie series expansions of the continuous- time nonlinear differential equations, again assuming that the inputs are constant between sampling times. This class of approximation includes the Euler discretization as a special case. There, derivatives appearing in the original continuous-time equations are replaced by approximating differences, so the structure of the original continuous-time equations is retained. The Euler discretization represents the most severely truncated Taylor-Lie series approximation; if more terms of the Taylor-Lie series are retained, the resulting discrete-time model exhibits a different (generally more complex) structure than the original continuous-time model. In addition, for sufficiently fast sampling rates, discretizations of nonlinear systems with relative order greater than one exhibit unstable zero dynamics (the nonlinear analog of nonminimum phase behavior), not present in the original model. This observation is closely related to the undesirable inverse response seen by Horton et al. (1991) in their compartmental model reduction results.

This observation is also closely related to the following simple example (Agarwal, 1992). For $\beta > 0$ and $\gamma > 0$, the Velhurst differential equation

$$\frac{dy}{dt} = \beta y - \gamma y^2, \qquad (1.17)$$

is used in modeling the growth of a population y, inhibited by the nonlinear decay term $-\gamma y^2$. Like the Eaton-Rawlings reactor model (1.14), this equation may be solved analytically and has the solution

$$y(t) = \frac{y(0)}{c + (1-c)e^{-\beta t}}, \quad c = \frac{\gamma y(0)}{\beta}. \qquad (1.18)$$

One possible discretization of Eq. (1.17) is (Agarwal, 1992, p. 120)

$$\frac{y(k) - y(k-1)}{T} = [\beta - \gamma y(k-1)]y(k). \qquad (1.19)$$

This difference equation may also be solved analytically, having the solution

$$y(k) = \frac{y(0)}{c + (1-c)(1-\beta T)^k}. \qquad (1.20)$$

Motivations and Perspectives

The qualitative behavior of this solution depends strongly on the discretization time T. If $0 < T < 1/\beta$, this solution monotonically approaches the same limit as the continuous-time solution $y(t)$, but if $T > 1/\beta$, the solution of the difference equation is oscillatory. In the numerical integration of differential equations, solutions of the resulting difference equations that are not close in character to the desired solution of the differential equation are called *phantom* or *ghost* solutions. The key point is that approximate discretizations may or may not preserve the qualitative character of the continuous-time model, particularly in the case of very large models like those discussed in Sec. 1.1.1.

1.1.4 Empirical models

Empirical models describe an approximate relationship between measured input and output variables. It has already been noted that fundamental models often use phenomenological models developed from empirical data as components. Alternatively, it is also possible to attempt to describe the dynamic behavior of the complete physical system directly from input/output data. This approach has one major advantage and two significant disadvantages, relative to fundamental model development. The advantage is that empirical modeling offers explicit, direct control over the structure of the resulting model. Thus, if this model structure is selected for compatibility with the intended application (e.g., a specific control system design methodology), utility is guaranteed. Conversely, one significant disadvantage of empirical models is that they generally lack the direct physical interpretation that characterizes fundamental models. This characteristic represents a limitation with respect to both the interpretation and the validation of the resulting empirical model. The second significant disadvantage of empirical models is that, because they are based exclusively on finite collections of observed data, predictions outside the range of these datasets may be highly unreliable.

The following example illustrates this point. Fig. 1.2 shows a plot of seven points $(k, y(k))$, corresponding to uniform samples of the exponential decay $y(t) = \exp(-t)$, taken at times $t_k = 0, 1, \ldots, 6$. These points are indicated with black triangles in the plot, and the open circles represent the predictions of a simple empirical model of the form

$$\hat{y}(k) = a_1 + b_1 t_k + c_1 t_k^2. \tag{1.21}$$

The coefficients for this model were obtained by forcing an exact agreement between the exponential decay $y(t_k)$ and the model predictions $\hat{y}(k)$ for $k = 0$, 1, and 2. Despite this exact agreement, it is clear that the extrapolation behavior of the polynomial prediction defined by Eq. (1.21) is disastrous. Examples of this sort are often cited to illustrate both the consequences of over-fitting (i.e., asking for too close agreement between observed data and the prediction model) and the poor extrapolation behavior of empirical models based on simple functions like polynomials.

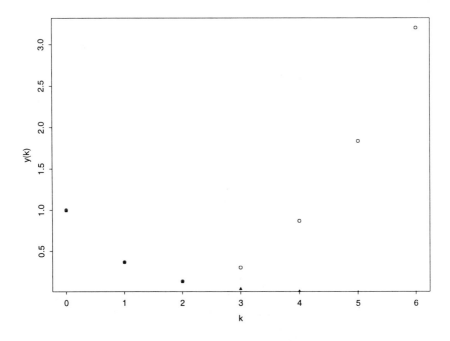

Figure 1.2: Exponential decay and first empirical approximation

Conversely, Fig. 1.3 shows a plot of the same seven data points, again indicated with black triangles, now fit to the alternative empirical model

$$\hat{y}(k) = \frac{1}{a_1 + b_1 t_k + c_1 t_k^2}. \tag{1.22}$$

The same exact agreement criterion is used as before, and the model predictions are again shown as open circles, but now the extrapolation behavior of the model appears to be quite good, in marked contrast to the previous example.

The point of this example is *not* to argue in favor of overfitting: model parameters should always be estimated from large enough datasets that the number of data points exceeds the number of empirical model parameters, preferably by a substantial factor. The key point is that, instead of viewing the first example as representative of empirical model behavior, it should be viewed as a natural consequence of poor model structure selection. Specifically, the exponential decay $y(t)$ asymptotically approaches zero as $t \to \infty$, and polynomial functions are inherently incapable of approaching any horizontal or vertical asymptote. Thus, in the first example there is an inherent qualitative mismatch between the behavior of the "true system" [i.e., the exponential decay $y(t)$] and the "model" [i.e., the quadratic approximation defined in Eq. (1.21)]. It should be clear on reflection that increasing either the size of the dataset used to estimate the model parameters or the order of the polynomial will not substantially alter

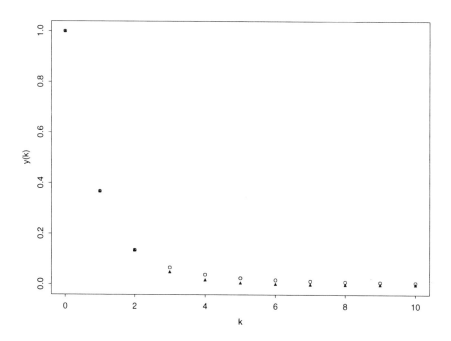

Figure 1.3: Exponential decay and second empirical approximation

these conclusions. In contrast, the second example involves a model of identical complexity (i.e., three parameters in both models), fit by exactly the same procedure. In this case, however, the model structure [i.e., the rational function defined in Eq. (1.22)] exhibits asymptotic behavior that is qualitatively consistent with that of the exponential decay. Consequently, the extrapolation behavior of this example is dramatically better than that of the first example. *To re-iterate: the primary point of this example is to emphasize the importance of careful model structure selection in empirical modeling.*

A more detailed discussion of the empirical modeling problem is given in Chapter 8, but it is useful to introduce the following points here. The class of linear ARMAX models (*autoregressive moving average models with exogenous inputs*) has been widely studied in the statistical time-series literature and is discussed further in Chapter 2. These models are of the form

$$\hat{y}(k) = \sum_{i=1}^{p} a_i \hat{y}(k-i) + \sum_{i=0}^{q} b_i u(k-i) + \sum_{i=1}^{r} c_i e(k-i) + e(k), \qquad (1.23)$$

where $\hat{y}(k)$ represents a prediction of the observed output $y(k)$ at time k, $u(k)$ is the observed input at time k, and $e(k) = \hat{y}(k) - y(k)$ is the model prediction error at time k. A detailed discussion of the mechanics of fitting ARMAX models to observed input/output data $(u(k), y(k))$ is given in the book by Box et al. (1994). In broad terms, this process involves the following four steps:

1. Select a model class $\mathcal{C}$ (here, the ARMAX class)
2. Determine the order parameters p, q, and r
3. Estimate the model parameters a_i, b_i, and c_i
4. Validate the resulting model.

More generally, the model class $\mathcal{C}$ may be taken as either an extension of the ARMAX class, like the class of ARIMA models discussed in Chapter 2 (Sec. 2.4.3), or some restricted class of ARMAX class like the ARX class, obtained by setting $q = 0$. Steps 2 and 3 of this procedure are similar in character (i.e., both involve parameter determination), but different enough in mechanics to warrant separate consideration. In particular, Step 3 is often based on a least squares fitting procedure applied to observed input/output data, whereas Step 2 often involves an evaluation of the change in these results as p, q, and r are varied. Finally, Step 4 typically involves either the search for systematic structure in the prediction error sequence $\{e(k)\}$, assessment of the model's predictive capability with respect to some *other* input/output dataset that was not used in the model building process, or assessment in terms of other "reasonableness" criteria (e.g., stability).

This book is primarily concerned with nonlinear extensions of the ARMAX model class, defined by

$$\hat{y}(k) = F(\hat{y}(k-1), \ldots, \hat{y}(k-p), u(k), \ldots, u(k-q),$$
$$e(k-1), \ldots, e(k-r)) + e(k). \quad (1.24)$$

This model class will be designated NARMAX (Nonlinear ARMAX) models, following the terminology of Billings and co-workers (Billings and Voon, 1983, 1986a,b; Zhu and Billings, 1993). Chapter 4 is devoted to a detailed discussion of this model class, emphasizing its breadth: most of the discrete-time dynamic models considered in this book may be represented as NARMAX models by the appropriate choice of the function $F(\cdot)$ and the order parameters p, q, and r. Further, it is important to note that this observation does not represent a restriction of scope, since most of these model structures were not originally proposed as NARMAX models but arose naturally from other considerations, generally either application-specific or estimation-related. The same basic steps are involved in obtaining NARMAX models as those listed for ARMAX model development, although the implementation of this general procedure may differ substantially in detail. In particular, one important difference is the wider range of possible subsets of the NARMAX class available in Step 1 of this procedure.

1.1.5 Gray-box models

Gray-box models attempt to overcome the limitations of fundamental models (structural complexity) and empirical models (sole reliance on empirical data) by combining both empirical data and fundamental knowledge. This idea is

Motivations and Perspectives

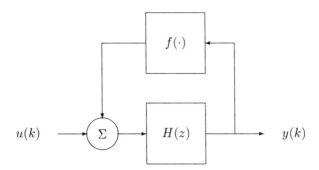

Figure 1.4: The Lur'e model structure

illustrated by Tulleken (1993), who considered the development of linear multivariable dynamic models for an industrial distillation process, comparing two approaches. First, he applied standard procedures to obtain empirical linear models from observed input/output data, but rejected almost all of them because certain aspects of their qualitative behavior were deemed unacceptable. In particular, the rejected models exhibited unphysical open-loop instabilities, steady-state gains of the wrong algebraic sign, or other qualitative behavior inconsistent with the physical system. Tulleken then considered a modified model identification procedure in which both stability and agreement with the known signs of the steady-state gains were imposed as explicit constraints. Although these modifications led to a somewhat more complex parameter estimation procedure, they also led to substantially better models.

Many different approaches to the gray-box modeling problem are possible, depending on the type of fundamental knowledge used, the way this knowledge is translated into modeling constraints, and the class of empirical models on which the approach is based. As a simple example of NARMAX model development, Pottmann and Pearson (1998) imposed exact steady-state agreement as a constraint in fitting empirical models to systems exhibiting output multiplicities. The motivation for this approach was two-fold: first, the *potential* steady-state behavior of even simple NARMAX models can be extremely complicated and second, fundamental steady-state knowledge is generally easier to develop than fundamental dynamic knowledge. These observations led to a consideration of the Lur'e model structure shown in Fig. 1.4, for which exact agreement with known steady-state behavior could be obtained easily; this model consists of a single static nonlinearity $f(\cdot)$ that appears as a feedback element around a linear dynamic model and is discussed further in Chapter 4. In the problem considered by Pottmann and Pearson, exact steady-state agreement determined the nonlinear function $f(\cdot)$ and imposed a constraint on the steady-state gain of the linear model. The remaining free parameters in this linear model were then chosen to best fit observed input/output data.

Other authors have also seen improved empirical modeling results when physically reasonable constraints are imposed explicitly. For example, Eskinat et al. (1993) also considered the use of steady-state gain constraints in linear model identification, along with other auxiliary information like estimates of cross-over frequency and amplitude estimates obtained with independent experiments. More generally, Johansen (1996) argues that purely empirical modeling, without incorporating prior knowledge like stability constraints, is an ill-posed problem in the following sense: first, it may fail to have a unique solution and second, any solutions found may not depend continuously on the available data.

1.1.6 Indirect empirical modeling

In practice, control systems configured from local, linear single-loop controllers (e.g., PID controllers) are quite popular. Simple linear models relating the variables on which these control systems are based are useful in configuring and tuning these control systems, and these models are frequently obtained by applying empirical modeling procedures to simulations of a detailed fundamental model. This general approach may be termed *indirect empirical modeling* to distinguish it from *direct empirical modeling* based on measured input/output data from the physical process. More specifically, indirect empirical modeling is an iterative process based on the following sequence of steps:

1. Development of a detailed fundamental model
2. Selection of an empirical model structure
3. Design of an excitation sequence $\{\mathbf{u}(k)\}$
4. Estimation of model parameters
5. Validation of the resulting model
6. Iteration, as required, to obtain a satisfactory final model.

This procedure inherits both advantages and disadvantages from the fundamental and empirical modeling approaches it combines. The primary advantages arise in steps 3, 5, and 6. Specifically, direct empirical modeling also requires the design of an input sequence (step 3) to be used in exciting the physical process, but practical operating constraints often impose significant restrictions on this sequence, a point discussed further in Chapter 8. Indirect empirical modeling permits greater flexibility in the choice of input sequences because the generation of the response data does not disrupt normal system operation, whereas direct empirical modeling often does. Similarly, more thorough model validation (step 5) is possible with indirect empirical modeling, including comparative evaluation of open-loop model responses and closed-loop responses under various control strategies, under different operating conditions, in different original system configurations, and in response to various simulated external disturbances. Finally, indirect empirical modeling facilitates the iterative validation and refinement of the empirical model (step 6). For example, a cycle of 50 ten-minute simulation runs to evaluate different input sequences, operating conditions, model structures, or control system configurations would not be unreasonable in most

Motivations and Perspectives 21

cases, but a comparable cycle of two-day plant tests every week for a year in a manufacturing facility would be difficult to imagine.

In addition, indirect empirical modeling has two other practical advantages. The first is that it permits full advantage to be taken of commercial developments in large-scale simulation programs (e.g., chemical process simulators, electromagnetic finite-element codes, etc.). The second advantage is that indirect empirical modeling does not suffer from the presence of unexpected external disturbances, measurement noise, outliers, or missing data that routinely arise in direct empirical modeling. Both of these advantages are noted in Mandler (1998), where the indirect modeling approach is illustrated for a variety of industrial gas separation and liquefaction processes. Conversely, indirect empirical modeling does suffer from three significant disadvantages. First, this approach directly inherits any limitations of the fundamental model on which it is based. For example, if perfect mixing is assumed in a reactor model but the physical reactor is not well-mixed, both the fundamental model and all simulation-based empirical approximations of it will suffer the consequences of this inappropriate modeling assumption. In contrast, direct empirical modeling can sometimes provide clues to the existence of phenomena like mixing imperfections. Second, like direct empirical models, the results of the indirect empirical modeling procedure also generally lack any physical interpretation. Consequently, the problems of interpretation and validation of the resulting models are comparable to those obtained by direct empirical modeling. Third, indirect empirical modeling requires the same initial model structure and input sequence choices as direct empirical modeling, and the resulting model also inherits the consequences of this choice. Alternatively, because input/output datasets are generally much easier to generate from a fundamental process model than from a physical process, indirect empirical modeling does provide greater flexibility in exploring different choices of model structure and input sequence, offsetting this last disadvantage somewhat.

1.2 Inherently nonlinear behavior

Linear dynamic models admit much analysis and have provided a practical basis for the development of many different control system design methods. Further, the use of linear models is widely accepted by practitioners in many different fields. In view of these advantages for linear models, it is reasonable to ask "when is a nonlinear model really necessary?" One answer is that there are certain types of qualitative behavior that linear models cannot exhibit. In cases where these types of qualitative behavior are important, nonlinear models are required to describe them.

The following sections introduce six different types of inherently nonlinear qualitative behavior and some of the physical situations in which they arise. It is important to note that both continuous-time and discrete-time models can exhibit all of these types of qualitative behavior. Subsequent chapters present detailed discussions of the relationship between these phenomena and different discrete-time model structure choices (e.g., nonlinear moving-average models can exhibit superharmonic generation but not subharmonic generation). In addition, the list of six inherently nonlinear phenomena described here will be expanded as additional types of qualitative behavior are introduced in subsequent chapters. Here, the main point is that all six of the following phenomena represent observable input/output behavior that is indicative of the need for a nonlinear dynamic model.

1.2.1 Harmonic generation

Oscillatory phenomena arise frequently in both electrical and mechanical engineering applications and have been widely studied there (Nayfeh and Mook, 1979; Scott, 1970). Such phenomena arise less commonly in chemical engineering applications, but there has been a great deal of interest in certain oscillatory oxidation-reduction reactions in recent years (Hudson and Mankin, 1981), and oscillatory phenomena also arise in some electrochemical systems (Krischner et al., 1993). In addition, the world abounds in uncontrollable disturbances that are at least approximately periodic (e.g., diurnal temperature variations, vibrations induced by rotating shafts, etc.). If these oscillations are strictly periodic, it is a standard result (Shilov, 1974) that they may be represented by a Fourier series. In particular, suppose $u(t)$ is periodic with period T, implying $u(t+T) = u(t)$ for all t; the Fourier series expansion of $u(t)$ is given by

$$u(t) = u_0 + \sum_{n=1}^{\infty} A_n \cos(2\pi n f t + \phi_n), \qquad (1.25)$$

where $f = 1/T$ is the *fundamental frequency* of the oscillations, corresponding to the lowest frequency sinusoidal component appearing in the Fourier series expansion. Analogously, if T is the smallest nonzero value for which the condition $u(t+T) = u(t)$ holds for all t, it is called the *fundamental period* of $u(t)$. The necessity for this terminology arises from the fact that if $u(t)$ is periodic with period T, it is also periodic with period mT for all integers m: the *fundamental period* T is unique, but "the period of oscillation" is not.

The simplest periodic oscillation is the sinusoid $u(t) = A\cos(2\pi f t + \phi)$. This model describes oscillations of amplitude A that are symmetric about zero, oscillating between $-A$ and $+A$; the phase angle ϕ specifies the value of $u(t)$ at the reference time $t = 0$ since $u(0) = A\cos\phi$. An important point discussed further in Chapter 2 is that a *linear, time-invariant* system may be completely characterized by its *frequency response* $H(f)$, provided this response is known for all f. Specifically, the response of such a linear system to a unit-amplitude sinusoidal input of frequency f is

$$y(t) = |H(f)|\cos[2\pi f t + \angle H(f)]. \qquad (1.26)$$

Here, $|H(f)|$ is the magnitude of the complex-valued frequency response at the frequency f and $\angle H(f)$ is the phase angle at the frequency f. It is useful to note

Motivations and Perspectives

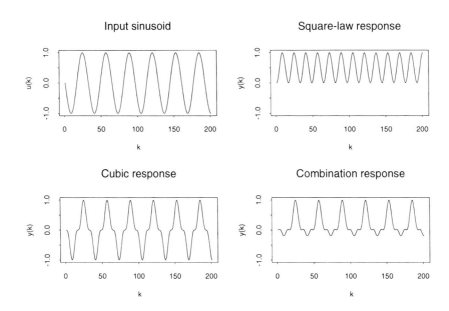

Figure 1.5: Examples of superharmonic generation

two features of this result, both of which are important characteristics of linear systems: first, the response of a linear, time-invariant system to a sinusoidal input is itself sinusoidal, and second, this response is of the same frequency f as the input sinusoid.

The response $y(t)$ of a nonlinear system to a sinusoidal input $u(t)$ can take one of the following three forms:

1. Superharmonic generation
2. Subharmonic generation
3. Nonperiodic responses.

The term *superharmonic generation* refers to the generation of higher harmonic terms $n > 1$ in the Fourier series expansion of $y(t)$. In this case, the response remains periodic with period T but is no longer sinusoidal in shape. This phenomenon is the simplest and most common of the three listed here and is therefore discussed first; subharmonic generation is discussed in Sec. 1.2.2 and nonperiodic responses are discussed in Sec. 1.2.3. Superharmonic generation is illustrated by the following three examples, summarized graphically in Fig. 1.5. The upper left plot shows a sinusoidal input signal $u(t)$ of unit amplitude and frequency f and the upper right plot shows the response of the *square-law nonlinearity* $y(t) = u^2(t)$ to this input. It follows from standard trigonometric identities that this response may be rewritten as

$$y(t) = \frac{A^2}{2} + \left(\frac{A^2}{2}\right)\cos(2\omega t + 2\phi), \qquad (1.27)$$

where $\omega = 2\pi f$ is the *angular frequency* of the sinusoidal input. Alternatively, note that this response may be viewed as a combination of a zero-frequency sinusoid (i.e., the constant term) and a *second harmonic* term at frequency $f' = 2f$. This frequency doubling is clearly visible when the upper two plots in Fig. 1.5 are compared.

The response of the cubic nonlinearity $y(t) = u^3(t)$ may also be obtained from standard trigonometric identities as

$$y(t) = \frac{3A^3}{4} \cos(\omega t + \phi) + \left(\frac{A^3}{4}\right) \cos(3\omega t + 3\phi). \tag{1.28}$$

This response is shown in the lower left plot in Fig. 1.5 and differs from the square-law response in three important respects. First, the cubic nonlinearity does not introduce any constant offset, a phenomenon known as *rectification* and associated with either even-symmetry or asymmetric nonlinearities. Second, the cubic nonlinearity response includes a term at the fundamental frequency f that is not present in the square-law response. This behavior is characteristic of odd-symmetry nonlinearities and implies that the fundamental period is not shortened, as it was in the case of the square-law nonlinearity. Third, the harmonic term introduced by the cubic nonlinearity occurs at $f' = 3f$, in contrast to the frequency doubling seen for the square-law nonlinearity.

Finally, the lower right plot in Fig. 1.5 shows the sinusoidal response of the combination nonlinearity

$$y(t) = 0.4u^2(t) + 0.6u^3(t). \tag{1.29}$$

This response includes all of the phenomena seen previously: the mean is nonzero (indicative of rectification), the fundamental period is unchanged, but the shape is highly non-sinusoidal (indicative of superharmonic generation). Further, the asymmetry of this response reflects the asymmetry of the nonlinearity defined in Eq. (1.29).

All three of the nonlinearities considered in this discussion belong to the general family of *static nonlinearities* or *memoryless nonlinearities*, defined by a nonlinear relationship $y(t) = N[u(t)]$ that depends on the argument $u(t)$ at time t alone. That is, the response of a static nonlinearity to a time-varying input signal $u(t)$ is instantaneous and does not depend on input values $u(t - \tau)$ at other times, on the rate of change of the input at time t, or other dynamic characterizations. For example, any real-valued function $f : R^1 \to R^1$ defines a static nonlinearity, while dynamic transformations like $N[u(t)] = u(t)du(t)/dt$ do not. An important characteristic of static nonlinearities is that they preserve periodicity, i.e.

$$u(t + T) = u(t) \Rightarrow y(t + T) = N[u(t + T)] = N[u(t)] = y(t). \tag{1.30}$$

Thus, it follows as a corollary that static nonlinearities are incapable of subharmonic generation. In general, as the preceding examples have demonstrated, static nonlinearities will cause superharmonic generation, often simply called

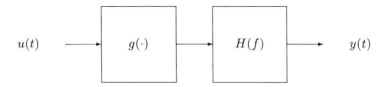

Figure 1.6: Structure of the Hammerstein model

harmonic generation because it is much more commonly observed in practice than subharmonic generation. In contrast, it follows from Eq. (1.26) that linear systems are incapable of either (super)harmonic generation or subharmonic generation.

A static nonlinearity $N[\cdot]$ is called *even* if $N[-x] = N[x]$ and it is called *odd* if $N[-x] = -N[x]$; a static nonlinearity that is neither even nor odd is called *asymmetric*. One advantage of this distinction is that if $N[\cdot]$ is even, its response to a sinusoidal input of frequency f may be written as

$$y(t) = y_0 + \sum_{n=1}^{\infty} A_{2n} \cos(2n\omega t + \phi_{2n}), \qquad (1.31)$$

consisting of a constant term and even harmonics at frequencies $2nf$. Similarly, if $N[\cdot]$ is odd, its sinusoidal response may be written as

$$y(t) = A_1 \cos(2\pi f t + \phi_1) + \sum_{n=1}^{\infty} A_{2n+1} \cos([2n+1]\omega t + \phi_{2n+1}), \qquad (1.32)$$

consisting of the fundamental and odd harmonics at frequencies $[2n+1]f$. Note that even nonlinearities exhibit rectification and cause a shortening of the fundamental period, since the lowest frequency oscillatory component appearing in Eq. (1.31) is of frequency $2f$. Conversely, odd nonlinearities generally preserve the period of the input sequence, since the lowest frequency component appearing in Eq. (1.32) is of frequency f. The most general static nonlinearities are asymmetric and may be decomposed uniquely into even and odd components: $N(x) = N_e(x) + N_o(x)$ where $N_e(x)$ is even and $N_o(x)$ is odd.

The results just presented for linear dynamic models and static nonlinearities may be used to obtain detailed expressions for the responses of *block-oriented nonlinear models* to sinusoidal excitations. These models consist of the interconnection of static nonlinearities and linear dynamic elements, and they arise naturally in describing certain types of physical systems, a point illustrated in Sec. 1.3. Probably the best known block-oriented nonlinear model is the *Hammerstein model* shown in Fig. 1.6 and consisting of the cascade connection of a static nonlinearity $g(\cdot)$ and a linear dynamic model, specified here by its transfer function $H(f)$. Hammerstein models are appealing because they combine two familiar process characteristics: a nonlinear steady-state gain curve $g(\cdot)$, and linear dynamics $H(f)$. In particular, note that if the steady-state gain of the linear system is $H(0) = 1$, then the steady-state response y_s of the Hammerstein

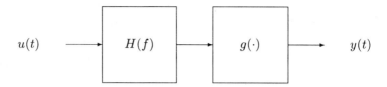

Figure 1.7: Structure of the Wiener model

model is simply $y_s = g(u_s)$, where u_s is the steady-state input value. Because of its relative simplicity and straightforward interpretation, the Hammerstein model is quite popular in both the engineering and the biomedical literature.

Like the Hammerstein model, the *Wiener model* also consists of the cascade connection of a static nonlinearity and a linear dynamic model, but in the opposite order. This structure is shown in Fig. 1.7, where $H(f)$ again represents the transfer function of the linear model, and $g(\cdot)$ is the static nonlinearity. Because of the close structural relationship that exists between these two model classes, it will be convenient to refer to Hammerstein and Wiener models constructed from the same components as *dual* models. Specifically, if $\mathcal{H}$ is the Hammerstein model represented in Fig. 1.6, then the corresponding dual Wiener model $\mathcal{W}$ is that represented in Fig. 1.7, and conversely. In the cascade connection of *linear models*, this ordering does not matter: the result of either ordering is the same, more complex, linear model. Here, however, profound differences arise: Wiener models are *not* the same as Hammerstein models in their qualitative behavior. This point is important, since if $H(0) = 1$, the *steady-state behavior* of the Hammerstein model is identical to that of its dual Wiener model. The differences in these models therefore lie entirely in their *transient behavior* and these differences can be fairly dramatic, as illustrated in Sec. 1.3.

The general nature of these differences may be seen by comparing the sinusoidal responses for the Hammerstein model and its dual Wiener model, constructed from the square-law nonlinearity and a linear dynamic model with arbitrary frequency response $H(f)$. Combining Eqs. (1.26) and (1.27), it follows that the response of the Hammerstein model is

$$y_H(t) = \frac{A^2 H(0)}{2} + \left(\frac{A^2 |H(2f)|}{2}\right) \cos[4\pi f t + \angle H(2f)], \qquad (1.33)$$

and the response of the Wiener model is

$$y_W(t) = \frac{A^2 |H(f)|^2}{2} + \left(\frac{A^2 |H(f)|^2}{2}\right) \cos[4\pi f t + 2\angle H(f)]. \qquad (1.34)$$

Note that the magnitude of the constant term is different for these two models, as are both the magnitude and phase of the second harmonic term.

1.2.2 Subharmonic generation

The phenomenon of *subharmonic generation* refers to a lengthening of the fundamental period of oscillation by a nonlinear system. This phenomenon is both less common and more complicated than that of superharmonic generation. In particular, it follows from the results presented in Sec. 1.2.1 that subharmonic generation is possible only in systems based on *dynamic nonlinearities*. For example, because linear dynamic models can exhibit neither superharmonic nor subharmonic generation and static nonlinearities can exhibit only superharmonic generation, neither the Hammerstein models nor the Wiener models described in Sec. 1.2.1 can exhibit subharmonic generation. An example of a model that *can* exhibit subharmonic generation is the Duffing equation

$$\frac{d^y(t)}{dt^2} + \omega_0^2 y(t) + sy^3(t) = u(t), \quad (1.35)$$

driven by the sinusoidal input $u(t) = r\cos\omega t$. For $s = 0$, this equation reduces to the linear simple harmonic oscillator, exhibiting a sinusoidal response whose amplitude diverges as $\omega \to \omega_0$. When $s \neq 0$, the Duffing equation can exhibit both superharmonic and subharmonic responses, but the exact nature of these responses depend on the model parameters ω_0 and s, the amplitude r of the sinusoidal input $u(t)$, and its angular frequency ω.

Perhaps the simplest of these phenomena to understand is the generation of subharmonics of order 1/3, corresponding to a period lengthening by a factor of three. This result may be obtained by first *assuming* a subharmonic solution of the form

$$y(t) = A\cos\left(\frac{\omega t}{3} + \phi\right), \quad (1.36)$$

and then substituting this expression into Eq. (1.35). It follows from Eq. (1.28) that the nonlinear term $sy^3(t)$ in Eq. (1.35) yields two components: one at angular frequency $\omega/3$ and another at angular frequency ω. If the amplitude and frequency of the input are chosen correctly, the term at frequency ω will balance with the excitation $u(t)$ on the right-hand side of Eq. (1.35), and the term of frequency $\omega/3$ will balance the other two terms (i.e., the linear simple harmonic oscillator terms) on the left-hand side of the equation. This solution for the Duffing equation and the analogous solution for the Helmholtz equation obtained by replacing the cubic nonlinearity sy^3 with the quadratic nonlinearity sy^2 have been discussed in detail, along with perturbations of these solutions (Gravador et al., 1995). For any $s > 0$ and any excitation amplitude $r > 0$, a solution of the form (1.36) exists with phase $\phi = 0$ and amplitude

$$A = 2\left(\frac{r}{2s}\right)^{1/3}, \quad (1.37)$$

provided the excitation frequency ω satisfies the condition

$$\omega = 3\sqrt{\omega_0^2 + \frac{3sA^2}{2}}. \quad (1.38)$$

Note that this frequency condition depends on both the model parameters ω_0 and s, and the excitation amplitude r through Eq. (1.37). This example illustrates that subharmonic generation is a rather subtle phenomenon, requiring a delicate balance between the excitation input $u(t)$ and the system parameters, in contrast to superharmonic generation discussed in Sec. 1.2.1, which always occurs when static nonlinearities are present.

To illustrate that subharmonic generation is actually observed in practice, it is instructive to consider two specific physical systems where this phenomenon arises. The first example is that of the inverted pendulum, consisting of a curved steel spring with a mass attached to the top, driven by a periodic torque applied at the base. An experimental implementation of this system for undergraduate physics laboratory experiments has been described in detail (Duchesne et al., 1991). The dynamics are well approximated by the following second-order nonlinear ordinary differential equation

$$ML\ddot{\theta} + \gamma\dot{\theta} + k\theta - MgL\sin\theta = \tau_0 \cos\omega_d t. \qquad (1.39)$$

Here, M is the mass attached to the end of the spring, L is the length of the spring, $\theta(t)$ is the angle of inclination of the top of the spring, γ is a constant describing the damping inherent in the system, τ_0 is the amplitude of the applied sinusoidal torque, and ω_d is its frequency. Note that the nonlinearity of this equation enters through the last term on the left-hand side, which has the Taylor series expansion

$$\sin\theta = \theta - \frac{\theta^3}{3!} + \frac{\theta^5}{5!} - \frac{\theta^7}{7!} + \cdots \qquad (1.40)$$

For very small angular excursions θ corresponding to low-amplitude excitations (i.e., sufficiently small τ_0), well away from the natural resonant frequency ω_0, Eq. (1.39) should be well approximated by the linear damped harmonic oscillator. As the amplitude of the response increases, a more reasonable approximation requires the cubic term in Eq. (1.40), yielding the Duffing equation (1.35) as an approximate description of the inverted pendulum dynamics. In fact, a very strong subharmonic resonance of order 1/3 is observed in the experimental system (Duchesne et al., 1991), consistent with the behavior of the Duffing equation.

The second example of subharmonic generation in a physical system is considerably more complex. One of the examples discussed briefly in Sec. 1.1.1 is a fundamental model of sea-ice dynamics in the Sea of Okhotsk (Yang and Honjo, 1996). In addition to the existence of a subsurface layer that remains nearly frozen the entire year (called the Okhotsk Dichothermal Layer or ODTL), one of the interesting features of the Sea of Okhotsk is that the interannular variations in sea-ice coverage there are among the greatest observed anywhere in the world. The proposed fundamental model for these dynamics is driven by surface heat fluxes, precipitation, river runoff, evaporation, inflows from the Pacific Ocean and the Sea of Japan, and outflow to the Pacific Ocean. These driving terms are assumed periodic with a one year period, but it is found that both the temperature and the salinity of the predicted ODTL exhibit strong subharmonic resonances, with a five year period. This behavior is explained in terms of mixing restrictions caused by the presence of the ODTL, once again emphasizing the potential importance of non-ideal mixing in real physical systems. In addition, these subharmonic responses are reflected in the predicted sea-ice thickness, suggesting a possible mechanistic explanation for the large interannular variations seen in the Sea of Okhotsk.

1.2.3 Chaotic response to simple inputs

Subharmonic generation corresponds to a lengthening of the fundamental period T of a periodic input to some integer multiple nT for $n > 1$. It is also possible for nonlinear systems to exhibit *nonperiodic responses* to periodic inputs, corresponding to the limit of subharmonic behavior as $n \to \infty$. One of the best known examples of nonperiodic responses to periodic inputs is *chaos*, which may be loosely defined as "highly irregular behavior" exhibited by deterministic systems. In chapter 2 of their book, Guckenheimer and Holmes (1983) present reasonably detailed discussions of four simple models that exhibit chaotic behavior:

1. Van der Pol's equation, describing a nonlinear electronic circuit
2. Duffing's equation, describing nonlinear mechanical oscillations
3. The Lorenz equations, approximating nonlinear fluid convection
4. A discrete-time model of a ball bouncing on a sinusoidally oscillating table.

All four of these mathematical models are extremely simple, and the first three are ordinary differential equations involving low-order polynomial nonlinearities, but the dynamic behavior they can exhibit is extremely complex. A particularly simple example of a physical system that can exhibit chaotic behavior is the inverted pendulum discussed in Sec. 1.2.2. This point is illustrated in (Duchesne et al., 1991, Fig. 4), which shows the measured position of the end of the pendulum versus time in response to a 0.37 Hz sinusoidal excitation at the base. While neither Guckenheimer and Holmes nor Duchesne et al. give rigorous mathematical definitions of chaos, both discussions clearly illustrate the notion: the systems considered exhibit responses that appear erratic, strongly dependent on initial conditions, and deterministic but not predictable in the long term. A detailed survey of both the underlying mathematical notions of chaos in continuous-time systems and its applications to problems of chemical engineering interest is available (Doherty and Ottino, 1988). Specific examples considered include fluid flow under a variety of conditions and chaotic chemical reactions. Similar behavior has also been observed in industrial scale polymerization reactors (Teymour and Ray, 1992).

It is a standard result (Guckenheimer and Holmes, 1983) that a continuous time, nonlinear model of at least third order is required to exhibit chaotic responses to initial conditions. This observation may appear to be in conflict with the behavior of the second-order pendulum model, but in fact there is no conflict. Specifically, note that Eq. (1.39) requires a state-space of dimension 4 to represent both the pendulum dynamics and the sinusoidal input $u(t)$ as the response of an unforced system to a specified initial condition.

That is, the complete state-space model is of the form $\dot{\mathbf{x}}(t) = \mathbf{f}(\mathbf{x}(t))$ where $\mathbf{x}(t) = [\theta(t), \dot{\theta}(t), u(t), \dot{u}(t)]^T$, and

$$\mathbf{f}(\mathbf{x}) = \begin{bmatrix} x_2 \\ (k/ML^2)x_1 - (g/L)\sin x_1 + (\gamma/ML^2)x_2 + (1/ML^2)x_3 \\ x_4 \\ -\omega_d^2 x_3 \end{bmatrix}. \quad (1.41)$$

The initial condition $\mathbf{x}(0) = [0, 0, \tau_0, 0]^T$ corresponds to the pendulum being driven at frequency ω_d and amplitude τ_0.

In discrete-time, chaotic responses to simple input sequences can be observed in simpler models. For example, one of the classic examples of a chaotic discrete-time system is the *logistic equation* (Tong, 1990)

$$y(k) = ay(k-1)[1 - y(k-1)]. \quad (1.42)$$

Viewed as a dynamic model, this equation describes the response of a system to an initial condition $y(0) \in (0, 1)$ for all $k \geq 1$. The constant a lies in the range $0 < a \leq 4$, and the behavior of this model becomes more complex as a increases. This point may be seen in Fig. 1.8, which shows 100 samples of $y(k)$ for $y(0) = 1/7$ and $a = 1, 2, 3$, and 4. Note that the qualitative behavior of this model changes significantly as a increases. For $a = 1$, $y(k)$ decays monotonically to zero, but for $a = 2$, it increases monotonically to a steady-state value of $y_s = 1/2$. Further, for $a = 3$, $y(k)$ exhibits an oscillatory approach to the steady-state value $y_s = 2/3$, and for $a = 4$, the behavior of $y(k)$ is highly erratic and does not approach *any* steady-state value. This behavior—seen in the lower right-hand plot in Fig. 1.8—is chaos.

1.2.4 Input-dependent stability

Both the phenomena of subharmonic generation and chaotic responses to periodic excitations exhibit significant dependence on the amplitude and frequency of the driving input $u(t)$. Another closely related phenomenon is *input-dependent stability* in which bounded inputs sometimes yield bounded outputs and sometimes yield unbounded outputs. In contrast, if a system is linear, it is either stable or unstable, and this characterization holds for all inputs. Some (but not all) nonlinear systems can exhibit stable responses to certain inputs and unstable responses to others.

A specific physical example of this type of behavior is that of hydrocarbon oxidation in the 600 to 900 Kelvin temperature range (Gaffuri et al., 1997). These reactions are responsible for various phenomena in internal combustion engines (e.g., autoignition and engine knock) and have therefore been the subject of considerable experimental research. Gaffuri et al. develop a detailed fundamental model of the complex reaction kinetics involved in these oxidations and compare the results with several different experimental reactor configurations, obtaining reasonable agreement. One of these configurations is a static reactor, consisting of a pre-heated, evacuated vessel, into which the reaction mixture is

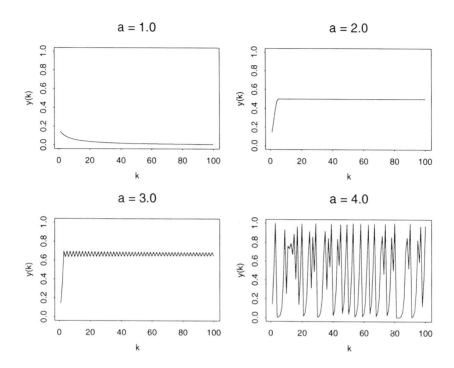

Figure 1.8: Four responses of the logistic equation

rapidly introduced at time $t = 0$. Thus, the initial temperature of the reactor may be regarded as a "thermal impulse" $u(t) = T_0 \delta(t)$, and the subsequent temperature evolution of the reaction mixture may be viewed as the observed response $y(t)$. The authors give a plot of the predicted temperature responses for propane oxidation at different initial temperatures (Gaffuri et al., 1997, Fig. 8). Three distinct types of response are seen in this plot, corresponding to slow combustion, cool flames, and ultimately *ignition*. This last phenomenon may be regarded as a form of physical instability, and the key point here is that it only occurs in response to input excitations of sufficiently large amplitude.

A number of physically significant, continuous-time dynamic model classes exist that exhibit this type of input-dependent stability behavior. One of the simplest is the class of continuous-time bilinear models, defined by state-space models of the form

$$\dot{\mathbf{x}}(t) = \mathbf{A}\mathbf{x}(t) + \mathbf{B}\mathbf{u}(t) + \sum_{i=1}^{m} u_i(t) \mathbf{N}_i \mathbf{x}(t). \tag{1.43}$$

These models have been developed for an extremely wide variety of applications, ranging from electrical networks to fluid flow to cardiovascular regulation. Further, many of the basic notions of linear control theory have been extended to this nonlinear model class (Mohler, 1991). One characteristic feature of the bilinear model class is that it exhibits input-dependent stability, analogous to the ignition phenomenon just described. To see this point, consider the response of Eq. (1.43) with $m = 1$ to a step input of amplitude α. It follows that the state-vector $\mathbf{x}(t)$ evolves according to the equivalent *linear* model

$$\dot{\mathbf{x}}(t) = [\mathbf{A} + \alpha \mathbf{N}_1]\mathbf{x}(t) + \mathbf{B}\mathbf{u}(t). \qquad (1.44)$$

Except in the linear case $\mathbf{N}_1 \equiv \mathbf{0}$, if this linear model is stable for $\alpha = 0$, it will generally become unstable for α of sufficiently large amplitude.

Similarly, in the development of fundamental models for chemical processes, it has been noted that the resulting state-space models often exhibit a *control-affine* form (Kantor, 1987). In the single-input, single-output case, this model structure is defined by the equation

$$\dot{\mathbf{x}}(t) = \phi[\mathbf{x}(t)] + \psi[\mathbf{x}(t)]u(t) \qquad y(t) = h[\mathbf{x}(t)], \qquad (1.45)$$

where $\phi[\cdot]$ and $\psi[\cdot]$ are functions mapping R^n into itself and $h : R^n \to R^1$. This utility of this model arises from the fact that the control input $u(t)$ is most commonly a flow rate and that many effects depend linearly on flow rates. As a specific example, note that the mass flow rate at which a chemical species A flows into or out of a well-mixed vessel is of the form $c_A u$, where c_A is the concentration of species A in the mixture and u is the volumetric flow rate of the mixture, directly determined by the setting of a flow control valve. While the restriction to control-affine models does exclude a wide range of possible state-space models, the subclass that remains can exhibit an extremely wide range of qualitative behavior (Pearson and Ogunnaike, 1997). Here, the key point is that one particular form of nonlinear behavior to be expected from this model class is that of input-dependent stability. The reasoning is analogous to that for the bilinear model (1.44): if the model is stable for $u(t) = 0$, it is likely to become unstable for step inputs of sufficiently large amplitude. Conversely, this situation may not arise if the control input $u(t)$ is restricted on physical grounds either to be non-negative or to lie within some bounded range of values.

1.2.5 Asymmetric responses to symmetric inputs

Besides subharmonic generation, chaotic responses to simple inputs, and input-dependent stability, many other forms of input-dependent qualitative behavior are also possible in nonlinear systems. One of the simplest of these input-dependent phenomena is the generation of asymmetric responses to symmetric input changes. Specifically, one of the defining characteristics of linearity is *homogeneity*: if a system S is linear, then $S[\lambda u(t)] = \lambda S[u(t)]$ for all real multipliers λ. This general characterization is discussed in some detail in subsequent chapters, but here it is enough to note that one special case for which it holds is $\lambda = -1$. Restricted to this special case alone, homogeneity reduces to the concept of odd symmetry introduced in Sec. 1.2.1 for static nonlinearities: $\mathcal{N}[-x] = -\mathcal{N}[x]$. Thus, it follows as a corollary that *failure to exhibit odd symmetry represents evidence of nonlinearity*.

Motivations and Perspectives 33

Eek (1995) uses this idea as an informal validity test for the linear crystallizer models he obtains from model reduction and linearization of his original fundamental model. Specifically, he presents graphical comparisons of three model responses. One is the response of a reduced, linearized fundamental model to a positive step input $u(t)$, which will be denoted $y_L(t)$ here. The second response, denoted $y_+(t)$, is the response of the original nonlinear fundamental model to this same step input. The third response is $-y_-(t)$ where $y_-(t)$ is the response of the nonlinear model to the negative step input $-u(t)$. A single plot with $y_L(t)$, $y_+(t)$ and $-y_-(t)$ gives a useful informal measure of the fundamental model's dynamic nonlinearity. In particular, if the crystallizer dynamics were exactly described by the reduced, linearized model, these three lines would coincide; the extent to which they differ gives a graphical indication of the degree of nonlinearity of the process dynamics. In fact, such plots can be extremely useful since they summarize three comparisons:

1. The difference between $y_L(t)$ and $y_+(t)$
2. The difference between $y_L(t)$ and $-y_-(t)$
3. The difference between $y_+(t)$ and $-y_-(t)$.

The first two of these comparisons may be viewed as nonlinearity measures, while the third gives a measure of the asymmetry of the nonlinearity.

Often, physical systems exhibit pronounced asymmetries when they are operating near some fundamental constraint. This point is illustrated in the high-purity distillation column example considered in Sec. 1.3: the magnitude of the manipulated variable change required to move the top concentration *toward* the 100% purity limit (e.g., from 99.5% to 99.6%) is much greater than that required to make the same change in the opposite direction (e.g., from 99.5% to 99.4%). Further, it is possible for some physical systems to exhibit even stronger *qualitative asymmetries* in response to symmetric input changes. As a specific example, the exothermic CSTR model considered in Sec. 1.4 exhibits a monotonic response to step *decreases* in the control input $u(t)$, but exhibits an oscillatory response to step *increases* in $u(t)$ of the same magnitude.

1.2.6 Steady-state multiplicity

In linear systems, steady-state input values u_s and output values y_s are related by a single steady-state gain: $y_s = K u_s$. Unless $K = 0$, this relationship is one-to-one: given the input u_s, the output y_s is uniquely determined, and vice versa. The case $K = 0$ is somewhat pathological (i.e., $y_s = 0$ for all u_s), but it can occur in nontrivial linear systems, a point discussed further in Chapter 2. This behavior is an extreme example of *input multiplicity*, one of two closely related notions:

- ▶ a system is said to exhibit *input multiplicity* if multiple steady-state input values u_s can correspond to a single steady-state output value y_s
- ▶ a system is said to exhibit *output multiplicity* if multiple steady-state output values y_s can correspond to a single steady-state input value u_s.

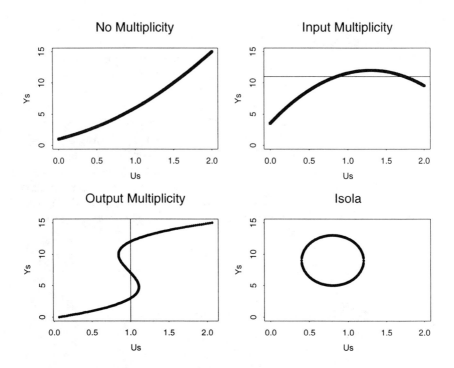

Figure 1.9: Four possible steady-state loci

Other than the pathological example ($K = 0$) just discussed, linear systems cannot exhibit either input or output multiplicities. In contrast, nonlinear systems can exhibit either, neither, or both of these types of behavior, as illustrated in Fig. 1.9. "Mildly nonlinear" models usually do not exhibit either multiplicity, having a nonlinear but monotonic steady-state locus like the curve shown in the upper left plot in Fig. 1.9 (note, however, that nonlinear dynamic models *can* exhibit *linear* steady-state behavior, a topic discussed in detail in Chapter 3). The relationship between monotonicity steady-state uniqueness is fundamental and is discussed further in Chapter 4, but for now it is enough to note that either a horizontal line or a vertical line drawn anywhere on the upper left graph in Fig. 1.9 will intersect the curve in precisely one point. Therefore, given any steady-state input value u_s in the range shown, the corresponding steady-state output value y_s is well-defined and unique, and vice versa.

The upper right plot in Fig. 1.9 illustrates the notion of input multiplicity defined above. Here, the curve of y_s versus u_s is not monotonic: y_s increases with increasing u_s for $0 \leq u_s \leq 1.3$, but then y_s *decreases* with increasing u_s for $u_s > 1.3$. Consequently, for any y_s less than the maximum value (here, $y_s = 12$ at $u_s = 1.3$), there are *two* possible values of u_s. This point is illustrated by the horizontal line $y_s = 11$, which is seen to intersect the steady-state locus at two distinct points. Conversely, note that for any steady-state output $y_s > 12$, there is *no* corresponding steady-state input u_s. This situation represents the most common form of steady-state multiplicity, occurring in systems that exhibit either "best case" or "worst case" operating conditions within a specified operating range. A simple illustrative example might be "little Albert's joy as a function of the quantity of candy eaten on his birthday": there is an optimum, somewhere between "Waah!! Gimme more!!" and "Waah!! I'm gonna be sick!!" As a more process-oriented example, consider a reactor producing some desired product B from raw material A, with reaction kinetics of the form $A \to B \to C$ where C is an undesirable byproduct. Produced in a continuously stirred tank reactor (CSTR), the concentration of product B exhibits a maximum as a function of the residence time of the reactor: if the residence time is too short, there is not enough time to form much of the desired product B, but if the residence time is too long, the product B has time to degrade to the undesirable product C.

The lower left plot in Fig. 1.9 illustrates the notion of output multiplicity. Here again, the steady-state response locus is not monotonic, but this time the locus is "S-shaped" rather than "N-shaped" as in the previous example. Viewed as a function of the steady-state output y_s, the steady-state input u_s exhibits monotonic behavior for $y_s \lesssim 5$ and for $y_s \gtrsim 10$, but for intermediate values of y_s, u_s exhibits both a local maximum and a local minimum. Between these two turning points, each steady-state input value u_s corresponds to three possible steady-state output values. This point is illustrated by the horizontal line $u_s = 1.0$, which intersects the steady-state locus at three distinct points: one at $y_s \sim 3$, another at $y_s \sim 7$, and a third one at $y_s \sim 12$. Generally, output multiplicity seems to be somewhat rarer than input multiplicity, but it has been observed in a variety of processes, including polymerization reactors (Hamer et al., 1981; Schmidt et al., 1984; Teymour and Ray, 1989, 1992), catalytic crackers (Arbel et al., 1995), and distillation columns (Bekiaris et al., 1993; Güttinger et al., 1997; Jacobsen and Skogestad, 1991).

Finally, the lower right plot in Fig. 1.9 illustrates the notion of an *isola* (Hlavacek and van Rompay, 1981), which is an "isolated" steady-state locus or, more generally, an isolated *branch* of a steady-state locus. That is, an isola represents a closed curve like the ellipse shown in the lower right plot in Fig. 1.9. Note that here, both input multiplicities and output multiplicities are present simultaneously. Consequently, although isolas do occur in a variety of physical systems (Russo and Becquette, 1995; Schmidt et al., 1984; Teymour and Ray, 1989, 1992), they tend to be rarer than either input multiplicities or output multiplicities alone.

1.3 Example 1: distillation columns

Separation processes in general and distillation processes in particular represent one of the most important unit operations in the chemical process industries. As with many other industrial process units, the degree to which a distillation column exhibits nonlinear dynamics depends strongly on its operating conditions [see the discussion on pp. 1140–1142 of Ogunnaike and Ray (1994) for an illustration of this point]. In particular, a *high-purity* distillation column attempts to separate one component almost completely from a mixture of two or more components; because the separation becomes more difficult as the desired product purity increases, the column dynamics become progressively more nonlinear. For example, the papers by Chien and Ogunnaike (1992) and Sriniwas et al. (1995) consider the control of a 27 tray methanol-ethanol column where the objective is to achieve 99% product purity at both the top and bottom of the column. These results are based on a first-principles model (Weischedel and McAvoy, 1987) consisting of 56 nonlinear ordinary differential equations coupled by a number of algebraic equations included to approximate energy balances, fluid dynamics, vapor-liquid equilibrium relations, and other important physical phenomena. The basic control strategy considered is the LV structure in which reflux and steam flows are used to control the two product compositions. Because direct composition measurements are more difficult than temperature measurements, the detailed control scheme infers composition from temperatures measured near the top and bottom of the column.

The authors consider ±1% and ±5% changes in the reflux and steam flow rates and give plots showing the response of both upper and lower tray temperatures to all four of these manipulated variable changes. A typical example is the change in the normalized upper tray temperature y_1 in response to a change in the reflux flow rate u_1: in dimensionless units, a 5% increase in u_1 causes an increase in y_1 of approximately 0.4, while a 5% decrease in u_1 causes a decrease in y_1 of approximately 0.1. In addition, both of these responses are significantly faster than the responses to a ±1% change in the reflux flow rate: 80% risetimes on the order of 10 minutes are observed for the ±5% flow rate changes, compared to the order of 50 minutes for the ±1% changes. Qualitatively similar results are seen in the other three model responses considered, and they give a clear indication of the nonlinearity of high-purity distillation column dynamics.

1.3.1 Hammerstein models

Because of its practical importance, many authors have considered the problem of distillation control and, as a consequence, many different nonlinear models have been investigated as empirical approximations of distillation column dynamics. For example, Eskinat et al. (1991) compare the performance of Hammerstein models and linear models for both a simulated high-purity distillation column and an experimental heat exchanger. The Hammerstein model structure is popular in part because it combines two elements with which practicing engineers are already familiar: a nonlinear steady-state gain curve and a linear dynamic model. In both applications, the authors find that the Hammerstein model gives a much better representation of the process dynamics than do linear models. In the case of the distillation column, a 25-tray binary column is considered, operating at a top product purity specification of 99.5%. The Hammerstein models considered consist of a polynomial $g(\cdot)$ of either third or fourth order, followed by first-order linear dynamics. These models are identified empirically from a simulation of the 27 ordinary differential equations

Motivations and Perspectives

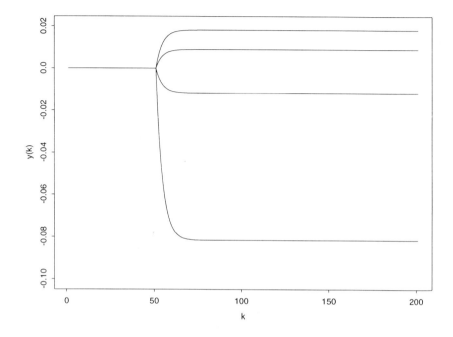

Figure 1.10: Hammerstein model step response

describing the individual tray mass balances, together with mass balances for the reboiler/bottom section and the accumulator/condenser section. As in the example discussed previously (Sriniwas et al., 1995), changes in reflux ratio to increase the purity of the top product above 99.5% have dramatically smaller gains than changes of the same magnitude in the opposite direction.

Two positive and two negative step responses are shown in Fig. 1.10 for the fourth-order Hammerstein model presented in Eskinat et al. (1991) The linear dynamics on which this Hammerstein model is based are described by the transfer function

$$H(z) = \frac{0.243z^{-1}}{1 - 0.757z^{-1}}, \qquad (1.46)$$

and the static nonlinearity is the fourth-order polynomial

$$g(x) = 1.04x - 14.11x^2 - 16.72x^3 + 562.75x^4. \qquad (1.47)$$

Note that the steady-state gain of the linear model is $H(1) = 1$, so the polynomial $g(\cdot)$ may be interpreted as the steady-state gain curve. The specific responses shown in Fig. 1.10 correspond to $\pm 1\%$ and $\pm 5\%$ changes in the magnitude of the input variable.

Based on their comparisons, Eskinat et al. (1991) conclude that this Hammerstein model gives a better description of column dynamics than the best linear model. This result is not surprising, given the dramatic differences in the effective gain for positive and negative step changes in the reflux flow rate. However, these authors observe that at higher product purities (e.g., 99.9% and above), the differences in gain are so pronounced that the best fourth-order Hammerstein models they were able to obtain still exhibit unacceptably poor performance. In addition, they also present an estimated steady-state gain curve for this distillation column, clearly illustrating the physically mandated "hard saturation" at 100% product purity. This curve illustrates one reason polynomial Hammerstein models are necessarily inadequate at sufficiently high product purities: finite-order polynomials cannot approach constant asymptotes. An alternative would be to consider nonpolynomial Hammerstein models (e.g., incorporating hyperbolic tangents, piecewise linear functions, or other choices) that could better represent this saturation phenomenon.

Hammerstein models are not entirely adequate for another reason, however, and the authors also note this point (Eskinat et al., 1991). Specifically, the *dynamic character*—that is, the effective time constant—of the response is also a function of the reflux flow rate. It is not difficult to show that the influence of the nonlinearity $g(\cdot)$, whatever its functional form, will be to transform a step of magnitude α into another step, of magnitude $g(\alpha)$. Although the *magnitude and sign* of the Hammerstein model step response can depend on the magnitude and sign of the input step, its *dynamic character* cannot. In particular, for the Hammerstein models considered by Eskinat, Johnson, and Luyben, the effective time constant for *any* step input is determined by the linear model coefficient $a_1 = -0.747$ that defines the single pole in $H(z)$. The "slower response to smaller steps" observed in high-purity columns is thus beyond the inherent capacity of the Hammerstein model dynamics.

Despite these particular limitations, the Hammerstein model is an interesting one and it is considered in more detail in Chapter 5. In particular, it turns out that Hammerstein models based on polynomial nonlinearities or other analytic functions may be represented as Volterra series with a particularly convenient structure. In addition, the Hammerstein model is closely related to both the Wiener model introduced in Sec. 1.2.1 and the Uryson model introduced in Sec. 1.4.5. Further, polynomial Hammerstein models may also be represented as polynomial NARMAX models; in particular, the NARMAX representation for the Hammerstein model defined in Eqs. (1.46) and (1.47) is

$$\begin{aligned} y(k) &= 0.757 y(k-1) + 0.253 u(k-1) - 3.429 u^2(k-1) \\ &\quad -4.063 u^3(k-1) + 136.75 u^4(k-1). \end{aligned} \qquad (1.48)$$

Overall, though it is somewhat limited in its flexibility, the behavior of the Hammerstein model often provides a useful baseline for better understanding the inherent behavior of these other model classes.

1.3.2   Wiener models

One increasingly popular alternative to the Hammerstein is the Wiener model. As noted in Sec. 1.2.1, this model structure consists of the same two components

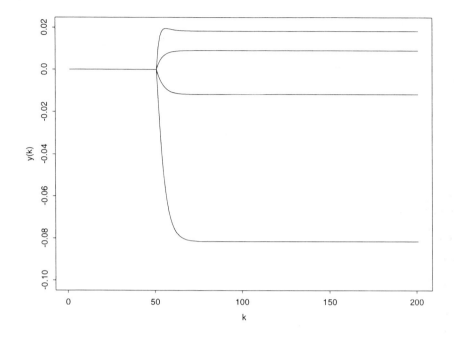

Figure 1.11: Wiener model step response

as the Hammerstein model, but connected in the opposite order, with the static nonlinearity *following* the linear dynamic subsystem. Further, it was also noted in Sec. 1.2.1 that if the steady-state gain of this linear subsystem is 1, both of these models exhibit the same steady-state behavior. This point is seen clearly in a comparison of the four Hammerstein model step responses shown in Fig. 1.10 with the four Wiener model step responses shown in Fig. 1.11. Specifically, Fig. 1.11 shows the $\pm 1\%$ and $\pm 5\%$ step responses for the Wiener model constructed from the first-order linear system defined in Eq. (1.46) and the fourth-order polynomial defined in Eq. (1.47). As noted, these step responses settle out to the same steady-state values as the corresponding Hammerstein step responses plotted in Fig. 1.10. The *dynamic character*, however, is quite different: note the slight "second order-like" overshoot observed in the $+5\%$ step response of the Wiener model.

Although Eskinat, Johnson, and Luyben do not consider this Wiener model in their study of distillation column dynamics, two points are worth noting. First, the performance of Wiener and Hammerstein models has been compared for a very similar distillation column model using a numerical suitability measure (Menold, 1996; Menold et al., 1997b). The results of this comparison suggest the Wiener model structure may be inherently better suited to approximating these dynamics than the Hammerstein model. The second point is that Wiener models have been used with some success *implicitly* in the control of high-purity distilla-

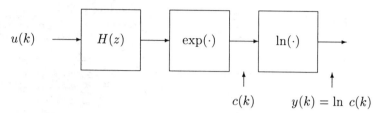

Figure 1.12: An implicit Wiener model

tion columns. In particular, one popular practice is to develop a linear dynamic model relating the manipulated variable of interest (e.g., steam or reflux flow rate) to either the *logarithm* of product composition $c(k)$ or some other, closely related transformation of $c(k)$ (Chien and Ogunnaike, 1992; Eskinat et al., 1991; Mejdell and Skogestad, 1991). The resulting model structure is shown in the block diagram in Fig. 1.12: the relationship between $u(k)$ and $y(k) = \log c(k)$ is described by a linear dynamic model, corresponding to the *implicit* construction of a Wiener model that relates $u(k)$ and $c(k)$. Note that the nonlinearity appearing in this model is the *inverse* of the transformation nonlinearity, since the Wiener model relates the variables $u(k)$ and $c(k) = \exp[y(k)]$. Constructing a linear model of "log composition" is therefore equivalent to constructing a Wiener model of composition based on an exponential nonlinearity.

This idea is closely related to the concept of *extensive variable control* (Georgakis, 1986). The basis for this concept is the observation that the available manipulated variables in industrial process control applications are often thermodynamic *extensive variables* like flow rates that scale with the total material volume. In contrast, measured variables are often thermodynamic *intensive variables* like temperatures that *do not* scale with the total material volume. It is sometimes possible, however, to find static nonlinear transformations mapping variables that are not thermodynamically extensive to those that are. Probably the simplest example is provided by the level in a storage tank of nonuniform cross-section: the relationship between inlet and outlet flow rates and liquid *level* is nonlinear, but the relationship between these flow rates and the total *volume* of liquid in the tank is linear. If the geometry of the tank is known, the total liquid volume may be computed from the liquid level by the appropriate nonlinear transformation. *The point is that control strategies of this type often give rise to implicit "block-oriented" nonlinear models like the Hammerstein and Wiener models considered here.* These models are discussed further in Chapter 5 in connection with the closely related class of Volterra models.

1.3.3 A bilinear model

Stromberg et al. (1995) consider the local dynamics of a distillation column, developing two-input, two-output first-order linear models at two different operating points. The basis of this study is a 15-tray experimental column, and the linear models are obtained directly from observed input/output data. Examining the results, the authors find that while some model parameters are fairly consistent between the two models, others vary significantly. To obtain a single global model, approximately valid for both operating points, they retain the model parameters that are essentially the same in the two models, but replace those that vary with *linear interpolations over the range of one of the output variables*. The result is the following *discrete-time bilinear* model

$$y_1(k) = a_1 y_1(k-1) + b_{11} u_1(k-2) + b_{12} u_2(k-2) + c_1,$$
$$y_2(k) = a_2 y_2(k-1) + [b_{210} + b_{211} y_2(k-1)] u_1(k-4)$$
$$+ [b_{220} + b_{221} y_2(k-1)] u_2(k-2) + c_2. \quad (1.49)$$

The inputs and outputs in this model are essentially the same as those considered by Sriniwas et al. (1995): u_1 is a normalized reflux flow rate, u_2 is a normalized reboiler steam flow rate, and y_1 and y_2 are tray temperatures near the top and bottom of the column, respectively.

The class of discrete-time bilinear models may be obtained by applying the Euler discretization to continuous-time bilinear models, replacing derivatives with first differences. Like their continuous-time counterpart, discrete-time bilinear models exhibit certain useful connections with discrete-time *linear* models. Some of these connections are exploited in Chapter 3 where bilinear models are discussed in detail. The important point to note here is that the general dynamic character of these models is quite different from that of the Hammerstein and Wiener models discussed in Secs. 1.3.1 and 1.3.2. In particular, the stability of Hammerstein and Wiener models is determined entirely by the stability of the linear models on which they are based; in contrast, bilinear models like Eq. (1.49) exhibit instability for input changes of sufficiently large magnitude. The practical importance of this model characteristic depends on two factors: first, the possibility or impossibility of observing input-dependent stability in the physical process of interest and second, the range over which the model will be used in the intended application. For example, if the range of inputs u_1 and u_2 is restricted enough in Eq. (1.49), the model exhibits stable responses. The primary practical consequence of this observation is the importance of *explicitly* restricting the range of operation considered for this model. Analogous considerations arise for other model classes.

1.3.4 A polynomial NARMAX model

Sriniwas et al. (1995) compare the performance of linear dynamic models obtained by standard empirical modeling procedures with that of nine polynomial NARMAX models, all identified from simulations of the 27-tray fundamental model described at the beginning of this section. The linear models each consist of 2 × 2 transfer function matrices, and each element of these matrices is specified by four parameters (a steady-state gain, one zero location, and two pole locations), resulting in a total of 16 parameters. The general character of these linear models depends strongly on the input sequences from which they are identified. For example, the steady-state gains relating the reboiler flow rate to the upper tray temperature differs by an order of magnitude in two different linear models obtained from qualitatively similar pseudorandom input sequences. In addition, some of these models exhibit nonminimum-phase behavior not seen in the fundamental model but others do not. Both this strong input sequence dependence and the evident nonlinear qualitative behavior of the step responses obtained from the fundamental model motivated the consideration of polynomial NARMAX models.

The NARMAX models considered in this study are comparable in complexity to these linear models. A specific example is designated "Model 5.5" and is defined by the following equations

$$y_1(k) = 0.0012 + 0.98 y_1(k-1) - 0.18 u_1(k-1)$$
$$+ 1.1 y_1(k-3) u_2(k-1) - 1.8 y_2(k-1) u_1(k-1)$$
$$y_2(k) = 0.0018 + 0.92 y_2(k-1) - 0.22 u_1(k-1)$$
$$+ 30.4 y_2^2(k-1) u_2(k-1) - 1.7 u_2^2(k-1). \tag{1.50}$$

Note that this model involves 10 parameters; other models considered are more complex, involving up to 23 parameters. Both the linear and nonlinear empirical models are used to design model-predictive controllers, evaluated for the original fundamental model (i.e., closed-loop simulations are performed). Not surprisingly, better closed loop performance is obtained with the nonlinear models. Conversely, the general qualitative behavior of polynomial NARMAX models tends to be rather exotic, including input-dependent stability, steady-state multiplicities, chaotic regimes, and other phenomena, most of which are not seen in the Hammerstein, Wiener, or bilinear model classes considered in Secs. 1.3.1, 1.3.2, or 1.3.3. This point is discussed at some length in Chapter 4.

1.4 Example 2: chemical reactors

The primary point of the distillation column example considered in Sec. 1.3 was to illustrate the variety of nonlinear discrete-time dynamic models that have been investigated as approximations of distillation column dynamics. The motivation for the following discussion is different: various approximations of CSTR dynamics are compared to illustrate some of the ways observable qualitative behavior can be used in the selection of discrete-time dynamic model structures. A brief survey of the variety of dynamics seen in the CSTR model is given by Ray and Jensen (1980), based on both detailed bifurcation analysis of the fundamental model equations and experimental observations in laboratory reactors. Depending on the reaction kinetics considered, the CSTR model can exhibit many different forms of nonlinear behavior, including input multiplicity, output multiplicity, isolas, and impulse or step responses showing persistent oscillations or chaos.

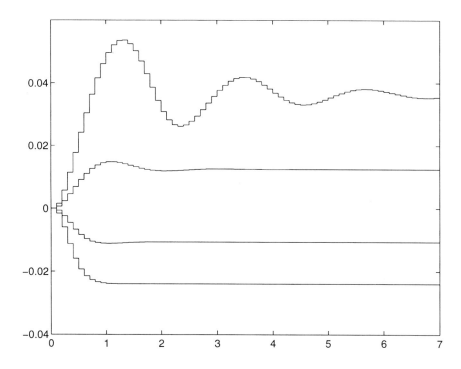

Figure 1.13: Qualitatively asymmetric CSTR response

In the example considered here, it is assumed that reactive species A flows into a perfectly mixed vessel where it undergoes an irreversible, exothermic reaction $A \to B$. This example has been discussed by a number of authors (Nahas et al., 1992; Menold, 1996; Menold et al., 1997a; Pearson, 1995; Seborg and Henson, 1997), and it is interesting here for two reasons. First, this model exhibits the dramatically asymmetric step response shown in Fig. 1.13. These responses relate changes in the cooling water flow rate in the reactor jacket to changes in the concentration of species A in the reactor. Like the high-purity distillation column responses discussed in Sec. 1.3, this response is strongly asymmetric, but here the asymmetry is *qualitative*: responses to negative steps are monotonic, typical of a first-order linear system, whereas responses to positive steps are oscillatory, typical of an underdamped second-order system.

The second reason this CSTR model is interesting here is that it can also exhibit output multiplicity. In particular, exothermic reactions generate heat at rates that depend on both the reaction kinetics and the operating conditions of the CSTR. Similarly, these operating conditions also determine the rate at which heat is removed from the reactor, and in steady-state operation, the rates of heat generation and heat removal must be equal. At fixed operating conditions, the heat removal rate depends approximately linearly on the reactor temperature, but the heat generation rate exhibits a strongly nonlinear temperature dependence. Because of this nonlinearity, these two curves can ex-

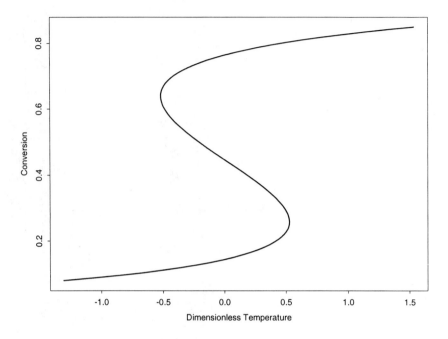

Figure 1.14: Output multiplicity in an exothermic CSTR

hibit multiple intersections, implying that the reactor can exhibit more than one steady-state operating condition. This point is illustrated in Fig. 1.14, showing reactor conversion x_1 (i.e., percentage of species A converted to species B in the reactor) as a function of the reactor jacket temperature u. Both the steady-state and the dynamic behavior of this reactor are described by the heat and mass balance equations (Uppal et al., 1974)

$$\frac{dx_1}{dt} = -x_1 + Da(1-x_1)\exp\left\{\frac{x_2}{1+x_2/\phi}\right\}$$
$$\frac{dx_2}{dt} = -x_2 + BDa(1-x_1)\exp\left\{\frac{x_2}{1+x_2/\phi}\right\} + C(u-x_2) \qquad (1.51)$$

where x_2 represents the dimensionless reactor temperature. In this particular example, model parameters have the same values as those considered by Pottmann and Pearson (1998), namely $Da = 0.072$, $\phi = 20.0$, $B = 8.0$, and $C = 0.3$, and the steady-state locus may be determined analytically by setting the time derivatives to zero in Eq. (1.51). The following sections consider six different popular empirical model structures in terms of their ability to exhibit the qualitative asymmetry of response shown in Fig. 1.13 and the output multiplicity illustrated in Fig. 1.14.

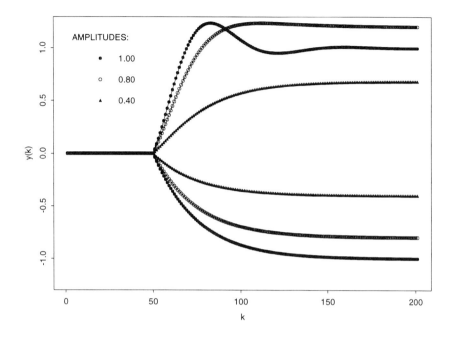

Figure 1.15: CSTR Wiener model responses

1.4.1 Hammerstein and Wiener models

Because they are simple and popular, it is reasonable to first explore the suitability of Hammerstein and Wiener models as approximations of the CSTR dynamics just described. It is easy to demonstrate that Hammerstein models cannot exhibit the qualitative asymmetry shown in Fig. 1.13: the effect of the nonlinearity $g(\cdot)$ in this model is to transform the original step of amplitude α into another step, of amplitude $g(\alpha)$. The qualitative character of the Hammerstein model step response (i.e., monotone vs. oscillatory) is therefore determined entirely by the step response of its linear part. In particular, while the steady-state gain of the Hammerstein model can be markedly asymmetric (as seen in Fig. 1.10), its qualitative character cannot change with the magnitude or sign of the step input.

In contrast, the step response shown in Fig. 1.11 demonstrates that such qualitative asymmetry is possible for Wiener models, although it does appear that this behavior is near the limit of the Wiener model's qualitative capability. That is, the step responses shown in Fig. 1.15 for a second Wiener model *do* exhibit the qualitative behavior seen in the CSTR model for $+10\%$ and -10% flow rate changes. In this particular example, steps of amplitude $A = \pm 0.4$, ± 0.8, and ± 1.0 are shown, and it may be seen that the steps of amplitude ± 1.0 exhibit exactly the desired response; the response of the "intermediate" amplitude steps

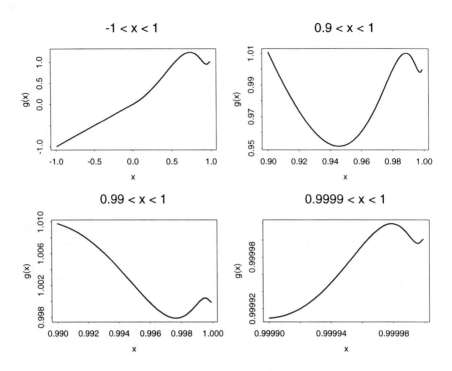

Figure 1.16: Wiener model nonlinearity over four ranges

is discussed further at the end of Sec. 1.4.5. The Wiener model considered here consists of a first-order linear system followed by a rather complicated static nonlinearity, described explicitly elsewhere (Pearson, 1995). Mathematically, this static nonlinearity is very badly behaved, much like $g(x) = sin(1/x)$, the notorious *topologist's sine curve* (Munkres, 1975). That is, this function exhibits fractal-like behavior, as illustrated in Fig. 1.16. There, four plots of $g(x)$ are shown: the first covers the entire domain of definition $-1 < x < 1$, while the other three cover the progressively more restricted domains $0.9 < x < 1$, $0.99 < x < 1$, and $0.9999 < x < 1$. It is clear that this function exhibits qualitatively similar but finer scale oscillations in progressively smaller neighborhoods of the point $x = 1$. Although this example clearly illustrates the differences between Wiener and Hammerstein models, it is also clear that such pathologically behaved functions do not provide a practical basis for constructing discrete-time process models of moderate complexity.

Neither the Hammerstein model nor the Wiener model can exhibit the output multiplicity seen in the CSTR example. To see this point, consider the Hammerstein and Wiener models constructed from the static nonlinearity $g(\cdot)$ and a linear dynamic model with steady-state gain K_0. The responses y_H and y_W of these models to the constant input $u(k) = u_s$ are unique, given by

$$y_H = K_0 g(u_s) \quad \text{and} \quad y_W = g(K_0 u_s). \qquad (1.52)$$

Motivations and Perspectives

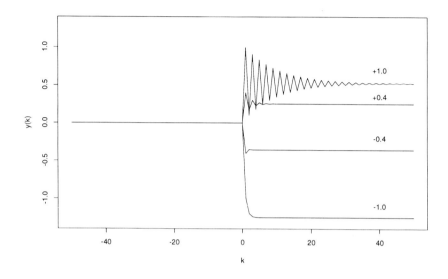

Figure 1.17: Bilinear model step responses

Hence, neither model can exhibit output multiplicity. Conversely, if $g(\cdot)$ is a non-monotonic function, both models can exhibit input multiplicity.

1.4.2 Bilinear models

Overall, the results presented in Sec. 1.4.1 illustrate two points. First, Hammerstein models are inherently incapable of exhibiting either of the two forms of qualitative behavior considered here and second, Wiener models are completely incapable of one of these forms of behavior and only marginally capable of the other. Bilinear models are discussed in detail in Chapter 3, where it is shown that these models cannot exhibit either input multiplicity or output multiplicity. Conversely, the following example illustrates that this model class is flexible enough to easily exhibit the qualitative asymmetry seen in Fig. 1.13. In particular, the following example exhibits the four step responses shown in Fig. 1.17:

$$y(k) = -0.35y(k-1) + u(k-1) - 0.55u(k-1)y(k-1). \tag{1.53}$$

The step amplitudes are ±1.0 and ±0.4, as indicated on the plot. For steps of amplitude ±1.0, these responses exhibit the same qualitative behavior as the CSTR example, but the behavior changes somewhat for the intermediate amplitude steps. In particular, note that for steps of amplitude ±0.4, both the positive and negative responses are somewhat oscillatory, although this character is more pronounced for the positive response.

The question of whether bilinear models are good approximations of CSTR dynamics will depend on the particular application considered. Specifically, since bilinear models cannot exhibit output multiplicity, they are inherently unsuitable in applications where the required range of model validity is large enough to encompass more than one steady-state operating point (u_s, y_s). Conversely, bilinear models may represent reasonable *local* approximations, valid in some neighborhood of a single steady-state operating point. Because bilinear models are capable of exhibiting the qualitative asymmetry of the CSTR dynamics that linear models cannot exhibit, local bilinear approximations may offer significant improvement over more traditional linear approximations. However, it is also important to note that bilinear models exhibit input-dependent stability, a point discussed in detail in Chapter 3. Hence, the following general behavior can be expected. Linear dynamic models should be adequate to describe local dynamics in some sufficiently small neighborhood of a steady-state operating point (u_s, y_s). For larger neighborhoods, bilinear models should offer significant improvements because of their ability to capture some of the nonlinear behavior of the CSTR dynamics. Ultimately, however, bilinear models will become inadequate for sufficiently large approximation neighborhoods, either because these neighborhoods include more than one steady-state operating point or because they exceed the stability limits of the bilinear model.

1.4.3 Polynomial NARMAX models

The family of polynomial NARMAX models is discussed in detail in Chapter 4 and is enormous, including both the bilinear model class and the polynomial Hammerstein model class as proper subsets. Any form of qualitative behavior that these subclasses can exhibit is therefore possible more generally in polynomial NARMAX models. A particularly simple polynomial NARMAX model exhibiting the asymmetric CSTR step response is

$$y(k) = ay^2(k-1) + bu(k-1). \tag{1.54}$$

Specifically, if $a < 0$ and $b > 0$, the response of this model to positive steps is oscillatory, whereas the response to negative steps is monotone. To accurately approximate the CSTR dynamics seen in Fig. 1.13, a more complex model would be required, but this example illustrates that the polynomial NARMAX class exhibits sufficient flexibility to capture the desired qualitative behavior. Unfortunately, this flexibility also admits a wide range of other qualitative phenomena, some of which may be highly undesirable in any given application. Specifically, like the bilinear model (1.53), the quadratic model defined by Eq. (1.54) also exhibits input-dependent stability, leading to the same issues regarding the range of model validity as in the case of bilinear models. In addition, the quadratic model can also exhibit more exotic phenomena like chaotic step responses, a point discussed further in Chapter 4.

In contrast to the class of bilinear models, more general polynomial NARMAX models can exhibit both input and output multiplicity, along with more complex behavior like isolas. In general, the characterization of this steady-state behavior is extremely difficult, essentially defining the field of classical algebraic geometry (Fulton, 1969). In special cases, however, simple characterizations are possible. A specific example is the Lur'e model introduced in Sec. 1.1.5, consisting of a linear dynamic model with a static nonlinearity appearing as a feedback element (see Fig. 1.4). This model structure is discussed further in Chapter 4, but here it is enough to note two points. First, if the static nonlinearity $f(\cdot)$ is a polynomial, the Lur'e structure defines a polynomial NARMAX model. Second, if attention is restricted to a bounded region of the (u_s, y_s) plane large enough to contain all three of the CSTR operating points in its region of multiplicity (e.g., the entire region shown in Fig. 1.14), it is possible to approximate the exact steady-state locus of the CSTR reasonably well by a cubic polynomial. Together, these observations permit the construction of highly structured polynomial NARMAX models that give reasonable approximations to both steady-state and dynamic behavior for the CSTR model (Pottmann and Pearson, 1998).

1.4.4 Linear multimodels

Traditionally, one of the most popular approaches to the analysis and control of nonlinear systems is linearization about some steady-state operating point (u_s, y_s). A logical extension of this idea is to construct *several* local linear models, each centered at some steady-state operating point (u_i, y_i), and combine the results into a global nonlinear model. The details of the procedure used for combining these local linear models are extremely important, but the basic approach has a number of advantages and appears to be the subject of growing interest (Murray-Smith and Johansen, 1997). A detailed discussion of this *linear multimodel* approach is given in Chapter 6, but the following three simple examples illustrate the basic idea and some of its inherent possibilities.

The first two examples were developed to capture the qualitative asymmetry of the CSTR model seen in Fig. 1.13 (Pearson, 1995)

$$y(k) = a|y(k-1)| + bu(k-1) \tag{1.55}$$
$$y(k) = ay(k-1) + bu(k-1) + c\theta_+[y(k-2)], \tag{1.56}$$

where $\theta_+(x) = x$ if $x \geq 0$ and 0 otherwise. The rationale for the first of these models was the observation that positive step responses behave qualitatively like a first-order linear model with a negative autoregressive coefficient, while negative step responses behave qualitatively like a first-order linear model with a positive autoregressive coefficient. This observation suggested the following linear multimodel, equivalent to Eq. (1.55)

$$y(k) = \begin{cases} ay(k-1) + bu(k-1) & y(k-1) \geq 0 \\ -ay(k-1) + bu(k-1) & y(k-1) < 0. \end{cases} \tag{1.57}$$

Choosing $a < 0$ and $b > 0$ yields a model that exhibits the correct general qualitative behavior. Alternatively, Eq. (1.56) attempts to capture the asymmetric CSTR dynamics by approximating the response to positive steps with a second-order, underdamped linear model and approximating the response to negative steps by a first-order, monotone linear model. This interpretation follows directly for the following linear multimodel representation for Eq. (1.56)

$$y(k) = \begin{cases} ay(k-1) + cy(k-2) + bu(k-1) & y(k-2) \geq 0 \\ ay(k-1) + bu(k-1) & y(k-2) < 0. \end{cases} \quad (1.58)$$

Again, choosing a, b, and c appropriately yields the correct general qualitative behavior.

Both of these models belong to the family of TARMAX models introduced in Chapter 3 and discussed further in Chapter 6. An important and unusual feature of this model class is that TARMAX models exhibit *positive-homogeneity*, a property discussed at length in Chapter 3. Specifically, note that if the input sequence $\{u(k)\}$ is scaled by any positive constant λ, the response of these models also scales by λ. Despite this "linear scaling behavior," the overall qualitative behavior of these models can be extremely complex. In particular, Eq. (1.56) can exhibit step responses with persistent oscillations, while Eq. (1.55) can exhibit chaotic step responses (Pearson, 1995).

The third example considered here is a linear multi-model developed to approximate the global dynamics of a simple CSTR model (Banerjee et al., 1997). In developing this model, several local operating points were chosen from the exact steady-state locus and linear approximations of the CSTR dynamics were developed around each of these operating points. An important question considered in this paper was the number of local models required, and in the specific example considered, three local models were found to be sufficient, but two local models were not. Defining a region of validity for each model then resulted in a piecewise-linear approximation of the steady-state locus. The global qualitative behavior of locally linear models like these depends strongly on the selection criterion, a point discussed in detail in Chapter 6. In general, however, this class of models is extremely flexible, capable of exhibiting input-multiplicity, output-multiplicity, and more complex behavior like isolas.

1.4.5 The Uryson model

Although it appears to be much less well-known than the Hammerstein model, the *Uryson model* (Billings, 1980) consists of m Hammerstein models connected in parallel, as shown in Fig. 1.18. That is, a common input sequence $\{u(k)\}$ is applied to m static nonlinearities $g_i(\cdot)$ for $i = 1, 2, ..., m$, and these m transformed inputs drive m linear models, defined by the transfer functions $H_i(z)$. The m linear model responses are then summed to obtain the overall model response $y(k)$. Because it includes the Hammerstein model as a special case, the Uryson model class exhibits a wider range of qualitative behavior. For example, unlike Hammerstein models, Uryson models can exhibit the asymmetric CSTR step response behavior shown in Fig. 1.13; also, unlike Wiener models, the Uryson models required to exhibit this behavior can be extremely simple.

Motivations and Perspectives

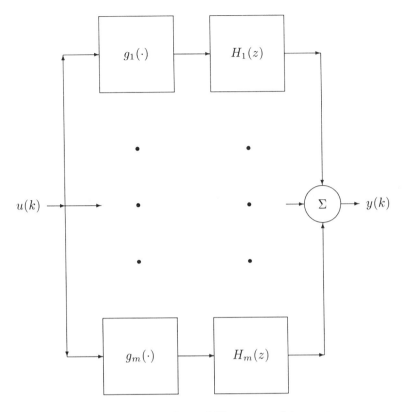

Figure 1.18: m-channel Uryson model structure

To see this point, consider a two-channel Uryson model based on the following pair of static nonlinearities:

$$g_1(x) = \begin{cases} x & x \geq 0 \\ 0 & x < 0, \end{cases} \qquad g_2(x) = \begin{cases} 0 & x \geq 0 \\ x & x < 0. \end{cases} \qquad (1.59)$$

Let $H_1(z)$ and $H_2(z)$ be any two linear dynamic models and consider the step response of the Uryson model defined by $g_1(\cdot)$, $g_2(\cdot)$, $H_1(z)$, and $H_2(z)$. For positive amplitudes, the step response of the Uryson model will simply be that of the linear model $H_1(z)$, since $g_1(u(k)) = u(k)$ and $g_2(u(k)) = 0$ for positive step inputs. Conversely, the step response of this Uryson model for negative steps will be identical to that of the linear model $H_2(z)$. Therefore, if $H_1(z)$ is a second-order linear model with an underdamped step response and $H_2(z)$ is a first-order linear model with a monotonic step response, the resulting Uryson model will exhibit the general type of step response shown in Fig. 1.19.

This particular Uryson model is an interesting one for two reasons. First, it represents a direct implementation of the semantic description given originally for the CSTR model step responses: positive responses look like those of a second-order linear model, while negative responses look like those of a first-

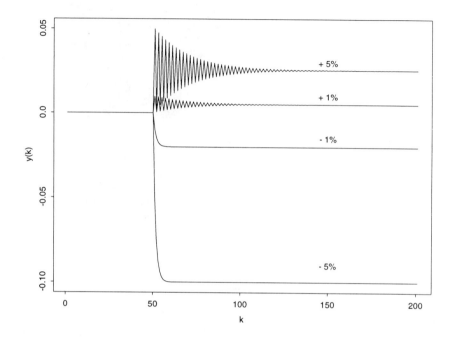

Figure 1.19: Uryson model step responses

order linear model. Second, like the models defined by Eqs. (1.55) and (1.56), this Uryson model is also positive-homogeneous. This behavior follows from the positive homogeneity of the functions $g_1(x)$ and $g_2(x)$: $g_i(\lambda x) = \lambda g_i(x)$ for any $\lambda > 0$. Further, this Uryson model may be also represented as a linear multi-model but in contrast to the previous two examples, the representation is infinite-dimensional (this point is discussed further in Chapter 6). Once again, this example illustrates that, in combining several "local linear models" to obtain a global nonlinear model, the details of this combination are critically important.

Further, although it was noted in Sec. 1.4.4 that linear multi-models can exhibit output multiplicity, Uryson models cannot. The steady-state behavior of the Uryson model defined in Fig. 1.18 is given by

$$y_s = \sum_{i=1}^{m} K_i g_i(u_s). \tag{1.60}$$

Here, u_s is the constant input sequence value and K_i is the steady-state gain of the ith linear model $H_i(z)$. As in the case of Hammerstein and Wiener models, the Uryson model can exhibit input multiplicity [i.e., Eq. (1.60) may yield the same value of y_s for more than one value of u_s], but output multiplicity is not possible since y_s is uniquely defined once u_s is specified.

Finally, it is worth briefly discussing one of the other qualitative differences between the various approximations of the CSTR dynamic asymmetry considered here. It was noted that the Wiener model required to exhibit this qualitative behavior is not simple, incorporating a rather badly-behaved static nonlinearity. In particular, this nonlinearity is extremely nonmonotonic; since this nonlinearity also determines the steady-state behavior of the model, it follows that the amplitude dependence of the step responses is also nonmonotonic. This point is seen clearly in the Fig. 1.15: the response to a step of amplitude $+1.0$ is more oscillatory than the response to a step of amplitude $+0.8$, but it settles out to a smaller steady-state value. In contrast, the three positive-homogeneous models considered here (the Uryson model and the first two linear multimodel examples) exhibit strict monotonicity in both positive and negative step responses. In fact, it might be argued that this monotonicity is also unrealistic and that it is more realistic to expect a smooth *qualitative* transition between highly oscillatory responses to positive steps of large amplitudes and strictly monotone responses to negative steps of large amplitude. Such arguments would favor the bilinear model, since the step response shown in Fig. 1.17 exhibits precisely this type of behavior.

1.5 Organization of this book

Sec. 1.1 briefly described the following five approaches to the development of dynamic models for complex physical systems:

1. Fundamental or first-principles modeling
2. Simplification and discretization of a fundamental model
3. Direct empirical modeling
4. Gray-box modeling
5. Indirect empirical modeling.

The first of these five modeling procedures arguably yields the best overall model, but the result is generally too complex to be used for model-based control system design, and it is not in the discrete-time form required for computer control. In contrast, the other four of these modeling procedures can lead to discrete-time dynamic models that are simple enough in structure to be useful for model-based computer control. One of the first steps in all four of these approaches is the specification of a class $\mathcal{C}$ of model structures to which the final model will belong. Further, the final model obtained by any of these four approaches can depend strongly on this choice, both in terms of its utility in the application for which it is developed, and in terms of its accuracy in representing the physical system dynamics.

The aim of this book is to address the following two fundamental questions. First, what classes $\mathcal{C}$ are available for consideration? Certainly, it is always possible to define new mathematical model classes, but it is usually more convenient to choose from those model classes for which parameter estimation algorithms, stability conditions, control system design procedures, and other useful results already exist. Second, given a particular model class $\mathcal{C}$, what types of qualitative behavior can it exhibit? Even partial answers to this question can provide extremely useful practical guidance in model structure selection. In particular, if the general qualitative behavior of the physical process of interest is known either from the behavior of fundamental model simulations or from practical experience, comparing this qualitative behavior with that of various model classes $\mathcal{C}_i$ can provide a systematic procedure for characterizing these model classes as reasonable or unreasonable for the application at hand.

To address these two questions, the remainder of this book is largely organized around specific discrete-time model classes, as follows. Chapter 2 begins with a reasonably detailed discussion of linear discrete-time models, undertaken for three reasons. First, this chapter introduces some important notation and some broadly applicable concepts that will appear repeatedly throughout subsequent chapters (e.g., the notion of a moving average model). The second reason for focusing initially on linear models is to review certain important qualitative features of linear models that are well-known and inform much of our intuition about dynamic phenomena. The third reason for focusing on linear models is to illustrate that certain violations of "low-order, time-invariant linear intuition" can occur if high-order (e.g., infinite-dimensional) or time-varying linear models are considered. Both of these less familiar linear model classes arise in practice and it is important to understand how their behavior differs from that of the nonlinear model classes considered in later chapters.

As a first step into the land of nonlinearity, Chapter 3 briefly discusses four classes of "nonlinear but almost linear" models. The first is the class of *bilinear models*, which illustrates the notion of *structural nonlinearity*: the model class is defined on the basis of a specific nonlinear structure. It is also possible to define nonlinear models *behaviorally* by specifying their qualitative input-output behavior. The class of *homogeneous models* discussed in Chapter 3 illustrates this idea: this class includes the class of linear models, but it also includes many other *nonlinear* models that share the behavioral feature of homogeneous response to arbitrary input sequences (e.g.., doubling the input doubles the output). The third and fourth model classes introduced in Chapter 3 are also defined behaviorally and both include the homogeneous model class as subsets: the class of *positive-homogeneous* models and the class of *static-linear* models. The key point of Chapter 3 is that it is possible to specify either the *structure* of the model and ask how it behaves, or the *behavior* of a model class and ask what structures exist within this class. Both of these questions are extremely difficult, so it is usually necessary to settle for fairly incomplete answers. Because the structural approach is somewhat easier, it is more popular and is the primary focus of this book, although the behavioral approach can yield some extremely useful insights, as subsequent discussions illustrate.

Chapter 4 presents a detailed discussion of the important class of NARMAX models introduced in Sec. 1.4.1. Like the class of bilinear models, this class is defined structurally, so one of the main topics of the chapter is the qualitative behavior of these models. Unfortunately, the general NARMAX class is too large to permit much direct analysis, but it is possible to obtain some surprisingly general insights into qualitative behavior by distinguishing between the broad subclasses of *nonlinear autoregressive models* and *nonlinear moving average models*. In particular, nonlinear moving average models are generally both better behaved and easier to analyze than nonlinear autoregressive models. In addition, many other structurally-defined subclasses of the NARMAX family have been explored by various authors, and some of these ideas are also presented in Chapter 4.

Many popular models (e.g., the Hammerstein and Wiener models discussed in this chapter) can be represented as *Volterra models*. Consequently, Chapter 5 is devoted to a detailed discussion of this model class, emphasizing its connections with other model classes. One of the key results presented in Chapter 5 is that the broad class of *block-oriented nonlinear models* is equivalent to the class of Volterra models, provided that the nonlinear functions involved are analytic. In addition, the class of *pruned Volterra models* is introduced, obtained by requiring certain model parameters to be zero (i.e., "pruning" the unconstrained Volterra model). As is seen in subsequent discussions, the question of how this pruning effects Volterra model behavior is closely related to the question of how Hammerstein model behavior differs from Wiener model behavior.

An entirely different approach to the development of nonlinear models is considered in Chapter 6, which introduces the class of *linear multimodels*. These models are *locally linear* but *globally nonlinear*, and are obtained by "piecing together" several linear dynamic models. The exact details of this piecing together are important, a point discussed in some detail in Chapter 6. The practical motivation for considering this class of models is that they can be extremely flexible, as noted in Sec. 1.4.4. In addition, the class of linear multimodels is closely related to some of the *positive homogeneous* models discussed in Chapter 3, and this connection is also explored in Chapter 6.

Chapter 7 is devoted to the description of *relationships* among the different model classes discussed in Chapters 2 through 6. Initially, relations based on structural inclusions are presented: model class X is a subset of model class Y but not of model class Z. Many of these inclusions are fairly easy to see once they have been pointed out but are not obvious initially. For example, Hammerstein models are members of the class of *structurally additive models* defined and discussed in Chapter 4, but Wiener models are not. As an aid in going beyond these basic inclusions, Chapter 7 also introduces the fundamental notions of *category theory*, a fairly abstract mathematical construct that provides an extremely useful basis for focusing on both relations between different model classes and the role of input sequences in determining model behavior. Specifically, category theory deals with *objects*, which will here be taken as "sets of sequences," and *morphisms*, which will be taken as "models mapping an input sequence into an output sequence." The primary advantages of introducing the formalism of category theory are first, that it provides a powerful framework for considering very general issues like model realizations, steady-state characterization, linearization, and discretization and second, that it requires a consideration of both dynamic models and their input sequences together.

Finally, Chapter 8 provides a brief but general summary of the main ideas and results presented in this book, with references back to more detailed discussions given in earlier chapters. The intent of this final chapter is to provide a useful first step in deciding which classes of nonlinear models seem most appropriate as a starting point in developing nonlinear model-based control strategies. As noted repeatedly throughout this chapter, this choice is extremely important in determining the relative ease or difficulty of the model identifications, model validations, controller designs, and performance evaluations that follow. In addition, brief introductions to the topics of data pretreatment, input sequence design and model parameter estimation are also given, along with references for more complete discussions of these important topics.

Chapter 2

Linear Dynamic Models

It was emphasized in Chapter 1 that low-order, linear time-invariant models provide the foundation for much intuition about dynamic phenomena in the real world. This chapter provides a brief review of the characteristics and behavior of linear models, beginning with these simple cases and then progressing to more complex examples where this intuition no longer holds: infinite-dimensional and time-varying linear models. In continuous time, infinite-dimensional linear models arise naturally from linear partial differential equations whereas in discrete time, infinite-dimensional linear models may be used to represent a variety of "slow decay" effects. Time-varying linear models are also extremely flexible: In the continuous-time case, many of the ordinary differential equations defining special functions (e.g., the equations defining Bessel functions) may be viewed as time-varying linear models; in the discrete case, the gamma function arises naturally as the solution of a time-varying difference equation.

Sec. 2.1 gives a brief discussion of low-order, time-invariant linear dynamic models, using second-order examples to illustrate both the "typical" and "less typical" behavior that is possible for these models. One of the most powerful results of linear system theory is that any time-invariant linear dynamic system may be represented as either a moving average (i.e., convolution-type) model or an autoregressive one. Sec. 2.2 presents a short review of these ideas, which will serve to establish both notation and a certain amount of useful intuition for the discussion of NARMAX models presented in Chapter 4. Sec. 2.3 then briefly considers the problem of characterizing linear models, introducing four standard input sequences that are typical of those used in linear model characterization. These standard sequences are then used in subsequent chapters to illustrate differences between nonlinear model behavior and linear model behavior. Sec. 2.4 provides a brief introduction to infinite-dimensional linear systems, including both continuous-time and discrete-time examples. Sec. 2.5 provides a similar introduction to the subject of time-varying linear systems, emphasizing the flexibility of this class. Finally, Sec. 2.6 briefly considers the nature of linearity, presenting some results that may be used to define useful classes of *nonlinear* models.

2.1 Four second-order linear models

Since this book is primarily concerned with nonlinear, discrete-time dynamic models, a thorough treatment of linear models is not attempted here. Instead, this section illustrates the range of behavior possible from low-order linear models by considering four specific examples, chosen because they exhibit a wide range of behavior and illustrate some important points. For a discussion of low-order linear models that is much more comprehensive and specific to process control applications, refer to part II of Ogunnaike and Ray (1994). For a more thorough discussion of the theory of linear, time-invariant, discrete-time models, refer to the first two chapters of Oppenheim and Schafer (1975).

Fig. 2.1 consists of four plots, each one showing the step response of a different second-order, linear, time-invariant discrete-time model. The step response shown in the upper left plot represents a typical underdamped response, and corresponds to the following model:

$$y(k) = -0.5y(k-1) - 0.8y(k-2) + 0.5u(k-1) + 0.5u(k-2). \quad (2.1)$$

Alternatively, this model may be represented in terms of its transfer function:

$$H(z) = \frac{0.5z^{-1} + 0.5z^{-2}}{1 + 0.5z^{-1} + 0.8z^{-2}}. \quad (2.2)$$

The oscillatory character of this step response results from the fact that this transfer function has complex conjugate poles at $z = -0.25 \pm 0.86i$. These poles lie inside the unit circle, so the model is stable, exhibiting a decaying oscillatory approach to its steady-state value. Also, note that the steady-state gain of this model is $H(1) \simeq 0.435$, which can be seen clearly in the plot of the unit step response.

The step response shown in the upper right plot in Fig. 2.1 is a little more unusual in its behavior. Specifically, note that the initial response of this model to the positive unit step is a *negative* excursion. Ultimately, this response also settles out to the same steady-state value seen in the first step response, but the initial negative excursion is called an *inverse response* and corresponds to the fact that this model has a zero outside the unit circle in the complex z plane. In particular, the transfer function defining this model is:

$$H(z) = \frac{-0.5z^{-1} + 1.5z^{-2}}{1 + 0.5z^{-1} + 0.8z^{-2}}, \quad (2.3)$$

which has the same pair of complex conjugate poles as the transfer function in Eq. (2.2), but different zeros. That is, the first transfer function exhibits a zero at $z = -1$, lying *on* the unit circle, while the transfer function in Eq. (2.3) exhibits a zero at $z = 3$. Note that the steady-state gain is exactly the same for these two transfer functions, so the two unit step responses settle out to the same final value.

Linear Dynamic Models

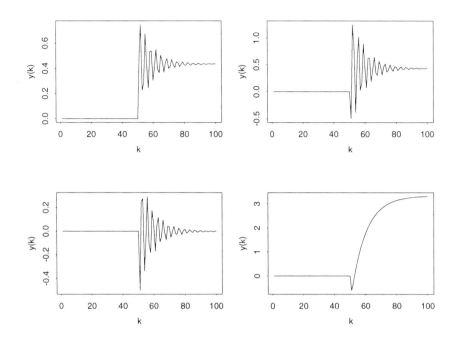

Figure 2.1: Four second-order step responses

The step response shown in the bottom left plot in Fig. 2.1 corresponds to the transfer function

$$H(z) = \frac{-0.5z^{-1} + 0.5z^{-2}}{1 + 0.5z^{-1} + 0.8z^{-2}}. \tag{2.4}$$

Note that this transfer function again exhibits the same complex conjugate pole pair as in the previous two examples, so the oscillatory character of the step response is also the same as in those two examples. Here, however, the steady-state gain of this model is *zero*, corresponding to the fact that the transfer function in Eq. (2.4) has a zero at $z = 1$. Physically, responses of this type can arise if there is a natural differentiation at work, with the response variable $y(k)$ dependent on *changes* in the input variable $u(k)$. A coil of wire placed in a magnetic field, for example, generates a voltage that is proportional to the *rate of change* of the magnetic field. This voltage can exhibit a significant response to magnetic field transients, but its steady-state response is identically zero, as in this discrete-time system example.

Finally, the bottom right plot in Fig. 2.1 shows the step response of the second-order linear model defined by the transfer function

$$H(z) = \frac{-0.6z^{-1} + .9z^{-2}}{1 - z^{-1} + 0.09z^{-2}}. \tag{2.5}$$

This transfer function exhibits two real poles at $z = 0.1$ and $z = 0.9$, both inside the unit circle in the complex z-plane, so the model is stable. As in the second example, however, this model exhibits a zero at $z = 3$, which lies outside the unit circle, thus giving this model an inverse response. In fact, this response is somewhat easier to see in this example than in the upper right plot because the predominant response is the monotonic increase expected for a second-order system with two real poles.

The principal point illustrated by the last three examples is that the nature of the *transient behavior* can be somewhat at odds with that of the steady-state behavior, even in the case of simple, low-order, time-invariant linear models. Specifically, note that the characteristic feature of an inverse-response system is that the *sign* of the high-frequency (i.e., transient) gain is *opposite* that of the low-frequency (i.e., steady-state) gain. Both of the examples on the right side of Fig. 2.1 illustrate this point, since the gain of the transient response is negative, whereas the steady-state gain is positive. Similarly, the example in the lower left plot illustrates that it is possible for the transient response to be arbitrarily behaved in a system whose steady-state response is identically zero. As model complexity grows, the number of ways in which the transient behavior can "disagree" with steady-state behavior grows correspondingly. In fact, it was noted in Chapter 1 that Hammerstein models and Wiener models can behave quite differently, *even though their steady-state behavior is identical*. This contrast between steady-state and transient behavior is an extremely important one that is revisited often in subsequent chapters.

2.2 Realizations of linear models

Linear, time-invariant, discrete-time models may be represented in a number of ways, including the following four:

1. State-space models
2. Autoregressive (AR) models
3. Moving average (MA) models
4. Autoregressive moving average (ARMA) models.

All four of these representations have extensions to both time-varying linear models and nonlinear models, so it is useful to include brief discussions of each of these basic linear model structures here for future reference. It is also important to note that these structures are equivalent in the case of linear, time-invariant discrete-time models, but this equivalence does not hold for the nonlinear extensions. This observation provides further motivation for briefly considering all four of these realizations and not just focusing on one of them. Because the simple models considered above are most simply represented as autoregressive moving average (ARMA) models, it will be convenient to start there and consider the four realizations listed above in reverse order.

Linear Dynamic Models

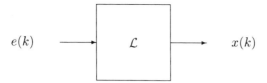

Figure 2.2: Block diagram of a time-series model

2.2.1 Autoregressive moving average models

The terms *autoregressive* and *moving average* have their origin in the statistical time-series analysis literature. There, the basic problem of interest is the development of a detailed model of an observed sequence of Gaussian random variables $\{x(k)\}$. To solve this problem, models of the general form shown in Fig. 2.2 are considered: a zero-mean, unit variance, independent, identically distributed (i.i.d.) Gaussian sequence $\{e(k)\}$ is assumed as an *unobserved* input to a linear, time-invariant model $\mathcal{L}$ whose output is the observed data sequence $\{x(k)\}$. The basis for this general model structure is discussed later in this chapter, but the point is that a substantial literature now exists concerned with the problem of determining the unknown model $\mathcal{L}$ from the available data $\{x(k)\}$ (Box et al., 1994; Priestley, 1981). The last three of the four general representations for $\mathcal{L}$ listed above have been considered extensively in this literature and the terminology for these representations has been widely adopted.

Specifically, an *autoregressive* time-series model of order p, abbreviated as $AR(p)$, has the form

$$x(k) = \sum_{i=1}^{p} a_i x(k-i) + b_0 e(k). \tag{2.6}$$

This model is called *autoregressive* because it may be viewed as a regression model of $x(k)$ in terms of its own past p values $x(k-i)$. Similarly, a *moving average* time-series model of order q, abbreviated as $MA(q)$, has the form:

$$x(k) = \sum_{j=0}^{q} b_j e(k-j). \tag{2.7}$$

The name *moving average* derives from the fact that if the coefficients b_j are nonnegative and sum to 1, this expression for $x(k)$ may be regarded as a weighted average of the random variables $e(k)$ through $e(k-q)$. Because this "averaging window" *moves* with the time index k, this representation is referred to as a *moving average* representation, even when the coefficients b_j do not sum to 1. Finally, combining both of these representations leads to the *autoregressive moving average* or $ARMA(p,q)$ model structure:

$$x(k) = \sum_{i=1}^{p} a_i x(k-i) + \sum_{j=0}^{q} b_j e(k-j). \tag{2.8}$$

Note that this model structure includes both the AR and MA model structures as special cases.

The ARMA model just described is a difference equation relating the unobserved Gaussian white noise sequence $\{e(k)\}$ to the observed data sequence $\{x(k)\}$, which may be viewed as the time-domain representation of a linear input/output model. Taking the z-transform of this difference equation yields the equivalent frequency-domain or transfer function representation

$$X(z) = H(z)E(z). \tag{2.9}$$

Here, $E(z)$ and $X(z)$ represent the z-transforms of the input and output sequences (Oppenheim and Schafer, 1975, ch. 2)

$$E(z) = \sum_{n=-\infty}^{\infty} e(n)z^{-n} \qquad X(z) = \sum_{n=-\infty}^{\infty} x(n)z^{-n}, \tag{2.10}$$

and $H(z)$ is the transfer function

$$H(z) = \frac{\sum_{j=0}^{q} b_j z^{-j}}{\sum_{i=1}^{p} a_i z^{-i}}. \tag{2.11}$$

The key point is that the $ARMA(p,q)$ representation given in Eq. (2.8) is equivalent to the transfer function representation (2.11) with p poles and q zeros.

The statistical time-series literature is primarily concerned with models that relate an observed data sequence $\{x(k)\}$ to an *unobserved* stochastic process $\{e(k)\}$, sometimes called an *endogenous* variable, intended to explain the irregular variation of $\{x(k)\}$. In addition, the statistical time-series literature also considers the class of autoregressive moving average models with exogenous inputs or *ARMAX* models. These models are defined by adding one or more "extra" or *exogenous* input sequences besides $\{e(k)\}$. Thus, appending the exogeneous input sequence $\{v(k)\}$ to the $ARMA(p,q)$ model defined in Eq. (2.8) yields the following ARMAX model:

$$x(k) = \sum_{i=1}^{p} a_i x(k-i) + \sum_{j=0}^{q} b_j e(k-j) + \sum_{\ell=0}^{r} c_\ell v(k-\ell). \tag{2.12}$$

In contrast to the endogenous sequence $\{e(k)\}$, the exogenous input sequence $\{v(k)\}$ represents an *observable* data sequence, included to explain the influence of related variables on the total variation of $\{x(k)\}$.

Model-based control applications generally require a *prediction model* that relates variations in a control input sequence $\{u(k)\}$ to the variations in an observable process response sequence $\{y(k)\}$. In particular, linear prediction models are of the form:

$$\hat{y}(k) = \sum_{i=1}^{p} a_i \hat{y}(k-i) + \sum_{j=0}^{q} b_j u(k-j), \tag{2.13}$$

Linear Dynamic Models

where $\hat{y}(k)$ is a prediction of the response $y(k)$ based on the available input data $\{u(k)\}$. Defining the prediction error sequence $e(k) = \hat{y}(k) - y(k)$, this prediction model may be rewritten as the $ARMAX(p,q)$ model

$$y(k) = \sum_{i=1}^{p} a_i y(k-i) + \sum_{j=0}^{q} b_j u(k-j) + \sum_{\ell=0}^{r} c_\ell e(k-\ell), \qquad (2.14)$$

for $r = p$, $c_0 = -1$, and $c_\ell = a_\ell$ for $\ell = 1, 2, ..., p$. The point is, if the prediction error sequence $\{e(k)\}$ is well approximated by a stationary stochastic process, the prediction model (2.13) is equivalent to the relabeled $ARMAX(p,q)$ model (2.14). In effect, this relabeling reverses the roles of the exogenous and endogenous inputs, consistent with the differences of focus between the control community and the time-series community.

Because this book deals exclusively with the qualitative input/output behavior of prediction models, this role reversal is taken to its extreme limit and $e(k)$ is assumed to be identically zero for all k. *Throughout the rest of this book, the term* ARMAX(p,q) *model refers to the prediction model (2.13) with the circumflex over* y(k) *omitted for simplicity.* That is, the linear models considered here will be of three types:

1. The $AR(p)$ model, Eq. (2.6) with $e(k) \to u(k)$ and $x(k) \to y(k)$
2. The $MA(q)$ model, Eq. (2.7) with $e(k) \to u(k)$ and $x(k) \to y(k)$
3. The $ARMA(p,q)$ model, Eq. (2.8) with $e(k) \to u(k)$ and $x(k) \to y(k)$

The primary motivation for this subtle change in notation is that the time-series community draws useful distinctions between *autoregressive* and *moving average* models, but only for the endogenous variable sequence $\{e(k)\}$. One of the main points of Chapter 4 is that these distinctions are also extremely important for the class of NARMAX prediction models of primary interest here.

Before proceeding to a discussion of the relationship between autoregressive and moving average linear models, it is important to say something about the role of initital conditions in these models. In particular, note that the response of a linear moving average model is completely determined by the input sequence $\{u(k)\}$. In contrast, to compute the response of a linear autoregressive or autoregressive moving average model, it is necessary to specify p initial values for $\{y(k)\}$. That is, if $\{u(k)\}$ is defined for $k \geq 0$, to compute $y(k)$ for $k \geq 0$, it is necessary to specify $y(-1), y(-2), ..., y(-p)$ *a priori. Unless otherwise specified, this book will make the "standard" assumption that* y(k) *is identically zero for all negative* k *in autoregressive and autoregressive moving average models.* Since linear models are often used to represent "deviation variables"—i.e., deviations from some set of "nominal" (e.g., steady-state) operating conditions—this assumption generally means that the system of interest is in this nominal state for $k < 0$. There are circumstances—for example, the case of completely bilinear models discussed in Chapter 3—under which this specification is not appropriate, but any specification other than the "standard" one just given will be noted explicitly. This issue is also discussed in Chapter 6 in connection with the class of linear multimodels.

2.2.2 Moving average and autoregressive models

It is a standard result that a linear, time-invariant, discrete-time dynamic model is completely characterized by its impulse response $\{h(k)\}$. Specifically, given $\{h(k)\}$, the model's response to any input sequence $\{u(k)\}$ is given by the discrete convolution

$$y(k) = \sum_{i=0}^{\infty} h(i)u(k-i). \qquad (2.15)$$

The key to proving this result lies in the fact that if $\{u(k)\}$ is any discrete-time input sequence, it may be rewritten as

$$u(k) = \sum_{i=-\infty}^{\infty} u(i)\delta(k-i), \qquad (2.16)$$

where $\delta(j)$ is the Kronecker delta function

$$\delta(j) = \begin{cases} 1 & j = 0 \\ 0 & j \neq 0. \end{cases} \qquad (2.17)$$

If $\mathbf{L}$ is any linear, time-invariant, discrete-time dynamic model, its response to the input sequence $\{u(k)\}$ may thus be expressed as

$$\begin{aligned} y(k) &= \mathbf{L}\left[\sum_{i=-\infty}^{\infty} u(i)\delta(k-i)\right] \\ &= \sum_{i=-\infty}^{\infty} u(i)\mathbf{L}[\delta(k-i)] \\ &= \sum_{i=-\infty}^{\infty} u(i)h(k-i) \\ &= \sum_{j=-\infty}^{\infty} h(j)u(k-j), \end{aligned} \qquad (2.18)$$

where the last line has been obtained by substituting $j = k - i$ into the previous line. Note that this expression defines an explicit *infinite-dimensional moving average representation* for the linear model. If the model is a member of the $MA(q)$ class for some finite q, it follows immediately from substitution of $e(k) = \delta(k)$ into Eq. (2.7) that $h(k)$ is nonzero only for $k = 0, 1, ..., q$. For this reason, moving average models are also frequently called *finite impulse response* (FIR) models.

For a *causal* linear, time-invariant dynamic model, the impulse response $\{h(k)\}$ is identically zero for $k < 0$, because the system does not respond before it is stimulated. Closely related to this observation is the fact that some systems exhibit an inherent delay $d > 0$; the impulse response of causal systems with a

Linear Dynamic Models

delay d is identically zero for $k < d$. This point is important in practice since many physical system models (in particular, most chemical process models) have an inherent delay $d \geq 1$. It is often possible to incorporate this inherent delay into a model with $d = 0$ by simply setting the appropriate model coefficients to zero. As a specific example, consider the following model:

$$y(k) = \sum_{i=1}^{p} a_i y(k-i) + \sum_{j=0}^{q} b_j u(k-d-j), \qquad (2.19)$$

which may be viewed either as an $ARMA(p,q)$ model with an inherent delay d, or as a special case of the following $ARMA(p, q+d)$ model:

$$y(k) = \sum_{i=1}^{p} a_i y(k-i) + \sum_{j=0}^{p+q} b'_j u(k-j), \qquad (2.20)$$

with $b'_0 = b'_1 = \cdots = b'_{d-1} = 0$. Generally, the latter view is adopted in this book, although sometimes this approach is inadequate, as the next result demonstrates.

In addition to the (generally infinite-order) moving average representation defined directly by the impulse response via Eq. (2.15), it is also possible to derive an autoregressive representation, also generally infinite dimensional. Specifically, a causal, time-invariant linear dynamic model with zero inherent delay may be represented in the form

$$y(k) = \sum_{i=1}^{\infty} a_i y(k-i) + b_0 u(k), \qquad (2.21)$$

where

$$a_i = \frac{1}{h(0)} \left\{ h(i) - \sum_{j=1}^{i-1} a_j h(i-j) \right\} \qquad b_0 = h(0). \qquad (2.22)$$

This result is easily verified by simply substituting $u(k) = \delta(k)$ into Eqs. (2.21) and (2.22), thus yielding the result $y(k) = h(k)$. Since the impulse response provides a complete characterization for this model, it follows that the representation given by Eqs. (2.21) and (2.22) is unique.

The assumption of zero inherent delay is important here since it is sufficient to guarantee that $h(0) \neq 0$. For causal, linear, time-invariant models with inherent delay $d > 0$, $h(0) = 0$ and the above results must be modified to

$$y(k) = \sum_{i=1}^{\infty} a_i y(k-i) + b_0 u(k-d), \qquad (2.23)$$

where

$$a_i = \frac{1}{h(d)} \left\{ h(d+i) - \sum_{j=1}^{i-1} a_j h(d+i-j) \right\} \qquad b_0 = h(d). \qquad (2.24)$$

2.2.3 State-space models

As discussed in Chapter 1, another useful representation for linear, discrete time models is the *state-space form* (Priestley, 1988, p. 19). In particular, if $q \leq p$, the linear $ARMA(p,q)$ model may be represented as the state-space model

$$\mathbf{x}(k) = \mathbf{A}\mathbf{x}(k-1) + \mathbf{b}u(k) \qquad y(k) = \mathbf{c}^T\mathbf{x}(k), \qquad (2.25)$$

where $\mathbf{x}(k)$ is a time-varying p-vector and $\mathbf{A}$ is the $p \times p$ matrix

$$\mathbf{A} = \begin{bmatrix} 0 & 0 & \cdots & 0 & a_p \\ 1 & 0 & \cdots & 0 & a_{p-1} \\ 0 & 1 & \cdots & 0 & a_{p-2} \\ \cdot & \cdot & \cdots & \cdot & \cdot \\ \cdot & \cdot & \cdots & \cdot & \cdot \\ \cdot & \cdot & \cdots & \cdot & \cdot \\ 0 & 0 & \cdots & 1 & a_1 \end{bmatrix}. \qquad (2.26)$$

The constant p-vectors $\mathbf{b}$ and $\mathbf{c}$ are defined by

$$\begin{aligned} \mathbf{b} &= [0, 0, \cdots, b_q, b_{q-1}, \cdots, b_1, b_0]^T, \\ \mathbf{c} &= [0, 0, \cdots 0, 0, \cdots, 0, 1]^T. \end{aligned} \qquad (2.27)$$

[The notation used here is slightly different from that in Priestley (1988, p. 19), but the results are the same.] Note that the initial condition $\mathbf{x}(0) = \mathbf{0}$ represents the "standard" initialization described earlier for ARMA models.

The state-space representation is convenient mathematically and emphasizes a number of useful connections between linear algebra and the theory of linear dynamical systems. For example, the impulse response of this model may be written as

$$h(k) = \sum_{n=1}^{p} \gamma_n \lambda_n^k, \qquad (2.28)$$

where $\{\lambda_n\}$ are the p eigenvalues of the matrix $\mathbf{A}$. Note that these eigenvalues depend only on the autoregressive coefficients $\{a_i\}$ of the $ARMA(p,q)$ model; dependence of the impulse response on the moving average coefficients $\{b_j\}$ arises from the coefficients $\{\gamma_n\}$.

2.2.4 Exponential and BIBO stability of linear models

This last observation provides the basis for an extremely useful stability analysis of $ARMA(p,q)$ models. Specifically, note that the eigenvalues of $\mathbf{A}$ can be either real or complex, but if they are complex, they must appear in complex conjugate pairs because the impulse response $h(k)$ is real. Therefore, define the two index sets I and J, associated with the real and complex eigenvalues of this matrix, and rewrite Eq. (2.28) as

$$h(k) = \sum_{n \in I} \gamma_n \lambda_n^k + \sum_{n \in J} [\gamma_n \lambda_n^k + \gamma_n^*(\lambda_n^*)^k], \qquad (2.29)$$

Linear Dynamic Models

where $*$ denotes complex conjugation. This result simplifies further if the complex eigenvalues are written in angular form as

$$\lambda_n = \mu_n e^{i\phi_n}, \tag{2.30}$$

where $\mu_n > 0$. Then, for $n \in J$, it follows that

$$\begin{aligned}\gamma_n \lambda_n^k + \gamma_n^*(\lambda_n^*)^k &= [\gamma_n e^{ik\phi_n} + \gamma_n^* e^{-ik\phi_n}]\mu_n^k \\ &= [\alpha_n \cos k\phi_n + \beta_n \sin k\phi_n]\mu_n^k, \end{aligned} \tag{2.31}$$

where the representation in terms of sin and cos follows from standard relations between trigonometric functions and complex exponentials. Combining all of these results leads to the general expression

$$h(k) = \sum_{n \in I} \gamma_n \lambda_n^k + \sum_{n \in J} [\alpha_n \cos k\phi_n + \beta_n \sin k\phi_n]\mu_n^k. \tag{2.32}$$

To obtain a stability characterization for linear models, define

$$\rho = \max_{n \in I}\{|\lambda_n|\} \quad \nu = \max_{n \in J}\{\mu_n\} \quad \omega = \max\{\rho, \nu\}. \tag{2.33}$$

It follows from the triangle inequality for absolute values that

$$\begin{aligned}|h(k)| &\leq \sum_{n \in I} |\gamma_n||\lambda_n|^k + \sum_{n \in J} [|\alpha_n| + |\beta_n|]\mu_n^k \\ &\leq \left[\sum_{n \in I} |\gamma_n|\right]\rho^k + \left[\sum_{n \in J} [|\alpha_n| + |\beta_n|]\right]\nu^k \\ &\leq \left[\sum_{n \in I} |\gamma_n| + \sum_{n \in J} [|\alpha_n| + |\beta_n|]\right]\omega^k. \end{aligned} \tag{2.34}$$

Therefore, this linear model is stable if $\omega < 1$, which is equivalent to the condition that all of the poles of the associated transfer function $H(z)$ lie inside the unit circle. This result is even stronger, however, since it shows explicitly that this model is *exponentially stable*. That is, not only does the impulse response decay to zero, it decays at least as rapidly as $C\omega^k$ as $k \to \infty$, where C is the constant appearing in brackets in the final expression in Eq. (2.34). This observation is significant, since "stability" may be defined in a number of different ways. A particularly important stability notion used throughout this book is that of *bounded-input, bounded-output* (BIBO) stability: a system is BIBO stable if $|u(k)| < M$ for all k implies $|y(k)| < N$ for all k for some finite constants N and M. For finite-dimensional, linear, time-invariant, discrete-time models, BIBO stability is equivalent to exponential stability. However, as subsequent discussions illustrate, BIBO stability does *not* imply exponential stability for infinite-dimensional, time-invariant linear models, for time-varying linear models, or for nonlinear models.

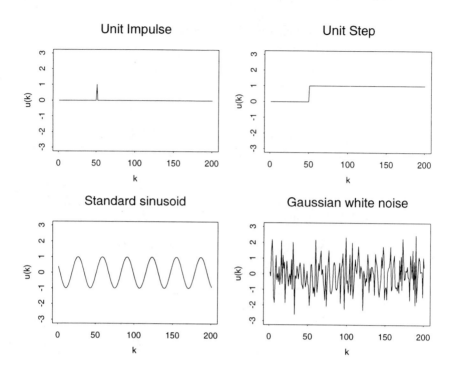

Figure 2.3: The four standard input sequences

2.3 Characterization of linear models

Although the topic of input sequence design is beyond the scope of this book, it is useful to briefly consider a few standard sequences that are capable of providing complete (or nearly complete) characterizations of linear models. It will be demonstrated (fairly dramatically in some cases) that these input sequences *do not* provide complete characterizations of *nonlinear* models, but they remain useful because they generally provide useful *partial* characterizations of nonlinear models (e.g., evidence of nonlinearity). Specifically, because they are all popular characterizations of linear systems, this book will consider four particular input sequences extensively in subsequent examples. These four standard sequences are listed in Table 2.1 and plotted in Fig. 2.3.

As the previous discussion has emphasized, impulse responses provide a complete characterization of causal, linear, time-invariant dynamic models. Informationally, step responses are equivalent to impulse responses, since if $\{u(k)\}$ is a unit step, $u(k) - u(k-1)$ is a unit impulse; therefore, if $y(k)$ is the unit step response for a linear, time-invariant model, $y(k) - y(k-1)$ is its impulse response. Similarly, responses to sinusoidal input sequences are also equivalent to impulse responses, *provided they are available for all angular frequencies* $-\pi \leq \omega \leq \pi$. This result follows from the fact that the *frequency response* $H(\omega)$ of such a

Table 2.1: The four standard input sequences

No.	Input sequence	Definition	
1	Unit impulse	$u(k) =$	$\begin{cases} 0 & -50 \leq k < 0 \\ 1 & k = 0 \\ 0 & 0 < k \leq 150 \end{cases}$
2	Unit step	$u(k) =$	$\begin{cases} 0 & -50 \leq k < 0 \\ 1 & 0 \leq k \leq 150 \end{cases}$
3	Standard sinusoid	$u(k) =$	$\sin\left(\frac{k\pi}{16}\right)$ for $-50 \leq k \leq 150$
4	Gaussian white noise	$u(k) \sim$	$N(0,1)$ for $-50 \leq k \leq 150$

model is defined as the discrete Fourier transform of its impulse response $h(k)$:

$$H(\omega) = \sum_{k=-\infty}^{\infty} h(k)e^{-ik\omega}. \qquad (2.35)$$

Because the Fourier transform is invertible, these two descriptions are equivalent; in particular, note that $h(k)$ may be recovered from $H(\omega)$ via

$$h(k) = \frac{1}{2\pi} \int_{-\pi}^{\pi} H(\omega)e^{ik\omega} d\omega. \qquad (2.36)$$

In fact, since $h(k)$ is real, it follows from Eq. (2.36) that $H(-\omega) = H^*(\omega)$, implying that $h(k)$ can be reconstructed from $H(\omega)$ over the reduced range $0 \leq \omega \leq \pi$. Intuitively, the frequency response is a complex quantity whose magnitude $|H(\omega)|$ quantifies the amplification or attenuation of a sinusoidal input of angular frequency ω, and whose phase $\angle H(\omega)$ quantifies the phase advance or retardation of this sinusoidal input. Specifically, for the input

$$u(k) = A\cos(k\omega + \theta), \qquad (2.37)$$

the output sequence $y(k)$ is given by

$$y(k) = A|H(\omega)|\cos[k\omega + \theta + \angle H(\omega)]. \qquad (2.38)$$

Time-invariant linear models may also be characterized by their response to stochastic input sequences. In particular, if the linear model is stable and the input sequence is a *stationary Gaussian stochastic process*, the output sequence will also be a stationary, Gaussian stochastic process (Papoulis, 1965). In this case, the frequency response $H(\omega)$ of the linear model may be recovered from the following result:

$$H(\omega) = \frac{S_{uy}(\omega)}{S_{uu}(\omega)}. \qquad (2.39)$$

Here, $S_{uy}(\omega)$ is the cross-spectrum between the input sequence $\{u(k)\}$ and the output sequence $\{y(k)\}$, defined as the Fourier transform of the cross-correlation function $R_{uy}(\ell)$

$$S_{uy}(\omega) = \sum_{\ell=-\infty}^{\infty} R_{uy}(\ell) e^{-i\ell\omega} \qquad R_{uy}(\ell) = E\{u(k) y(k+\ell)\}. \qquad (2.40)$$

Similarly, $S_{uu}(\omega)$ is the autospectrum of the input sequence, defined as the Fourier transform of the autocorrelation function $R_{uu}(\ell)$

$$S_{uu}(\omega) = \sum_{\ell=-\infty}^{\infty} R_{uu}(\ell) e^{-i\ell\omega} \qquad R_{uu}(\ell) = E\{u(k) u(k+\ell)\}. \qquad (2.41)$$

The qualitative interpretation of these spectra and correlations are discussed briefly in Chapter 8 (Sec. 8.4), but the key point here is that these quantities provide a basis for the input/output characterization of linear, time-invariant, discrete-time dynamic models.

To provide a simple "linear baseline" for comparison with more complex models as they are introduced, responses to the four standard input sequences listed in Table 2.1 are illustrated in Fig. 2.4 for the following $AR(1)$ model:

$$y(k) = 0.95 y(k-1) + u(k-1). \qquad (2.42)$$

Note that this model exhibits a single pole inside the unit circle at $z = 0.95$, indicating a stable response, although a slow one, given the proximity of this pole to the unit circle. Further, because the autoregressive coefficient $a_1 = 0.95$ is positive, the impulse response is monotonic, comparable to that of a continuous-time linear first-order model. This behavior is seen clearly in the unit impulse response shown in the upper left plot in Fig. 2.4.

The transfer function for this model follows directly from Eq. (2.42) as

$$H(z) = \frac{z^{-1}}{1 - 0.95 z^{-1}}. \qquad (2.43)$$

The steady-state gain of this model is $H(1) = 20$, consistent with the step response shown in the upper right plot in Fig. 2.4, which approaches this limit as $k \to \infty$. The frequency response is related to the transfer function via the substitution $z = e^{i\omega}$, from which it follows that the magnitude of this model's response to the "standard sinusoid" of frequency $\omega = \pi/16$ should be approximately 5.06. This result is in reasonable agreement with the response

Linear Dynamic Models

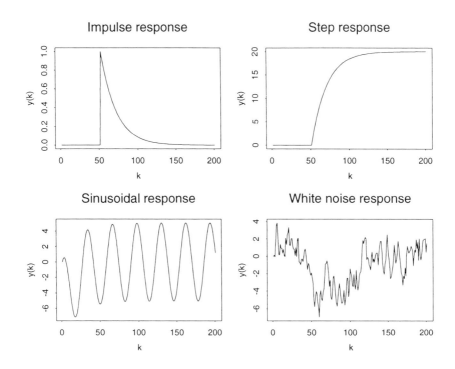

Figure 2.4: Standard responses for $AR(1)$ model

shown in the lower left plot in Fig. 2.4. More generally, note that the magnitude of the frequency response $|H(\omega)|$ decreases monotonically from the value 20 at $\omega = 0$ to approximately 0.5 at $\omega = \pi$. This general behavior means that the model is "low pass" in nature, amplifying low frequency inputs more than high frequency ones. For Gaussian white noise input sequences, this behavior means that the output sequence is smoother or less irregular than the input sequence. This behavior is seen clearly here in comparing the lower right plot in Fig. 2.4 with the lower right plot in Fig. 2.3.

Finally, it is important to say something about the practical handling of transient responses in the simulation of autoregressive models. Specifically, note that in the first two standard input sequences are explicitly intended to elicit transient behavior. In contrast, the last two sequences are assumed to be of infinite duration, extending back to $k = -\infty$ so that any initial transients have decayed to zero. In other words, in considering the response of any *stable* model to these last two sequences, it will *generally* be assumed that the sequence has been acting on the system long enough that any startup transient associated with switching the input sequence on has died out. In the first-order model responses shown in Fig. 2.4, this assumption is *not* valid, and the presence of a startup transient may be seen in the lower left plot (i.e., the sinusoidal response). In some cases (e.g., the sinusoidal response of an infinite-dimensional linear model), it may be possible to compute the *transient free* response exactly. In other cases, however, the most practical way to proceed is to start the simulation at some time in the remote past (say, $k = -1800$) and only retain the results for the range of interest ($-50 \leq k \leq 150$). A comparison of the general nature of the initial and final portions of such a simulation should provide confirmation that any startup transients have decayed to insignificance. Unless explicitly noted otherwise, the simulation results presented in this book for sinusoidal and white noise responses of models involving autoregressive terms were obtained by simulating 2000 values with an initial condition $y(k) = 0$ for $k \leq 0$ and retaining the final 200 values of this sequence.

2.4 Infinite-dimensional linear models

One of the points made in Sec. 2.2 is that the impulse response of a finite-dimensional, linear, time-invariant discrete-time model may be represented as a finite sum of exponentials. This result is closely related to the fact the transfer function $H(z)$ for such a model may be represented as the ratio of polynomials of finite orders q and p. Analogous results hold in continuous time: the impulse response of a linear, time-invariant model described by a finite collection of ordinary differential equations may also be represented as a finite sum of exponentials. Similarly, the continuous-time transfer function $H(s)$ may also be represented as a ratio of finite-order polynomials. Because our "linear intuition" genererally grows from a study of such models, it is easy to form the misconception that *all* linear, time-invariant models behave this way. The primary objective of the brief discussion given here is to illustrate that *infinite-dimensional* linear, time-invariant models exist, arise in practice, and can behave very differently from their finite-dimensional counterparts.

2.4.1 Continuous-time examples

In the discussion of fundamental models presented in Chapter 1, it was noted that linear partial differential equations can normally be expressed as equivalent *infinite-dimensional* state-space models. In particular, this point was illustrated for the case of the simple diffusion model, repeated here for convenience

$$\partial \psi / \partial t = \frac{\kappa}{\rho C_p} \nabla^2 \psi. \tag{2.44}$$

The most important qualitative difference between infinite-dimensional models like this one and more familiar finite-dimensional linear models is the significantly wider range of possible qualitative behavior in the infinite-dimensional case. For example, solutions to the diffusion equation commonly involve the complementary error function (Abramowitz and Stegun, 1972, ch. 7):

$$\operatorname{erfc} x = \frac{2}{\sqrt{\pi}} \int_x^\infty e^{-t^2} dt. \tag{2.45}$$

As a specific example, Johnson and Newman (1971) develop a mathematical model of an electrochemical desalination system based on porous carbon electrodes. The current density i obtained in response to a step change ΔV in the applied electrode potential is determined by setting up and solving a diffusion equation. The final result is of the form (Johnson and Newman, 1971, Eq. (23))

$$i = \left(\frac{\Delta V}{R}\right) e^{t/\tau} \text{ erfc } \sqrt{t/\tau}. \tag{2.46}$$

The principal point is that this type of response is *functionally* more complicated than that possible from finite-dimensional, time-invariant linear models.

Not surprisingly, the transfer functions associated with infinite-dimensional, time-invariant linear models are also generally more complex in structure than their more familiar finite-dimensional counterparts. For example, an electrical transmission line may be described by a pair of first-order linear partial differential equations. Solving these equations leads to the following expression for the impedance of an open-circuited transmission line of length a (Scott, 1970):

$$Z(s) = Z_0(s) \coth \gamma_0(s) a, \tag{2.47}$$

where

$$Z_0(s) = \sqrt{\frac{ls}{cs-g}} \quad \text{and} \quad \gamma_0(s) = \sqrt{ls(cs-g)}. \tag{2.48}$$

Here, s is the standard Laplace transform variable, l is the inductance per unit length of the transmission line, c is the capacitance per unit length, and g is the conductance per unit length. Once again, the point of this example is to illustrate the range of dynamic behavior possible from infinite-dimensional linear models: frequency-dependent behavior of this type cannot be represented as a ratio of polynomials in s.

Finally, it is instructive to consider one more example. Diffusion phenomena are important in electrochemistry (Bockris and Reddy, 1970), and one particularly interesting example is the following. Oldham and Spanier (1974) consider the time-evolution of an electro-reducible species A, initially present at a uniform concentration C_0 throughout a volume with a hanging mercury sphere electrode of small radius R. Transport is assumed to occur solely by radial diffusion with a diffusion coefficient D, and the relationship between the concentration of species A and the electrode current i is described by the following *semi-integral equation*:

$$\frac{d^{-1/2} i(t)}{dt^{-1/2}} = KC_0 \left[\xi(t) + \frac{\sqrt{D}}{R} \frac{d^{-1/2} \xi(t)}{dt^{-1/2}} \right], \tag{2.49}$$

where $\xi(t) = [C_0 - C(0,t)]/C_0$ is a normalized measure of the concentration of species A at the surface of the mercury electrode. The semi-integral symbol $d^{-1/2}/dt^{-1/2}$ is discussed at length in the book by Oldham and Spanier, who present a reasonably complete discussion of fractional-order calculus. The fundamental idea is very closely related to transfer functions $H(s)$ that are not simple ratios of polynomials in s. Specifically, recall that if the Laplace transform of $x(t)$ is $X(s)$, the Laplace transform of time integral of $x(t)$ is $s^{-1}X(s)$: analogously, the Laplace transform of the semi-integral of $x(t)$ is $s^{-1/2}X(s)$. Once again, the point of this example is that distributed-parameter phenomena like diffusion give rise to some extremely interesting dynamic effects that have no counterpart in finite-dimensional linear models. In fact, a fractional-order extension of the traditional PID controller structure has been recently proposed for systems whose dynamics exhibit this type of nonrational frequency response (Podlubny, 1999).

2.4.2 Discrete-time slow decay models

The discretization results for continuous-time linear state-space models presented in Chapter 1 also extend to infinite-dimensional models. Therefore, phenomena analogous to those just described in continuous-time also exist in discrete-time. In particular, one of the hallmarks of infinite-dimensional linear time-invariant systems is nonexponential decay, as the following example illustrates.

Even in the infinite-dimensional case, it is true that a linear, time-invariant, discrete-time dynamic model is completely characterized by its impulse response $h(k)$. Here, consider the family of models characterized by impulse responses of the form:

$$h_\alpha(k) = \frac{1}{(k+1)^\alpha}, \qquad (2.50)$$

for some $\alpha > 1$. Fig. 2.5 emphasizes the initial portion of four decay transients: $h_\alpha(k)$ for $\alpha = 1.5$, 1.7, and 2.0, together with the exponential decay of the first-order autoregressive model

$$y(k) = \gamma y(k-1) + u(k), \qquad (2.51)$$

for $\gamma = 0.8$. All four of these impulse responses decay from an initial value of $h(0) = 1$ to zero as $k \to \infty$. From this figure, it appears that $h_\alpha(k)$ decays much faster than the first-order model response for all three of these values of α. Asymptotically, however, the opposite is true, as seen in Fig. 2.6, which shows these same four responses for k between 20 and 30. It is clear from these plots that $h_\alpha(k)$ decays faster with increasing α, but the first-order model response has fallen below $h_\alpha(k)$ by $k = 22$ for $\alpha = 1.5$ and by $k = 27$ for $\alpha = 1.7$. Extrapolating these results, it is clear that the first-order impulse response will fall below $h_{2.0}(k)$ for some k slightly larger than 30.

In fact, it is easy to prove that, for any $\alpha > 1$, the impulse response $h_\alpha(k)$ ultimately decays more slowly than *any* exponential decay. Specifically, recall from the discussion in Sec. 2.2 that a response $r(k)$ decays exponentially if there exists some real constant $0 < \omega < 1$ and an integer N such that $|r(k)| < C\omega^k$ for

Linear Dynamic Models

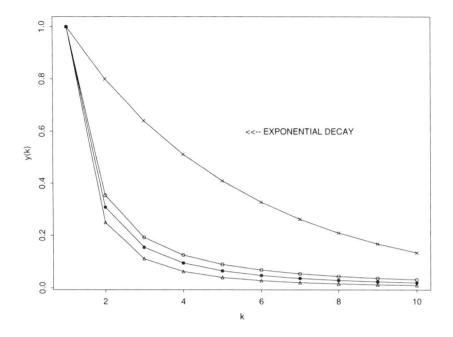

Figure 2.5: Initial comparison of four decays

all $k > N$. To show that $h_\alpha(k)$ does not satisfy such a bound, consider any real constant $0 < \omega < 1$ and any value $\alpha > 1$. Further, suppose that $|h_\alpha(k)| < C\omega^k$, that is:

$$(1+k)^{-\alpha} < C\omega^k. \tag{2.52}$$

Raising both sides of this inequality to the power $-1/\alpha$ yields

$$1+k > C^{-1/\alpha}\omega^{-k/\alpha} \equiv Ae^{\lambda k}, \tag{2.53}$$

where $\lambda = \ln[\omega^{-1/\alpha}] > 0$. (Recall that transforming both sides of an inequality by a negative power reverses the inequality sign.) It is a standard result (Abramowitz and Stegun, 1972, ch. 4) that for any positive integer n and any positive real x,

$$e^x > 1 + \frac{x^n}{n!}, \tag{2.54}$$

a result that follows directly from the Taylor series expansion for e^x. Substituting this result into the inequality (2.53) for $n = 2$ yields the quadratic inequality

$$1+k > A(1+\frac{\lambda^2 k^2}{2}). \tag{2.55}$$

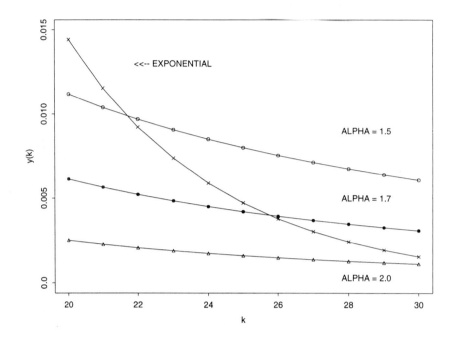

Figure 2.6: Later comparison of four decays

Note that because the quadratic term on the right-hand side of the inequality increases faster with increasing k than the linear term on the left-hand side of this inequality, this condition can only be satisfied for sufficiently small k. In particular, condition (2.51) can only be satisfied for $k < N$ where N depends on A and λ or, equivalently, α and ω. This result establishes, for any $\alpha > 1$ and $0 < \omega < 1$, that the original inequality (2.52) can only be satisfied for bounded values of k, since for k sufficiently large, the impulse response $h_\alpha(k)$ will exceed any decaying exponential $C\omega^k$. Because stable $ARMA(p,q)$ models are *exponentially stable* for any finite p and q, it follows that this slow decay behavior is an inherently infinite-dimensional phenomenon.

Conversely, although this linear model is not exponentially stable, it is easy to demonstrate BIBO stability. This result follows from the fact that the response of this model to arbitrary input sequences $\{u(k)\}$ is given by the convolution Eq. (2.15). Specifically, if $|u(k)| \leq M$ for all k, it follows that

$$|y(k)| \leq \sum_{i=0}^{\infty} |h(i)||u(k-i)|$$
$$\leq M \sum_{i=0}^{\infty} \frac{1}{(i+1)^\alpha} = M\zeta(\alpha), \qquad (2.56)$$

where $\zeta(\alpha)$ is the Riemann zeta function (Abramowitz and Stegun, 1972, p.

Linear Dynamic Models

807). This function is finite for all $\alpha > 1$, thus establishing the BIBO stability of the model whose impulse response is given by Eq. (2.50).

2.4.3 Fractional Brownian motion

One extremely important practical application of infinite-dimensional, time-invariant, linear discrete-time models is in the statistical modeling of fluctuations with *long range dependence* (Beran, 1994; Brockwell and Davis, 1991). Such fluctuations seem to arise in a number of different fields, and a very large literature has grown up around their analysis, both in statistics under the name long memory processes (Beran, 1994; Brockwell and Davis, 1991), and in physics under the name $1/f$-noise (Keshner, 1982). A general review of these ideas is well beyond the scope of this book, but for a comprehensive introduction, refer to Beran (1994). There, he treats five datasets in considerable detail, including one of minimal water levels for the Nile River between the years 622 and 1281. He briefly discusses the historical significance of this dataset, which led to the formulation of the Hurst effect in hydrology in an effort to understand the long-range behavior (i.e., alternating periods of flooding and drought) of the Nile River. Subsequent efforts by Mandelbrot and coworkers led to the formulation of *fractional Brownian motion* processes as stochastic models for these fluctuations. The following paragraphs give a brief introduction to this notion.

A logical place to begin is with the definition of ordinary Brownian motion, which is a nonstationary, Gaussian stochastic process generated by the simple recursion relation

$$\beta_k = \beta_{k-1} + e_k. \tag{2.57}$$

Here, $\{e_k\}$ is an independent, identically distributed, zero-mean, Gaussian stochastic process with variance σ^2. Initializing this recursion relation with $\beta_0 = 0$, it follows by direct substitution that

$$\beta_k = \sum_{j=1}^{k} e_j, \tag{2.58}$$

for all $k \geq 1$, from which it further follows that

$$E\{\beta_k\} = 0 \quad \text{and} \quad E\{\beta_k^2\} = k\sigma^2. \tag{2.59}$$

Brownian motion holds an extremely important place in the theory of stochastic processes (Brockwell and Davis, 1991; Priestley, 1981), but it is enough to note here that the sequence $\{\beta_k\}$ has zero mean and a variance that grows linearly with time, consistent with the interpretation of Eq. (2.57) as an $AR(1)$ model with the unstable coefficient $a_1 = 1$ (sometimes referred to as a *unit root* process).

There are circumstances under which Brownian motion is a reasonable model for fluctuations that "drift" with time, and it provides the basis for the class of *autoregressive integrated moving average (ARIMA) models* frequently studied in time-series analysis (Box et al., 1994). To define these models, begin by defining the *difference operator* ∇ acting on a time-series $\{x_k\}$

$$\nabla x_k = (1 - z^{-1})x_k = x_k - x_{k-1}, \qquad (2.60)$$

where z^{-1} represents the backward shift operator. Applying this difference operator to Brownian motion yields

$$\nabla \beta_k = \beta_k - \beta_{k-1} = e_k. \qquad (2.61)$$

In other words, differencing Brownian motion yields Gaussian white noise, a stationary stochastic process. Often, differencing observed time series seems to correct for apparent nonstationarity, thus leading to the notion of ARIMA models. Specifically, ARIMA models replace the Gaussian white noise sequence $\{e_k\}$ in Eq. (2.61) with an $ARMA(p,q)$ sequence

$$\nabla z_k = x_k \qquad x_k = \sum_{i=1}^{p} a_i x_{k-i} + \sum_{i=0}^{q} b_i e_{k-i}. \qquad (2.62)$$

It is not difficult to show that this model may be reexpressed as

$$z_k = \sum_{i=1}^{p} a_i z_{k-i} + \sum_{i=0}^{q} b_i \beta_{k-i}, \qquad (2.63)$$

where $\{\beta_k\}$ is the Brownian motion or "integrated white noise" sequence defined above. This observation underlies the acronym ARIMA: autoregressive integrated moving average models.

As noted, ARIMA models sometimes provide reasonable descriptions of a time series, but there are circumstances in which the "drift" represented by ordinary Brownian motion is too severe. In these cases, *fractional Brownian motion* can provide a reasonable alternative. There, the difference operator ∇ is replaced with the *fractional difference operator* ∇^d, defined as

$$\nabla^d = (1 - z^{-1})^d = \sum_{j=0}^{\infty} \pi_j z^{-j}. \qquad (2.64)$$

The basis for this definition is the binomial expansion of the fractional power $(1-x)^d$, and the coefficients π_j are given by (Brockwell and Davis, 1991, p. 520)

$$\pi_j = \frac{\Gamma(j - d)}{\Gamma(j + 1)\Gamma(-d)}, \qquad (2.65)$$

where $\Gamma(x)$ is the gamma function (Abramowitz and Stegun, 1972), reducing to the factorial $\Gamma(x) = (x-1)!$ for positive integers. A fractional Brownian motion sequence $\{f_k\}$ is defined by replacing ∇ with ∇^d in Eq. (2.61), that is,

$$\nabla^d f_k = e_k, \qquad (2.66)$$

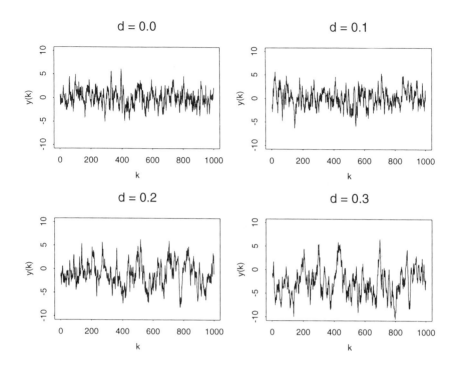

Figure 2.7: Fractional ARIMA models, $d = 0.0, 0.1, 0.2, 0.3$

where $\{e_k\}$ is Gaussian white noise, as before. Note that $\{f_k\}$ reduces to Gaussian white noise when $d = 0$, and to ordinary Brownian motion when $d = 1$. The sequence is well defined for all $d > -1$, but it is of primary interest for $0 < d < 0.5$.

To see the connection with infinite-dimensional linear models, note that f_k may be represented as the infinite-order moving average process (Brockwell and Davis, 1991, p. 521)

$$f_k = \sum_{j=0}^{\infty} \psi_j e_{k-j}, \qquad (2.67)$$

where the coefficients ψ_j are given by

$$\psi_j = \frac{\Gamma(j+d)}{\Gamma(j+1)\Gamma(d)}. \qquad (2.68)$$

The class of *fractional ARIMA models* is obtained by replacing the Brownian motion sequence $\{\beta_k\}$ appearing in Eq. (2.63) with the fractional Brownian motion sequence $\{f_k\}$.

Eight examples of fractional ARIMA model responses are shown in Figs. 2.7 and 2.8. In all cases, the basic model is

$$z_k = 0.8 z_{k-1} + f_k, \qquad (2.69)$$

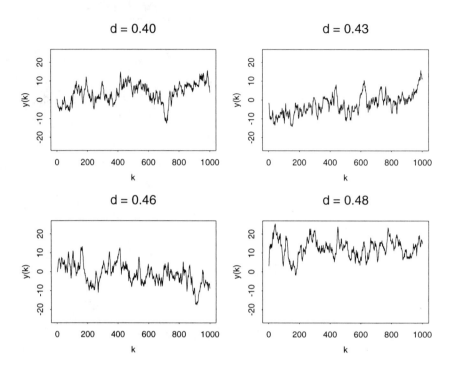

Figure 2.8: Fractional ARIMA models, $d = 0.40, 0.43, 0.46, 0.49$

where $\{f_k\}$ is a fractional Brownian motion sequence of order d. In Fig. 2.7, the upper left plot shows this response for $d = 0$, corresponding to the repsonse of a first-order linear model to a Gaussian white noise input sequence. The other three plots in Fig. 2.7 show the corresponding ARIMA sequences for $d = 0.1$, 0.2, and 0.3; note the appearance of what look like long term drifts or possibly cycles in these sequences, particularly for $d = 0.3$. These tendencies are even more pronounced in Fig. 2.8, where the ARIMA sequences for $d = 0.40, 0.43$, 0.46, and 0.48 are plotted. Also, note the wider range of variation in Fig. 2.8 relative to Fig. 2.7; in particular, note that the difference in scales in these two plots. Relative to the first-order autoregressive sequence shown in the upper left plot in Fig. 2.7, these sequences exhibit clear evidence of *long memory effects*, reflecting the fact that the moving average model coefficients ψ_j decay slowly with j.

2.5 Time-varying linear models

The main point of Sec. 2.4 was to illustrate that *infinite-dimensional* time-invariant linear models can behave very differently from the more familiar finite-dimensional ones. Similar conclusions hold for the class of *time-varying linear models*, defined as linear dynamic models whose parameters vary with time.

Linear Dynamic Models

An important feature of these models is that this time variation is *independent* of variations in the input $u(t)$ or $u(k)$. That is, if the parameters in a "linear model" were to depend, directly or indirectly, on either the model inputs or the model outputs, the resulting model would be *nonlinear*. A fairly well-developed theory is available for time-varying linear models (Halanay and Ionescu, 1994), but a detailed discussion of this theory lies beyond the scope of this book. Instead, a few simple examples are presented to illustrate the range of behavior these models can exhibit. Time-varying linear models may be either finite-dimensional or infinite-dimensional, but only the finite-dimensional case is considered here for simplicity. Similarly, this book generally has little to say about the class of time-varying *nonlinear* models because it is too large and too difficult to analyze. It is sometimes possible, however, to represent the responses of *time-invariant, nonlinear* models to specific input sequences as the responses of an *equivalent time-varying linear model*, as examples presented in subsequent chapters illustrate.

2.5.1 First-order systems

Two particularly simple examples of time-varying linear models are the (unforced) first-order continuous-time model

$$\frac{dy}{dt} = -a(t)y(t),$$

$$\Rightarrow y(t) = y(0)\exp\left\{-\int_0^t a(\tau)d\tau\right\}, \quad (2.70)$$

and the corresponding (unforced) first-order discrete-time model

$$y(k) = a(k)y(k-1),$$

$$\Rightarrow y(k) = y(0)\prod_{i=1}^k a(i) = y(0)\exp\left\{\sum_{i=1}^k \ln a(i)\right\}. \quad (2.71)$$

As in the more familiar time-invariant case, note the close relationship between the continuous-time and discrete-time models. Also, note that the solution of the discrete-time equation will be particularly useful, since it provides a basis for constructing time-varying linear models with specified impulse responses. Specifically, consider a model of the form

$$y(k) = a(k)y(k-1) + u(k), \quad (2.72)$$

and note that the unit impulse response is given by

$$h(k) = \prod_{i=1}^k a(i), \quad \text{for } k > 0 \qquad h(0) = 1. \quad (2.73)$$

This result follows from the fact that for an impulse input $u(k)$, Eq. (2.73) reduces to the unforced model (2.71) for $k > 0$. The following four examples illustrate the utility of this result.

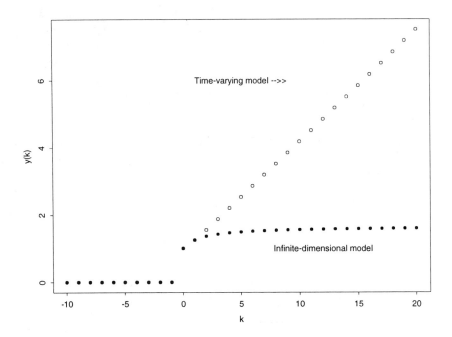

Figure 2.9: A comparison of two "slow decay" model responses

The first example illustrates the difference between time-varying linear models and the infinite-dimensional, time-invariant linear models. In particular, both of these model classes can exhibit slow decay phenomena that finite-dimensional, time-invariant models cannot exhibit. To gain insight into the difference between these two model classes, it is useful to compare two models, one from each class, that exhibit the same impulse response

$$h(k) = \frac{1}{(k+1)^2}. \tag{2.74}$$

It is not difficult to show that this impulse response may be realized by the first-order, time-varying linear model (2.72) with coefficients $a(k)$ given by

$$a(k) = \frac{k^2}{(k+1)^2}. \tag{2.75}$$

Recall that this same impulse response was exhibited by the time-invariant, infinite-dimensional linear model considered in Sec. 2.4. The model itself is not the same, however, as seen by comparing the step responses shown in Fig. 2.9 for both models.

In particular, simple explicit expressions may be developed for both of these step responses, and these expressions clearly illustrate the difference between these two models. For the time-varying case, it follows immediately from Eq. (2.72) that $y(0) = u(0) = \alpha$ for a step of amplitude α. For $k \geq 1$, it may be shown by induction that the step response is

$$y(k) = \frac{\alpha}{(k+1)^2} \sum_{n=1}^{k+1} n^2. \tag{2.76}$$

The finite sum appearing in this expression may be evaluated analytically (Gradshteyn and Ryzhik, 1965, no. 0.121.2),

$$\sum_{n=1}^{k+1} n^2 = \frac{(k+1)(k+2)(2k+3)}{6}. \tag{2.77}$$

Hence, the step response for the time-varying model is given by

$$y(k) = \alpha \left[\frac{k+2}{k+1}\right] \left(\frac{2k+3}{6}\right) \to \alpha \left(\frac{2k+3}{6}\right) \text{ as } k \to \infty. \tag{2.78}$$

An important corollary of this result is that the time-varying model is *not* BIBO stable: the response to a finite amplitude step is unbounded. This result is somewhat counterintuitive since $|a(k)| < 1$ for all k, which might *appear* to generalize the time-invariant linear stability condition $|a| < 1$. The difficulty lies in the fact that there is no *uniform* bound on $|a(k)|$: that is, there is no $M < 1$ such that $|a(k)| \leq M$ for all k. If such a bound did exist, it would follow that the time-varying model was *exponentially stable*, thus further implying BIBO stability. In contrast, it follows from the results presented in Sec. 2.4.2 that the infinite-dimensional time-invariant model whose impulse response is given by Eq. (2.74) *is* BIBO stable. In particular, the step response of this model is given directly by the convolution sum

$$\begin{aligned} y(k) &= \sum_{n=0}^{\infty} h(n) u(k-n) \\ &= \sum_{n=0}^{k} \frac{\alpha}{(n+1)^2} \\ &\leq \alpha \zeta(2) = \frac{\pi^2 \alpha}{6} \simeq 1.64 \alpha. \end{aligned} \tag{2.79}$$

The dramatic difference in these step responses is seen in Fig. 2.9.

The second time-varying linear model example considered here exhibits *superexponential decay*, in contrast to the *subexponential decay* seen in the previous example Specifically, consider the following impulse response

$$h(k) = \frac{1}{(k+1)!}, \tag{2.80}$$

which decays *much* faster than any exponential. (In particular, it follows from Stirling's approximation (Abramowitz and Stegun, 1972) that $n! \simeq n^n$ for large

n, rapidly becoming much larger than c^n for any fixed c.) It follows from Eq. (2.73) that the first-order, time-varying model (2.72) will exhibit this impulse response if $a(k)$ is taken as:

$$a(k) = \frac{1}{k+1}. \tag{2.81}$$

It is interesting to contrast this solution with that obtained for the analogous *continuous-time* model. Specifically, consider the time-varying model (2.70) with $a(t) = 1/(t+1)$: It follows immediately from the expression for the solution for this equation that

$$y(t) = \frac{y(0)}{t+1}. \tag{2.82}$$

In other words, simply "translating" the continuous-time model (2.70) into the discrete-time model (2.71) transforms a superexponential decay model into a subexponential decay model. This observation should serve as a warning for nonlinear models, where the situation is *much* more complex. The key point is that the behavioral similarity that exists between *time-invariant, linear* discrete-time and continuous-time models *does not* extend to either time-varying or nonlinear models.

It is also interesting to note that the response $y(t)$ given in Eq. (2.82) satisfies the *nonlinear* continuous-time ordinary differential equation

$$\frac{dy(t)}{dt} = -y^2(t), \tag{2.83}$$

subject to the initial condition $y(0) = 1$. Both this model and the un-forced time-varying model (2.70) may be viewed as the responses of externally forced differential equation models to impulse inputs that effectively impose the initial condition $y(t) = y(0)$ at $t = 0$. The key point here, then, is that the response of a physical system to a single input is not generally enough to determine its nature: For example, is a time-varying linear model more appropriate, or is a nonlinear model more appropriate? For this reason, it is important to examine responses to a number of different excitations.

The third time-varying linear model example considered here is closely related to this last one. Specifically, the following model exhibits superexponential *instability*. Defining $a(k)$ as

$$a(k) = k + 1, \tag{2.84}$$

it follows that the impulse response is

$$h(k) = (k+1)! \tag{2.85}$$

This example illustrates that unstable models, which are necessarily *exponentially unstable* in the finite-dimensional, time-invariant case, can "explode" much more rapidly in the case of time-varying linear models. An even more dramatic

Linear Dynamic Models

form of instability is possible in nonlinear models, known as *finite escape time*, in which the solution diverges in a finite number of steps. It should be clear that for bounded input sequences $\{u(k)\}$ and bounded model coefficients $\{a(k)\}$, time-varying linear models cannot exhibit finite escape times.

The fourth and final time-varying model example is based on the following recursion relation for the nth-order exponential integral $E_n(x)$ (Abramowitz and Stegun, 1972, ch. 5)

$$E_n(x) = \int_1^\infty t^{-n} e^{-xt} dt. \qquad (2.86)$$

Specifically, for fixed x, $y(n) = E_{n+1}(x)$ satisfies (Wimp, 1984, p. 14)

$$y(n) = \left[-\frac{x}{n}\right] y(n-1) + \frac{e^{-x}}{n}. \qquad (2.87)$$

Clearly, this recursion relation corresponds to the time-varying linear model defined in Eq. (2.72) with model coefficient

$$a(k) = -\frac{x}{k}, \qquad (2.88)$$

and the external input

$$u(k) = \frac{e^{-x}}{k}, \qquad (2.89)$$

for some fixed constant x. The real point of this example is to illustrate the range of behavior observable in time-varying linear models. In fact, Wimp (1984) lists many recursion relations for computing numerical values of special functions that may be regarded as first- and second-order, time-varying linear models. Ultimately, most of the special functions of mathematical physics emerge as solutions of these equations.

2.5.2 Periodically time-varying systems

In continuous-time, an important special case of the class of time-varying linear systems is the class of *periodically time-varying systems* (Richards, 1983). As a simple example, consider the following first-order unforced periodic system:

$$\frac{dy}{dt} + (a - 2q \cos \omega t) x = 0. \qquad (2.90)$$

The solution of this equation follows directly from Eq. (2.70):

$$y(t) = y(0) e^{-at} e^{(2q/\omega) \sin \omega t}. \qquad (2.91)$$

For $q = 0$, this system reduces to the first-order time-invariant linear model considered earlier, but for $q \neq 0$, the general behavior changes substantially.

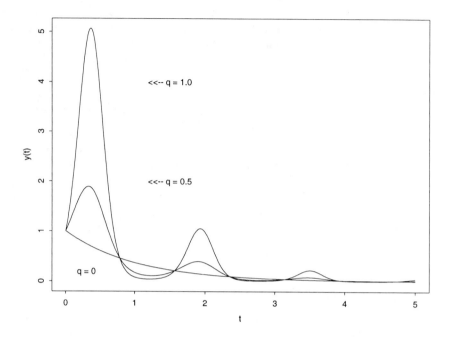

Figure 2.10: A first-order periodic model: three responses

This point may be seen from the solutions shown in Fig. 2.10 for $y(0) = 1$, $a = 1$, $\omega = 4$, and $q = 0$, 0.5, and 1.0, all plotted on the same axes.

Within the class of periodically time-varying continuous-time linear models, the most detailed results seem to be available for Hill's equation, a second-order periodic linear differential equation of the form (Richards, 1983):

$$\frac{d^2y}{dt^2} + [a - 2q\psi(t)]y = 0, \qquad (2.92)$$

where $\psi(t)$ is a periodic function. One of the best known special cases of Hill's equation is the Mathieu equation, obtained for $\psi(t) = \cos 2t$; much is known about the general behavior of this equation, although the solution is not expressible in terms of elementary functions (Abramowitz and Stegun, 1972; Richards, 1983). For $q = 0$, Hill's equation describes the simple harmonic oscillator (e.g., small amplituded oscillations of a pendulum or a lossless spring-mass system) and has the solution

$$y_0(t) = A \sin \sqrt{a} t + B \cos \sqrt{a} t. \qquad (2.93)$$

For $q \neq 0$, this equation can exhibit *parametric amplification*, in which the magnitude of the response is much larger than those observed for $q = 0$. Physically, these responses result when the system of interest extracts energy from the source of the sinusoidal variation $\psi(t)$. In fact, similar responses may be seen in the first-order system considered above: note how much larger the maximum response is for this system when $q = 1.0$ than when $q = 0$. Since the natural response of Hill's equation is oscillatory in the absence of periodic parameter variations (i.e., when $q = 0$), it turns out that the relationship between the period of the natural response and that of $\psi(t)$ is important.

A more tractable special case of Hill's equation is the Meissner equation (Richards, 1983), for which $\psi(t)$ alternates periodically between ± 1. Specifically, the solution to this equation may be obtained by piecing together the solutions (2.93) obtained for the constant coefficients $a + 2q$ and $a - 2q$. Generalizations of this equation can also be solved analytically, for which $\psi(t)$ is again piecewise constant, but assumes more than two values per period. Similarly, it is also possible to solve the "sawtooth" version of Hill's equation, obtained by taking $\psi(t) = 1 - 2t/T$ on $[0, T]$ and replicating it periodically outside this interval. By a change of variable, it is possible to transform this special case of Hill's equation into Stoke's equation (Richards, 1983, p. 33):

$$\frac{d^2 y}{dt^2} + \left(\frac{t}{\beta^2}\right) y = 0. \tag{2.94}$$

The advantage of this transformation is that the solution of Stoke's equation may be expressed analytically in terms of Bessel functions of order $\pm 1/3$. The point of this example is two-fold. First, it once again illustrates the strong connections that exist between many of the known special functions of mathematical physics and the class of time-varying linear models. Second, this example illustrates that it is sometimes possible to relate periodically time-varying models to other, non-periodic, time-varying linear models.

Discrete-time analogs of these periodically time-varying linear models have also been investigated, including the class of *cyclostationary stochastic processes* (Adams and Goodwin, 1995; Bloomfield et al., 1994). For a *stationary stochastic process* $\{x(k)\}$, it is a standard result that the mean $\bar{x}$ is constant and the autocorrelation function $R_{xx}(k_1, k_2)$ depends only on the difference $k_1 - k_2$. In contrast, for a cyclostationary process, the mean is periodic with period P and the autocorrelation function satisfies $R_{xx}(k_1 + P, k_2 + P) = R_{xx}(k_1, k_2)$. One natural model for periodically correlated sequences is the *periodic ARMA model* or PARMA model (Adams and Goodwin, 1995; Bloomfield et al., 1994), defined as follows (Bloomfield et al., 1994):

$$\begin{aligned} y(nP + \nu) &= \sum_{k=1}^{p(\nu)} a_k(\nu) y(nP + \nu - k) \\ &+ \sum_{k=0}^{q(\nu)} b_k(\nu) u(nP + \nu - k). \end{aligned} \tag{2.95}$$

Here, ν is called the *seasonal index*, and it lies in the range $1 \leq \nu \leq P$. This model may be viewed as a linear ARMAX model whose order parameters $p(\nu)$ and $q(\nu)$ and coefficients $a_k(\nu)$ and $b_k(\nu)$ exhibit "seasonal" dependence. Restricting consideration to $q(\nu) = 0$ for all ν defines the subclass of *periodic autoregressive models* (McLeod, 1994), whereas the restriction $q(\nu) = 0$ defines the *periodic moving average models* (Bentarzi and Hallin, 1994).

Note that all of these results reduce to the more familiar problem of ARMAX modeling of stationary processes (or time-invariant linear systems) if $P = 1$. To see the difference between this case and the more general case, consider the following simple example, somewhat analogous to the first-order continuous-time model considered at the beginning of this section. Specifically, consider the periodic autoregressive model for $P = 2$ defined by

$$y(2n + \nu) = a(\nu)y(2n + \nu - 1) + b_0 u(2n + \nu). \tag{2.96}$$

Here, $a(\nu)$ assumes two values, one for $\nu = 1$ and another for $\nu = 2$, so this model may be re-written as

$$y(2n + 1) = a(1)y(2n) + b_0 u(2n + 1)$$
$$y(2n + 2) = a(2)y(2n + 1) + b_0 u(2n + 2). \tag{2.97}$$

A sufficient condition for the stability of this model is $|a(1)a(2)| < 1$ (Bloomfield et al., 1994); as a specific example, note that the model defined by $a(1) = 1/2$, $a(2) = -3/2$ is stable, despite the fact that $|a(2)| > 1$. The standard four responses for this particular periodic autoregressive model are shown in Fig. 2.11. The complex nature of these responses—in contrast with the time-invariant first-order linear model considered in Sec. 2.3—provides yet another dramatic illustration of the differences that exist between the time-invariant and the time-varying classes of linear models.

Finally, it is worth noting that the model just described may be reformulated as follows. Combining the two equations in (2.97) yields the following pair of equivalent second-order time-varying linear models

$$y(2n + 2) = [a(1)a(2)]y(2n) + b_0 u(2n + 2) + [a(2)b_0]u(2n + 1)$$
$$y(2n + 1) = [a(1)a(2)]y(2n - 1) + b_0 u(2n + 1) + [a(1)b_0]u(2n). \tag{2.98}$$

Note that this reformulation has moved the time variation from the autoregressive model coefficients to the moving average model coefficients; in fact, the sequence $\{y(k)\}$ may be further reexpressed as the response of the *second-order, linear, time-invariant model*

$$y(k) = [a(1)a(2)]y(k - 2) + b_0 v(k), \tag{2.99}$$

to the *periodically modulated input sequence*

$$v(k) = \begin{cases} u(k) + a(1)u(k-1) & k \text{ odd} \\ u(k) + a(2)u(k-1) & k \text{ even.} \end{cases} \tag{2.100}$$

This interpretation clarifies the stability condition noted above; in particular, the linear time-invariant model (2.99) satisfies the necessary and sufficient stability conditions discussed in Chapter 4 for second-order models if and only if $|a(1)a(2)| < 1$. This interpretation also provides an explanation for the "complex" behavior seen in the step and sinusoidal responses shown in Fig. 2.11: the modulated input sequences $\{v(k)\}$ for these examples are *not* steps and sinusoids, respectively, but oscillatory sequences with period 2.

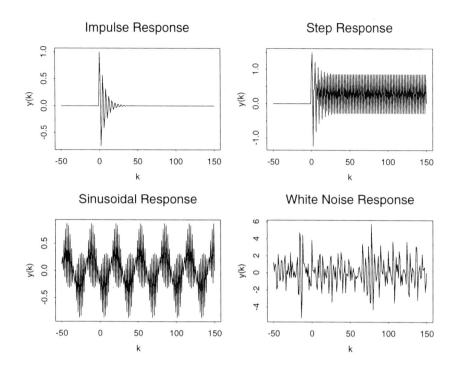

Figure 2.11: A first-order PAR model: standard four responses

2.6 Summary: the nature of linearity

Since this chapter is primarily intended to provide a linear baseline for subsequent discussions of nonlinear models, it seems appropriate to conclude with a brief discussion of the notion of linearity itself. In terms of its *behavior*, a mathematical function $f(x)$ is linear if it satisfies the superposition condition

$$f(a_1 x_1 + a_2 x_2) = a_1 f(x_1) + a_2 f(x_2), \qquad (2.101)$$

for all real constants a_1, a_2, and all x_1, x_2 in its domain $\mathcal{D}$. This condition generalizes easily from R^1 to arbitrary Euclidean spaces R^n, and to sequence spaces like ℓ_2 or ℓ_∞. In fact, *bounded input, bounded output (BIBO) stable systems* may be viewed as transformations mapping a bounded input sequence $u \in \ell_\infty$ into a bounded output sequence $y \in \ell_\infty$. Taking this view, a BIBO system is *linear* if it satisfies condition (2.101) for arbitrary ℓ_∞ input sequences x_1 and x_2.

It is also possible to define linearity as a *structural restriction* on the function $f(x)$. If the domain of the function is R^n, linearity implies

$$f(\mathbf{x}) = \sum_{i=1}^{n} c_i x_i, \qquad (2.102)$$

where $\{c_i\}$ is a set of n real constants and x_i represents the ith component of the argument $\mathbf{x}$. This definition also extends naturally to sequence spaces like ℓ_2 or ℓ_∞, taking the limit as $n \to \infty$. Here, it is necessary to restrict the coefficients c_i so that the sum converges for all $\mathbf{x}$ in the domain of $f(\cdot)$. In the case of linear BIBO systems, the argument $\mathbf{x}$ represents the input sequence and the coefficients c_i are directly given by the system's impulse response.

Mathematically, it yields a certain amount of useful insight to decompose condition (2.101) into two somewhat weaker conditions. Specifically, $f(x)$ is *additive* if it satisfies Cauchy's equation (Aczel and Dhombres, 1989),

$$f(x_1 + x_2) = f(x_1) + f(x_2), \qquad (2.103)$$

and *homogeneous* if it satisfies the scaling relation (Aczel and Dhombres, 1989; Saaty and Bram, 1964; Saaty, 1981),

$$f(ax) = af(x), \qquad (2.104)$$

for all real a and all x in the domain of $f(\cdot)$. Hence, if $f(\cdot)$ satisfies Eqs. (2.103) and (2.104) it also satisfies Eq. (2.101) and is therefore linear. The domain of $f(\cdot)$ is critically important, and it is useful to distinguish two cases: scalar functions $f: R \to R$, and multivariable functions $f: R^n \to R$ for $n > 1$. In particular, note that the only homogeneous function $f: R \to R$ is the linear function $f(x) = cx$ for some real constant c. To see this point, consider $f(1)$: if $f(1) = 0$, it follows by homogeneity that $f(x) = xf(1) = 0$ for all x. Similarly, if $f(1) \neq 0$, then $f(x) = xf(1) = cx$ where $c = f(1)$, so the result again holds. *Thus, for scalar functions, homogeneity and linearity are equivalent, a result that does not extend to multivariable functions.*

The case of scalar additive functions is somewhat more complex. Clearly, the linear function $f(x) = cx$ satisfies Eq. (2.103), but it is known that certain very badly behaved nonlinear functions also satisfy this condition. In particular, *nonlinear* solutions of Cauchy's equation are nowhere-continuous functions whose graph is dense in the plane (Aczel and Dhombres, 1989, ch. 2). Consequently, any function $f: R \to R$ satisfying Eq. (2.103) is of the form $f(x) = cx$ for some real constant c if it satisfies any of the following conditions:

1. $f(x)$ is continuous at any point $x_0 \in R$
2. $f(x)$ is monotonic on any measurable interval in R
3. $f(x)$ is bounded above or below on any measurable interval in R.

Further, it can be shown that all solutions of Cauchy's equation satisfy $f(x) = cx$ for some real constant c for all *rational* points $x = m/n$. Therefore, any scalar function that is likely to arise in either fundamental or empirical modeling of real-world phenomena will satisfy Cauchy's equation if and only if it is linear.

Cauchy's equation may be extended to functions $f: R^n \to R$ for arbitrary n and solutions for this case may be obtained by extending the scalar result. That is, define the scalar functions $f_k(x_k)$:

$$f_k(x_k) = f(0, 0, \ldots, x_k, \ldots, 0), \qquad (2.105)$$

where x_k appears in the kth argument of the original function $f(\cdots)$. It is not difficult to show (Aczel and Dhombres, 1989, p. 34) that if $f : R^n \to R$ satisfies Cauchy's equation, it also satisfies the equation:

$$f(x_1, x_2, \ldots, x_n) = \sum_{k=1}^{n} f_k(x_k), \tag{2.106}$$

where the functions $f_k(\cdot)$ all satisfy Cauchy's scalar equation (2.103). Thus, if $f : R^n \to R$ satisfies Cauchy's equation, it is necessarily of the form

$$f(x_1, x_2, \ldots, x_n) = \sum_{k=1}^{n} c_k x_k, \tag{2.107}$$

provided any one of the following additional conditions is satisfied:

1. $f(\mathbf{x})$ is continuous at any point $\mathbf{x}_0 \in R^n$
2. $f(\mathbf{x})$ is monotonic on any measurable subset of R^n
3. $f(\mathbf{x})$ is bounded above or below on any measurable subset of R^n.

The power of this result is that it links *additive behavior* defined by condition (2.103) for any function $f : R^n \to R$ to the *linear structure* defined in Eq. (2.107), subject only to the exclusion of the somewhat pathological exceptions noted above. As a specific application, consider the class of NMAX models discussed in Chapter 4

$$y(k) = F(u(k), u(k-1), \ldots, u(k-q)). \tag{2.108}$$

If the input/output behavior of this model is required to satisfy Cauchy's equation, then $u(k) \to y(k)$ and $v(k) \to z(k)$ implies that $u(k) + v(k) \to y(k) + z(k)$. Thus, it follows that the function $F : R^{q+1} \to R$ defining this model must also satisfy Cauchy's equation, reducing it to a linear function of the form given in Eq. (2.107). Hence, the NMAX model is in fact the linear moving average model

$$y(k) = \sum_{j=0}^{q} b_j u(k-j), \tag{2.109}$$

provided that the pathological exceptions noted above are excluded (as they always will be in modeling physical phenomena).

Extension of these results to more general NARMAX models is also possible, but a new issue arises. Specifically, in the NMAX example considered here, additive input/output behavior directly implies that the unknown function $F(\cdots)$ must satisfy Cauchy's equation. In contrast, in models with autoregressive dependence, since the arguments of $F(\cdots)$ are only indirectly determined by the input sequence $\{u(k)\}$, it does not necessarily follow that $F(\cdots)$ must satisfy Cauchy's equation on all of its domain. Therefore, to extend the argument that "linear behavior implies linear structure" to more general NARMAX models, it is necessary to impose some type of "controllability" condition to guarantee that additive input/output behavior does indeed imply that the defining function satisfies Cauchy's equation. To see the essential nature of this controllability issue, consider the following simple example. Suppose it is known that a model of the form

$$y(k) = F(y(k-1), u(k)) \tag{2.110}$$

exhibits additive input/output behavior, but only for $|u(k)| \leq 1$. It is not difficult to see that $F(\cdot, \cdot)$ can be taken as any function of the form

$$F(x_1, x_2) = a\phi_1(x_1) + (1-a)\phi_2(x_2), \tag{2.111}$$

for $0 < a < 1$ with $\phi_i(x)$ given by

$$\phi_i(x) = \begin{cases} x & |x| \leq 1 \\ g_i(x) & |x| > 1, \end{cases} \tag{2.112}$$

where the functions $g_i(x)$ are completely arbitrary. This point is illustrated in some detail in Chapter 6, where similar results are developed for a special class of linear multimodels (the TARMAX model family).

In contrast to the requirement that a function $f : R^n \to R$ satisfy Cauchy's equation, the homogeneity condition becomes significantly less restrictive for $n > 1$. For example, it is easy to show that the following nonlinear function from R^2 to R^1 is homogeneous:

$$f(x_1, x_2) = \frac{x_1 x_2}{x_1 + x_2}. \tag{2.113}$$

Returning to the view of BIBO stable systems as maps from ℓ_∞ into itself, it is possible to define a class of *homogeneous systems*, analogous to and including the class of linear systems. That is, homogeneous BIBO systems may be defined as those systems whose input-output map is a homogeneous mapping from ℓ_∞ into itself. This idea is explored in considerable detail in Chapter 3 and revisited periodically throughout the rest of the book.

Finally, it is important to comment on a possible source of confusion in later discussions. The term *additive* for functions satisfying Cauchy's equation has a long history in mathematics, but more recently this term has also been introduced in the statistics and time-series literature, with a very different meaning. This newer definition refers to the *structure* of a model and not its behavior. In fact, the class of nonlinear models based on this newer notion of additivity is an extremely interesting one that is discussed at length in Chapter 4, in marked contrast to the class of additive nonlinear models considered here. To distinguish these two very different notions, condition (2.103) will generally be called *functional additivity*, and the newer notion of additivity considered in Chapter 4—the statistical and time-series notion—will be called *structural additivity*.

Chapter 3

Four Views of Nonlinearity

The review of linear models presented in Chapter 2 was intended to provide a baseline, establishing notation and reviewing some important aspects of this reference class against which *nonlinear* models are necessarily judged. This chapter demonstrates that nonlinearity can be approached from at least two fundamentally different directions. The first of these directions is *structural*—by far the most common approach—in which a class of nonlinear models is defined by specifying the basic structure of all elements of that class. The Hammerstein, Wiener, and Uryson model classes discussed in Chapter 1 illustrate this approach. The structurally defined model class considered in this chapter is the class of *bilinear models* discussed in Sec. 3.1, which may be viewed as "almost linear" for reasons that will become apparent.

Alternatively, it is possible to adopt *behavioral* definitions of nonlinear model classes, although this approach is generally more difficult. There, a particular type of input/output behavior is specified and subsequent analysis seeks model structures that can exhibit this behavior. In practice, this approach tends to be difficult because it is often not clear how to construct explicit examples that exhibit specified qualitative behavior. Of necessity, then, the primary focus of this book is structurally defined model classes like the bilinear models discussed in Sec. 3.1, although three behaviorally defined model classes are considered in some detail in Secs. 3.2 through 3.4. The first of these is the class of *homogeneous models*, obtained by relaxing one of the two defining conditions for linearity: homogeneous models do not obey the superposition principle of linear systems, but they are invariant under scaling of the input sequence by arbitrary real constants. Relaxing these conditions further and requiring only that this scaling hold for *positive* constants leads to the class of *positive-homogeneous models*, described in Sec. 3.3. Alternatively, requiring *linearity* to hold *but only for constant input sequences* leads to the class of *static-linear* models, described in Sec. 3.4. As these and subsequent discussions illustrate, some remarkably general results may be obtained concerning the structure of these three model classes.

3.1 Bilinear models

Chapter 1 (Sec. 1.2.4) introduced the class of continuous-time bilinear models as a special case of the more general class of control-affine models. It was noted that both of these model classes have been employed frequently in the modeling of physical systems. Continuous-time bilinear models have been applied to describe certain classes of electrical networks, mechanical links, heat transfer, nuclear fission, aircraft dynamics, fluid flow, chemical kinetics, and a variety of biological phenomena (Mohler, 1991). Detailed stability results are available, building on linear systems theory and providing a basis for the design of control systems for bilinear models. In fact, continuous-time bilinear models are sometimes viewed as "nearly linear" (Bruni et al., 1974) because of their close structural and behavioral connections with the class of linear models.

Similar observations hold for *discrete-time bilinear models*, but it is important to note that these models may be defined in at least two different ways; further, although they are related, these two definitions are *not* equivalent. The first of these definitions retains the general form of the continuous-time bilinear state-space model, essentially replacing derivatives with differences. Specifically, the *discrete-time state-space bilinear model structure* is defined by

$$\mathbf{x}(k+1) = \mathbf{A}\mathbf{x}(k) + u(k)\mathbf{N}\mathbf{x}(k) + \mathbf{b}u(k), \quad y(k) = \mathbf{c}^T\mathbf{x}(k). \tag{3.1}$$

A significant disadvantage of this definition is that the general case does not appear to exhibit a convenient input-output representation. Specifically, consider the two-dimensional example defined by the following choices of $\mathbf{A}$, $\mathbf{N}$, $\mathbf{b}$ and $\mathbf{c}$:

$$\mathbf{A} = \begin{bmatrix} 0 & 1 \\ a_1 & a_2 \end{bmatrix} \quad \mathbf{N} = \begin{bmatrix} n_{11} & n_{12} \\ n_{21} & n_{22} \end{bmatrix} \quad \mathbf{b} = \begin{bmatrix} b_1 \\ b_2 \end{bmatrix} \quad \mathbf{c} = \begin{bmatrix} 0 \\ 1 \end{bmatrix}. \tag{3.2}$$

The two components of this state vector evolve according to the following pair of difference equations

$$\begin{aligned} x_1(k) &= x_2(k-1) + u(k-1)[n_{11}x_1(k-1) + n_{12}x_2(k-1) + b_1] \\ x_2(k) &= a_1 x_1(k-1) + a_2 x_2(k-1) \\ &\quad + u(k-1)[n_{21}x_1(k-1) + n_{22}x_2(k-1) + b_2]. \end{aligned} \tag{3.3}$$

To reduce these equations to an input-output representation, it would be necessary to eliminate the internal state variable $x_1(k)$ from this pair of equations and there is no obvious way to do this in the general case. Conversely, if $n_{11} = 0$, it is possible to obtain the following input-output representation:

$$\begin{aligned} y(k) &= a_2 y(k-1) + a_1 y(k-2) + b_2 u(k-1) + a_1 b_1 u(k-2) \\ &\quad + n_{22} u(k-1) y(k-1) + a_1 n_{12} u(k-2) y(k-2) + n_{21} u(k-1) y(k-2) \\ &\quad + n_{21} b_1 u(k-1) u(k-2) + n_{21} n_{12} u(k-1) u(k-2) y(k-2). \end{aligned} \tag{3.4}$$

An important feature of this result is the appearance of the last two nonlinear terms, involving $u(k-1)u(k-2)$ and $u(k-1)u(k-2)y(k-2)$; in particular, the presence of these terms are not suggested by the name "bilinear model." More generally, Baheti et al. (1980) note that if the matrices $\mathbf{A}$ and $\mathbf{N}$ have the form:

$$\mathbf{A} = \begin{bmatrix} 0 & 1 & 0 & \cdots & 0 \\ 0 & 0 & 1 & \cdots & 0 \\ \vdots & \vdots & \vdots & \vdots & \vdots \\ 0 & 0 & 0 & \cdots & 1 \\ a_n & a_{n-1} & a_{n-2} & \cdots & a_1 \end{bmatrix} \quad \mathbf{N} = \begin{bmatrix} 0 & 0 & 0 & \cdots & 0 \\ 0 & 0 & 0 & \cdots & 0 \\ \vdots & \vdots & \vdots & \vdots & \vdots \\ 0 & 0 & 0 & \cdots & 0 \\ \eta_n & \eta_{n-1} & \eta_{n-2} & \cdots & \eta_1 \end{bmatrix},$$

the state-space bilinear model has an input-output representation of the form

$$y(k) = \sum_{i=1}^{n} a_i y(k-i) + \sum_{i=1}^{n} \gamma_i u(k-i)$$
$$+ \sum_{i=1}^{n} \eta_i y(k-i) u(k-n) + \sum_{i=1}^{n} \nu_i u(k-i) u(k-n), \quad (3.5)$$

where γ_i and ν_i are constants determined by the model parameters $\{a_j\}$ and $\{b_j\}$. In the context of the previous example, note that this restriction on the matrix $\mathbf{N}$ corresponds to $n_{12} = 0$, eliminating the $u(k-1)u(k-2)y(k-2)$ term.

The second definition of discrete-time bilinear models is based directly on the following input-output representation:

$$y(k) = \sum_{i=1}^{p} a_i y(k-i) + \sum_{i=0}^{q} b_i u(k-i)$$
$$+ \sum_{i=1}^{P} \sum_{j=1}^{Q} c_{i,j} y(k-i) u(k-j). \quad (3.6)$$

These models have been considered by a number of different authors (Granger and Anderson, 1978; Priestley, 1988; Rao, 1981; Rao and Gabr, 1984; Tong, 1990), typically in the context of statistical time-series models. One motivation for considering these models is their ability to exhibit "burst-like" random fluctuations when driven by white Gaussian input sequences (Priestley, 1988; Rao and Gabr, 1984). Following Rao (1981), the model defined by Eq. (3.6) will be denoted $BL(p, q, P, Q)$. As in the case of the linear ARMA models discussed in Chapter 2, note that the lower limit on the moving average sum in this model is $i = 0$, rather than $i = 1$ or $i = d$ for some assumed delay d. As before, this assumption is made primarily to avoid having to treat certain models, or model components, that do not exhibit delays as special cases. Generally, it will make little difference whether this lower limit is assumed to be 0, 1, or d for any arbitrary fixed integer; in cases where such differences are important, this point is noted explicitly. *Because this book is primarily concerned with input-output model representations and because the model class defined by Eq. (3.6) is amenable to so much analysis, this definition of the term "bilinear model" will be adopted in all subsequent discussions unless it is explicitly noted otherwise.*

In the context of physical system modeling, this class of discrete-time bilinear models was also introduced in Chapter 1. Specifically, Sec. 1.3.3 described a bilinear model developed as an interpolation between two local linear models for a distillation column (Stromberg et al., 1995), and Sec. 1.4.2 described a simple model from this class developed to emulate the qualitative asymmetry seen in the CSTR dynamics considered there (Pearson, 1995). Another illustration of the practical utility of this class of discrete-time bilinear models is given in Bartee and Georgakis (1994), who consider the problem of modeling a fluid catalytic cracking unit (FCCU), one of the key process units found in petroleum refineries. A critical control consideration in operating these units is the concentration of carbon monoxide leaving the regenerator section, which is restricted by environmental regulations. Bartee and Georgakis consider the problem of controlling this concentration by manipulating the lift air flow rate into the unit, comparing the response of a detailed first-principles simulation with that of two approximations: an empirically identified linear model, and an empirically identified bilinear model. These models are compared for an increasing sequence of six consequetive step changes in the lift air flow rate. For the first step in this sequence, the linear model provides an excellent approximation of the FCCU dynamics, whereas the bilinear model approximation is not nearly as good. In contrast, for the other five step changes in this input sequence, the bilinear model provides a much better approximation of the FCCU dynamics than the linear model does. In particular, the linear model's performance degrades consistently with each successive step change in the input, predicting unphysical negative carbon monoxide concentrations after the third step change. The bilinear model's performance generally exhibits the opposite behavior, improving consistently with each successive step change and never predicting negative carbon monoxide concentrations. The key point of this example is that, although linear models can provide excellent approximations to process dynamics over a sufficiently narrow operating range, bilinear models have often been found to provide a reasonable alternative over wider operating ranges. Conversely, as noted in Chapter 1 (Sec. 1.4.2), bilinear models also generally have a restricted range of utility, and one of the objectives of the discussion of these models given here is to illustrate the basic nature of this restriction.

Since the class of bilinear models reduces to the class of linear ARMA models if $c_{i,j} = 0$ for all i and j, it is clear that these coefficients determine the qualitative nature of the behavior of the bilinear model class. What is less obvious is that there exist qualitatively distinct subclasses of the bilinear family, depending on whether all or only some of these coefficients are nonzero. Specifically, the following subfamilies are of significant interest:

1. Linear models: $c_{i,j} = 0$ for all i, j
2. Diagonal models: $c_{i,j} = 0$ if $i \neq j$
3. Superdiagonal models: $c_{i,j} = 0$ if $i < j$
4. Subdiagonal models: $c_{i,j} = 0$ if $i > j$
5. Completely bilinear models: $b_i = 0$ for all i.

It is sometimes also useful to refer to *strictly superdiagonal models* and *strictly subdiagonal models* in describing models for which $c_{i,j} = 0$ if $i \leq j$ and $c_{i,j} = 0$ if $i \geq j$, respectively. Generally, however, strict superdiagonality or subdiagonality is implied when using the unmodified terms, as it is in the examples discussed here. Some of the differences between the first four of these five bilinear model classes are illustrated in Sec. 3.1.1, and the relationship between these classes is considered at the end of that discussion. The fifth class is considered in detail in Sec. 3.1.2.

3.1.1 Four examples and a general result

The differences between the linear, diagonal, superdiagonal, and subdiagonal model classes may be seen by comparing the following four examples:

$$y(k) = 0.99y(k-1) + u(k-1) \tag{3.7}$$
$$y(k) = 0.99y(k-1) + u(k-1) - 0.4y(k-2)u(k-1) \tag{3.8}$$
$$y(k) = 0.99y(k-1) + u(k-1) - 0.4y(k-1)u(k-1) \tag{3.9}$$
$$y(k) = 0.99y(k-1) + u(k-1) - 0.4y(k-1)u(k-2). \tag{3.10}$$

The first of these models corresponds to the linear part of the other three, which are all bilinear. The second model is superdiagonal, the third is diagonal, and the fourth is subdiagonal. The differences between these models are illustrated by the differences in their responses to the four standard input sequences defined in Chapter 2. For the impulse and step responses, the initial condition $y(k) = 0$ for $k \leq 0$ is assumed for all four models, and for the sinusoidal and Gaussian white noise inputs, it is assumed that $y(-N) = 0$ for some value $-N$ in the distant past (i.e., N large).

The impulse responses of these four models are shown in Fig. 3.1. In fact, the first three of these impulse responses are identical, illustrating a general characteristic of diagonal and superdiagonal models: their impulse response is identical to that of the linear model obtained by setting $c_{i,j} = 0$. This point is most easily seen for diagonal models like Eq. (3.9): since $u(k) \neq 0$ only for $k = 0$, it follows that the only possible nonlinear contribution to the impulse response of this model is $-0.4y(0)u(0)$, which is zero since $y(0) = 0$. This observation generalizes to any diagonal diagonal model for which $y(0) = 0$ since the only possible nonlinear contributions are of the form $c_{i,i}y(0)u(0)$. Similarly, for superdiagonal models, the only possible nonlinear contributions to the impulse response are of the form $c_{i,j}y(j-i)u(0)$; if $y(k) = 0$ for $k < 0$, this nonlinear contribution is zero since $c_{i,j} = 0$ for $i < j$ in a superdiagonal model. As a specific example, note that the nonlinear contribution to the impulse response of the superdiagonal model defined by Eq. (3.8) is $-0.4y(-1)u(0)$ for $k = 1$, which is zero since $y(-1) = 0$. An important conclusion to be drawn from this example is that the impulse response of diagonal and subdiagonal models provides no evidence of the model's nonlinearity, independent of the amplitude of the impulse considered. This result stands in marked contrast to the case of linear models, which are completely characterized by their impulse responses. Finally, for the subdiagonal model defined by Eq. (3.10), the bilinear term is

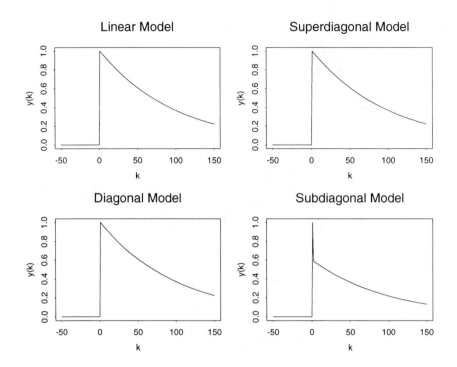

Figure 3.1: Bilinear model impulse responses

$-0.4y(k-1)u(k-2)$ and is nonzero for $k = 2$. The effect of this term on the impulse response is seen clearly in the lower right-hand plot in Fig. 3.1.

Fig. 3.2 compares the step responses for these models, and here the situation is quite different. In all cases, the bilinear models exhibit *much* faster step responses than the linear model, shown in the upper left plot. In fact, it is possible to describe the general behavior of these step responses in terms of equivalent linear models. Specifically, for $k \geq Q$, the response of the bilinear model (3.6) to a step of amplitude α may be rewritten as

$$y(k) = \sum_{i=1}^{r} \gamma_i y(k-i) + \sum_{i=0}^{q} b_i u(k-i), \qquad (3.11)$$

where $r = \max\{p, P\}$ and the autoregressive coefficients γ_i are given by

$$\gamma_i = \begin{cases} a_i + \alpha \sum_{j=1}^{Q} c_{i,j} & 1 \leq i \leq p, P \\ a_i & P < i \leq p \\ \alpha \sum_{j=1}^{Q} c_{i,j} & p < i \leq P. \end{cases} \qquad (3.12)$$

Note that the order r of this equivalent linear model depends on the parameters of the original bilinear model, and that the coefficients γ_i depend on both the original model parameters $c_{i,j}$ and the amplitude α of the step input. A key

Four Views of Nonlinearity 99

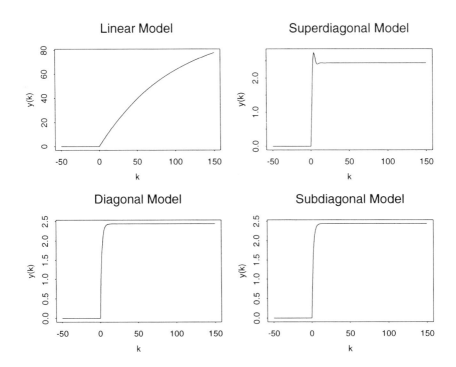

Figure 3.2: Bilinear model step responses

observation is that if the sum of the coefficients $c_{i,j}$ is nonzero, the bilinear model will exhibit instability for inputs of sufficiently large magnitude $|\alpha|$. That is, for some α sufficiently large in magnitude, the poles of this equivalent linear system will move outside the unit circle, causing an exponentially unstable step response. This behavior is generic for bilinear models: except for the linear case, these models are not BIBO stable, but exhibit instability for sufficiently large amplitude input sequences. This point is important and is discussed further in Sec. 3.1.3.

Returning to Fig. 3.2, the step response for the superdiagonal model (3.8) shown in the upper right plot exhibits a slight overshoot, reflecting the fact that its equivalent linear model is second-order, that is

$$y(k) = 0.99y(k-1) - 0.4\alpha y(k-2) + u(k-1). \tag{3.13}$$

The poles of this linear system are given by

$$z = \frac{0.99 \pm \sqrt{0.9801 - 1.6\alpha}}{2}, \tag{3.14}$$

from which it is not difficult to show that the response will be stable and monotonic (i.e., the poles will be real and inside the unit circle) if $-0.025 < \alpha \leq 0.612$. A stable oscillatory response will be obtained (i.e., the poles will be complex

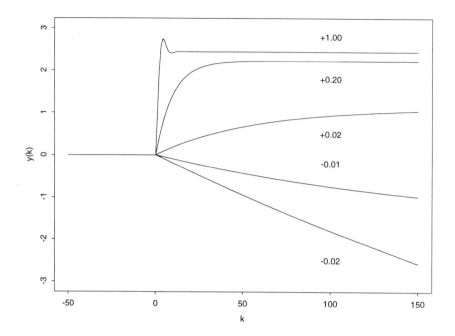

Figure 3.3: Family of step responses, superdiagonal model

conjugates lying inside the unit circle) for $0.612 < \alpha < 1.084$. If $\alpha < -0.025$ or $\alpha > 1.084$, the bilinear model's step response will be unstable. Step responses for $\alpha = -0.02, -0.01, +0.02, +0.20$, and $+1.000$ are shown in Fig. 3.3 and illustrate this general behavior. Note in particular that this model can exhibit the same qualitative behavior as the CSTR example discussed in Chapter 1: oscillatory responses to positive steps and monotonic responses to negative steps.

In contrast to this superdiagonal model, both the diagonal model (3.9) and the subdiagonal model (3.10) exhibit monotonic step responses, as seen in the bottom two plots in Fig. 3.2. Further, these responses are identical for $k > 1$ since the bilinear term reduces to $-0.4\alpha y(k-1)$ for a step of amplitude α. Thus, for $k > 1$, both step responses are described by an equivalent first-order linear model with autoregressive coefficient $0.99 - 0.4\alpha$. It follows, then, that the diagonal and subdiagonal step responses will be stable for $-0.025 < \alpha < 4.975$.

In many respects, the response of these models to sinusoidal input sequences is similar, as seen in Fig. 3.4. In all cases, the nonlinearity of these models is clearly evident in the nonsinusoidal character of the response. Note particularly the "overshoot" seen on the positive peaks of the superdiagonal response. For the diagonal and subdiagonal responses shown in the bottom two plots, the general shape is quite similar: the positive peaks of the response are fairly flat, whereas the negative peaks are much sharper. In contrast, note that the

Four Views of Nonlinearity   101

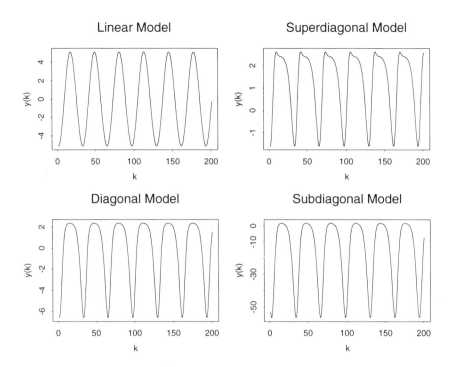

Figure 3.4: Bilinear model sinusoid responses

magnitudes of the negative peaks differ by about a factor of ten. Further, these responses are highly amplitude-dependent, a point that is revisited in Sec. 3.1.3. In particular, for sufficiently large amplitude inputs, bilinear models can exhibit *unstable subharmonic generation*, but in general these models exhibit only the superharmonic generation that is responsible for the nonsinusoidal character of the responses shown in Fig. 3.4. A more detailed discussion of this point is given in Sec. 3.1.2.

In the time-series setting, the input sequence $\{u(k)\}$ is typically assumed to be an independent, identically distributed zero mean, Gaussian sequence (i.e., Gaussian white noise). In this context, it turns out that superdiagonal models are the easiest to analyze, diagonal models are significantly harder to analyze, subdiagonal models are *very* hard to analyze, and the general bilinear model is essentially dismissed as intractible. The significant differences between these model classes may be seen by considering the simplest stochastic characterization problem, that of computing $E\{y(k)\}$ for an independent, identically distributed, zero mean input sequence $\{u(k)\}$. This problem requires the evaluation of the sum

$$E\{\sum_{i=1}^{P}\sum_{j=1}^{Q} c_{i,j} y(k-i) u(k-j)\} = \sum_{i=1}^{P}\sum_{j=1}^{Q} c_{i,j} E\{y(k-i) u(k-j)\}, \quad (3.15)$$

and the difficulty of this task depends on the model type. For a strictly superdiagonal model, $c_{i,j} = 0$ if $i \leq j$ and the expectations required to evaluate the right-hand side of this equation are of the form $E\{y(k-i)u(k-j)\}$ for $i > j$. In words, this expectation expresses the (non-normalized) correlation between the input at time $k-j$ and the output at the *earlier* time $k-i$. Since the input values are statistically independent at different times and the model is causal, it follows that $y(k-i)$ and $u(k-j)$ are statistically independent. Hence

$$E\{y(k-i)u(k-j)\} = E\{y(k-j)\}E\{u(k-i)\} = 0, \quad (3.16)$$

from which it follows that $E\{y(k)\} = 0$. This result extends to the diagonal case *provided* there is no direct feedthrough term (i.e., $b_0 = 0$), since the presence of such a term would introduce a correlation between $u(k-i)$ and $y(k-i)$. Conversely, for subdiagonal models, the expectations $E\{y(k-i)u(k-j)\}$ in Eq. (3.15) generally do not vanish, since they involve the input at time $k-j$ and the output at some *later* time $k-i$. Consequently, the expected value of the response $y(k)$ of a subdiagonal or general bilinear model is generally nonzero for an i.i.d., zero-mean input sequence. This observation represents a stochastic extension of the rectification phenomenon discussed in Chapter 1: persistent fluctuations of the input sequence about zero result in a constant offset (i.e., nonzero mean) in the response.

The responses of the four models defined in Eqs. (3.7) through (3.10) to Gaussian white noise input sequence are shown in Fig. 3.5. In this case, these results are not obtained from the standard 201-sample sequence defined in Chapter 2, but rather from a much longer sequence (2000 samples) that better illustrates both the general nature of the model responses and the differences between them. In particular, careful examination of these plots illustrates some of the dramatic differences between these four models. The linear model response shown in the upper left plot exhibits some evidence of long-term drift-like behavior, consistent with the fact that this model output is "almost Brownian motion" (i.e., the autoregressive coefficient is $\phi = 0.99$ vs. $\phi = 1$ for Brownian motion). In marked contrast, the superdiagonal model response shown in the upper right plot exhibits two large, highly localized *bursts*; similar behavior is exhibited by a superdiagonal example of Priestley (1988, p. 52), who notes that such models have been explored for analyzing time series from earthquakes and underground explosions. Also, note that these superdiagonal responses are consistent with the result $E\{y(k)\} = 0$ established in the preceeding discussion. The plot of the diagonal model response shown in the lower left plot in Fig. 3.5 also exhibits a mean value of zero, but here the general character of the response is quite different from the superdiagonal case. In particular, note that the diagonal model response is positive most of the time, exhibiting values between $y(k) \simeq 0$ and $y(k) \simeq 3$, but it also exhibits isolated negative bursts, occurring frequently enough to achieve an overall mean value of zero. The subdiagonal response shown in the lower right plot appears to exhibit somewhat similar behavior to the diagonal model, but two important differences should be noted. First, the scale of this response is an order of magnitude larger than that of the diagonal model response, exhibiting isolated values between $y(k) \simeq -100$

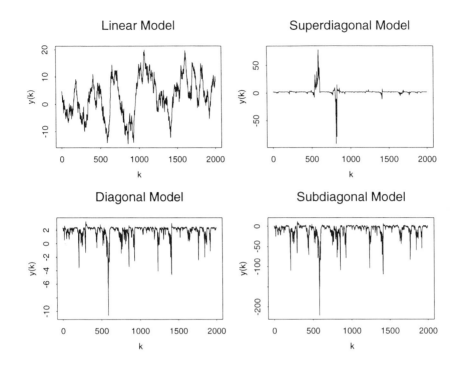

Figure 3.5: Bilinear model white noise responses

and $y(k) \simeq -200$. The second difference is that the subdiagonal model clearly exhibits a nonzero (specifically, negative) mean value, again in agreement with the results presented in the previous paragraph.

These examples give some idea of the range of behavior that the class of bilinear models is capable of representing. The focus has been on specific subclasses of this model family because, by exploiting their structure, it is possible to draw more detailed conclusions about the behavior of these subclasses than is possible for the general family of bilinear models. It turns out, however, that *any* bilinear model is essentially equivalent to a subdiagonal model, as the following discussion illustrates. First, consider the diagonal model defined in Eq. (3.9) and note that it may be rewritten as

$$y(k) = 0.99y(k-1) + v(k-2) - 0.4y(k-1)v(k-2), \qquad (3.17)$$

where $v(k) = u(k+1)$. In terms of $v(k)$, note that the model defined in Eq. (3.17) is *subdiagonal*. Similar results may be obtained for the superdiagonal model defined in Eq. (3.8), rewriting it as

$$y(k) = 0.99y(k-1) + w(k-3) - 0.4y(k-2)w(k-3), \qquad (3.18)$$

where $w(k) = u(k+2)$. Note that in both of these examples, bilinear models that are not subdiagonal have been converted to subdiagonal form by *advancing* the input sequence. Although this operation is not physically realizable in real time, it does illustrate that the inherent behavior of both the superdiagonal and diagonal model examples can be obtained from a closely related subdiagonal model, provided the input sequence is also modified slightly. In fact, the basic result of the following theorem is that the qualitative input/output behavior of *any* bilinear model may be obtained from a closely related subdiagonal model for an appropriately advanced input sequence. Thus, the subdiagonal model class is equivalent to the general bilinear model class. This observation explains the difficulty of understanding the qualitative behavior of the subdiagonal class relative to the more restrictive diagonal and superdiagonal classes.

Theorem:

Any bilinear model from the general class $BL(p, q, P, Q)$ may be converted into a subdiagonal model by replacing $u(k)$ by $v(k) = u(k + L)$ for some finite $L \geq 0$.

Proof:

Clearly, if the model is subdiagonal, the result holds trivially for $L = 0$. If the model is diagonal, taking $L = 1$ will result in a subdiagonal model in $BL(p, q+1, P, Q+1)$ by simply replacing the linear coefficients b_i with b_{i+1} and the bilinear coefficients $c_{i,j}$ with $c_{i,j+1}$ in Eq. (3.6). For superdiagonal models, taking $L = P$, replacing b_i with b_{i+P} and replacing $c_{i,j}$ with $c_{i,j+P}$ yields a subdiagonal model in $BL(p, q+P, P, Q+P)$, although simpler representations may be possible for some $L < P$. In fact, this last result holds for general bilinear models.

□

3.1.2 Completely bilinear models

Granger and Anderson (1978) define *completely bilinear models* as bilinear models for which all of the linear coefficients b_j are identically zero. Rao (1981) terms these models *homogeneous*, although it is important to note that this terminology derives from an analogy with differential equations and is unrelated to the notion of homogeneity discussed later in this chapter. Rao notes that if $\{u(k)\}$ is an i.i.d. sequence (specifically, Gaussian white noise), then the completely bilinear model is either asymptotically equivalent to a *deterministic* model, or it exhibits the nonstationarity characteristic of unstable models. This observation suggests that completely bilinear models are somewhat pathological in their behavior, a conclusion supported by the following observations.

Although it is not the most general case, consideration of the completely bilinear model with $P = p$ is instructive. This model may be rewritten as

$$y(k) = \sum_{i=1}^{p} \left[a_i + \sum_{j=1}^{Q} c_{i,j} u(k - j) \right] y(k - i). \tag{3.19}$$

Expressed in this form, the completely bilinear model represents a very special case of the class of *random coefficient autoregressive models* (Nicholls and Quinn, 1982), obtained by permitting the coefficients in a linear autoregressive model to be random variables. This class of models is an interesting one that is known to exhibit burst phenomena like those discussed earlier in connection with superdiagonal models. Normally, however, random coefficient autoregressive models also include a second, independent input sequence $\{e(k)\}$ that serves as a source of persistent excitation. In particular, note that in Eq. (3.19), if $y(k) = 0$ for $k < 0$, then $y(k) = 0$ identically for all k, a situation that does not arise in more typical random coefficient autoregressive models. This observation represents yet another disadvantage of the class of completely bilinear models, since the necessity of specifying a nonzero initial condition is frequently undesirable in practice. In addition, it will be seen in Chapter 5 that completely bilinear models do not exhibit Volterra series representations, in contrast to the other subclasses of bilinear models considered here. For all of these reasons, the class of completely bilinear models does not appear to be a particularly promising one for practical applications.

Despite these limitations, however, completely bilinear models are important in practice because they arise naturally in two practically important situations. The first of these situations is as *transient solutions* of the general bilinear model (3.6). That is, suppose $y(k)$ satisfies Eq. (3.6) and $z(k)$ is a nonzero solution of the corresponding completely bilinear model:

$$z(k) = \sum_{i=1}^{p} a_i z(k-i) + \sum_{i=1}^{P} \sum_{j=1}^{Q} c_{i,j} z(k-i) u(k-j). \qquad (3.20)$$

It follows immediately that $w(k) = y(k) + z(k)$ is also a solution of Eq. (3.6). Further, note that $z(k) = 0$ for all k is one possible solution of Eq. (3.20) and, if $z(k)$ is required to satisfy the initial conditions $z(k) = 0$ for $k = -1, \ldots, -r$ where $r = \max\{p, P\}$, this zero solution is unique for all $k \geq 0$. Conversely, if $z(k) \neq 0$ for any initial k, it follows that a nonzero solution of Eq. (3.20) exists. An example of a stable transient solution is shown in Fig. 3.6, which shows the response of the superdiagonal model considered in Sec. 3.1.1 to 2000 samples of the standard sinusoidal input sequence defined in Chapter 2. The final 201 samples corresponds to the upper right plot in Fig. 3.4, but the initial portion of this plot shows the presence of a very large transient response to the initial condition $y(k) = 0$ that decays rapidly to zero.

To see the nature of this transient solution more clearly, consider the completely bilinear, diagonal model

$$z(k) = az(k-1) + cz(k-1)u(k-1) = [a + cu(k-1)]z(k-1). \qquad (3.21)$$

If $z(0) = 0$, it follows immediately that $z(k) = 0$ for all $k > 0$, so necessary and sufficient conditions for the existence of a nonzero solution are $z(0) \neq 0$. Further, note that if $z(k)$ satisfies Eq. (3.21), then so does $\lambda z(k)$ for any real λ. It therefore suffices to consider the *nontrivial prototype* $z^*(k)$ defined by

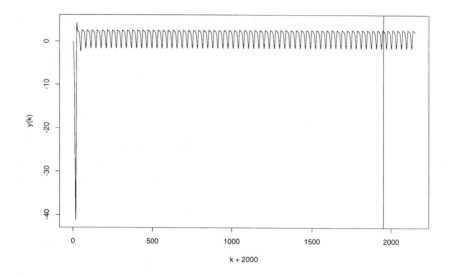

Figure 3.6: Superdiagonal model, stable transient response

requiring $z(0) = 1$, since the solution for $z(0) = z_0$ may be obtained directly as $z(k) = z_0 z^*(k)$. Finally, note that if the input sequence $\{u(k)\}$ is fixed, Eq. (3.21) is equivalent to the time-varying *linear* model

$$z^*(k) = \alpha(k) z^*(k-1) \qquad z^*(0) = 1, \tag{3.22}$$

where $\alpha(k) = a + cu(k-1)$. From the results presented in Chapter 2 (Sec. 2.5.1), it follows that $z^*(k)$ is given by

$$z^*(k) = \prod_{i=1}^{k} [a + cu(k-i)]. \tag{3.23}$$

Sufficient conditions for the stability and instability of the nontrivial prototype are easily obtained from this solution. In particular, if

$$|a + cu(k)| \leq \gamma < 1, \tag{3.24}$$

for all k, it follows that $|z^*(k)| \leq \gamma^k \to 0$ as $k \to \infty$. Conversely, if

$$|a + cu(k)| \geq \gamma > 1, \tag{3.25}$$

for all k, it follows that $|z^*(k)| \geq \gamma^k \to \infty$ as $k \to \infty$. *More generally, nontrivial solutions of the completely bilinear model (3.20) may be viewed as transient solutions of the general bilinear model (3.6) that normally either decay asymptotically to zero or diverge.*

Four Views of Nonlinearity 107

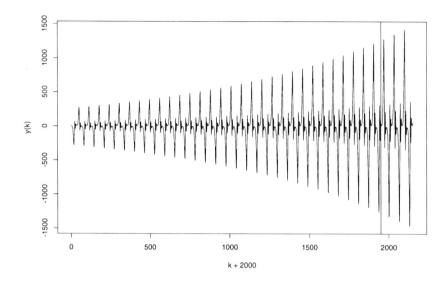

Figure 3.7: Superdiagonal model, unstable transient response

The second important situation where completely bilinear models arise naturally is in connection with the response of general bilinear models to periodic input sequences. As noted in Chapter 1, a useful characterization of nonlinear models involves their response to periodic input sequences, and a particularly important question is whether a given model preserves the periodicity of these input sequences. To address this question, consider an input sequence of period P (i.e., $u(k + P) = u(k)$ for all k) and define $z(k) = y(k + P) - y(k)$. If $y(k)$ satisfies Eq. (3.6), it follows directly that $z(k)$ satisfies the associated completely bilinear model (3.20) and the question of whether $y(k)$ is periodic with period P or not reduces to the question of whether or not $z(k) \equiv 0$. Hence, *a general bilinear model can exhibit subharmonic generation or nonperiodic responses to periodic inputs if and only if the associated completely bilinear model can exhibit such solutions.*

In view of the interpretation of the solution $z(k)$ as a transient response of the general bilinear model given in the previous discussion, it appears that subharmonic or nonperiodic responses to periodic inputs generally lie beyond the range of qualitative behavior that bilinear models can exhibit. The one notable exception is the possibility of *unstable subharmonic responses*, illustrated in Figs. 3.7 and 3.8. Specifically, Fig. 3.7 shows the response of the same superdiagonal model considered in Fig. 3.6 to a sinusoidal input of the same frequency but twice the amplitude. The instability of this response is clear from Fig. 3.7, and the fact that this response is approximately subharmonic of order $1/2$ may be seen in Fig. 3.8, where the final 201 samples of this response are overlaid with an appropriately scaled replica of the input sinusoid.

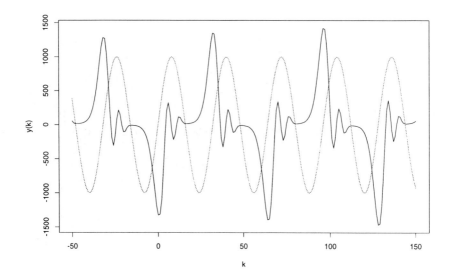

Figure 3.8: Approximately subharmonic character of the unstable response

3.1.3 Stability and steady-state behavior

The results presented in Secs. 3.1.1 and 3.1.2 have demonstrated clearly that bilinear models can exhibit input-dependent stability. As noted in Chapter 1, such qualitative behavior is inherently nonlinear since the stability or instability of linear models depends entirely on the model parameters and holds, or fails to hold, for all possible input sequences. In contrast, the step response results presented in Sec. 3.1.1 demonstrated that the stability of bilinear models generally depends on the amplitude of the step input considered. In particular, recall that for step inputs, the bilinear model may be rewritten as an equivalent linear model whose coefficients depend on the amplitude of the step input; hence, for step inputs of sufficient amplitude, most bilinear models become unstable. Similarly, the unstable subharmonic generation results presented in Sec. 3.1.2 demonstrated that the response of a bilinear model to a sinusoidal input of fixed frequency can be either stable or unstable, depending on the amplitude of that input. In fact, this amplitude-dependent stability behavior is generic to the class of bilinear models, as the following result demonstrates (Lee and Mathews, 1994). First, define λ_i to be the p roots of the polynomial

$$z^p - \sum_{i=1}^{p} a_i z^{p-i} = 0, \qquad (3.26)$$

and suppose $|u(k)| \leq M$ for all k. Lee and Mathews show that the bilinear model (3.6) is stable if the following conditions hold:

$$|\lambda_i| < 1 \text{ for } i = 1, 2, \ldots, p, \tag{3.27}$$

$$\sum_{i=1}^{P} \sum_{i=1}^{Q} |c_{i,j}| < \frac{\prod_{i=1}^{p}(1 - |\lambda_i|)}{M}. \tag{3.28}$$

Note that Eq. (3.27) corresponds to the well-known necessary and sufficient conditions for the stability of the linear model that results when $c_{i,j} = 0$ for all i, j. Viewed in this light, condition (3.28) represents an additional constraint on the bilinear model sufficient to guarantee stability for all input sequences satisfying the fixed bound $|u(k)| \leq M$. In particular, note that this condition may be viewed as a constraint on the severity of the nonlinearity that becomes more stringent as the bound M on the input sequences increases. Also, note that this result is strictly weaker than BIBO stability since the input sequences are required to satisfy a *fixed* bound and not just *some* bound. This distinction is important, since the results noted in the previous paragraph illustrate that bilinear models are *not* BIBO stable, in general.

Conversely, it is also important to note that the stability of bilinear models can depend strongly on the *type* of input considered. Specifically, recall from the discussion in Sec. 3.1.1 that the impulse response of a diagonal or superdiagonal model is identical to that of the model obtained by setting $c_{i,j} = 0$. Consequently, these impulse responses cannot exhibit amplitude-dependent stability. In fact, this result extends to general bilinear models, by the following argument. For an impulse of amplitude α, the response of the general bilinear model is given by

$$y(k) = \sum_{i=1}^{p} a_i y(k-i) + \sum_{i=0}^{q} b_i u(k-i) + \nu(k), \tag{3.29}$$

where $\nu(k)$ is the perturbation term

$$\nu(k) = \alpha \sum_{i=1}^{P} \sum_{j=1}^{Q} y(k-i)\delta(k-j)$$

$$= \begin{cases} \alpha \sum_{i=1}^{P} \sum_{j=1}^{Q} y(j-i) & k = 1, \ldots, Q \\ 0 & k > Q. \end{cases} \tag{3.30}$$

Since $y(k)$ is finite for all finite k even if the model is unstable, it follows that $\nu(k)$ is bounded, so $y(k) \to \pm\infty$ as $k \to \infty$ if and only if the linear part of the bilinear model is unstable. This observation establishes the important point that the stability of a nonlinear model will depend, in general, on both the *magnitude* and the *type* of the inputs considered.

It is useful to conclude this discussion of bilinear models with a brief examination of their steady-state behavior, which is quite easy to characterize.

Setting $u(k) = u_s$ and $y(k) = y_s$ for all k in Eq. (3.6) leads immediately to the following equation for the steady-state response y_s

$$y_s = \frac{\bar{b} u_s}{1 - \bar{a} - \bar{c} u_s}. \qquad (3.31)$$

Here, the constants $\bar{a}$, $\bar{b}$, and $\bar{c}$ are defined as

$$\bar{a} = \sum_{i=1}^{p} a_i \quad \bar{b} = \sum_{i=0}^{q} b_i \quad \bar{c} = \sum_{i=1}^{P} \sum_{i=1}^{Q} c_{i,j}. \qquad (3.32)$$

Note that for a linear model, $\bar{c} = 0$ and Eq. (3.31) reduces to the standard expression for the steady-state gain. Conversely, note that this gain expression holds *whenever $\bar{c} = 0$, regardless of whether the system is linear or not.* That is, $\bar{c} = 0$ is not sufficient to imply $c_{i,j} = 0$ for all i and j; instead, this condition defines a bilinear model that is also a member of the family of *static linear models*, discussed in detail in Sec. 3.4.

It follows from Eq. (3.31) that the bilinear model cannot exhibit output multiplicity since, given any steady-state input value u_s, the steady-state output value y_s is determined uniquely by Eq. (3.31). The only exception is the singular case $1 - \bar{a} - \bar{c} u_s = 0$. For linear models, this condition reduces to $\bar{a} = 1$, corresponding to a model that is not BIBO stable (i.e., the "unit root" or "integrating" case in time-series modeling (Brockwell and Davis, 1991)); also, note that if $\bar{a} = 1$, the steady-state response of the linear model is not unique, since $y(k) = y_s$ satisfies the general linear model equation with $u(k) \equiv 0$ for any real y_s. In fact, these conclusions apply more generally to bilinear models with $\bar{c} = 0$; if $\bar{c} \neq 0$, this singularity condition implies $u_s = (1 - \bar{a})/\bar{c}$ and the steady-state equation requires $\bar{b} u_s = 0$, implying either $\bar{a} = 1$ or $\bar{b} = 0$. Note that $\bar{b} = 0$ implies $y_s = 0$, generalizing the class of linear models with zero steady-state gains.

Bilinear models cannot generally exhibit input multiplicity, either, since it is possible to invert Eq. (3.31) to obtain

$$u_s = \frac{1 - \bar{a}}{\bar{b} + \bar{c} y_s}. \qquad (3.33)$$

Here again, it is necessary to exclude the singular case $\bar{b} + \bar{c} y_s = 0$; for linear models or bilinear models with $\bar{c} = 0$, this singularity condition again reduces to $\bar{b} = 0$. If $\bar{c} \neq 0$ and this singularity condition is satisfied, two cases must be considered. First, if $y_s = 0$, it again follows that $\bar{b} = 0$. Second, if $y_s \neq 0$ then $\bar{b} = -\bar{c} y_s$; combining this condition with Eq. (3.31) and simplifying then implies $\bar{a} = 1$. Overall, then, the steady-state behavior of bilinear models falls into one of three cases:

1. $\bar{a} = 1 \Rightarrow$ steady-state $(0, y_s)$, arbitrary y_s
2. $\bar{b} = 0 \Rightarrow$ steady-state $(u_s, 0)$, arbitrary u_s
3. $\bar{a} \neq 1, \bar{b} \neq 0 \Rightarrow$ unique steady-state (u_s, y_s), Eqs. (3.31) and (3.33).

Note that these results apply for all $\{c_{i,j}\}$, including the linear case $c_{i,j} \equiv 0$.

3.2 Homogeneous models

As noted in the introduction to this chapter, the topic of nonlinear dynamic models may be approached from either a structural or a behavioral perspective. The discussion of bilinear models given in Sec. 3.1 illustrates the structural approach, by far the most common, and one that will be revisited many times in subsequent chapters. This section and the next two illustrate the behavioral approach, in part to demonstrate that such an approach is feasible. Specifically, this section defines and briefly describes the class of *homogeneous* nonlinear models, and the next two sections define and discuss two larger classes of nonlinear models, each containing the class of homogeneous models as a proper subset. Although it is generally more difficult than the structural approach, the behavioral approach does lead to some extremely interesting insights that will be developed further in later chapters.

3.2.1 Homogeneous functions and homogeneous models

Recall from the discussion at the end of Chapter 2 that a function $\mathbf{f}(\mathbf{x})$ mapping R^n to R^m is called *homogeneous* if it satisfies the condition $\mathbf{f}(\lambda \mathbf{x}) = \lambda \mathbf{f}(\mathbf{x})$ for all $\mathbf{x} \in R^n$ and all real λ. The class of *homogeneous models* is defined by replacing the Euclidean spaces R^n and R^m in this definition with the appropriate space of input/output sequences, either ℓ_∞ or ℓ_∞^0, as in the treatment of the linear problem given in Chapter 2. More specifically, suppose $\mathcal{M}$ is a (generally nonlinear) model mapping the input sequence $\{u(k)\} \in \ell_\infty$ into the output sequence $\{y(k)\} \in \ell_\infty$. This model will be called *homogeneous* if it maps $\{\lambda u(k)\}$ into $\{\lambda y(k)\}$ for all real λ. Alternatively, the input and output spaces ℓ_∞ may be replaced with ℓ_∞^0 in this definition.

It is clear that the class of *linear* models belongs to the class of homogeneous models, since the scaling behavior just described is one of the hallmarks of linear models. Given only the definition, however, it is not obvious that any *nonlinear* members of this class of models exist. This observation represents the key limitation of the behavioral approach to defining and analyzing nonlinear dynamic models. In fact, nonlinear members of this class *do* exist, including the following example

$$y(k) = \frac{u(k-1)u(k-2)u(k-3)}{u^2(k-1) + u^2(k-2)}. \tag{3.34}$$

Note that this response is well-defined if and only if $u(k-1)$ and $u(k-2)$ are not both zero; to overcome this limitation, L'Hospital's rule (Klambauer, 1975, p. 245) may be applied to obtain the consistent limiting response for $u(k-1) = u(k-2) = 0$:

$$y(k) = \frac{u(k-3)}{2}. \tag{3.35}$$

Further, it is interesting to note that both the impulse and step responses of the nonlinear model defined by Eqs. (3.34) and (3.35) are identical to those of

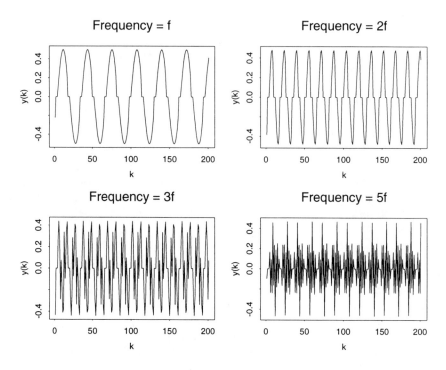

Figure 3.9: Frequency dependence of homogeneous model response

the linear model defined by Eq. (3.35) alone. Although impulse and step inputs are standard ones for characterizing linear models—either response provides a complete characterization—it should be clear from the discussion of bilinear models given in Sec. 3.1 and this example that these two inputs are *not* adequate to characterize nonlinear models, in general. In addition, the response to any constant sequence $u(k) = u_s$ for all k is also linear, and this behavior is a characteristic of the class of homogeneous models; this point is important and is considered further in Sec. 3.4.

The nonlinear nature of the homogeneous model (3.34) may be seen from its responses to input sequences that exhibit greater variation over an extended period. As a specific example, note that the response to the periodic input sequence $u(k) = \alpha \sin \omega k$ is given by

$$y(k) = \left[\frac{2\cos(\omega/2)\sin\omega(k-3/2)}{1 - \cos\omega \cos\omega(2k-3)} \right] \alpha \sin\omega(k-3), \tag{3.36}$$

for any $\alpha \neq 0$ and $\omega \neq 0$. Plots of this response for $\alpha = 1$ and four different values of ω are shown in Fig. 3.9. Note that because this model is homogeneous, the nature of this response is independent of the input amplitude α, but the frequency dependence is quite pronounced, as seen in this figure.

In particular, note the marked *crossover distortion* that occurs whenever the response $y(k)$ passes through the value zero. This effect occurs because $y(k) = 0$

Four Views of Nonlinearity 113

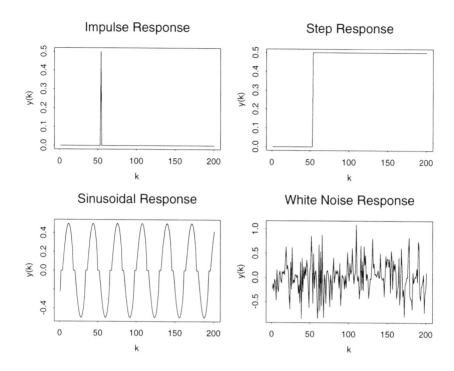

Figure 3.10: Four standard homogeneous model responses

whenever $u(k-1) = 0$, $u(k-2) = 0$, or $u(k-3) = 0$; since $u(k) = 0$ twice in each period of the input sinusoid, $y(k) = 0$ will occur for two sets of three successive points in each period. This behavior is seen in the upper two plots in Fig. 3.9, which show the response of this model for frequencies f and $2f$ where $f = 1/32$ is the frequency of the standard sinusoidal sequence introduced in Chapter 2. The lower two plots in Fig. 3.9 show the responses for frequencies $3f$ and $5f$, and these responses are seen to be more complicated, becoming more "noise-like" with increasing frequency. Because the homogeneous model considered here is a member of the moving average class of models defined and discussed in Chapter 4, it turns out that the response to a periodic input of any frequency is necessarily periodic with the same frequency. It is clear, however, that this response is highly nonsinusoidal, exhibiting a relatively complicated dependence on excitation frequency.

Fig. 3.10 gives the standard four input sequence responses to facilitate comparison of this model with the other specific model examples discussed in this book. In addition, this figure may be viewed as a graphical summary of the general results presented above: the impulse and step responses presented in the upper two plots of this figure give no indication of the model's nonlinearity, being completely indistinguishable from the related linear model $y(k) = 0.5u(k-3)$. The lower two plots, however, do provide clear evidence of this nonlinearity, as seen in the nonsinusoidal character of the response to a sinusoidal input and, perhaps less obviously, the non-Gaussian character of the response to the Gaussian white noise input. It is not difficult to show that the response of this model to any zero-mean, independent, identically distributed input sequence is itself a zero-mean sequence. Beyond this observation, however, the analysis rapidly becomes difficult because of the model's nonlinearity.

One important feature of the class of nonlinear homogeneous functions is that they are necessarily multivariable maps, as the following theorem demonstrates. Further, it follows as an immediate corollary that nonlinear homogeneous models of the general NMAX form discussed in Chapter 4:

$$y(k) = \Phi(u(k), \ldots, u(k-M)), \tag{3.37}$$

require $M > 0$. Similarly, it also follows from this result that the nonlinear autoregressive model

$$y(k) = f(y(k-1)) + bu(k), \tag{3.38}$$

cannot exhibit homogeneous behavior, aside from the linear case $f(y) = \alpha y$. In contrast, nonlinear models of the form

$$y(k) = f(y(k-1), y(k-2)) + bu(k) \tag{3.39}$$
$$\text{or} \quad y(k) = f(y(k-1), u(k)), \tag{3.40}$$

can exhibit nonlinear homogeneous dynamics. Differences between these and other nonlinear autoregressive models are explored further in Chapter 4.

Theorem:

If $f : R^1 \to R^1$ is homogeneous, it is linear.

Proof:

Suppose $f : R^1 \to R^1$ is homogeneous and $x \neq 0$; then,

$$\begin{aligned} f(x+y) = f([1+y/x]x) &= (1+y/x)f(x) \\ &= f(x) + (y/x)f(x) \\ &= f(x) + f(y). \end{aligned}$$

Thus, $f(\cdot)$ is both homogeneous and (functionally) additive, from which it follows that $f(\cdot)$ is linear, by the results discussed at the end of Chapter 2.

□

In view of this result, the simplest examples of nonlinear homogeneous maps are those from R^2 to R^1. The following examples will be useful in subsequent discussions and suggest immediate generalizations to R^n for arbitrary n:

$$f_1(x,y) = [x^3 + y^3]^{1/3} \qquad f_2(x,y) = \frac{xy}{x+y}$$

$$f_3(x,y) = \frac{x^3 + y^3}{x^2 + y^2} \qquad f_4(x,y) = \frac{x^3 - y^3}{xy}.$$

Further, note that restricting these functions to the subspace $x = y$ yields a function $f_i(x,x)$ defined on R^1. This function is necessarily homogeneous and, by the theorem presented above, therefore linear: $f_i(x,x) = \alpha x$ for some real α. Further, note that the constant α can assume any real value, including zero.

A generalization of homogeneity that arises in the classical theory of ordinary differential equations (Davis, 1962) is the following. For any real number ν, a function $f : R^m \to R^1$ is called *homogeneous of order* ν if $f(\lambda \mathbf{x}) = \lambda^\nu f(\mathbf{x})$. To be consistent with the notation introduced previously, the term *homogeneous function* with no further specification will always be taken to mean "homogeneous of order 1." As in the case $\nu = 1$, the range of possible homogeneous functions of order ν is highly restricted for scalar functions. Specifically, the following theorem generalizes the previous one.

Theorem:

If $f : R^1 \to R^1$ is homogeneous of order ν, then $f(x) = \alpha x^\nu$ for some real constant α.

Proof:

Let $\alpha = f(1)$ and consider any $x \in R^1$; it follows that

$$f(x) = f(x \cdot 1) = x^\nu f(1) = \alpha x^\nu.$$

$\square$

This result will be useful in subsequent developments, as will the following observation: if $f : R^1 \to R^1$ is homogeneous of order ν and invertible, then the inverse function $f^{-1}(\cdot)$ is necessarily homogeneous of order $1/\nu$. As a simple example, note that the function $f(x) = x^3$ is homogeneous of order 3, whereas its inverse $f^{-1}(x) = x^{1/3}$ is homogeneous of order $1/3$. As in the special case $\nu = 1$ already considered, the behavior of homogeneous functions of order ν becomes much more interesting in the multivariable case. This point is illustrated in Sec. 3.2.3 for the important special case $\nu = 0$.

3.2.2 Homomorphic systems

In chapter 10 of their signal processing book, Oppenheim and Schafer (1975) describe a class of nonlinear techniques called *homomorphic signal processing*. These techniques are based on the class of *homomorphic systems* shown in Fig. 3.11, consisting of two static nonlinearities and a linear dynamic model. More specifically, this structure consists of the scalar nonlinearity $\phi(\cdot)$, followed by the linear dynamic model defined by the discrete transfer function $H(z)$, followed finally by the inverse of the original function, $\phi^{-1}(\cdot)$. This structure is extremely special, for reasons that will become partially apparent in the discussions presented here. In fact, Oppenheim and Schafer consider a somewhat more general structure than this, but the structure indicated in Fig. 3.11 is sufficient

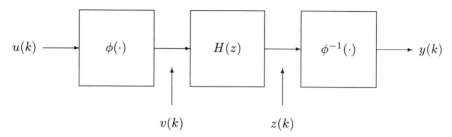

Figure 3.11: Representation of a homomorphic system

for the purposes of discussion here, and is extremely important in subsequent discussions in Chapter 7.

Here, the following additional restriction is imposed: besides being invertible, the function $\phi(\cdot)$ will also be homogeneous of order ν for some $\nu \neq 0$. Consequently, it follows that $\phi^{-1}(\cdot)$ is homogeneous of order $1/\nu$ from the results given in Sec. 3.2.1. The significance of these restrictions is that the resulting homomorphic model illustrated in Fig. 3.11 is then *homogeneous* (i.e., homogeneous of order 1). To see this point, consider the effect of multiplying an arbitrary input sequence $\{u(k)\}$ by λ. Defining $\{v(k)\}$ as the output of the first block in Fig. 3.11 [i.e., $v(k) = \phi(u(k))$], it follows that $\phi(\lambda u(k)) = \lambda^\nu \phi(u(k))$. Similarly, defining $z(k)$ as the output of the linear block in Fig. 3.11, it follows that scaling $u(k)$ by λ will then scale $z(k)$ by λ^ν. Since the output of the homomorphic system is $y(k) = \phi^{-1}(z(k))$ and this nonlinearity is homogeneous of order $1/\nu$, it follows that the ultimate response $y(k)$ will be scaled by λ when the input is scaled by this factor. Thus, the homomorphic system will exhibit a homogeneous response to arbitrary input sequences.

As a simple but specific example, suppose $\phi(x) = x^3$, an invertible function that is homogeneous of order 3; as noted previously, its inverse is $\phi^{-1}(x) = x^{1/3}$ and is homogeneous of order $1/3$. For the linear portion of this structure, consider the simple moving averge model $z(k) = v(k-1) + v(k-2)$. The input-output behavior of this homomorphic system may then be represented as

$$y(k) = [u^3(k-1) + u^3(k-2)]^{1/3} = f_1(u(k-1), u(k-2)), \quad (3.41)$$

where $f_1(\cdot, \cdot)$ is the first of the homogeneous function examples discussed in Sec. 3.2.1. As another example, consider the same linear moving average model, but take $\phi(x) = 1/x$, a function that is its own inverse and is homogeneous of order -1. The input-output behavior of the homomorphic system constructed from this nonlinearity is given by

$$y(k) = \frac{u(k-1)u(k-2)}{u(k-1) + u(k-2)} = f_2(u(k-1), u(k-2)), \quad (3.42)$$

where $f_2(\cdot, \cdot)$ is the second of the homogeneous function examples discussed in Sec. 3.2.1.

These two examples raise the interesting question of whether *all* homogeneous models could be represented as homomorphic systems. The answer to

Four Views of Nonlinearity   117

this question is no, a result that is demonstrated in Sec. 3.2.3 but which rests on the following observation. Suppose $h(k)$ is the impulse response of the linear system on which the homomorphic model is based, and consider the model's response to an impulse of amplitude α. The static nonlinearity $\phi(\cdot)$ maps this input sequence into a second impulse, of amplitude α^ν. Hence, the linear system's response is $\alpha^\nu h(k)$, implying the homomorphic model's output is

$$y(k) = [\alpha^\nu h(k)]^{1/\nu} = \alpha h^{1/\nu}(k). \tag{3.43}$$

Similar reasoning establishes that, if $s(k)$ is the step response of the linear system, the response of the homomorphic model is $\alpha s^{1/\nu}(k)$. The key point is that not all homogeneous models exhibit impulse and step responses of this form, a result demonstrated in Sec. 3.2.3. This point is also discussed further in Chapter 7.

Two other points are worth making here. First, note that the class of homogeneous functions of order ν is somewhat irregularly behaved for even orders ν. Specifically, note that in the scalar case, these maps are not invertible, since if $f(\cdot)$ is homogeneous of some even order ν, it follows immediately that $f(-x) = (-1)^\nu f(x) = f(x)$. Thus, positive and negative roots cannot be distinguished from $f(x)$ alone, implying $f(\cdot)$ is not invertible on R^1. The case $f(x) = x^2$ provides the classic example here: while $f^{-1}(x)$ is well-defined if $f(x)$ is restricted to the domain $x \geq 0$, this inverse is not defined on the whole space R^1. This point is important and closely related to the class of *positive homogeneous functions* discussed in Sec. 3.3.

Second, note that the homogeneity of the homomorphic system shown in Fig. 3.11 does not actually require the dynamic model to be linear, only homogeneous. For example, replacing the linear model that leads to Eq. (3.41) with the homogeneous model defined in Eqs. (3.34) and (3.35) yields the new homogeneous model

$$y_k = \frac{u(k-1)u(k-2)u(k-3)}{[u^6(k-1) + u^6(k-2)]^{1/3}}, \tag{3.44}$$

where $\phi(x) = x^3$ as before. This construction is closely related to the fact that the class of homogeneous models is closed under cascade connection, a point discussed at length in Chapter 7.

3.2.3 Homogeneous ARMAX models of order zero

The second systematic procedure for constructing homogeneous nonlinear dynamic models considered here is based on homogeneous functions of order zero, denoted H^0-functions for simplicity. These functions satisfy the scaling condition

$$F(\lambda x_1, \ldots, \lambda x_m) = F(x_1, \ldots, x_m), \tag{3.45}$$

for all real λ. The motivation for this development is the observation that the simplest homogeneous functions of order zero are the constant functions

$F(x_1,\ldots,x_m) = \alpha$. The class of *homogeneous ARMAX models of order 0* or H^0-ARMAX models is constructed as follows. Let Φ_i and Ψ_j be H^0-functions, each with arguments $y(k-1),\ldots,y(k-p), u(k),\ldots,u(k-q)$ for some $p,q \geq 0$, adopting the convention that the functions do not depend on $y(k-i)$ if $p=0$. Given these functions, consider the following class of models:

$$y(k) = \sum_{i=1}^{p} \Phi_i y(k-i) + \sum_{j=0}^{q} \Psi_j u(k-j). \tag{3.46}$$

Note that if $\Phi_i = a_i$ for all i and $\Psi_j = b_j$ for all j, this definition reduces to the class of linear ARMAX models discussed in Chapter 2. More generally, since Φ_i and Ψ_j are all H^0-functions, it follows that the class of models defined in Eq. (3.46) is homogeneous.

It follows as a corollary of the theorem presented in Sec. 3.2.1 that if the functions Φ_i and Ψ_j depend on a single argument they are constant, again leading to the class of linear ARMAX models. To obtain nonlinear homogeneous models, it is therefore necessary to consider functions involving at least two arguments. Specific examples of nonconstant H^0-functions with two arguments include the following:

$$F_1(x,y) = \frac{x-y}{x+y} \qquad F_2(x,y) = \frac{x^2-y^2}{x^2+y^2}$$

$$F_3(x,y) = \exp\left\{\frac{x^2-y^2}{x^2+y^2}\right\} \qquad F_4(x,y) = \frac{|x+y|}{\sqrt{x^2+y^2}}.$$

More generally, note that if $F : R^m \to R^1$ is any H^0-function and $G : R^1 \to R^1$ is *any real-valued function defined on all of* R^1, it follows that the composition $G(F(\cdots))$ is also a homogeneous function of order 0 mapping R^m into R^1. For this reason, the class of multivariable H^0-functions is quite broad.

To illustrate the range of behavior the family of H^0-ARMAX models can exhibit, consider the following simple but general example:

$$y(k) = \Phi_1[y(k-1), u(k-1)]y(k-1) + b_1 u(k-1), \tag{3.47}$$

where $\Phi_1(v,w)$ is an H^0 function, $b_1 \neq 0$, and $y(k) = 0$ for $k \leq 0$. Surprisingly, an explicit expression for the impulse response of this model can be derived with essentially no further knowledge about the function $\Phi(v,w)$. That is, the initial response to an impulse of amplitude α is $y(1) = b_1 \alpha$, independent of the function $\Phi_1(v,w)$ and for $k > 1$, the first term on the right-hand side of Eq. (3.47) is $\Phi_1[y(k-1), 0]y(k-1)$ since $u(k-1) = 0$ for $k > 1$. Because $\Phi_1(v,w)$ is H^0, the restriction $\Phi_1(v,0)$ is also H^0, so by the second theorem presented in Sec. 3.2.1, $\Phi_1(v,0) = \gamma$ for all v for some constant γ. Hence, the impulse response of the model defined in Eq. (3.47) is identical to that of the linear model

$$y(k) = \gamma y(k-1) + b_1 u(k-1), \tag{3.48}$$

independent of the behavior of the H^0 *function* $\Phi_1(v,w)$ *for* $w \neq 0$.

Four Views of Nonlinearity 119

To see this point, consider the following four examples. The linear model defined by Eq. (3.48) is designated Model 0, Model 1 is defined by taking $\Phi_1(v,w) = \gamma v^2/(v^2 + w^2)$, yielding

$$y(k) = \gamma \left[\frac{y^2(k-1)}{y^2(k-1) + u^2(k-1)}\right] y(k-1) + b_1 u(k-1), \quad (3.49)$$

and Model 2 is defined by taking $\Phi_1(v,w) = \gamma(v^2 - w^2)/(v^2 + w^2)$, yielding

$$y(k) = \gamma \left[\frac{y^2(k-1) - u^2(k-1)}{y^2(k-1) + u^2(k-1)}\right] y(k-1) + b_1 u(k-1). \quad (3.50)$$

Finally, define Model 3 according to Eq. (3.47) with $\Phi_1(v,w)$ given by

$$\Phi_1(v,w) = \begin{cases} \alpha & |v| < |w| \\ \beta & |v| \geq |w|. \end{cases} \quad (3.51)$$

To see that this function is homogeneous of order zero, note that $|v| \leq |w|$ if and only if $|\lambda v| \leq |\lambda w|$ for all $\lambda \neq 0$; thus, $\Phi_1(\lambda v, \lambda w) = \Phi_1(v,w)$. The H^0-ARMAX model defined by Eq. (3.47) for this choice of $\Phi_1(v,w)$ may be written more explicitly as

$$y(k) = \begin{cases} \alpha y(k-1) + u(k-1) & |y(k-1)| < |u(k-1)| \\ \beta y(k-1) + u(k-1) & |y(k-1)| \geq |u(k-1)|. \end{cases} \quad (3.52)$$

This representation illustrates that Model 3 belongs to the class of *linear multimodels* discussed in Chapter 6. A key feature of this example is that, whereas these local linear models are both extremely simple (i.e., first-order), they are *generally selected* since the choice of which local model is "valid" or "active" at any time k depends on both the immediate past input $u(k-1)$ and the immediate past output $y(k-1)$. The significance of this observation is discussed further in Chapter 6.

To illustrate the difference between these four models, Fig. 3.12 shows their unit step responses for the following parameter values:

Model 0: $\gamma = -0.8$,
Model 1: $\gamma = -0.8$,
Model 2: $\gamma = -0.8$,
Model 3: $\alpha = 2.0$, $\beta = -0.8$.

For these parameter values, all four models exhibit identical impulse responses, each consisting of a decaying oscillation. The upper left plot in Fig. 3.12 shows the step response of the linear reference model (Model 0), which exhibits the "ringing" behavior typical of a first-order linear model with negative autoregressive coefficient γ. Note that Model 1 (upper right plot) exhibits qualitatively similar behavior, although the steady-state gain is larger than for the linear model (0.77 vs. 0.56). It is not difficult to show that the step response of Model 2 is $y(k) = u(k-1)$, independent of the parameter γ, so this model exhibits *no*

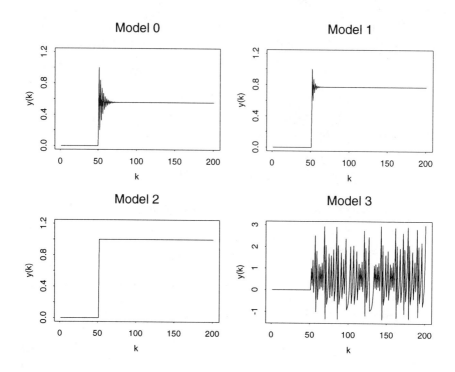

Figure 3.12: A comparison of four step responses

ringing in its step response, in contrast to Models 0 and 1. The most dramatic contrast, however, is provided by Model 3, which exhibits a chaotic step response that never settles to a steady-state value. *In fact, this response establishes the claim made in Sec. 3.2.2: because linear dynamic models cannot exhibit chaotic step responses, it follows that Model 3 is not a homomorphic model.* Specifically, recall that the unit step response of a homomorphic model based on the static nonlinearity $\phi(x) = x^\nu$ is given by $s^{1/\nu}(k)$, where $s(k)$ is the step response of the linear system on which the homomorphic model is based. For this response to be chaotic, it would follow that the linear system's step response $s(k)$ would also have to be chaotic and this is not possible.

Fig. 3.13 shows four responses of Model 3 to the standard sinusoidal input sequence introduced in Chapter 2. The difference between these plots is the parameter values assumed in Eq. (3.52), which are indicated above the plots in the form (α, β). It is clear from these plots that the sinusoidal response of this model can range from one that is essentially indistinguishable from a linear model response (lower left plot, $\alpha = 2.0, \beta = 0.8$) to one that is extremely irregular and "noise-like" (lower right plot, $\alpha = 2.0, \beta = -0.8$). Intermediate cases (e.g., the two upper plots) can exhibit responses that are clearly nonsinusoidal—thus demonstrating the model's nonlinearity—but "more regular" than the response seen in the lower right plot.

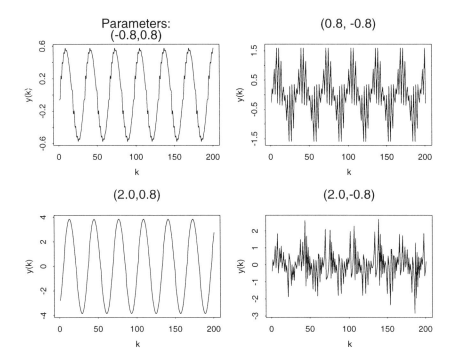

Figure 3.13: A comparison of four sinusoidal responses

3.3 Positive homogeneous models

The class of homogeneous dynamic models was defined in Sec. 3.2 by requiring invariance under scaling of the input sequence. This requirement may be relaxed in either of two ways, both of which lead to interesting classes of nonlinear dynamic models. The first of these relaxations is to require invariance under *positive* input scalings and the consequences of this relaxation is considered here. Alternatively, the homogeneity requirement may be relaxed by requiring only that it hold for *constant input sequences*. The consequences of this second relaxation of homogeneity are explored in Sec. 3.4.

3.3.1 Positive-homogeneous functions and models

The defining condition for the class of *positive-homogeneous models* is

$$u(k) \to \lambda u(k) \quad \Rightarrow \quad y(k) \to \lambda y(k), \tag{3.53}$$

for all $\lambda > 0$. Here, $\{u(k)\}$ is the input sequence driving the system and $y(k)$ is its response to this input sequence. Since any homogeneous model is also positive-homogeneous, positive-homogeneity is a strictly weaker requirement than homogeneity, implying that the class of positive-homogeneous models is larger than the class of homogeneous models. Consequently, positive-homogeneous models seem to appear in applications more frequently than homogeneous models. For example, note that the *RMS-to-DC converter* is an electronic component that may be viewed as an analog computation of the root-mean-square amplitude of a time-varying signal $x(t)$. Digital implementations of this device are also possible, and these implementations may be viewed as nonlinear dynamic models of the form

$$y(k) = \sqrt{\frac{1}{M+1} \sum_{i=0}^{M} u^2(k-i)}. \tag{3.54}$$

It is easy to see that this system satisfies the positive-homogeneity condition (3.53), consistent with the interpretation of $y(k)$ as a measure of the amplitude of the input sequence $\{u(k)\}$. It is also interesting to note that, if the input sequence is constrained to be nonnegative, the digital RMS-to-DC converter defined in Eq. (3.54) may also be viewed as a homomorphic system, based on the quadratic nonlinearity $\phi(x) = x^2$ and an unweighted moving average filter. Generalizing to arbitrary nonnegative weights leads to the class of *Pythagorean models of order* M, defined by

$$y(k) = \sqrt{\sum_{i=0}^{M} \alpha_i u^2(k-i)}, \tag{3.55}$$

and denoted $P(M)$. The name of this class of systems derives from its relationship to the Pythagorean theorem: $y(k)$ represents the length of the hypotenuse in R^{M+1} of the "right hypertriangle" whose sides are defined by $u(k-i)$.

More generally, positive-homogeneous models arise in connection with phenomena that depend on the *intensity* of a of a variable quantity, but not on details like its phase or its short-term time history. For example, note that positive-homogeneity is a characteristic of the p-norm

$$||\mathbf{x}||_p = \left[\sum_{i=1}^{m} |x_i|^p \right]^{1/p}, \tag{3.56}$$

representing a map from R^m to R. This result holds for any p, including the ∞-norm

$$||\mathbf{x}||_\infty = \max\{|x_1|, \ldots, |x_m|\}. \tag{3.57}$$

obtained in the limit as $p \to \infty$. A physically analogous phenomenon is Joule heating, which depends on the magnitude of the current flow through a conductor but not on the direction of the current flow. Also, note that if $f : R^m \to R$ is any homogeneous function, it follows that $|f(\mathbf{x})|$ is a positive-homogeneous function. Similarly, if $h : R^m \to R$ is a homogeneous function of order 2, it follows that $\sqrt{h(\mathbf{x})}$ is positive-homogeneous.

Unlike the more restrictive homogeneous case discussed previously, *nonlinear* positive homogeneous functions *do* exist on R^1, as the following example illustrates:

$$\theta(a, b; x) = ax + b|x|. \tag{3.58}$$

Here, a and b are any real constants, and it is not difficult to show that $\theta(a, b; \lambda x) = \lambda \theta(a, b; x)$ for any $\lambda > 0$. In fact, the following theorem establishes that all positive homogeneous functions on R^1 are of this form.

Theorem:

On R^1, if $f(x)$ is positive homogeneous, then $f(x) = \theta(a, b; x)$ for some real constants a and b.

Proof:

Note that

$$f(x) = \begin{cases} xf(1) & x \geq 0 \\ x[-f(-1)] & x < 0, \end{cases}$$

$$= \left(\frac{f(1) - f(-1)}{2}\right) x + \left(\frac{f(1) + f(-1)}{2}\right) |x| = \theta(a, b; x),$$

where $a = [f(1) - f(-1)]/2$ and $b = [f(1) + f(-1)]/2$.

□

Four specific examples of this family of functions are shown in Fig. 3.14. The top left plot shows the linear identity function $\theta(1, 0; x) = x$ and the top right plot shows the nonlinear absolute value function $\theta(0, 1; x) = |x|$. The bottom left plot shows the nonnegative *threshold nonlinearity*

$$\theta(\frac{1}{2}, \frac{1}{2}; x) = \theta_+(x) \equiv \begin{cases} x & x \geq 0 \\ 0 & x < 0. \end{cases}$$

Not shown is the analogous nonpositive threshold nonlinearity

$$\theta(\frac{1}{2}, -\frac{1}{2}; x) = \theta_-(x) \equiv \begin{cases} 0 & x \geq 0 \\ x & x < 0. \end{cases}$$

In the general case, the function $\theta(a, b; x)$ is piecewise linear, continuous at $x = 0$, but with different slopes for positive and negative arguments, as shown in the bottom right plot in Fig. 3.14.

In general, note that

$$\theta(\alpha, \beta; x) = \begin{cases} (\alpha + \beta)x & x \geq 0 \\ (\alpha - \beta)x & x < 0. \end{cases} \tag{3.59}$$

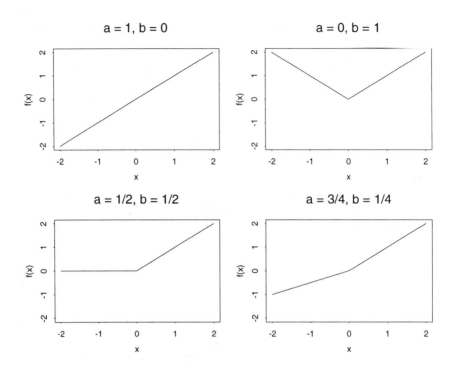

Figure 3.14: Four examples of positive-homogeneous functions

This representation is useful because it leads immediately to both conditions for $\theta(a, b; x)$ to be invertible and a simple expression for this inverse when it exists. In particular, it is a standard result that a continuous function is invertible on any closed, bounded interval if and only if it is strictly monotonic (Klambauer, 1975, p. 181). This condition is met if and only if the following two conditions are satisfied:

1. The slopes $\alpha + \beta$ and $\alpha - \beta$ must both be nonzero
2. The slopes $\alpha + \beta$ and $\alpha - \beta$ must both have the same sign.

These conditions reduce to the following requirements:

$$\alpha \neq 0 \qquad |\beta| < |\alpha|, \tag{3.60}$$

and the representation for $\theta(\alpha, \beta; x)$ given in Eq. (3.59) leads immediately to the following expression for the inverse:

$$\theta^{-1}(\alpha, \beta; x) = \begin{cases} x/(\alpha + \beta) & x \geq 0 \\ x/(\alpha - \beta) & x < 0. \end{cases} \tag{3.61}$$

In fact, it is apparent from this result that the inverse function is itself a function $\theta(\mu, \nu; x)$, for μ and ν satisfying

$$\mu + \nu = 1/(\alpha + \beta) \qquad \mu - \nu = 1/(\alpha - \beta). \tag{3.62}$$

Solving these equations leads to the following explicit result for the inverse function:

$$\theta^{-1}(\alpha, \beta; x) = \theta\left(\frac{\alpha}{\alpha^2 - \beta^2}, \frac{-\beta}{\alpha^2 - \beta^2}; x\right). \tag{3.63}$$

The family of functions $\theta(\alpha, \beta; x)$ is positive-homogeneous with respect to the argument x and *linear* with respect to the parameters α and β. These observations lead to the following results:

$$\theta(\alpha, \beta; \lambda x) = \lambda \theta(\alpha, \beta; x) \text{ for all } \lambda \geq 0$$
$$\theta(\alpha, 0; x) = \alpha x$$
$$\theta(0, \beta; x) = \beta|x|$$
$$\theta(\alpha, \alpha; x) = 2\alpha\theta_+(x)$$
$$\theta(\alpha, -\alpha; x) = 2\alpha\theta_-(x)$$

$$\sum_{i=1}^{m} \theta(\alpha_i, \beta_i; x) = \theta\left(\sum_{i=1}^{m} \alpha_i, \sum_{i=1}^{m} \beta_i; x\right). \tag{3.64}$$

As will be seen in subsequent examples, these results are useful in simplifying expressions involving the function $\theta(\alpha, \beta; x)$.

3.3.2 PH^0-ARMAX models

Just as homogeneous functions of order zero led naturally to the class of homogeneous ARMAX models of order zero introduced in Sec. 3.2.3, it is also possible to define an interesting class of positive-homogeneous models based on *positive homogeneous functions of order zero*, denoted PH^0-functions for simplicity. These functions are mappings $F : R^m \to R$ satisfying the requirement that, for all $\lambda > 0$

$$F(\lambda x_1, \lambda x_2, \ldots, \lambda x_m) = F(x_1, x_2, \ldots, x_m). \tag{3.65}$$

In analogy with the class of H^0-ARMAX models defined in Sec. 3.2.3, define the class of *positive-homogeneous ARMAX models of order zero* (or, more simply, PH^0-ARMAX models) by the equation:

$$y(k) = \sum_{i=1}^{p} \Phi_i y(k-i) + \sum_{j=0}^{q} \Psi_j u(k-j), \tag{3.66}$$

where Φ_i and Ψ_j are PH^0-functions in the arguments $y(k-1), \ldots, y(k-p), u(k), \ldots, u(k-q)$. It follows immediately that the resulting class of dynamic models is positive-homogeneous.

Clearly, the H^0-functions considered in Sec. 3.2.3 are also PH^0-functions, but the larger PH^0 class also includes functions like $\phi_1(x, y) = x/\sqrt{x^2 + y^2}$ that is invariant under positive scalings of x and y, but antisymmetric with respect to x. Similarly, $\phi_2(x,y) = \exp[\phi_1(x,y)]$ is also a PH^0 function since it is of the form $F(G(x,y))$ where $G(x,y)$ is a PH^0 function. Finally, note that the following PH^0 function is both extremely simple and extremely discontinuous:

$$\phi_3(x,y) = \mathrm{sgn}(x+y) = \begin{cases} -1 & x+y < 0 \\ 0 & x+y = 0 \\ 1 & x+y > 0. \end{cases}$$

To illustrate the general behavior of this model class, consider the following four examples:

1. *Model 1:*

$$y(k) = \left[\frac{\alpha y(k-1)}{\sqrt{y^2(k-1) + u^2(k-1)}}\right] y(k-1) + u(k-1),$$

2. *Model 2:*

$$y(k) = \left[\frac{\alpha u(k-1)}{\sqrt{y^2(k-1) + u^2(k-1)}}\right] y(k-1) + u(k-1),$$

3. *Model 3:*

$$y(k) = \alpha[\mathrm{sgn}(u(k-1) + y(k-1))]y(k-1) + u(k-1),$$

4. *Model 4:*

$$y(k) = \alpha y(k-1) + [\mathrm{sgn}(u(k-1) + y(k-1))]u(k-1).$$

First, consider the responses of these models to an impulse of magnitude γ at $k = 0$, assuming $y(k) = 0$ for $k \leq 0$. For Models 1, 2, and 3, it follows that $y(1) = \gamma$. For Model 1, the response for $k > 1$ is $y(k) = \alpha|y(k-1)|$; hence, the response to positive impulses is indistinguishable from that of the linear model $y(k) = \alpha y(k-1) + u(k-1)$, whereas for negative impulses, an immediate reversal of the sign of the response occurs at $k = 2$, followed by an exponential decay from this positive value. For Model 2, $y(k)$ is proportional to $u(k-1)$ for $k > 1$, so the response to any impulse input is identically zero for $k > 1$. Model 2 thus exhibits exactly the same response to impulses—either positive or negative—as the trivial linear model $y(k) = u(k-1)$. Models 3 and 4 are interesting for a number of reasons, not least of which is their deceptive simplicity. In fact, these models can be represented as members of the class of linear multimodels discussed in detail in Chapter 6. The impulse response for Model 3 is identical to that for Model 1, but Model 4 exhibits highly unusual behavior: its responses to positive and negative impulse inputs is *identical*, equal in both cases to the response of the linear model $y(k) = \alpha y(k-1) + u(k-1)$ to a *positive* impulse of magnitude $|\gamma|$. Both of these conclusions may be verified easily by examining the equations defining Models 3 and 4.

The responses of Models 1 and 2 to step inputs are shown in Fig. 3.15, with the upper plots showing responses to positive steps and the lower plots showing

Four Views of Nonlinearity   127

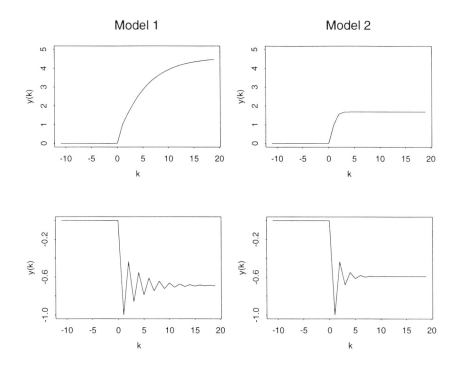

Figure 3.15: Step responses, Models 1 and 2

responses to negative steps. Note that, although the effect of changing the direction of the input step has the opposite sense of the qualitatively asymmetric CSTR response considered in Chapter 1, the behavior is completely analogous: positive step inputs result in monotonic responses, whereas negative step inputs result in oscillatory responses. Also, note that Model 2 exhibits a much more rapid approach to its steady-state value than Model 1 for step inputs of either sign. Like the impulse response, the general qualitative behavior of the step response for Model 3 is essentially the same as that for Model 1 and is therefore not shown here. The response of Model 4 to positive steps is idetentical to that of Model 3, but the response to negative steps is dramatically different, as seen in Fig. 3.16. Specifically, the negative step response exhibits persistent oscillations that never settle out to a constant steady-state response.

3.3.3 TARMAX models

Both the H^0-ARMAX models introduced in Sec. 3.2.3 and the PH^0-ARMAX models just described are defined by replacing the constants appearing in the linear ARMAX model with homogeneous or positive-homogeneous functions of order 0. The product of these functions with $u(k-i)$ or $y(k-i)$ then yields either a homogeneous or a positive-homogeneous nonlinear function of the arguments, implying homogeneous or positive-homogeneous behavior for

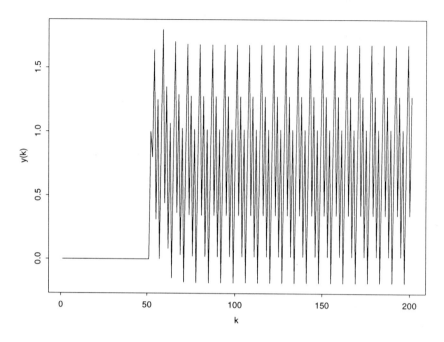

Figure 3.16: Negative step response, Model 4

the resulting dynamic model. Because scalar functions can be both positive-homogeneous and nonlinear, it is also possible to define the following class of positive-homogeneous dynamic models:

$$\begin{aligned} y(k) &= \sum_{i=1}^{p} \theta(a_i, c_i; y(k-i)) + \sum_{i=0}^{q} \theta(b_i, d_i; u(k-i)) \\ &= \sum_{i=1}^{p} a_i y(k-i) + \sum_{i=1}^{p} c_i |y(k-i)| \\ &\quad + \sum_{i=0}^{q} c_i u(k-i) + \sum_{i=0}^{q} d_i |u(k-i)|. \end{aligned} \quad (3.67)$$

This equation defines the class of TARMAX models (theta-function NARMAX models), a subset of the class of *structurally additive models* defined and discussed in Chapter 4.

These models exhibit an extremely interesting structure that can be exploited to gain insight into the realtionship between internal structure and input/output behavior. In addition, the TARMAX model family belongs to the class of linear multimodels considered in Chapter 6, where the TARMAX models are discussed further. For now, it is enough to illustrate some of the interesting behavior these models can exhibit by briefly considering the following two examples, introduced in Chapter 1 (Sec. 1.4.4). Both models were described in Pearson (1995) for their ability to match the qualitative dynamics of the exothermic CSTR example discussed in Chapter 1, and both may be represented in terms of the positive-homogeneous $\theta(a, b; x)$ functions:

$$y(k) = a\theta(0, 1; y(k-1)) + u(k-1) \tag{3.68}$$

for the first example and

$$y(k) = b\theta(1, 0; y(k-1)) + c\theta(\frac{1}{2}, \frac{1}{2}; y(k-2)) + u(k-1) \tag{3.69}$$

for the second. The coefficients appearing in these models are $a = -0.70$, $b = 0.60$, and $c = -0.80$, and in both cases, positive steps are oscillatory in character, while negative ones are monotonic.

The general behavior of these models is illustrated in Fig. 3.17, which shows the standard four responses for the second-order model defined by Eq. (3.69); the responses for the first-order model defined by Eq. (3.68) are virtually identical and are therefore not shown. Note that the impulse response looks very much like the nonminimum phase examples discussed in Chapter 2: the response is initially positive, then switches to a negative value, and ultimately decays monotonically toward zero. In contrast, the step response is identical to that of the minimum-phase linear model

$$y(k) = 0.6y(k-1) - 0.8y(k-2) + u(k-1). \tag{3.70}$$

Not surprisingly, the sinusoidal and white noise responses are more complex and provide stronger reason to suspect nonlinearity. In particular, note the asymmetry of the sinusoidal response, reflecting the model's positive-homogeneous nonlinear behavior: the positive half cycle of the sinusoidal response exhibits small "ripples" that are not seen in the negative half cycles of the response, and the negative half cycles appear generally "sharper" than the positive half cycles. Careful examination of the Gaussian white noise responses reveal similar asymmetries.

3.4 Static-linear models

The class of positive-homogeneous models introduced in Sec. 3.3 represents one possible extension of the class of homogeneous models described in Sec. 3.2. Another possible extension is the class of *static-linear* models, defined by requiring that the linearity conditions hold *only for constant input sequences*. The consequences of this assumption are quite interesting, particularly since they lead to a family of nonlinear models with members in almost all of the structurally-defined nonlinear model classes considered in this book.

3.4.1 Definition of the model class

Consider a model $\mathcal{M}$ that maps an input sequence $\{u(k)\} \in \ell_\infty$ into an output sequence $\{y(k)\} \in \ell_\infty$ and restrict consideration to input sequences of the form

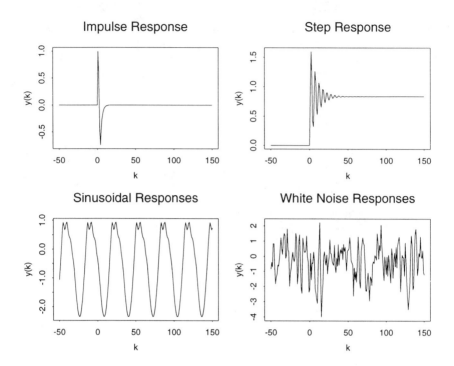

Figure 3.17: Four standard responses of a TARMAX model

$u(k) = u_s$ for all k. The model $\mathcal{M}$ is static linear if it satisfies the linearity condition

$$\mathcal{M}[\alpha u_s^1 + \beta u_s^2] = \alpha \mathcal{M}[u_s^1] + \beta \mathcal{M}[u_s^2], \quad (3.71)$$

for all possible constant input sequences u_s^1 and u_s^2. As in the case of homogeneous models, this definition may be extended to the class ℓ_∞^0 of input sequences, defined only for $k \geq 0$.

It is important to note that the response $y(k)$ to a constant input sequence need not be constant itself to satisfy the static linearity conditions, as the following simple example illustrates:

$$y(k) = u(k) \left[\frac{1 + u^2(k-1)}{1 + u^2(k-2)} \right] \sin(\omega k). \quad (3.72)$$

Although this model is time-varying, it is also static-linear, as the following argument illustrates. For $u(k) = u_s$, the response is a sinusoid of fixed frequency ω and constant amplitude u_s, so the response of this model to the constant input sequence $u(k) = \alpha u_s^1 + \beta u_s^2$ is a sinusoid of frequency ω and amplitude $\alpha u_s^1 + \beta u_s^2$. Since the phase of these sinusoids is the same, it follows that this response may be written as $y(k) = \alpha y_1(k) + \beta y_2(k)$, where $y_1(k)$ and $y_2(k)$ are the responses to the constant input sequences u_s^1 and u_s^2, respectively.

The class of static-linear models includes the entire class of homogeneous models as a subset. To see this point, suppose $\mathcal{M}$ is a homogeneous model and note that even if its response to constant input sequences is not constant, this response may be characterized by an infinite collection of scalar mappings $\{h^k : R^1 \to R^1\}$, each defined by

$$h^k(u_s) = y(k). \tag{3.73}$$

Since $\mathcal{M}$ is homogeneous, its response to constant input sequences is homogeneous, implying the mappings $h^k(\cdot)$ are homogeneous scalar functions. By the theorem given in Sec. 3.2.1, it follows that these mappings are necessarily linear. Therefore, it follows that

$$\begin{aligned} u_s = \alpha u_s^1 + \beta u_s^2 \Rightarrow y(k) &= h^k(\alpha u_s^1 + \beta u_s^2) \\ &= \alpha h^k(u_s^1) + \beta h^k(u_s^2) \\ &= \alpha y^1(k) + \beta y^2(k), \end{aligned} \tag{3.74}$$

where $y^1(k)$ is the response of $\mathcal{M}$ to the constant input sequence u_s^1 and $y^2(k)$ is the response to u_s^2. Consequently, any homogeneous model $\mathcal{M}$ is also static-linear.

Conversely, it is important to note that though the class of static-linear models includes the class of homogeneous models, this inclusion is proper. For example, consider the following model:

$$y(k) = \alpha \left[\frac{1 + u^2(k-2)}{1 + u^2(k-1)}\right] y(k-1) + \beta u(k-1). \tag{3.75}$$

For $u(k) = u_s$, this model exhibits the same steady-state behavior as the linear model

$$y(k) = \alpha y(k-1) + \beta u(k-1). \tag{3.76}$$

For nonconstant input sequences, however, the response of this model is neither linear nor homogeneous, as seen in Fig. 3.18. There, responses of this model are shown for a sinusoidal input sequence at four different amplitudes. Specifically, the upper left plot shows the response of this model to the standard sinusoidal input sequence introduced in Chapter 2, and the upper right plot shows the response to twice this input sequence. Similarly, the lower plots show the responses obtained for amplitudes 3 and 4. The nonlinearity of this model is evident from the nonsinusoidal shape of these responses, while the nonhomogeneity of this model is immediately clear from the pronounced change in shape that occurs as the input amplitude increases. In addition, note the dramatic increase in the amplitude of the responses, ranging from approximately 7 to 28 to 80 to 180; in contrast, note that if this model were homogeneous, the amplitude of the last response would be approximately 28.

An important point emphasized by the existence of static-linear models with nonlinear dynamic behavior is that the steady-state behavior of a discrete-time

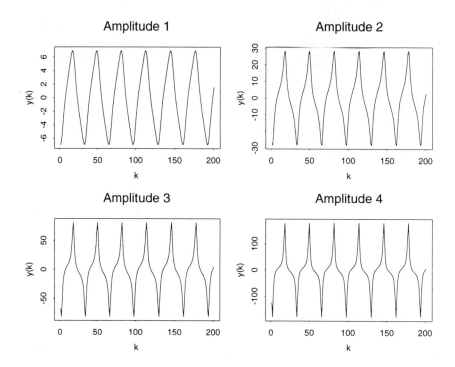

Figure 3.18: Four static-linear model responses

model can be completely unrelated to its dynamic behavior. In a sense, this observation may be regarded as a generalization of the unusual behavior of linear models with inverse responses: there, the steady-state gain and the high-frequency gain have opposite signs, resulting in a strong disparity between transient and steady-state behavior. An even stronger disparity exists in dynamic models whose steady-state gain is zero, like some of the examples discussed in Chapter 2. In fact, nonlinear generalizations of these models—that is, nonlinear dynamic models whose steady-state response is identically zero—also exist and are, by default, members of the class of static-linear dynamic models.

3.4.2 A static-linear Uryson model

Since the steady-state behavior of a Hammerstein model is defined by the static nonlinearity on which it is based, Hammerstein models cannot exhibit static-linearity except in the degenerate case of linear models. Uryson models consist of r Hammerstein models connected in parallel, and this augmentation is enough to allow static-linearity if $r > 1$. Specifically, consider the two-channel model obtained by combining the static nonlinearities

$$f_1(x) = \alpha_1 x + \alpha_3 x^3 \qquad f_2(x) = \alpha_1 x - \alpha_3 x^3, \tag{3.77}$$

with any two *distinct* linear dynamic models $H_1(z)$ and $H_2(z)$. If these linear models both have steady-state gains of 1, it follows that the steady-state behavior of the Uryson model is given by the equation

$$y_s = f_1(u_s) + f_2(u_s) = 2\alpha_1 u_s. \tag{3.78}$$

Because this steady-state response is linear, it follows that the Uryson model is static-linear, even though the dynamic behavior is nonlinear as long as $H_1(z) \neq H_2(z)$. This point is illustrated in Fig. 3.19, which shows responses of this model to the sinusoidal input $u(k) = A \cos 2\pi f t$ for the fourth-order linear moving average models

$$H_1(z) = \frac{z^{-1} - z^{-2} + z^{-3} + z^{-4}}{2}, \quad H_2(z) = \frac{z^{-1} + z^{-2} + z^{-3} - z^{-4}}{2}. \tag{3.79}$$

Specifically, the upper left plot shows the response of this Uryson model to the standard sinusoid input defined in Chapter 2, with frequency $f = 1/32$ and amplitude $A = 1$, and the other three plots show the effects of increasing f or A, or both.

These effects are particularly interesting, for three reasons. First, as several previous examples have demonstrated, sinusoidal input sequences are often effective in providing a simple visual indication of a model's nonlinearity. Second, because sinusoids approach a constant limit as $f \to 0$, it follows that the nonlinear behavior of static-linear models should exhibit a pronounced frequency dependence. Third, the amplitude-dependence of a static-linear model's behavior may be taken as a measure of its distance from homogeneity. All of these points are illustrated clearly in Fig. 3.19. In particular, note that the amplitude A and the frequency f are both low enough that the model's response to this input is approximately linear, as seen in the approximately sinusoidal shape of the response shown in the upper left plot. Quadrupling the input amplitude results in the highly nonsinusoidal response shown in the upper right plot, giving clear evidence of both nonlinearity and inhomogeneity. Similarly, doubling the frequency also results in a more highly nonsinusoidal response as shown in the lower right plot, illustrating the greater departure from linearity predicted at higher frequencies. Finally, increasing both the frequency and the amplitude leads to the even more highly nonlinear response shown in the lower right plot.

3.4.3 Bilinear models

It is also possible for the bilinear models introduced in Sec. 3.1 to exhibit static-linearity even when their dynamic behavior is highly nonlinear. In fact, it is easy to obtain an explicit characterization of the class of static-linear bilinear models. If $u(k) = u_s$ for all k, it follows that the response of the general bilinear model defined in Eq. (3.6) satisfies

$$y_s = \sum_{i=1}^{p} a_i y_s + \sum_{i=0}^{q} b_i u_s + \sum_{i=1}^{P} \sum_{j=1}^{Q} c_{i,j} y_s u_s. \tag{3.80}$$

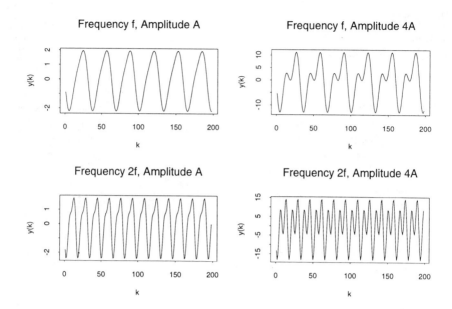

Figure 3.19: Sinusoidal responses of the Uryson model

This model will be static-linear if and only if the following condition is satisfied:

$$\sum_{i=1}^{P}\sum_{j=1}^{Q} c_{i,j} = 0. \tag{3.81}$$

Clearly, this condition is satisfied by linear models, for which $c_{i,j} = 0$ for all i and j, but it is also satsified for certain nonlinear models. For example, the general bilinear model

$$\begin{aligned} y(k) = &\, ay(k-1) + bu(k-1) \\ &+ cy(k-2)u(k-1) - cy(k-1)u(k-2), \end{aligned} \tag{3.82}$$

is nonlinear but static-linear. Analogous results are given in Chapter 6 for the TARMAX class of positive-homogeneous linear multimodels introduced in Sec. 3.3.3, and in Chapter 5 for the class of Volterra models.

An especially interesting static-linear model results from the choice $a = 0.8$, $b = 1$, and $c = -0.4$ in Eq. (3.82). Responses of this model to step inputs of amplitude ± 1 and ± 2 are shown in Fig. 3.20 and these responses clearly show the qualitative asymmetry exhibited by the exothermic CSTR example introduced in Chapter 1. That is, the response to a step of amplitude $+2$ is highly oscillatory, whereas the response to the mirror-image step of amplitude -2 is strictly monotonic. Despite this strong *transient* nonlinearity, note that the steady-state behavior is completely characterized by the linear relation $y_s = 5u_s$. This behavior is seen clearly in the uniform spacing of the asymptotic values to which all four of these step responses settle.

Four Views of Nonlinearity   135

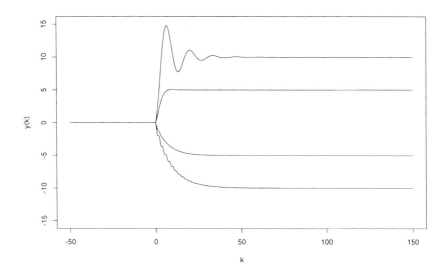

Figure 3.20: Bilinear model step responses

The nonlinearity of this example is also seen clearly in the amplitude dependence of its response to Gaussian white noise inputs. This behavior is illustrated in Fig. 3.21, which shows four plots. The upper left plot is the response of the linear reference model obtained by setting $c = 0$ in Eq. (3.82) to zero-mean, Gaussian white noise of amplitude (i.e., standard deviation) 0.5. For comparison, the upper right plot shows the response of the static-linear model to this same input sequence. Though they are not identical, these two responses are quite similar; in particular, the static-linear model's response gives no obvious indication of the model's nonlinearity. The lower left plot shows the response of the bilinear model to a white noise input sequence of amplitude 1.0 and the lower right plot shows the corresponding response to an input of amplitude 1.5. The nonlinearity of the model is clear from the qualitative change seen in these responses with increasing input amplitude. In particular, note the burst phenomena seen toward the end of the data record in the lower right plot.

3.4.4 Mallows' nonlinear data smoothers

An important topic in time series analysis and digital signal processing is the theory of smoothing filters (Hamming, 1983; Mallows, 1980; Oppenheim and Schafer, 1975). The most detailed results are available for the design and characterization of *linear* smoothing filters, which are generally based on weighted averages of the values in a moving data window of finite duration. Increasingly, however, interest in nonlinear smoothing filters is growing, in part because of their improved resistance to *outliers* or *anomalous data points*, relative to linear

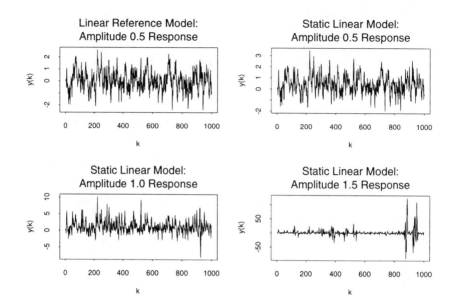

Figure 3.21: A comparison of four white noise responses

filters (refer to Chapter 8 for a further discussion of this point). Mallows (1980) considers the behavior of a class of nonlinear smoothing filters that are members of the NMAX model family discussed in Chapter 4 and that satisfy several restrictions, motivated by data analysis considerations. Two of these restrictions are the following. First, the smoother S is *location invariant*, implying that $S[x(k) + c] = S[x(k)] + c$, for any sequence $\{x(k)\}$ and any real constant c. Second, Mallows assumes S is *centered*, implying $S[\mathbf{0}] = \mathbf{0}$, where $\mathbf{0}$ denotes the zero sequence $x(k) \equiv 0$. Taken together, these two conditions imply that the smoother S preserves constant sequences or, in the terminology introduced in this chapter, S is a static-linear model with steady-state gain 1.

Mallows describes a number of smoothing filters that satisfy these criteria. One of these filters is the noncausal linear smoother defined by the transfer function

$$H(z) = \frac{-3z^{-2} + 12z^{-1} + 17 + 12z - 3z^2}{35}. \tag{3.83}$$

Mallows notes that cubic polynomials are invariant under this smoothing filter. An extremely important nonlinear smoothing filter is the *median filter*, defined by

$$S[\{x(k)\}] = \text{median}\{x(k - M), \ldots, x(k + M)\}. \tag{3.84}$$

The importance of this filter lies in its ability to reject isolated outliers while still preserving both constant portions of a sequence and step transitions between these constant portions. In addition, the median filter also preserves any monotonically increasing or decreasing sequence. Mallows describes other nonlinear smoothing filters based on extensions of the median filter, and related ideas remain important topics of research in the signal processing literature (see, for example, Flaig et al. (1998) and the references cited there). The key point here is that this specific form of static-linearity—that is, invariance of constant sequences under filtering—is both a natural restriction in data smoothing applications and a practically important one. Despite this very specific restriction, the class of nonlinear smoothing filters that satisfies it is both large and growing.

3.5 Summary: the nature of nonlinearity

It has been noted many times before, but it is worth repeating that the term *nonlinear* is in some ways an inherently poor one, defining an entire class of systems in terms of a crucial property that they lack. Recall from the discussion at the end of Chapter 2 that the class of linear models may be defined either behaviorally or structurally. Behaviorally, linear models may be viewed as a class of maps from ℓ_∞ into itself (i.e., *endomorphisms* of ℓ_∞) that satisfy the additivity and homogeneity conditions discussed in Sec. 2.6. Structurally, linear models may be viewed as members of the parametric family of $ARMAX(p,q)$ models discussed in Sec. 2.2. One of the key points of Sec. 2.6 is that these two descriptions are essentially equivalent. In contrast, one of the key points of this chapter is that the class of nonlinear models cannot be *completely* defined either structurally or behaviorally. For this reason, probably the most common basis for discussions of nonlinear dynamic models is to restrict consideration to a particular subclass that *can* be defined either structurally or behaviorally. For example, the class of bilinear models discussed in Sec. 3.1 represents a structurally defined mild relaxation of linearity, whereas the class of homogeneous models discussed in Sec. 3.2 represents a behaviorally defined mild relaxation of linearity. Without question, structural approaches are more popular because they provide *explicit* definitions of model classes and thus a clear starting point for investigations of qualitative behavior, empirical model identification, model-based control system design, and various other topics of interest. However, examining behaviorally defined model classes can provide a great deal of intuition about the overall nature of nonlinear behavior.

As an illustration of this point, consider the following question: Can bilinear models exhibit homogeneous behavior? To address this question, consider the bilinear model response

$$y_\gamma(k) = \sum_{i=1}^{p} a_i y_\gamma(k-i) + \sum_{i=0}^{q} b_i \gamma u(k-i) + \sum_{i=1}^{P} \sum_{j=1}^{Q} c_{i,j} \gamma y(k-i) u(k-j). \tag{3.85}$$

to the scaled input sequence $\{\gamma u(k)\}$ and define the sequence

$$z_\gamma(k) = y_\gamma(k) - \gamma y_1(k). \tag{3.86}$$

Note that if the bilinear model defined by the coefficients $\{a_i\}$, $\{b_i\}$, and $\{c_{i,j}\}$ does exhibit homogeneous behavior, then $z_\gamma(k) = 0$ for all γ. To show that this result does not hold, proceed by substituting Eq. (3.85) into Eq. (3.86) to obtain an explicit model for $z_\gamma(k)$:

$$z_\gamma(k) = \sum_{i=1}^{p} a_i z_\gamma(k-i) + \sum_{i=1}^{P}\sum_{j=1}^{Q} c_{i,j}\gamma z_\gamma(k-i)u(k-j)$$
$$+ \gamma(\gamma-1)\sum_{i=1}^{P}\sum_{j=1}^{Q} c_{i,j} y_1(k-i)u(k-j). \tag{3.87}$$

Note that $z_\gamma(k) = 0$ for all k only if the last term in this equation is identically zero for all γ and all possible input sequences $\{u(k)\}$. *Since this condition can only hold if $c_{i,j} = 0$ for all i and j, it follows that the only models that are both bilinear and homogeneous are linear.* This result stands in marked contrast to that presented in Sec. 3.4 for static-linear models, where it is sufficient that the *sum* of the bilinear model coefficients $c_{i,j}$ be zero for the model to be static-linear. This difference illustrates how much stronger the homogeneity requirement is than the static-linearity requirement.

This last example also serves as a nice illustration of the profound differences that can exist between different classes of nonlinear models. That is, although it was by no means obvious beforehand, this result establishes that the structurally defined family of bilinear models and the behaviorally defined family of homogeneous models represent small steps away from linearity in directions that may be loosely regarded as orthogonal. In addition, this example raises the following interesting general question that will be examined in a number of specific instances throughout the rest of this book:

> Can qualitative behavior $\mathcal{Q}$ be achieved with a nonlinear model of structure $\mathcal{S}$?

In practical terms, this question is an extremely important one, since one of the primary objectives in developing empirical models is to match an observed or anticipated qualitative behavior (e.g., presence or absence of output multiplicity, chaotic regimes, oscillatory responses) to a model whose structure permits the solution of some problem of interest (e.g., implementation of a model-based controller).

Finally, one other important practical point about nonlinear models is that the number of structurally defined model classes that have been investigated is large and growing. Further, these model classes are not all treated in the same parts of the literature, so notation, working assumptions, and applications can vary enormously. It is therefore often unclear how different model classes relate to one another. The example just presented shows that two specific nonlinear model classes—the structurally defined class of bilinear models and the behaviorally defined class of homogeneous models—are mutually exclusive. At the other extreme, models of class $\mathcal{C}$ may be a proper subset of some other

Four Views of Nonlinearity   139

class $\mathcal{D}$. As a specific illustration, note that the class of homogeneous models introduced in Sec. 3.2 is a proper subset of the class of positive-homogeneous models introduced in Sec. 3.3. Although this relationship is fairly clear from the definition, the fact that the class of homogeneous models is also a subset of the class of static-linear models defined in Sec. 3.4 is somewhat less obvious. In intermediate cases, the relationships between two different model classes may be harder to characterize. For example, the existence of nonlinear but static-linear TARMAX models is demonstrated in Chapter 6, establishing that the intersection between the positive-homogeneous and static-linear model classes contains nonlinear, nonhomogeneous members. Taken together, the asymmetric character of models that are positive-homogeneous but not homogeneous, and the symmetric character of linear models would seem to suggest that the only positive-homogeneous, static-linear models would be elements of the homogeneous model class; the examples presented in Chapter 6 demonstrate that this is not the case. More generally, it is often possible to gain useful insights by asking the following question:

> What class of models $\mathcal{M}$ is defined by the intersection of the classes $\mathcal{C}$ and $\mathcal{D}$?

Specializations of this question will be examined throughout the rest of this book.

Because it does provide the most popular and arguably the most useful basis for defining specific nonlinear model classes, most of the rest of this book will focus on *structurally defined* nonlinear model classes. Specifically, Chapter 4 presents a detailed treatment of the large class of NARMAX models, along with several of its important subclasses, all defined in terms of further structural restrictions (e.g., nonlinear moving average models, polynomial NARMAX models, structurally additive models). Chapter 5 then treats the class of Volterra models, including structurally restricted subclasses like the analytic Hammerstein and Wiener models. Similarly, Chapter 6 introduces the family of linear multimodels, presenting and comparing a number of different (and nonequivalent) definitions. The inherent idea underlying the class of linear multimodels is that of a *local linear model*, combined with some way of switching between local model regimes. One interpretation of this idea leads ultimately to a class of positive-homogeneous models (the class of TARMAX models) discussed in detail in Chapter 6. Interestingly, another interpretation of this idea leads to the class of bilinear models, as noted in Chapter 1 (Sec. 1.3.3). Specifically, recall that linear interpolation between two local linear models resulted in a simple bilinear approximation to distillation column dynamics (Stromberg et al., 1995). Here, the key point is that the basic idea of combining local linear models into a global nonlinear one can be implemented in two very reasonable but very different ways, thus leading to models that are members of mutually exclusive families: the family of positive-homogeneous TARMAX models described in Sec. 3.3.3 and the family of bilinear models described in Sec. 3.1. Clearly, the connections between different approaches to nonlinear modeling can be extremely subtle and sometimes very counterintuitive.

Chapter 4

NARMAX Models

As noted In Chapter 1, the general class of NARMAX models is *extremely* broad and includes almost all of the other discrete-time model classes discussed in this book, linear and nonlinear. Because of its enormity, little can be said about the qualitative behavior of the NARMAX family in general, but it is surprising how much can be said about the behavior of some important subclasses. In particular, there are sharp qualitative distinctions between *nonlinear moving average models* (NMAX models) and *nonlinear autoregressive models* (NARX models). Since these representations are equivalent for linear models (c.f. Chapter 2), this observation highlights one important difference between linear and nonlinear discrete-time dynamic models. Consequently, this chapter focuses primarily on the structure and qualitative behavior of various subclasses of the NARMAX family, particularly the NMAX and NARX classes. More specifically, Sec. 4.1 presents a brief discussion of the general NARMAX class, defining five important subclasses that are discussed in subsequent sections: the NMAX class (Sec. 4.2), the NARX class (Sec. 4.3), the class of (structurally) additive NARMAX models (Sec. 4.4), the class of polynomial NARMAX models (Sec. 4.5), and the class of rational NARMAX models (Sec. 4.6). More complex NARMAX models are then discussed briefly in Sec. 4.7 and Sec. 4.8 concludes the chapter with a brief summary of the NARMAX class.

4.1 Classes of NARMAX models

The general class of NARMAX models was defined and discussed in a series of papers by Billings and various co-authors (Billings and Voon, 1983, 1986a; Leontartis and Billings, 1985; Zhu and Billings, 1993) and is defined by the equation

$$y(k) = F(y(k-1), \ldots, y(k-p), u(k), \ldots, u(k-q),$$
$$e(k-1), \ldots, e(k-r)) + e(k). \qquad (4.1)$$

NARMAX Models

Because this book is primarily concerned with the qualitative input/output behavior of these models, it will be assumed that $e(k) = 0$ identically as in the class of linear ARMAX models discussed in Chapter 2. This assumption reduces Eq. (4.1) to

$$y(k) = F(y(k-1), \ldots, y(k-p), u(k), \ldots, u(k-q)). \tag{4.2}$$

Again, as in the case of linear ARMAX models, note that for models with $p > 0$ [i.e., those involving autoregressive terms $y(k-j)$], it is necessary to specify initial conditions on $y(k)$ to fully define the model. Here, the standard assumption $y(k) = 0$ for $k < 0$ will always be made unless explicitly stated to the contrary.

It is important to note that the possibility of "direct feedthrough" is permitted in this model by including the undelayed input term $u(k)$. Although most dynamic models considered in the process control literature only include the *past* input terms $u(k-d)$ through $u(k-q)$ for some $d \geq 1$, inclusion of the current input $u(k)$ permits the static nonlinearity $y(k) = \phi(u(k))$ to be viewed as a NARMAX model. This generalization will simplify subsequent discussions, particularly in Chapter 7 where relationships between different model classes are considered. Finally, it is also important to note that the time index k is *not* included as an argument in the function $F(\cdots)$ in Eq. (4.2). This restriction excludes the linear time-varying models considered at the end of Chapter 2 and their nonlinear generalizations.

The model classes of principal interest here represent structural restrictions of the general model defined in Eq. (4.2), obtained either by constraining the form of the function $F(\cdots)$ or the choice of its arguments. More specifically, this chapter focuses on six subclasses of the NARMAX family, defined and discussed briefly here. These model classes are each represented as a set in the Venn diagram shown in Fig. 4.1, which illustrates their interrelationships. Specifically, the model classes represented in Fig. 4.1 are the following:

1. The linear ARMAX models discussed in Chapter 2
2. The NMAX models discussed in Sec. 4.2
3. The NARX* models discussed in Sec. 4.3
4. The additive NARMAX models discussed in Sec. 4.4
5. The polynomial NARMAX models discussed in Sec. 4.5
6. The rational NARMAX models discussed in Sec. 4.6.

At the center of this diagram is a triangular region representing the linear ARMAX family discussed in Chapter 2, for which the autoregressive and moving average representations are equivalent. The dashed lines emanating from the vertices of this triangle divide the nonlinear portion of the NARMAX class into three disjoint subsets: the nonlinear moving average (NMAX) models, the nonlinear autoregressive (NARX*) models, and the larger class of "other" or "general" NARMAX models that do not belong to either the NARX* or the NMAX classes. All three of these model classes are obtained by specifying the arguments of the function $F(\cdots)$: NMAX models depend on present and past

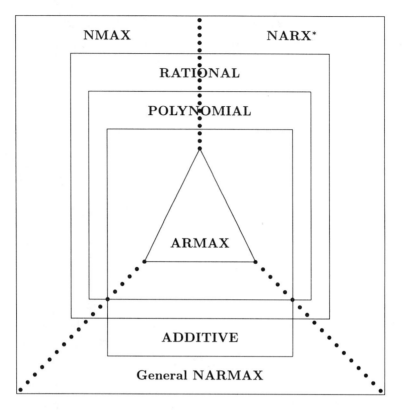

Figure 4.1: Some important subsets of the NARMAX family

input values $u(k-j)$, NARX* models depend on the current input value $u(k)$ and past output values $y(k-j)$, and general NARMAX models depend on both inputs $u(k-j)$ and outputs $y(k-j)$ for various time lags j. The family of NARX models is a proper subset of the NARX* class, defined by further restricting the function $F(\cdots)$ to be of the form $G(\cdots) + b_0 u(k)$ where $G(\cdots)$ is a function of only past outputs $y(k-j)$. In terms of qualitative behavior, there is no loss of generality in replacing the current input $u(k)$ in either the NARX class or the NARX* class with the delayed input $u(k-d)$ for $d \geq 1$. It is important to note, however, that the range of possible qualitative behavior can change in important ways if $F(\cdots)$ is permitted to depend on more than one input value $u(k-j)$. The class of general NARMAX models includes those that do not satisfy either of the restrictions defining the NMAX or NARX* model classes.

Frequently, when the term NARMAX model appears in the process control literature, it is tacitly assumed to mean the class of polynomial NARMAX models, defined by restricting the function $F(\cdots)$ to be a multivariate polynomial. This model class is represented by a rectangular box containing the triangular ARMAX region and intersecting all three of the broad classes just described (i.e., the NMAX, NARX*, and general NARMAX regions indicated by dashed lines). The polynomial NARMAX model class is a proper subset of the larger class of rational NARMAX models, represented by the rectangular box containing both the polynomial NARMAX and linear ARMAX portions of Fig. 4.1. Rational NARMAX are defined by restricting the function $F(\cdots)$ to be the ratio of two polynomials; like the polynomial NARMAX models, note that this model family intersects all three of the NMAX, NARX*, and general NARMAX regions shown in the diagram. Finally, the class of additive NARMAX models (NAARX models) is also represented as a rectangle in Fig. 4.1, again including the linear ARMAX models as a special case. This model family imposes the restriction of *structural additivity* discussed at the end of Chapter 2 on the function $F(\cdots)$. As indicated in Fig. 4.1, this model family intersects all of the other model families considered here: polynomial NARMAX models, rational NARMAX models, NMAX models, NARX* models, and general NARMAX models. The primary objective of this chapter is to illustrate the behavioral differences that exist between these structurally-defined model classes.

4.2 Nonlinear moving average models

As discussed in Chapter 2, linear ARMAX models may be represented in either of two forms: as autoregressive models, dependent on past *outputs* $y(k - i)$, or as moving average models, dependent on past *inputs* $u(k - j)$. By analogy, it is useful to define the complementary classes of *nonlinear autoregressive (NARX*)* models and *nonlinear moving average (NMAX)* models. Although these two representations are both equivalent and complete in the linear case (i.e., any linear ARMAX model may be represented in either of these two forms), they are neither equivalent nor complete in the nonlinear case. In particular, there are *dramatic* differences in the qualitative behavior of NARX* and NMAX models, as subsequent examples will illustrate. Further, it is important to note that this partitioning is not complete: the NARMAX class includes members that have neither NARX* nor NMAX representations. Still, these two subclasses are large enough and important enough in practice to be worth investigating in detail.

More specifically, the NMAX model class considered here is defined by setting $p = 0$ in Eq. (4.2), thereby constraining the NARMAX model to depend on *current and past inputs only*. This restriction results in a model of the form:

$$y(k) = g(u(k), \ldots, u(k - q)), \qquad (4.3)$$

where $g(\cdots)$ may be chosen as any function mapping R^{q+1} into R^1. Though it does not alter the qualitative behavior of the model in any important way, it will be convenient to assume generally that $g(0, 0, \ldots, 0) = 0$ so that if the input sequence is identically zero, the output sequence is also identically zero. Because *linear* moving average models exhibit impulse responses that are identically zero after a finite number of samples, they are also called *finite impulse response models*. It is easily seen from Eq. (4.3) that this same behavior is exhibited by NMAX models: if $u(k)$ is an impulse centered at $k = 0$ and if $g(0, 0, \ldots, 0) = 0$, then $y(k) = 0$ for all $k > q$.

The class of NMAX models is large and it includes many other models and model classes discussed in this book. For example, the first nonlinear homogeneous model example discussed in Chapter 3 [Sec. 3.2, Eq. (3.34)] is a nonlinear moving average model of order $q = 3$. Similarly, the class of Pythagorean models introduced in Sec. 3.3 is a subset of the NMAX model class, as is Mallows' class of nonlinear smoothing filters introduced in Sec. 3.4. Finally, the class of finite Volterra models introduced in Chapter 5 belongs to the NMAX family, including as special cases the finite Hammerstein models, the finite Wiener models, the finite Uryson models, and the finite projection-pursuit models defined there. These examples demonstrate that the NMAX model class is capable of superharmonic generation, asymmetric responses to symmetric input changes, homogeneity, positive-homogeneity, and static linearity. Conversely, models of this class *cannot* exhibit certain other types of nonlinear qualitative behavior, most notably subharmonic generation, input-dependent stability, and chaotic responses to simple input sequences (e.g., impulses, steps, and sinusoids). Consequently, the NMAX model class may be regarded as an especially well-behaved subset of the general NARMAX class, particularly in comparison with the NARX model class discussed in Sec. 4.3. The following discussion first considers the behavior of two simple NMAX models (Sec. 4.2.1) and then establishes some general behavioral results for the NMAX class (Sec. 4.2.2).

4.2.1 Two NMAX model examples

To illustrate the general range of qualitative behavior seen in the NMAX model family, it is instructive to consider the following two simple examples. The first example is the second-order Volterra model

$$y(k) = (1/4)[u(k) + u(k-1) + u(k-2) + u(k-3)] - u(k)u(k-4). \quad (4.4)$$

The response of this model to a unit impulse is easily seen to be

$$y(k) = \begin{cases} 1/4 & k = 0, 1, 2, 3 \\ 0 & \text{otherwise.} \end{cases} \quad (4.5)$$

This impulse response is identical to that of the linear FIR model obtained by simply omitting the nonlinear term $u(k)u(k-4)$ and it illustrates the point made previously about the finite duration of NMAX model impulse responses. Also, since this response is identical to that of the linear part of the model, it offers no evidence of the model's nonlinearity. As the next example further illustrates, this behavior is typical of NMAX models: they are generally not well characterized by their impulse responses, in marked contrast to the linear case, which is completely characterized by the impulse response.

Step responses are generally (though not always) more informative, particularly if a family of step responses is considered. This point is illustrated in Fig. 4.2, which shows the responses of this model to steps of amplitude ± 0.5 and ± 1. As with the impulse responses, these step responses all reach their steady-state values in finite time (here, 5 time steps), again consistent with the

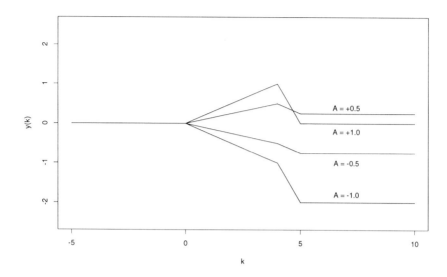

Figure 4.2: A family of four step responses

discussion in Sec. 4.2. In constrast to the impulse response, however, these step responses give clear evidence of both the nonlinearity of the model and the asymmetric nature of this nonlinearity. In particular, note the markedly nonmonotonic character of these step responses: the model's response to a step of amplitude +0.5 settles out to a value of +0.25, while the response to a step of amplitude +1.0 settles out to the value 0.

The second example considered here is the median filter introduced in Chapter 3 (Sec. 3.4.4). In practice, this filter is most commonly used as an off-line data smoothing filter, employing the noncausal implementation described in Chapter 3. Because this implementation is not strictly a member of the NMAX class considered here, this discussion will consider the following causal implementation:

$$y(k) = \text{median}\{u(k), \ldots, u(k-q)\}, \tag{4.6}$$

where q is an even integer. Note that the median may be defined in terms of the following rank-ordering of the original data sequence $\{u(k), \ldots, u(k-q)\}$:

$$u_{(0)} \leq u_{(1)} \leq \cdots \leq u_{(q-1)} \leq u_{(q)}. \tag{4.7}$$

Specifically, $u_{(0)}$ is the smallest element of the original data sequence, $u_{(q)}$ is the largest element, and the median is the middle value, $u_{(q/2)}$.

It was noted that the impulse response of the Volterra model considered in the previous example is not very informative since it gives no evidence of the model's nonlinearity. Here, the impulse response is even less informative:

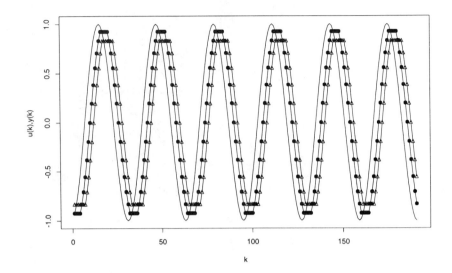

Figure 4.3: Sinusoidal responses of the median filter

for $q \geq 2$, the median filter's response to any impulse input is identically zero. In fact, this feature illustrates an extremely useful characteristic of the median filter in data analysis: it is completely resistant to isolated outliers in the data. Another useful feature of the median filter is that it preserves monotonic sequences, including steps. In the nonlinear digital filtering literature, such invariant sequences are called *root sequences* and are the subject of significant interest (Astola et al., 1987). In the causal example considered here, monotonic sequences are delayed by $q/2$ time steps, but the key point is that their shape is unchanged. In certain data filtering operations, this behavior is extremely desirable, but in the context of NMAX model characterization, this invariance means that the median filter cannot be distinguished from the trivial linear model $y(k) = u(k-q/2)$ on the basis of monotonic input sequences. This observation has important practical implications for nonlinear model identification that are discussed further in Chapters 7 and 8.

As an interesting corollary to the median filter's preservation of monotonic sequences, the response of this NMAX model to a sinusoidal input is *almost* sinusoidal, at least for low-frequency sinusoids. Specifically, note that sinusoidal input sequences are monotonic everywhere except at their minima and maxima. Consequently, the response of the median filter to a sinusoidal input sequence $\{u(k)\}$ will be $y(k) = u(k - q/2)$ except near these minima and maxima. This point is illustrated in Fig. 4.3, which shows overlaid plots of the standard sinusoidal input sequence defined in Chapter 2 (light line), the response of a median filter with $q = 6$ (solid circles) and the response of a median filter with $q = 10$ (open triangles). Note that the responses are characterized by a flat-

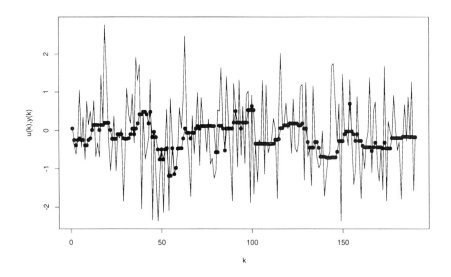

Figure 4.4: White noise response of the median filter

tening of the sinusoidal peaks by an amount that depends on the frequency of the sinusoid and the width parameter (q) of the median filter. Conversely, certain high-frequency sinusoids are eliminated completely, as in the example $u(k) = \sin(k\pi/2) = 0, 1, 0, -1, 0, \ldots$, for any even $q > 0$

The most dramatic evidence of the median filter's nonlinearity is seen in its response to the standard Gaussian white noise input sequence defined in Chapter 2. The median filter response for $q = 10$ is shown in Fig. 4.4 as the solid circles overlaid on the input sequence, which is shown as the light line. The profound difference in these sequences is due to the irregularity (alternatively, the extreme nonmonotonicity) of the Gaussian white noise sequence. The effect of the median filter is to significantly reduce both the magnitude and frequency of these variations.

It is also interesting to note that the median filter is homogeneous. Positive-homogeneity is easy to see, since scaling the input sequence by a positive value λ simply replaces the rank-ordered sequence $u_{(j)}$ defined in Eq. (4.7) by the scaled sequence $\lambda u_{(j)}$. Conversely, scaling the input sequence by a negative value *reverses* this rank-ordered sequence, but since the median is the central element in this rank-ordered list, it remains invariant under this order reversal. Hence, it follows that the median filter is invariant under arbitrary scalings of the input sequence, implying homogeneity. The static linearity of the median filter noted in Chapter 3 is a consequence of this homogeneity. Another consequence of homogeneity is that the median filter exhibits symmetric responses to symmetric input changes, in marked contrast to the Volterra model considered in the previous example.

4.2.2 Qualitative behavior of NMAX models

It is possible to give a fairly complete characterization of the general qualitative behavior of NMAX models on the basis of a few extremely weak assumptions. First, consider the steady-state behavior of the general NMAX model defined in Eq. (4.3). Specifically, note that if $u(k) = u_s$ for all k, it follows immediately that $y(k) = y_s$ for all k where

$$y_s = g(u_s, \ldots, u_s) \equiv \gamma(u_s). \qquad (4.8)$$

Further, since y_s is uniquely defined by Eq. (4.8) for any given value of u_s, NMAX models cannot exhibit output multiplicity. Similarly, if the function $\gamma(\cdot)$ is invertible, it follows that $u_s = \gamma^{-1}(y_s)$ and the NMAX model defined by Eq. (4.3) does not exhibit input multiplicity, either. As a very special case, note that if the function $\gamma(\cdot)$ is linear, as in the case of the median filter example discussed in the previous section (there, $\gamma(u_s) = u_s$), the NMAX model belongs to the static-linear model class introduced in Chapter 3. Conversely, if the function $\gamma(\cdot)$ is not invertible, it is possible for Eq. (4.8) to exhibit more than one solution u_s for a single value of y_s, implying that the model defined by Eq. (4.3) exhibits input multiplicity. As a specific example, this behavior is observed in the Volterra model considered in Sec. 4.2.1. There, $\gamma(u_s) = u_s - u_s^2$ and Eq. (4.8) exhibits two distinct solutions u_s for any $y_s < 0.5$. As noted in Chapter 3, it is a standard result (Klambauer, 1975, p. 181) that if $\gamma(x)$ is continuous on a compact (that is, closed and bounded) set S, it is invertible on S if and only if it is strictly monotonic on S. That is, an invertible, continuous function $\gamma(x)$ on S must be either strictly increasing [$x < y$ implies $\gamma(x) < \gamma(y)$] or strictly decreasing [$x < y$ implies $\gamma(x) > \gamma(y)$] on S. Further, if the function $\gamma(x)$ is differentiable on S (implying continuity), it is strictly increasing if and only if the derivative $\gamma'(x)$ is strictly positive for all $x \in S$ and it is strictly decreasing if and only if $\gamma'(x)$ is strictly negative for all $x \in S$.

Note that the addition of *linear autoregressive terms* to an NMAX model—as in Robinson's AR-Volterra model discussed in Sec. 4.5—does not change the steady-state character of an NMAX model. Specifically, define the class of LARNMAX models (linear autoregressive - nonlinear moving average models with exogenous inputs) by the equation

$$y(k) = \sum_{i=1}^{p} a_i y(k-i) + g(u(k), \ldots, u(k-q)) \qquad (4.9)$$

and observe that the steady-state response y_s for this model satisfies

$$y_s = \sum_{i=1}^{p} a_i y_s + g(u_s, u_s, \ldots, u_s). \qquad (4.10)$$

Again defining the scalar function $\gamma(u_s) = g(u_s, u_s, \ldots, u_s)$, this expression may be rewritten as

$$y_s = \frac{\gamma(u_s)}{1 - \sum_{i=1}^{p} a_i}. \qquad (4.11)$$

Another general feature of nonlinear moving average models is that they are bounded-input, bounded-output (BIBO) stable. In fact, this result follows under the extremely weak assumption that the function $g(\cdots)$ defining the model is continuous. As shown in Sec. 4.3, the conditions required to establish analogous results for nonlinear autoregressive models are *much* more restrictive. Recall that the function $g: R^{q+1} \to R^1$ is continuous if, given $\epsilon > 0$, there exists $\delta > 0$ such that $||\mathbf{x} - \mathbf{y}|| < \delta$ implies $|g(\mathbf{x}) - g(\mathbf{y})| < \epsilon$ (Klambauer, 1975, p. 447). Also, note that if $g: R^{q+1} \to R^1$ is continuous, it is bounded on compact subsets of R^{q+1} (Klambauer, 1975, p. 449).

Theorem:

> Suppose $g: R^{q+1} \to R^1$ is a continuous function defining a nonlinear moving average (NMAX) model and $\{u(k)\}$ is a bounded sequence; then $\{y(k)\}$ is also a bounded sequence.

Proof:

> Since $\{u(k)\}$ is bounded, there exist constants a and b such that $a \leq u(k) \leq b$ for all k. Further, since $g: R^{q+1} \to R^1$ is continuous, it follows that $g(\mathbf{x})$ is bounded on the compact domain $[a,b]^{q+1}$. Therefore, $\{y(k)\}$ is also a bounded sequence.

> □

As a corollary, this result implies that continuous NMAX models cannot exhibit the input-dependent stability behavior discussed in Chapter 1. In addition, it is also possible to establish the following stronger result for continuous NMAX models: if the input sequence $\{u(k)\}$ approaches a constant limit, so does the output sequence $\{y(k)\}$. To prove this result, introduce the following terminology: a sequence $\{x(k)\}$ is called *asymptotically constant* if, given $\epsilon > 0$, there exists a real number c and an integer N such that $k > N$ implies $|x(k) - c| < \epsilon$.

Theorem:

> Suppose $g: R^{q+1} \to R^1$ is a continuous function defining a nonlinear moving average (NMAX) model and $\{u(k)\}$ is a bounded, asymptotically constant sequence, converging to the limit u_s. Define $y_s = g(u_s, u_s, \ldots, u_s)$; then $\{y(k)\}$ is a bounded, asymptotically constant sequence, converging to the limit y_s.

Proof:

> The boundedness of the sequence $\{y(k)\}$ follows from the previous theorem. To establish asymptotic constancy, it must be shown that,

given $\epsilon > 0$, there exists N such that $k > N$ implies $|y(k) - y_s| < \epsilon$. Since $g(\cdots)$ is continuous, there exists $\delta > 0$ such that if

$$\|[u(k), u(k-1), \ldots, u(k-q)] - [u_s, u_s, \ldots, u_s]\| < \delta,$$

then,

$$|g(u(k), u(k-1), \ldots, u(k-q)) - g(u_s, u_s, \ldots, u_s)| < \epsilon.$$

Since $\{u(k)\}$ is an asymptotically constant sequence with limit u_s, there exists N' such that $k - q > N'$ implies

$$\|[u(k), u(k-1), \ldots, u(k-q)] - [u_s, u_s, \ldots, u_s]\| < \delta.$$

Taking $N = N' + q$, it therefore follows that, for $k > N$

$$|y(k) - y_s| = |g(u(k), u(k-1), \ldots, u(k-q)) - g(u_s, u_s, \ldots, u_s)| < \epsilon.$$

□

This result strengthens the previous one, excluding the possibility of bounded but persistent fluctuations in response to asymptotically constant input sequences like the unit step. In particular, neither persistent periodic oscillations of fixed amplitude nor more complex bounded phenomena like chaos are possible in the response of continuous NMAX models to asymptotically constant inputs. Again, subsequent examples illustrate that such "exotic" behavior is possible for NARX models. Further, the following result establishes that NMAX models are also well behaved in terms of their responses to periodic input sequences and cannot generate subharmonic or nonperiodic (e.g., chaotic) responses.

Theorem:

If $\{u(k)\}$ is a period P input sequence P and $\mathcal{M}$ is any NMAX model, the output sequence $\{y(k)\}$ is also periodic with period P.

Proof:

By the assumptions of the theorem,

$$y(k) = g(u(k), u(k-1), \ldots, u(k-q))$$

for some non-negative integer q and some function $g : R^{q+1} \to R^1$. Thus, it follows that

$$y(k+P) = g(u(k+P), u(k+P-1), \ldots, u(k+P-q))$$
$$= g(u(k), u(k-1), \ldots, u(k-q)) = y(k).$$

□

NARMAX Models

Before leaving the topic of NMAX models, it is worth noting two other results. First, consider m different NMAX models, each of the same order q, defined by the equations

$$y_j(k) = g_j(u(k), \ldots, u(k-q)). \tag{4.12}$$

The *parallel combination* of these models is defined by:

$$y(k) = \sum_{j=1}^{m} y_j(k), \tag{4.13}$$

corresponding to the construction introduced in Chapter 1 to obtain the Uryson model from m Hammerstein models. That is, all m models are driven by the same input sequence $\{u(k)\}$ and their outputs are summed to obtain the overall model output $y(k)$. The key point here is that the parallel combination of m NMAX models of order q is also an NMAX model of order q, based on the overall mapping:

$$g(u(k), \ldots, u(k-q)) = \sum_{j=1}^{m} g_j(u(k), \ldots, u(k-q)). \tag{4.14}$$

This result can be useful in manipulating block diagrams describing more complex models built from NMAX components. An important observation is that the parallel connection of NMAX models does not increase the dynamic order q; this result stands in contrast to autoregressive models, where dynamic order is not preserved even in the linear case. (Note, for example, that the parallel connection of two first-order autoregressive linear models generally results in a *second-order* autoregressive linear model.) In fact, the parallel combination of two NARX models does not necessarily even exhibit a general NARMAX representation; this point is closely related to the failure of Wiener models to exhibit NARMAX representations discussed in Sec. 4.4.1.

Similarly, the second useful result is that the cascade connection of NMAX models is again an NMAX model, an observation that will be extremely important in Chapter 7. Specifically, consider the composite model relating $\{u(k)\}$ to $\{y(k)\}$ defined by the equations

$$y(k) = g_1(v(k), \ldots, v(k-r)) \tag{4.15}$$
$$v(k) = g_2(u(k), \ldots, u(k-q)). \tag{4.16}$$

Combining these equations yields the following NMAX model of order $m = q+r$:

$$y(k) = g_1(g_2(u(k), \ldots, u(k-q)), \ldots, g_2(u(k-r), \ldots, u(k-r-q)))$$
$$\equiv g_3(u(k), \ldots, u(k-q-r)). \tag{4.17}$$

In contrast to the parallel connection result just presented, it is interesting to note that the cascade connection of NARX models is quite similar to this NMAX result. This point is discussed further in Sec. 4.3.

4.3 Nonlinear autoregressive models

The nonlinear moving average (NMAX) models considered in the previous section were defined by setting $p = 0$ in Eq. (4.2). The dual of this model class is obtained by setting $q = 0$, leading to a class of models in which $y(k)$ depends in an arbitrarily nonlinear fashion on the past outputs $y(k-1)$ through $y(k-p)$ but only on the *current* input $u(k)$. A particularly important special case of this model class is that in which dependence on $u(k)$ is linear, that is

$$y(k) = F(y(k-1), \ldots, y(k-p)) + b_0 u(k). \tag{4.18}$$

This case is important for three reasons. First, many of the nonlinear autoregressive models that have been studied in the literature belong to this class, as do many of the examples discussed in this book. Second, control problems are fundamentally concerned with the design of input sequences $\{u(k)\}$ that will drive the output $y(k)$ to some target value y_T. This problem leads naturally to a study of the inverse model that relates $\{y(k)\}$ to $\{u(k)\}$, and for the model defined by Eq. (4.18), this inverse model is simply

$$u(k) = b_0^{-1}[y(k) - F(y(k-1), \ldots, y(k-p))]. \tag{4.19}$$

The third reason this model class is interesting is that more can be said about the qualitative behavior of the model defined by Eq. (4.18) than the more general case defined by the condition $q = 0$. As a specific example, consider the following connection of nonlinear autoregressive models:

$$\begin{aligned} y(k) &= F_1(y(k-1), \ldots, y(k-p)) + b_1 v(k) \\ v(k) &= F_2(v(k-1), \ldots, v(k-r)) + b_2 u(k). \end{aligned} \tag{4.20}$$

To obtain a NARMAX model relating the input sequence $\{u(k)\}$ of the composite model to its output sequence $\{y(k)\}$, an expression for the internal sequence $\{v(k)\}$ is required. For models defined by Eq. (4.18), this sequence may be recovered from $\{y(k)\}$ as in Eq. (4.19). Specifically, note that

$$\begin{aligned} F_2(v(k-1), \ldots, v(k-r)) = \\ F_2(b_1^{-1}[y(k-1) - F_1(y(k-2), \ldots, y(k-p-1))], \ldots, \\ b_1^{-1}[y(k-r) - F_1(y(k-r-1), \ldots, y(k-r-p))]) \end{aligned} \tag{4.21}$$

from which it follows that $y(k)$ may be written as

$$\begin{aligned} y(k) &= F_1(y(k-1), \ldots, y(k-p)) + b_1[F_2(v(k-1), \ldots, v(k-r)) + b_2 u(k)] \\ &\equiv G(y(k-1), \ldots, y(k-p-r)) + b_1 b_2 u(k). \end{aligned} \tag{4.22}$$

Hence, this cascade connection yields a third model that also satisfies Eq.(4.18), of order $p+r$. Note that this result generalizes that for the cascade connection of two linear ARX models.

NARMAX Models

For the three reasons just given, the class of models defined by Eq. (4.18) will be designated NARX models. The more general model structure defined by setting $q = 0$, that is

$$y(k) = F(y(k-1), \ldots, y(k-p), u(k)), \qquad (4.23)$$

will be designated as the NARX* class. In contrast to the cascade connection result just presented, note that if the component NARX models in Eq. (4.20) are replaced with NARX* models, reconstruction of the intermediate sequence $\{v(k)\}$ is no longer possible, in general. Not only does the resulting cascade connection lack a NARX* representation, it generally does not exhibit any NARMAX representation at all. Overall, the qualitative behavior of both the NARX and NARX* model classes is much richer than that of the NMAX class considered in Sec. 4.2, as the following discussions illustrate.

4.3.1 A simple example

Tong (1990) defines and discusses a number of different *threshold autoregressive models*, a class that is discussed in detail in Chapter 6 since it represents one possible definition of a *linear multimodel*. The basic idea underlying these definitions is to assume that the function $F(\cdots)$ appearing in Eq. (4.2) is *piecewise linear*, as in the following simple example:

$$y(k) = \begin{cases} 2y(k-1) + u(k) & |y(k-1)| \leq 2 \\ u(k) & |y(k-1)| > 2 \end{cases} \qquad (4.24)$$

which may be rewritten as the first-order NARX model

$$y(k) = f(y(k-1)) + u(k) \qquad f(x) = \begin{cases} 2x & |x| \leq 2 \\ 0 & |x| > 2. \end{cases} \qquad (4.25)$$

Alternatively, this model may be viewed as two different linear autoregressive models that have been "glued together" at the boundaries $y(k-1) = \pm 2$.

The standard four responses for this model are shown in Fig. 4.5. The unit impulse response is shown in the upper left plot and may be interpreted as follows. Initially, $y(k) = 0$ for $k < 0$, while $y(0) = 1$ and $y(1) = 2$, so the condition $|y(k-1)| \leq 2$ is satisfied for $k \leq 2$. Therefore, Eq. (4.24) may be simplified to the *unstable first-order linear model* $y(k) = 2y(k-1) + u(k)$ for $k \leq 2$. Since $y(2) = 4$, the condition $|y(k-1)| \leq 2$ is not satisfied for $k = 3$, so the threshold model "switches" to the stable linear model $y(k) = u(k)$, implying $y(3) = 0$ since $u(3) = 0$. For $k > 3$, the condition $|y(k-1)| \leq 2$ is once again satisfied, so the threshold model "switches back" to the initial unstable linear model. Here, however, both the previous response value $y(k-1)$ and the input $u(k)$ are identically zero, so the threshold model's impulse response remains zero for all $k > 3$. Even though this model is not a member of the NMAX family, it exhibits finite impulse response behavior, thus establishing that such behavior is possible—although highly atypical—for NARX models. More generally, note

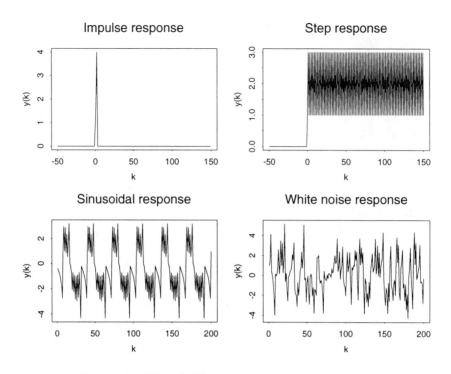

Figure 4.5: Threshold autoregressive model responses

that the duration of the response of this model to an impulse of amplitude α will depend on α. For example, if $\alpha \geq 2$, it follows that $y(1) = \alpha$, implying $y(k) = u(k) = 0$ for $k = 2$ and $y(k) = 2y(k-1) + u(k) = 0$ for $k > 2$. Consequently, the impulse response exhibits a one sample duration for $\alpha \geq 2$. For $0 < \alpha < 2$, the impulse response will be a sequence of the form

$$\{y(k)\} = \{\ldots, 0, \alpha, 2\alpha, \ldots, 2^r \alpha, 0, \ldots\} \qquad (4.26)$$

where r is the largest integer satisfying $2^r \alpha \leq 2$. Analogous results hold for negative impulse inputs.

The unit step response of this threshold autoregressive model is shown in the upper right plot in Fig. 4.5. This response may be understood on the basis of similar reasoning to that just presented for the impulse response. The primary difference, however, is that since $u(k)$ is nonzero for all $k \geq 1$, the threshold model does not remain in the domain of the unstable linear model, but is periodically driven into the domain of the stable linear model, where it is "reset" and returns to the domain of the unstable model. As a consequence, the step response oscillates with a period determined by the amplitude of the step input, at least for sufficiently small amplitude steps. This behavior is clearly seen in Fig. 4.6 which shows the responses to steps of amplitudes $\alpha = 0.1, 1, 2$, and 10. This example also illustrates that the asymptotic constancy results presented at the end of Sec. 4.2 for NMAX models do not hold for NARX

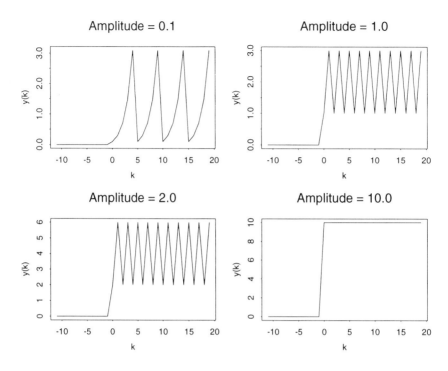

Figure 4.6: Amplitude dependence of step responses

models. In particular, for a step of amplitude α applied at time $k = 0$, note that $y(0) = u(0) = \alpha$. If $|\alpha| > 2$, it follows that $y(1) = u(1) = \alpha$ and, in fact, $y(k) = \alpha$ for all k. In other words, the response of this model to a step input of amplitude $|\alpha| > 2$ is simply the input step $u(k)$. However, if $|\alpha| \leq 2$, the "switching behavior" observed in the impulse response is observed here as well. The unstable model $y(k) = 2y(k-1) + u(k)$ will remain active until the instability amplifies the step amplitude α enough to exceed the threshold value $|y(k-1)| = 2$.

The response of this threshold autoregressive model to the standard sinusoidal input sequence is shown in the lower left plot in Fig. 4.5, illustrating the same reset phenomenon seen in the step response. Since $|u(k)| \leq 2$ for all k for the periodic input sequence considered here, the threshold model only remains in the domain of the stable model $y(k) = u(k)$ for one sample, returning to the domain of the unstable model $y(k) = 2y(k-1) + u(k)$ at the next sampling instant. The time required to "escape" from the limits $|y(k-1)| \leq 2$ depends on the input value to which this model was previously reset. The highly nonsinusoidal shape of this response is indicative of intense superharmonic generation; in fact, this particular model is also capable of subharmonic generation, as discussed in Tong (1990); for further discussion of the responses of NARX models to periodic input sequences, refer to Sec. 4.3.2. Finally, the white noise response shown in the lower right of Fig. 4.5 exhibits analogous, if much less regular, behavior. That is, because the input sequence assumes random values, the time the threshold model spends in the domain of the unstable model is also random. In addition, $|u(k)|$ may exceed the threshold value of 2 for this input sequence, so it is possible for the threshold model to remain in the domain of the stable linear model for more than one sampling period.

4.3.2 Responses to periodic inputs

One of the simplest characterizations of the NMAX model class developed in Sec. 4.2.2 was the periodic response of these models to periodic input sequences. In contrast, the characterization of NARX model responses to periodic input sequences is too difficult to treat in general, so discussion here is restricted to the following extremely special case:

1. The NARX model is of order 1
2. The function defining this model is $f(y) = gy^2$
3. The input sequence is of period 2.

Even with these highly restrictive assumptions, the analysis is somewhat complicated. One of the important ideas on which this analysis is based is the following: if the input sequence $\{u(k)\}$ is of period 2, it may be written as

$$u(k) = \left(\frac{u_0 + u_1}{2}\right) + \left(\frac{u_0 - u_1}{2}\right)(-1)^k \qquad (4.27)$$

where $u(k) = u_0$ for k even and $u(k) = u_1$ for k odd. Hence, if a first-order NARX model exhibits a period 2 response to a period 2 input sequence, these sequences may be written as

$$u(k) = a + b(-1)^k \qquad y(k) = c + d(-1)^k, \qquad (4.28)$$

for some constants a, b, c, and d, and these constants must satisfy the equation

$$c + d(-1)^k = f[c + d(-1)^{k-1}] + a + b(-1)^k. \qquad (4.29)$$

If this equation exhibits real solutions, they define stable, period 2 responses to the period 2 input, but if there are no real solutions, it follows that the model exhibits instability, subharmonic generation, or nonperiodic (e.g., chaotic) responses to period 2 input sequences.

To illustrate the mechanics of this approach, consider the first-order NARX model defined at the beginning of this discussion, that is

$$y(k) = gy^2(k-1) + u(k). \qquad (4.30)$$

Eq. (4.29) reduces to

$$c + d(-1)^k = g[c^2 - 2cd(-1)^k + d^2] + a + b(-1)^k \qquad (4.31)$$

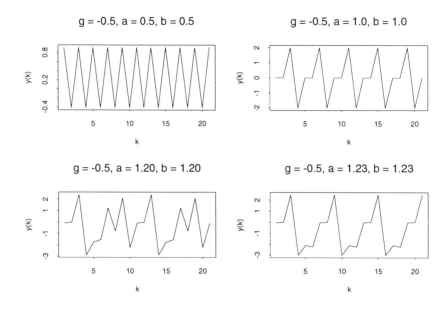

Figure 4.7: Four responses to period 2 inputs

and, if real solutions exist, the constant and variable terms must both match, leading to the simultaneous equations

$$c = gc^2 + gd^2 + a \qquad d = -2gcd + b. \tag{4.32}$$

The second of these equations may be solved for d to obtain

$$d = \frac{b}{1 + 2gc}. \tag{4.33}$$

Substituting this result into the first equation and simplifying ultimately yields the quartic equation for c

$$4g^2 c^4 + (4ag^2 - 3g)c^2 + (4ag - 1)c + a + gb^2 = 0. \tag{4.34}$$

Note that this equation can exhibit 0, 2, or 4 real roots, depending on the input excitation parameters a and b and the model parameter g.

Despite the extreme simplifications assumed here, the final result is fairly complex. Some useful insights may be obtained by considering the special cases illustrated in Figs. 4.7 and 4.8. In all cases, the model parameter is $g = -0.5$, and the plots show the influence of the input sequence parameters a and b; note that the input sequence assumes the values $a \pm b$. The upper left plot in Fig. 4.7 shows the period 2 response obtained for $a = b = 0.50$, illustrating that period 2 responses are possible for this model. Conversely, the upper right plot in Fig. 4.7 shows the period 4 response obtained for $a = b = 1.00$, illustrating that this model is also capable of subharmonic generation of order 1/2. Similarly, the

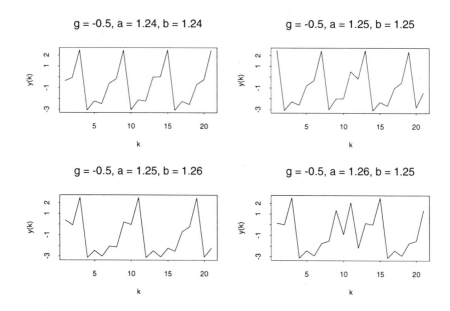

Figure 4.8: Four more responses to period 2 inputs

lower two plots in Fig. 4.7 show the responses obtained for $a = b = 1.20$ (left-hand side) and $a = b = 1.23$ (right-hand side). These responses exhibit periods 10 and 6, respectively, corresponding to subharmonic generation of orders 1/5 and 1/3.

The upper left plot in Fig. 4.8 shows the response of this quadratic NARX model for $a = b = 1.24$. In contrast to the previous examples, this response is *not* periodic but may be characterized as *mildly irregular*, exhibiting an *almost periodic* response. Changing the input parameters to $a = b = 1.25$ results in the *strongly irregular* response shown in the upper right plot in Fig. 4.8 and $a = 1.25$, $b = 1.26$ yields the strongly irregular response shown in the lower left. Interestingly, it can be shown that the response in the lower right, corresponding to $a = 1.26$, $b = 1.25$, is in fact periodic with period 12. Finally, taking $a = b = 1.26$ results in an unstable model response.

4.3.3 NARX model stability

The amplitude-dependent stability seen in the quadratic model responses just described appears to be typical of NARX models, just as it is typical of the bilinear models discussed in Chapter 3. Like the subharmonic generation also seen in NARX models, this behavior stands in marked contrast to that of NMAX models, which were shown in Sec. 4.2 to be BIBO stable under extremely mild conditions. Not surprisingly, general stability characterizations are substantially more difficult for NARX models than for NMAX models, but the following results nicely illustrate the general nature of the problem.

NARMAX Models

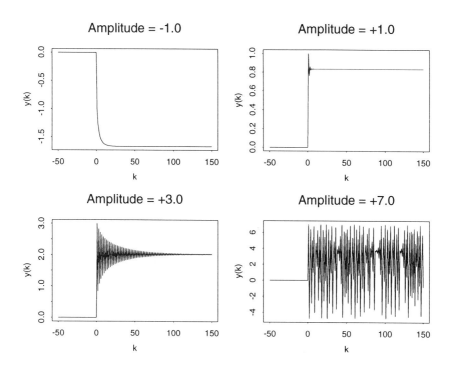

Figure 4.9: Amplitude dependence of NARX model behavior

First, consider the following sufficient condition for the *instability* of a first-order NARX model. That is, consider the model

$$y(k) = f(y(k-1)) + b_0 u(k), \quad (4.35)$$

and suppose that the function $f(\cdot)$ satisfies the following condition: there exist constants $R > 0$ and $\gamma \geq 1$ such that $|x| > R$ implies $|f(x)| > \gamma|x|$. Intuitively, this condition means that the function $f(\cdot)$ is radially unbounded, growing faster than $|x|$ for all x sufficiently large. In addition, assume for simplicity that $f(0) = 0$ and that $y(k) = 0$ for $k < 0$. Now consider the response of this model to an impulse of amplitude α at time $k = 0$. If $|b_0 \alpha| > R$, it follows that $y(0) = b_0 \alpha$ and, for $k > 0$,

$$|y(k)| = |f(y(k-1))| > \gamma|y(k-1)| \to \infty \quad \text{as } k \to \infty. \quad (4.36)$$

In other words, radial unboundedness of $f(\cdot)$ is a sufficient condition for amplitude-dependent instability in the first-order NARX model. As a specific application, note that $f(y) = gy^2$ satisfies this radial unboundedness condition. Fig. 4.9 shows the responses of this quadratic model for $g = -0.24$ to step inputs of amplitude -1.0, $+1.0$, $+3.0$ and $+7.0$. The qualitative behavior of these step responses changes from monotonic to slightly oscillatory to highly oscillatory to chaotic. Ultimately, for sufficiently large amplitudes, the step response becomes unstable, as predicted by Eq. (4.36).

To guarantee the stability of a NARX model, it is necessary to impose local growth conditions on the function $F : R^p \to R^1$ defining the model. In fact, BIBO stability conditions may be established for both the NARX* and the NARX model classes, but under *much* stronger conditions than those required to guarantee NMAX model stability. Both of these results essentially replace the continuity of the NMAX function $g(\cdot)$ with a *Lipschitz condition* on the function $F(\cdots)$ defining either the NARX or the NARX* model. The NARX result follows and the NARX* result is given in Sec. 4.3.5 where the NARX and NARX* model classes are compared. Further, since both results apply to *linear* autoregressive models, that discussion gives a brief comparison of both results with more familiar linear stability conditions.

The Lipschitz condition (Kreyszig, 1978; Wheeden and Zygmund, 1977) may be regarded as a stronger version of continuity. Specifically, consider a function $f : R^m \to R^1$ and let $K \subset R^m$ be a compact (i.e., closed and bounded) set. The function f satisfies the Lipschitz condition on K if there exists $\lambda > 0$ such that

$$|f(x_1, \ldots, x_m) - f(y_1, \ldots, y_m)| \leq \lambda \sum_{i=1}^{m} |x_i - y_i|, \qquad (4.37)$$

for all $\mathbf{x}, \mathbf{y} \in K$. It is not difficult to show that if f is Lipschitz on K for any $\lambda > 0$, then f is continuous on K. Further, note that if f satisfies the Lipschitz condition on K for some $\lambda > 0$, it also satisfies the Lipschitz condition for any $\lambda' > \lambda$. Thus, the Lipschitz condition for any given λ is a stronger requirement than continuity, and it becomes more stringent as λ is made smaller. Similarly, note that if K' is a second compact set and $K' \subset K$, then if f satisfies the Lipschitz condition with the constant λ on K, the condition is also satisfied on K'. Consequently, for any given λ, the Lipschitz condition becomes more stringent as the set K becomes larger. In particular, note that a function f can satisfy the Lipschitz condition on the entire space R^m, but this condition is very restrictive, a point that is discussed further at the end of this section. Still, this restriction provides sufficient conditions for the BIBO stability of NARX models, as the following result demonstrates.

Theorem:

Consider the NARX model:

$$y(k) = F(y(k-1), \ldots, y(k-p)) + b_0 u(k),$$

where $b_0 \neq 0$, $F(0, \ldots, 0) = 0$, and $F : R^p \to R^1$ satisfies the Lipschitz condition on all of R^p with a Lipschitz constant $\lambda < 1/p$. If $|u(k)| \leq M$ for all k, then $|y(k)| \leq |b_0| M / (1 - p\lambda)$ for all k.

Proof:

Proceed by induction, starting with the Lipschitz condition:

$$|y(k)| = |y(k) - 0| \leq |F(y(k-1), \ldots, y(k-p)) - F(0, \ldots, 0)| + |b_0||u(k)|$$

$$\leq \lambda \sum_{i=1}^{p} |y(k-i)| + |b_0| M.$$

NARMAX Models

If $|y(j)| \leq |b_0|M/(1-p\lambda)$ for all $j < k$, it follows that

$$|y(k)| \leq \frac{p\lambda |b_0| M}{1-p\lambda} + |b_0|M = \frac{|b_0|M}{1-p\lambda}.$$

Since $y(k) = 0 \leq |b_0|M/(1-p\lambda)$ for all $k < 0$, the inequality holds for all k.

□

It was noted that requiring f to satisfy a Lipschitz condition on the entire space R^p or R^{p+1} is extremely restrictive. More familiar terminology in the engineering literature is *sector-bounded nonlinearity*, which describes functions $f : R^m \to R^1$ such that $f(0, \ldots, 0) = 0$ and satisfying the Lipschitz condition on all of R^m. As a specific example, for $m = 1$ these conditions imply:

$$|f(x)| \leq \lambda |x|, \tag{4.38}$$

for all x and some $\lambda > 0$. *Note that no nonlinear polynomial satisfies these conditions.* Indeed, any polynomial of order $n > 1$ satisfies the radial unboundedness condition discussed previously, establishing that polynomial $NARX(1)$ models are not BIBO stable.

It should also be noted that this theorem establishes *sufficient* conditions for BIBO stability, not necessary conditions. For example, consider the threshold autoregressive model defined in Eq. (4.24). There, if the input sequence $\{u(k)\}$ satisfies the bound $|u(k)| < M$ for all k and some $M > 0$, it is not difficult to show that $|y(k)| < 3M^*$ for all k provided $|y(0)| < 3M^*$. Here, $M^* = max(M, 2)$ and the result follows by induction, assuming it holds for some k and considering the two possible values $y(k+1)$ could take, depending on whether $|y(k)| \leq 2$ or $2 < |y(k)| < 3M^*$. In either case, it is easily shown that $|y(k+1)| < 3M^*$, establishing the BIBO stability of the threshold autoregressive model. Note, however, that since the function $f(x)$ defined in Eq. (4.25) is not even continuous, it does not satisfy the conditions of the NARX stability theorem.

4.3.4 Steady-state behavior of NARX models

In contrast to the stability and periodic response results, the steady-state characterization of NARX models is quite similar to that of NMAX models. In particular, note that if $u(k) = u_s$ and $y(k) = y_s$ for all k, it follows from Eq. (4.18) that these two constants are related by the equation

$$y_s = F(y_s, \ldots, y_s) + b_0 u_s. \tag{4.39}$$

To simplify this result, define the scalar function

$$\phi(y_s) = y_s - F(y_s, \ldots, y_s) \tag{4.40}$$

and rewrite Eq. (4.39) as

$$\phi(y_s) = b_0 u_s. \tag{4.41}$$

This result may be viewed as the dual of the NMAX result presented in Sec. 4.2.2: given a steady-state response y_s, the corresponding steady-state input u_s is uniquely defined by $u_s = b_0^{-1}\phi(y_s)$, provided $b_0 \neq 0$. *Consequently, it is not possible for NARX models to exhibit input multiplicity.* Similarly, if the function $\phi(\cdot)$ is invertible, the steady-state output is also unique and is given by $y_s = \phi^{-1}(b_0 u_s)$. As in the discussion of NMAX model steady-states given in Sec. 4.2.2, note that if $\phi(\cdot)$ is continuous, it is invertible if and only if it is strictly monotonic (the continuity assumption is important here, a point illustrated at the end of this section). Conversely, if $\phi(\cdot)$ is not invertible, it is possible for the corresponding NARX model to exhibit output multiplicity.

As a specific example of output multiplicity for a NARX model, consider the EXPAR model (Tong, 1990), discussed further in Sec. 4.4.2:

$$y(k) = [\alpha_1 + \beta_1 \exp\{-\delta y^2(k-1)\}]y(k-1) + u(k-1). \tag{4.42}$$

The steady-state response of this model is given by the solution of the transcendental equation

$$[1 - \alpha_1 - \beta_1 \exp\{-\delta y_s^2\}]y_s = u_s. \tag{4.43}$$

If u_s is sufficiently large and if $|\alpha_1| < 1$, it follows that $y_s \simeq u_s/(1-\alpha_1)$, since the exponential term becomes negligible for sufficiently large y_s. Consequently, this particular model can only exhibit output multiplicities when $|u_s|$ is sufficiently small. Hence, consider the simplest case: $u_s = 0$. It follows by inspection that one solution is $y_s = 0$, but two other solutions also exist. Specifically, the term in brackets will vanish for the two nonzero values:

$$y_s = \pm\sqrt{\frac{\ln|\beta_1/(1-\alpha_1)|}{\delta}}, \tag{4.44}$$

provided $\beta_1/(1-\alpha_1) < -1$ and $\delta > 0$. These conditions are met by the specific EXPAR model considered in Sec. 4.4.2 [Eq. (4.87)], for which $\beta_1/(1-\alpha_1) = -2$ and $\delta = 1$. Hence, in addition to the steady-state at $y_s = 0$, two others exist for $u_s = 0$, at $y_s \simeq \pm 0.833$. Note that this general behavior is qualitatively quite similar to that discussed in Chapter 1 for the exothermic CSTR example.

Again as in the case of NMAX models, note that the addition of linear moving average terms to a NARX model does not alter the steady-state behavior in any fundamental way. That is, define the class of NARLMAX models (nonlinear autoregressive - linear moving average models with exogenous inputs) by the equation

$$y(k) = F(y(k-1), \ldots, y(k-p)) + \sum_{j=0}^{q} b_j u(k-j). \tag{4.45}$$

The steady-state input/output relationship for these models is

$$\phi(y_s) = \left(\sum_{j=0}^{q} b_j\right) u_s \tag{4.46}$$

NARMAX Models

which exhibits exactly the same multiplicity as Eq. (4.41). *In particular, note that neither of these models can exhibit input multiplicities.*

Finally, two additional points should be noted concerning the steady-state behavior of NARX models. First, even if a NARX model is BIBO stable, applying a constant input sequence $u(k) = u_s$ for all k need not imply that $y(k) = y_s$ for all k. In particular, if $\phi(\cdot)$ is not invertible, Eq. (4.41) may fail to exhibit any real roots for a given value of u_s. In such cases, no steady-state response exists. The second point is closely related but more subtle: the responses to the input sequences $u(k) = u_s$ for all k and $u(k) = u_s$ for all $k \geq 0$ can be radically different, even if the model is BIBO stable and a unique steady-state response y_s exists for the steady-state input u_s. As a specific example, the steady-state locus for the threshold autoregressive model defined in Eq. (4.24) is given by

$$y_s = \psi(u_s) = \begin{cases} -u_s & |u_s| \leq 2 \\ u_s & |u_s| > 2. \end{cases} \quad (4.47)$$

This steady-state relationship is unique (specifically, the function $\psi(x)$ is its own inverse, implying $u_s = \psi(y_s)$), illustrating a point noted earlier: *discontinuous* functions need not be strictly monotonic to be invertible. In particular, note that $\psi(x)$ is discontinuous and invertible but it is not monotonic. More important in this example, recall that the step response of this model exhibits persistent, bounded oscillations whose frequency depends on the amplitude of the imput step. Hence, although this model is BIBO stable and exhibits a unique steady-state response y_s for all possible steady-state inputs u_s, the response of a step of amplitude u_s does not settle out to the value y_s, in general.

4.3.5 Differences between NARX and NARX* models

The classes of NARX and NARX* models were both defined at the beginning of Sec. 4.3, and most of the discussion given here has been concerned with the more restrictive NARX class. The following discussion illustrates the behavioral differences between the NARX and NARX* model classes. This discussion begins with four simple examples of models that belong to the NARX* class but not to the NARX class. Then, comparisons of steady-state behavior and stability results are given and finally, the notion of *zero dynamics* is introduced and differences between the zero dynamics of NARX models and NARX* models are examined. It is appropriate to point out that the structural difference between these two model classes is closely related to the difference between the general NARMAX class and the additive NARMAX class discussed in Sec. 4.4.

The first NARX* model considered here is the superdiagonal bilinear model:

$$y(k) = ay(k-1) + bu(k) + cy(k-1)u(k). \quad (4.48)$$

Note that this model does not belong to the NARX class because the input $u(k)$ enters through the bilinear term $cy(k-1)u(k)$. Although most bilinear models are not members of the NARX* class, either, this example does illustrate that the NARX* class includes a small subset of superdiagonal bilinear models (specifically, those involving terms of the form $y(k-i)u(k)$ for some $i > 0$).

Another NARX* model is the homogeneous example considered in Chapter 3, repeated here for convenience:

$$y(k) = \left[\frac{y^2(k-1) - u^2(k)}{y^2(k-1) + u^2(k)}\right] u(k). \tag{4.49}$$

Note that this model belongs to the first-order NARX* class and it illustrates an important point: first-order NARX* models can exhibit nonlinear homogeneous behavior, but first-order NARX models cannot. To see this point, consider the response of the general first-order NARX model

$$y(k) = f(y(k-1)) + \beta u(k) \tag{4.50}$$

to an impulse of amplitude α. It follows that $y(1) = \beta\alpha$ and $y(2) = f(\beta\alpha)$. If this model is homogeneous, it follows that $y(2)$ must be a homogeneous function of the impulse amplitude α, implying either that the system is trivial ($\beta = 0$) or that the scalar function $f(x)$ is homogeneous and therefore linear.

The third NARX* example considered here is the positive-homogeneous model

$$y(k) = \sqrt{\alpha^2 y^2(k-1) + \beta^2 u^2(k)}, \tag{4.51}$$

which may be represented as a homomorphic system based on a first-order linear model and the nonlinear function $\phi(x) = x^2$. Although it is true that the NARX model class includes nonlinear, positive-homogeneous members [e.g., the NARX model $y(k) = a_1|y(k-1)| + u(k)$ is positive-homogeneous], no nonlinear *homomorphic* models belong to the NARX class. This result is closely related to the mutually exclusive character of Hammerstein and Wiener models discussed in Sec. 4.4.1. To see this point, consider the homomorphic model based on a first-order linear system and an invertible static nonlinearity $\phi[x]$. The input-output relation for this model is given by

$$y(k) = \phi^{-1}\left[a_1 \phi[y(k-1)] + b_0 \phi[u(k)]\right]. \tag{4.52}$$

If this model has a NARX representation, there must exist a function $f(x)$ and a constant β such that Eq. (4.50) is also satisfied. Now, suppose this requirement holds and define the variables

$$x_1 = a_1 \phi[y(k-1)] \qquad x_2 = b_0 \phi[u(k)] \tag{4.53}$$

and the functions

$$\begin{aligned} g_1(x_1) &= f(y(k-1)) &&= f(\phi^{-1}[x_1/a_1]) \\ g_2(x_2) &= \beta u(k) &&= \beta \phi^{-1}[x_2/b_0]. \end{aligned} \tag{4.54}$$

The requirement that $u(k)$ and $y(k)$ satisfy both Eqs. (4.50) and (4.52) may then be expressed as

$$\phi^{-1}[x_1 + x_2] = g_1(x_1) + g_2(x_2). \tag{4.55}$$

This relationship is called the *Pexider equation* (Aczel and Dhombres, 1989, p. 42) and, as discussed further in Sec. 4.4.1, it exhibits only affine solutions. Hence, it follows that the only homomorphic systems based on first-order linear models that also belong to the NARX model class are the first-order linear models themselves. This result extends to NARX models of arbitrary order p by the same reasoning presented in Sec. 4.4.1 for the Hammerstein and Wiener model result.

The fourth and final NARX* example presented here is the following variation of the previous model:

$$y(k) = \sqrt{\alpha^2 y^2(k-1) + \beta^2 |u(k)|}. \tag{4.56}$$

This model is neither positive-homogeneous nor homomorphic, but like the previous example it also lacks a NARX representation. This result may be derived by similar reasoning to that used in obtaining the previous result, first defining the auxilliary input $v(k) = \sqrt{|u(k)|}$. The model defined by Eq. (4.56) may then be represented as the cascade connection of this static nonlinearity (i.e., $g(x) = \sqrt{|x|}$) with the model defined by Eq. (4.52).

These four examples offer some insight into the structural differences between the NARX and NARX* model classes. As a consequence of these differences, these model classes also exhibit certain important behavioral differences. For example, it was established in Sec. 4.3.4 that NARX models can exhibit output multiplicity, but not input multiplicity. Consequently, it follows that NARX models cannot exhibit phenomena like isolas that involve both input and output multiplicity. Conversely, NARX* models *can* exhibit isolas, as the following example illustrates. Consider the isola defined by the steady-state relationship

$$(u_s - u_0)^2 + (y_s - y_0)^2 = r^2, \tag{4.57}$$

which defines a circle of radius r centered at (u_0, y_0) in the (u_s, y_s) plane. More specifically, consider the case $u_0 = y_0 = r = 1$, which may be rearranged to yield

$$y_s = \frac{1}{2} + \frac{1}{2}y_s^2 - u_s + \frac{1}{2}u_s^2. \tag{4.58}$$

Next, define the first-order NARX* model

$$y(k) = \frac{1}{2} + \frac{1}{2}y^2(k-1) - u(k) + \frac{1}{2}u^2(k) \tag{4.59}$$

and note that the steady-state behavior of this model is given by Eq. (4.58).

Sufficient conditions for the stability of NARX models were presented in Sec. 4.3.3, and similar stability results may be established for the larger class of NARX* models. Because the NARX* class contains the NARX class as a proper subset, it follows that the general NARX* stability results are necessarily more conservative than the NARX stability results. In particular, note that the Lipschitz condition required in the following theorem is more restrictive than that required in the NARX stability theorem.

Theorem:

Consider the NARX* model:

$$y(k) = f(y(k-1), \ldots, y(k-p), u(k)),$$

where $f : R^{p+1} \to R^1$ satisfies $f(0, \ldots, 0) = 0$ and satisfies the Lipschitz condition on all of R^{p+1} with a Lipschitz constant $\lambda < 1/p$. If $|u(k)| \leq M$ for all k, then $|y(k)| \leq \lambda M/(1-p\lambda)$ for all k.

Proof:

The proof proceeds by induction, from the Lipschitz condition:

$$|y(k)| = |y(k) - 0| = |f(y(k-1), \ldots, y(k-p), u(k)) - f(0, 0, \ldots, 0)|$$

$$\leq \lambda \left\{ \sum_{i=1}^{p} |y(k-i)| + |u(k)| \right\}$$

$$\leq \lambda \sum_{i=1}^{p} |y(k-i)| + \lambda M.$$

Assume $|y(j)| \leq \lambda M/(1-p\lambda)$ for all $j < k$; then,

$$|y(k)| \leq \frac{p\lambda^2 M}{1-p\lambda} + \lambda M = \frac{\lambda M}{1-p\lambda}.$$

Since $y(k) = 0 \leq \lambda M/(1-p\lambda)$ for all $k < 0$, it follows that the inequality holds for all k.

□

It is interesting to compare these results with those available for *linear* autoregressive models. Since these models belong to both the NARX and the NARX* classes, both of the stability theorems apply. For $p = 1$, the linear autoregressive model $y(k) = a_1 y(k-1) + b_0 u(k)$ satisfies the Lipschitz condition on R^1 with $\lambda = |a_1|$, so the BIBO stability condition from the NARX theorem reduces to the standard result $|a_1| < 1$. For comparison, note that the conditions of the NARX* theorem require both $|a_1| < 1$ and $|b_0| < 1$. This additional constraint on the magnitude of b_0 is a consequence of assuming only that the arguments of the NARMAX function $F(\cdots)$ are $y(k-1)$ and $u(k)$ and *not* making the stronger assumption $F(y(k-1), u(k)) = f(y(k-1)) + b_0 u(k)$.

NARMAX Models

For $p = 2$, note that the Lipschitz constant for the linear model $y(k) = a_1 y(k-1) + a_2 y(k-2) + b_0 u(k)$ is $\lambda = max(|a_1|, |a_2|)$, so the NARX stability condition requires $|a_1|, |a_2| < 1/2$. Necessary and sufficient conditions for the stability of this system are (Elaydi, 1996, Ex. 4.1.3):

$$a_1 + a_2 < 1 \quad a_2 - a_1 < 1 \quad a_2 > -1. \tag{4.60}$$

It is not difficult to see that these conditions are strictly weaker than the NARX stability requirements. For example, note that $a_1 = 1/8$, $a_2 = -3/4$ satisfies conditions (4.60) but not the NARX stability criteria. As before, the NARX* stability criterion is even more restrictive, requiring both that $|a_1|, |a_2| < 1/2$ and that $|b_0| < 1/2$.

The stability of NARX models—and NARMAX models more generally—is an important topic, about which much has been written. A number of useful results are available for discrete-time dynamic models (e.g., Liapunov stability criteria and LaSalle's invariance principle) that can be applied to study the stability of particular examples or, in more favorable cases, entire classes of models. Because the topic of stability is too broad to attempt to treat in detail here and because it is well treated elsewhere (Elaydi, 1996; Wimp, 1984), discussions of stability are limited here to specific examples and broadly applicable results like the previous theorem.

Finally, one last qualitative characterization that illustrates the differences between NARX models and NARX* models is that of *zero dynamics*. Specifically, given a system $\mathcal{M}$ that maps an input sequence $\{u(k)\}$ into an output sequence $\{y(k)\}$, consider the set $\mathcal{Z}$ of all input sequences $\{u(k)\}$ such that $y(k) \equiv 0$. For the general NARX model, it follows immediately that the set $\mathcal{Z}$ consists of the single sequence

$$u(k) = -\frac{F(0, \ldots, 0)}{b_0} \quad \text{for all } k. \tag{4.61}$$

Since $F(0, \ldots, 0)$ is frequently zero, it follows that the zero dynamics of NARX models usually correspond to the zero sequence $u(k) \equiv 0$. In contrast, the situation for NARX* models is much more complicated, as the following four examples illustrate. First, for the model

$$y(k) = \alpha \cos[y(k-1)u(k)], \tag{4.62}$$

it follows that $\mathcal{Z}$ is the empty set unless $\alpha = 0$, in which case the system is trivial and $\mathcal{Z}$ contains all possible input sequences $\{u(k)\}$. Conversely, the model

$$y(k) = \alpha \sin[y(k-1)u(k)], \tag{4.63}$$

is not trivial for $\alpha \neq 0$ but $\mathcal{Z}$ again contains all possible input sequences, regardless of the value of α. An intermediate example is the model

$$y(k) = \alpha \cos \sqrt{y^2(k-1) + u^2(k)}, \tag{4.64}$$

for which the zero dynamics set consists of all sequences $\{u(k)\}$ taking the values $u(k) = (2m+1)\pi/2$ for some integer m. Note that these sequences may be constant, periodic with arbitrary periods, monotonic [e.g., $u(k) = (2k+1)\pi/2$] or random (e.g., a random binary sequence assuming the values $\pm \pi/2$), so long as $u(k)$ is always an odd multiple of $\pi/2$. Finally, the last example is the model

$$y(k) = \alpha u(k) \cos \sqrt{y^2(k-1) + u^2(k)}, \qquad (4.65)$$

which exhibits the same zero dynamic set as the previous example except for the addition of the identically zero sequence $u(k) \equiv 0$. The point of these examples is that, while the zero dynamics of NARX models are extremely simple, the zero dynamics of NARX* models may be arbitrarily complicated.

4.4 Additive NARMAX Models

It was noted at the end of Chapter 2 that the term additive can have at least two distinct meanings. The discussion there was principally concerned with the older notion of *functional additivity*, while the following discussion is concerned with the newer notion of *structural additivity*. This notion comes from the regression literature, where *additive models* are of the form (Hastie and Tibshirani, 1990; Simonoff, 1996)

$$F(x_1, x_2, \ldots, x_m) = \sum_{i=1}^{m} f_i(x_i). \qquad (4.66)$$

In the motivating regression problem, an unknown function $F(x_1, x_2, \ldots, x_m)$ of m variables is to be determined by fitting the equation

$$y = F(x_1, x_2, \ldots, x_m) + e, \qquad (4.67)$$

to measured values of $x_1, x_2, \ldots, x_m$ and y. In practice, there are many different ways of solving this problem, typically involving the minimization of some measure of the model prediction error e. Standard *parametric* procedures approach this problem by specifying a particular form for the function $F(\cdots)$ (e.g., a polynomial in m arguments), thus reducing the problem to one of determining values for some finite set of unknown parameters. Alternatively, *nonparametric procedures* are also available that attempt to "ask the data" what functional form is appropriate, specifying some measure of the "regularity" or "smoothness" for this functional form to obtain a well-posed problem (Hardle, 1990).

In practice, nonparametric procedures can work quite well for small m (e.g., $m = 1$ or 2), but they rapidly become intractible as the number of variables involved becomes large (Hardle, 1990). This fact motivates the use of the (structurally) additive models defined in Eq. (4.66). Specifically, this assumption reduces the nonparametric regression problem to a set of m one-dimensional problems, for which nonparametric procedures are best suited. This idea has been applied to the nonlinear time-series modeling problem (Chen and Tsay,

1993; Chen et al., 1995), leading to the class of (structurally) *additive NARMAX models* defined by requiring the function $F(\cdots)$ in Eq. (4.2) to be structurally additive:

$$y(k) = \sum_{i=1}^{p} f_i(y(k-i)) + \sum_{j=0}^{q} g_j(u(k-j)). \tag{4.68}$$

Chen and Tsay (1993) refer to these models as *nonlinear additive autoregressive models with exogenous inputs*, or *NAARX models*; here, the terms *NAARX models* and *additive NARMAX models* are used interchangably.

The class of NAARX models is particularly interesting both because it represents a simple generalization of the class of linear ARMAX models discussed in Chapter 2, and because it includes several other important model examples and model classes as special cases. First, the NAARX model class includes the Hammerstein model class as a proper subset but it *excludes* all nonlinear Wiener models, a result discussed in some detail in Sec. 4.4.1. Next, note that the NAARX model class includes all first-order NARX models and is therefore capable of all of the qualitative behavior exhibited by these models, including harmonic generation, subharmonic generation, chaotic responses to simple inputs, input-dependent stability, asymmetric responses to symmetric input changes, and output multiplicity. Further, the NARX* model defined by Eq. (4.59) is also structurally additive, establishing that this model class can also exhibit isolas and, as a corollary, input multiplicity. In addition, the class of positive-homogeneous TARMAX models introduced in Chapter 3 may be written in the form of Eq. (4.68), establishing that structurally additive models can exhibit positive-homogeneous and static-linear behavior.

The class of structurally additive NARMAX models also includes the class of Lur'e models introduced in Chapter 1. To establish this result, recall that a Lur'e model consists of a linear model with a static nonlinearity $f(\cdot)$ connected as a feedback element. The response $y(k)$ of the linear model may be written as

$$y(k) = \sum_{i=1}^{m} a_i y(k-i) + \sum_{i=1}^{m} b_i u(k-i) \tag{4.69}$$

where two simplifying assumptions have been made. First, and without loss of generality, it has been assumed that $p = q = m$ in the usual $ARMAX(p,q)$ notation; it is always possible to do this by taking $m = \max\{p,q\}$ and setting the model coefficients a_i or b_i to zero if $i > p$ or $i > q$, respectively. The second assumption made here is that $b_0 = 0$, implying the linear dynamic model exhibits no direct feedthrough. This assumption is usually met in practice and it greatly simplifies the analysis of the Lur'e model, whose output may then be written as

$$y(k) = \sum_{i=1}^{m} a_i y(k-i) + \sum_{i=1}^{m} b_i [u(k-i) - f(y(k-i))], \tag{4.70}$$

corresponding to Eq. (4.68) with $f_i(x) = a_i - b_i f(x)$ and $g_i(x) = b_i x$ for $i = 1, \ldots, m$.

The remainder of this section discusses two special cases of the additive NARMAX model class, a general stability result that extends those presented earlier for the NARX and NARX* model classes, and the steady-state behavior of structurally additive NARMAX models. Specifically, Sec. 4.4.1 compares the Hammerstein and Wiener model classes, establishing a number of useful results. These results emphasize some of the differences between the additive Hammerstein model class and the non-additive Wiener model class. Similarly, Sec. 4.4.2 illustrates the consequences of structural additivity in connection with the class of EXPAR models mentioned in Sec. 4.3.5. In both cases, behavioral differences arise due to the presence of *cross-terms* that couple inputs and/or outputs at different time lags $k-i$ and $k-j$; generally, these differences appear to be most pronounced in the transient responses of the models. Sec. 4.4.3 then presents sufficient conditions for the stability of structurally additive NARMAX models and Sec. 4.4.4 briefly considers the steady-state behavior of these models.

4.4.1 Wiener vs. Hammerstein models

One of the most important subsets of the additive model class is the family of Hammerstein models, consisting of a static nonlinearity $v(k) = \phi(u(k))$ followed by the linear dynamic model

$$y(k) = \sum_{i=1}^{p} a_i y(k-i) + \sum_{j=0}^{q} b_j v(k-j). \quad (4.71)$$

It follows immediately that the Hammerstein model may be represented as Eq. (4.68) with

$$f_i(x) = a_i x \qquad g_j(x) = b_j \phi(x). \quad (4.72)$$

Conversely, consider the Wiener model formed from the linear moving average model $v(k) = u(k) + u(k-1)$ followed by the quadratic nonlinearty $y(k) = v^2(k)$:

$$y(k) = [u(k) + u(k-1)]^2 = u^2(k) + u^2(k-1) + 2u(k)u(k-1). \quad (4.73)$$

The presence of the cross-term $u(k)u(k-1)$ in this NARMAX representation establishes that Wiener models *do not* belong to the NAARX class. More generally, it is possible to establish three important representation results for Wiener models. First, the only structurally additive Wiener models are the degenerate special cases of linear models and static nonlinearities, for which the Wiener and Hammerstein representations are equivalent. Second, an explicit NARMAX representation can be obtained for Wiener models based on *invertible* static nonlinearities and third, *not all Wiener models have NARMAX representations*. The remainder of this section demonstrates these three results.

The dual Wiener model to the Hammerstein model defined by Eqs. (4.68) and (4.72) may be viewed as a dynamic model relating the input sequence $\{v(k)\}$ appearing in Eq. (4.71) to the output $z(k) = \phi(y(k))$. If $p = 0$, the linear model

NARMAX Models

defined by Eq. (4.71) is a moving average model and the resulting Wiener model is an NMAX model with the following explicit representation:

$$z(k) = \phi\left(\sum_{i=0}^{q} b_i v(k-i)\right). \tag{4.74}$$

For this model to be structurally additive, it is necessary that

$$\phi\left(\sum_{i=0}^{q} b_i v(k-i)\right) = \sum_{i=0}^{q} g_i(v(k-i)), \tag{4.75}$$

for some functions $\{g_i(x)\}$. For $q = 0$, this condition is trivially satisfied by $g_0(x) = \phi(b_0 x)$ and the Wiener model reduces to this static nonlinearity. For $q = 1$, this condition reduces to the Pexider equation introduced in Sec. 4.3.5 (Aczel and Dhombres, 1989, p. 42):

$$\phi(x_1 + x_2) = g_1(x_1) + g_2(x_2). \tag{4.76}$$

It is known that the only solutions to this equation are of the form

$$\phi(x) = ax + b \quad g_1(x) = a_1 x + b_1 \quad g_2(x) = a_2 x + b_2. \tag{4.77}$$

where a, a_i, b, and b_i are real constants. For $q > 1$, this result extends to the following theorem.

Theorem:

The only solutions of the equation

$$\phi\left(\sum_{i=0}^{q} b_i x_i\right) = \sum_{i=0}^{q} g_i(x_i),$$

for $q > 1$ are the affine functions $g_i(x) = c_i x + d_i$ for $i = 0, 1, \ldots, q$, and $\phi(x) = cx + d$ where c_i, c, d_i, and d are all real constants.

Proof:

Proceed by induction, assuming the result holds for $q-1$ and showing that this implies it holds for q. For simplicity, define y as

$$y = \sum_{i=0}^{q-1} b_i x_i,$$

and define the function $\Phi(x_0, \ldots, x_q)$ as

$$\Phi(x_0, \ldots, x_q) = \phi\left(b_q x_q + \sum_{i=0}^{q-1} b_i x_i\right) - g_q(x_q)$$

$$= \phi(b_q x_q + y) - g_q(x_q)$$

$$= \sum_{i=0}^{q-1} g_i(x_i).$$

If the last sum in this expression can be reduced to a function of y alone, the last two lines of this equation reduce to the Pexider equation, thus proving that $\phi(x)$ and $g_q(x)$ are affine. To establish this result, note that $\Phi(x_0,\ldots,x_q)$ is independent of x_q and take $x_q = 0$ to obtain

$$\Phi(x_0,\ldots,x_{q-1},x_q) = \Phi(x_0,\ldots,x_{q-1},0)$$
$$= \phi\left(\sum_{i=0}^{q-1} b_i x_i\right) - g_q(0) \equiv \psi\left(\sum_{i=0}^{q-1} b_i x_i\right).$$

Combining these equations reduces the problem to

$$\psi\left(\sum_{i=0}^{q-1} b_i x_i\right) = \sum_{i=0}^{q-1} g_i(x_i).$$

Since the theorem is assumed to hold for $q-1$, it follows that $g_i(x) = c_i x + d_i$ for $i = 0, 1, \ldots, q-1$, and $\psi(x) = cx + d$ for some real constants c and d. In particular

$$\sum_{i=0}^{q-1} g_i(x_i) = c\left(\sum_{i=0}^{q-1} b_i x_i\right) + d = cy + d,$$

which is a function of y alone. The desired result thus follows, holding for q provided it holds for $q - 1$. Taking $q = 2$, the problem reduces to the Pexider equation for $q - 1 = 1$, so the general case follows by induction.

□

This theorem establishes that the only structurally additive Wiener NMAX models for $q > 0$ are affine models. For the autoregressive case, the same result holds, but the general situation is somewhat more complex. Specifically, note that the Wiener model representation

$$z(k) = \phi(y(k)) \qquad y(k) = \sum_{i=1}^{p} a_i y(k-i) + \sum_{i=0}^{q} b_i v(k-i), \qquad (4.78)$$

relating the input sequence $\{v(k)\}$ to the output sequence $\{z(k)\}$ is *not* a NARMAX model because it does not directly relate the current output $z(k)$ to past outputs $z(k-i)$ and current and past inputs $v(k-i)$. That is, the intermediate variable $y(k)$ appears in these equations and it is not obvious how to eliminate it, in general. A special case where $y(k)$ can be eliminated from these equations occurs when $\phi(\cdot)$ is invertible. There, $y(k) = \phi^{-1}(z(k))$, and the pair of equations (4.78) may be reduced to the NARMAX model for $z(k)$:

$$z(k) = \phi\left(\sum_{i=1}^{p} a_i \phi^{-1}(z(k-i)) + \sum_{i=0}^{q} b_i u(k-i)\right). \qquad (4.79)$$

NARMAX Models

As a specific example, consider the Wiener model based on the cubic nonlinearity $\phi(x) = x^3$ and the first-order linear model, $p = 1$, $q = 0$. The NARMAX representation for this model is

$$\begin{aligned}
z(k) &= [a_1 z^{1/3}(k-1) + b_0 u(k)]^3 \\
&= a_1^3 z(k-1) + 3a_1^2 b_0 z^{2/3}(k-1)u(k) \\
&\quad + 3a_1 b_0^2 z^{1/3}(k-1)u^2(k) + b_0^3 u^3(k).
\end{aligned} \quad (4.80)$$

This Wiener model is a member of the NARX* family but not the NARX family and, like the NMAX case considered above, it is clear that this model is not structurally additive. In fact, the results of the above theorem may be applied to Eq. (4.79) to show that this NARMAX model is neither structurally additive nor a member of the NARX class. Since the Hammerstein model may be viewed as a simple prototype of both the NARX and the NARX* classes, comparisons of Hammerstein and Wiener models may provide some useful insight into both the distinction between additive and nonadditive models, and that between the NARX and NARX* model classes.

In the most general case, Wiener models do not exhibit NARMAX representations at all, as the following example illustrates. Consider the Wiener model formed from the first-order linear model $x(k) = ax(k-1) + u(k)$ with $0 < a < 1$ and the saturation nonlinearity

$$y(k) = \begin{cases} -1 & x(k) < -1 \\ x(k) & -1 \leq x(k) \leq 1 \\ 1 & x(k) > 1. \end{cases} \quad (4.81)$$

The response of the linear part of this model to an impulse of amplitude γ is $x(k) = \gamma a^k$ for all $k \geq 0$. For $0 < \gamma < 1$, the saturation nonlinearity has no effect, so $y(k) = v(k)$, but for $\gamma > 1$, $y(k) = 1$ for $k = 0, \cdots, r$ where r is the largest integer such that $\gamma a^r \geq 1$. Now, suppose the Wiener model does have the NARMAX representation

$$y(k) = F(y(k-1), \ldots, y(k-p), u(k), \ldots, u(k-q)) \quad (4.82)$$

for some finite integers p and q and some function $F(\cdots)$. Next, choose γ such that $r = \max\{p, q\} + 1$ and note that the following two conditions must be satisfied simultaneously:

$$\begin{aligned}
y(r) &= 1 &&= F(1, \ldots, 1, 0, \ldots, 0) \\
y(r+1) &= \gamma a^{r+1} &&= F(1, \ldots, 1, 0, \ldots, 0).
\end{aligned} \quad (4.83)$$

It follows from the definition of r that $\gamma a^{r+1} < 1$, implying that these two conditions are contradictory. *Hence, this simple Wiener model has no NARMAX representation.*

Because Wiener and Hammerstein models are so similar in structure, differences in their qualitative behavior are of particular interest and are discussed at length throughout the rest of this book. In particular, note that if the steady-state gain of the linear block incorporated in both models is constrained to be 1 (a restriction that may be imposed without loss of generality by re-scaling the static nonlinearity), the steady-state response of a Hammerstein model and its dual Wiener model are identical. Therefore, differences in behavior between these models are only present in the transient response, as has been noted (Pearson, 1995). This observation represents a general characteristic of structurally additive models, a point discussed further in Sec. 4.4.4.

4.4.2 EXPAR vs. modified EXPAR models

In the limit of either small or large arguments, many functions may be well approximated by other, simpler functions. In the context of NARMAX models, note that if $y(k-i)$ and $u(k-j)$ are sufficiently small in magnitude, most nonlinear maps $F(\cdots)$ defining NARMAX models can be well approximated by linear maps. This observation is the basis for linearization of nonlinear models and implies that, for inputs of sufficiently small magnitude, NARMAX models can often be approximated by linear ARMAX models. At the other extreme, for sufficiently *large* magnitudes of the arguments $y(k-i)$ and $u(k-j)$, many nonlinear functions $F(\cdots)$ are also well approximated by other, simpler functions. Here, the key point is that if this simpler approximation is *additive*, the nonadditivity of the exact NARMAX model based on the map $F(\cdots)$ will only be evident for *intermediate amplitude* responses. This point is illustrated by the following example.

Jones (1978) discussed the general first-order NARX model

$$y(k) = f(y(k-1)) + u(k). \tag{4.84}$$

Since this model belongs to the NAARX class for any choice of the function $f(\cdot)$, it follows it is necessary to examine higher-order autoregressive models to understand the qualitative differences between autoregressive NAARX models and the larger (nonadditive) NARX class. Jones focused on the statistics of the response sequence $\{y(k)\}$ under the assumption that the input sequence $\{u(k)\}$ was a sequence of independent, identically distributed random variables. Specific examples included both threshold autoregressive models like the one considered in Sec. 4.3 and models involving exponential nonlinearities.

Haggan and Ozaki (1980) considered a generalization of the exponential class discussed by Jones, motivated by a study of the behavior of the ordinary differential equations that describe nonlinear oscillators. This class of exponential NARMAX models is also discussed by Tong (1990, p. 108) under the name EXPAR models, and is defined by

$$y(k) = \sum_{i=1}^{p} [\alpha_i + \beta_i \exp\{-\delta y^2(k-1)\}] y(k-i) + u(k). \tag{4.85}$$

Note that for $p = 1$ (the case considered by Jones), this model is a member of the NAARX family, but not for $p > 1$. Specifically, note that terms in the sum in Eq. (4.85) for $i > 1$ involve both $y(k-1)$ and $y(k-i)$. Interestingly, in his discussion of these models Tong notes that it is sometimes advatageous to replace $y(k-1)$ appearing in all of the exponential terms in Eq. (4.85) with $y(k-i)$; this modification results in a model that *is* a member of the NAARX family. Specifically, Tong proposes the term *modified EXPAR model* to describe the model class

$$y(k) = \sum_{i=1}^{p} [\alpha_i + \beta_i \exp\{-\delta y^2(k-i)\}] y(k-i) + u(k). \qquad (4.86)$$

It is instructive to briefly compare the behavior of these model classes with two simple examples. The first of these examples (Model 1) is the EXPAR model:

$$y(k) = [0.9 - 1.8 \exp\{-y^2(k-1)\}] y(k-2) + u(k), \qquad (4.87)$$

whereas the second is the modified EXPAR model (Model 2):

$$y(k) = [0.9 - 1.8 \exp\{-y^2(k-2)\}] y(k-2) + u(k). \qquad (4.88)$$

In the limit as $|y(k)|$ becomes small, the exponential term in Eqs. (4.87) and (4.88) approaches 1, so both models behave increasingly like the second-order linear model

$$y(k) = -0.9 y(k-2) + u(k). \qquad (4.89)$$

In particular, for sufficiently small impulse amplitudes γ, the impulse response of Models 1 and 2 will be well approximated by the impulse response of this linear model, which is readily determined to be

$$y(k) = \begin{cases} (-0.9)^{k/2} \gamma & k \text{ even,} \\ 0 & k \text{ odd.} \end{cases} \qquad (4.90)$$

Conversely, if $|y(k)|$ is sufficiently large, the exponential term in these nonlinear models is small, so they behave like the second-order linear model:

$$y(k) = +0.9 y(k-2) + u(k), \qquad (4.91)$$

whose impulse response is

$$y(k) = \begin{cases} (0.9)^{k/2} \gamma & k \text{ even,} \\ 0 & k \text{ odd.} \end{cases} \qquad (4.92)$$

Since the nonlinear Models 1 and 2 both approach the *same* linear models in these two extreme limits, it follows that qualitative differences between these models will only appear in the "intermediate amplitude" regime for which $\exp\{-y^2(k-1)\}$ is significantly different from both 0 and 1. Further, note that both of these limiting models are stable, so $|y(k-2)| \to 0$ as $k \to \infty$. Thus, the linear approximation defined in Eq. (4.89) ultimately dominates, for both

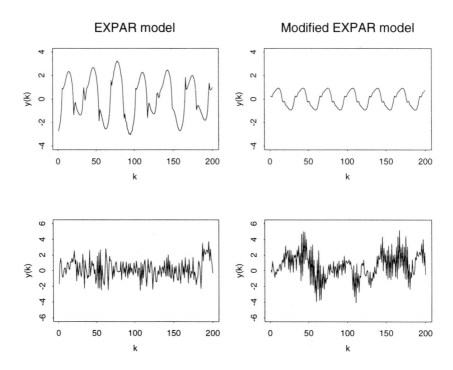

Figure 4.10: A comparison of EXPAR and modified EXPAR models

Model 1 and Model 2, implying that these models differ only in their transient responses and not in their asymptotic or "steady-state" responses, just as in the case of the Hammerstein and Wiener models discussed in Sec. 4.4.1. More specifically, since these two models differ only in the terms $\exp\{-y^2(k-1)\}$ vs. $\exp\{-y^2(k-2)\}$, it follows that these differences will be apparent only in "high-frequency, intermediate amplitude" ranges where $y(k-1)$ and $y(k-2)$ are significantly different and neither too small nor too large.

This last point may be seen clearly in the sinusoidal and white noise responses shown in the four plots in Fig. 4.10. The left-hand plots in this figure show the responses for the EXPAR model defined by Eq. (4.87), and the right-hand plots show the responses for the modified EXPAR model defined by Eq. (4.88). The upper plots compare the sinusoidal responses: note that the response of the modified EXPAR model exhibits both a more "regular" appearance than the unmodified response and a smaller magnitude. In particular, note that the modified EXPAR model appears to preserve the periodicity of the input sinusoid, whereas the unmodified model does not. The white noise responses are compared in the lower plots, and these also exhibit visually obvious differences in their qualitative character. Here, however, the opposite trend is observed: the response of the modified EXPAR model exhibits a response that is both larger in magnitude and somewhat less regular than that of the unmodified EXPAR model.

NARMAX Models

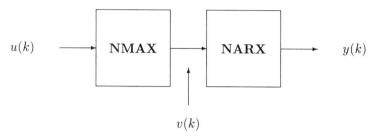

Figure 4.11: General representation for NAARX models

4.4.3 Stability of additive models

Recall from earlier discussions that the partitioning of the NARMAX model class between nonlinear autoregressive (NARX) and nonlinear moving average (NMAX) models is not complete, in contrast to the linear case. Among nonlinear model classes, this partitioning may be most nearly complete for the structurally additive models, which may be re-written as the following pair of equations:

$$y(k) = \sum_{i=1}^{p} f_i(y(k-i)) + v(k), \qquad (4.93)$$

$$v(k) = \sum_{j=0}^{q} g_j(u(k-j)). \qquad (4.94)$$

Since Eq. (4.93) defines an additive NARX model and Eq. (4.94) defines an additive NMAX model, it follows that *any* additive NARMAX model may be represented by the cascade structure shown in Fig. 4.11. Consequently, if the functions $\{g_j(x)\}$ defining the additive NMAX model (4.94) are continuous, it follows from the NMAX stability theorem presented in Sec. 4.2 that this part of the model will be BIBO stable. Hence, the overall NAARX model defined in Eq. (4.68) will be BIBO stable if the NARX model defined in Eq. (4.93) is BIBO stable. Sufficient conditions for the stability of NAARX models may therefore be obtained from the theorem presented in Sec. 4.3.3, but it is also possible to obtain a much stronger result by exploiting structural additivity.

Recall that the stability results given in Sec. 4.3.3 applied to the model:

$$y(k) = f(y(k-1), \ldots, y(k-p)) + b_0 u(k),$$

subject to the following three conditions:

1. $b_0 \neq 0$
2. $f(0, \ldots, 0) = 0$
3. $f : R^p \to R^1$ is Lipschitz on all of R^p with constant $\lambda < 1/p$.

The first of these conditions is satisfied by the additive NARX model (4.93) and the second is satisfied if $f_i(0) = 0$ for $i = 1, 2, \ldots, p$. This second condition may be relaxed, although it is not terribly restrictive. The third condition is satisfied if the functions $f_i(x)$ individually satisfy the Lipschitz condition on all of R^1 with $\lambda_i < 1/p$, but the following theorem establishes stability under much weaker conditions. As a specific example, note that this theorem is applicable to the class of modified EXPAR models discussed in Sec. 4.4.2.

Theorem:

Consider the additive NARX model

$$y(k) = \sum_{i=1}^{p} f_i(y(k-i)) + b_0 u(k),$$

where $b_0 \neq 0$, $y(k) = 0$ for all $k < 0$ and $f_i(0) = 0$ for $i = 1, \ldots, p$. If $f_i(x)$ satisfies a Lipschitz condition with constant λ_i on all of R^1, this model is BIBO stable if the associated linear model

$$z(k) = \sum_{i=1}^{p} \lambda_i z(k-i) + |b_0| v(k)$$

is stable, where $z(k) = 0$ for $k < 0$.

Proof:

Suppose $|u(k)| < \gamma$ for all k and some $\gamma > 0$. Then, for $k \geq 0$

$$|y(k)| \leq \sum_{i=1}^{p} |f_i(y(k-i))| + |b_0||u(k)|$$

$$\leq \sum_{i=1}^{p} \lambda_i |y(k-i)| + |b_0|\gamma,$$

Next, proceed by induction, assuming $|y(k-j)| \leq z(k-j)$ for $j \geq 1$ where $z(k)$ is the response of the above linear model to the step input $v(k) = \gamma$ for $k \geq 0$. It follows that

$$|y(k)| \leq \sum_{i=1}^{p} \lambda_i z(k-i) + |b_0|\gamma = z(k).$$

Since $|y(0)| = |b_0||u(0)| = |b_0|\gamma = z(0)$, and $y(k) = z(k) = 0$ for all $k < 0$, the induction hypothesis holds for $k = 1$, hence for all k. Consequently, if the linear model is stable, $|y(k)|$ is bounded.

□

NARMAX Models

In addition to the general representation result presented in Fig. 4.11, the following additive NMAX representation results are also useful. First, recall from Sec. 4.2 that the parallel combination of NMAX models results in another NMAX model of the same dynamic order. In the case of additive NMAX models, the parallel combination of m models defined by the functions $g_j^r(x)$ for $j = 0, 1, \ldots, q$ and $r = 1, 2, \ldots, m$ yields an additive NMAX model defined by the additive functions

$$\gamma_j(x) = \sum_{r=1}^{m} g_j^r(x). \tag{4.95}$$

The second interesting result to note here is that the class of additive NMAX models is equivalent to the class of NMAX Uryson models. To see this equivalence, first note that any additive NMAX model may be represented as the Uryson model shown in Fig. 4.12. Here, the functions $g_j(x)$ define the static nonlinearities on which the Uryson model is based, whereas the linear dynamic blocks are the pure delay terms z^{-j} for $j = 0, 1, \ldots, q$. The proof of the converse—that every Uryson model may be represented as an additive NMAX model proceeds by a simple rearrangement of sums (Menold, 1996). Specifically, note that any r-channel finite Uryson model (i.e., any member of the class $U_{(N,M)}^r$ defined in Chapter 5) may be represented as:

$$y(k) = \sum_{j=1}^{r} \left(\sum_{i=0}^{q} b_{ij} g_j(u(k-i)) \right) = \sum_{i=0}^{q} \phi_i(u(k-i)), \tag{4.96}$$

where $\phi_i(x) = \sum_{j=1}^{r} b_{ij} g_j(x)$.

4.4.4 Steady-state behavior of NAARX models

The following observation has been made several times in previous sections of this chapter: the restriction of additivity constrains the transient response much more strongly than it constrains the steady-state response. To see this point, consider the response of an arbitrary NARMAX model to the input sequence $u(k) = u_s$ for all $k \geq 0$. If the model under consideration exhibits a constant response $y(k) = y_s$ to this input sequence, this response will be a solution of the equation

$$y_s = F(y_s, \ldots, y_s, u_s, \ldots, u_s) = \phi(y_s, u_s). \tag{4.97}$$

For NMAX models, it follows that $\phi(y_s, u_s)$ is independent of y_s, whereas for NARX models, it follows that $\phi(y_s, u_s) = \psi(y_s) + b_0 u_s$. Restricting consideration to *additive* NARMAX models, however, it follows only that the function $\phi(y_s, u_s)$ must be of the form

$$\phi(y_s, u_s) = \psi(y_s) + \gamma(u_s). \tag{4.98}$$

Since the function $\phi(y_s, u_s)$ is of this form for both NMAX and NARX models, it follows immediately that *any steady-state response that can be exhibited by*

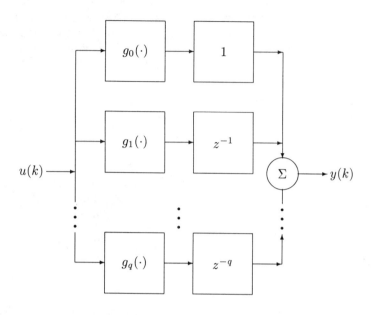

Figure 4.12: Uryson representation for additive NMAX models

either NMAX or NARX models can be matched by an additive model. Therefore, it follows that the qualitative effects of the structural additivity restriction will be seen either in transient responses or in more complex NARMAX models that are not members of either the NMAX or the NARX model classes.

4.5 Polynomial NARMAX models

Without question, one of the most popular classes of NARMAX models is the subset of *polynomial* NARMAX models. There, the nonlinear mapping $F(\cdots)$ defined in Eq. (4.2) is taken as a polynomial of order n in its $p+q+1$ arguments. For example, taking $n = 2$, and $p = q = 1$ leads to the polynomial NARMAX model

$$y(k) = y_0 + a_1 y(k-1) + b_0 u(k) + b_1 u(k-1)$$
$$+ c_{01} u(k) y(k-1) + c_{11} u(k-1) y(k-1)$$
$$+ d_{00} u^2(k) + d_{01} u(k) u(k-1) + d_{11} u^2(k-1) + e_{11} y^2(k-1). \quad (4.99)$$

Note that this model involves 10 parameters and includes an enormous range of special cases obtained by setting some of these parameters to zero. To illustrate this point, particular special cases of this model include the following:

NARMAX Models

1. The first-order autoregressive linear model:
$$y(k) = a_1 y(k-1) + b_1 u(k-1)$$

2. The static nonlinearity:
$$y(k) = y_0 + b_0 u(k) + d_{00} u^2(k)$$

3. The Hammerstein model formed from (1) and (2):
$$y(k) = y_0 + a_1 y(k-1) + b_1 u(k-1) + d_{11} u^2(k-1)$$

4. The linear moving average model:
$$y(k) = b_0 u(k) + b_1 u(k-1)$$

5. The Hammerstein model formed from (2) and (4):
$$y(k) = y_0 + b_0 u(k) + b_1 u(k-1) + d_{00} u^2(k) + d_{11} u^2(k-1)$$

6. The dual Wiener model to (5), of the form:
$$y(k) = y_0 + b_0 u(k) + b_1 u(k-1) + d_{00} u^2(k) \\ + d_{01} u(k) u(k-1) + d_{11} u^2(k-1)$$

7. The superdiagonal bilinear model:
$$y(k) = a_1 y(k-1) + b_0 u(k) + c_{01} u(k) y(k-1)$$

8. The diagonal bilinear model:
$$y(k) = a_1 y(k-1) + b_1 u(k-1) + c_{11} u(k-1) y(k-1)$$

9. Robinson's Volterra model, discussed in Sec. 4.5.1:
$$y(k) = b_0 u(k) + b_1 u(k-1) + d_{01} u(k) u(k-1)$$

10. Robinson's AR-Volterra model, also discussed in Sec. 4.5.1:
$$y(k) = a_1 y(k-1) + b_0 u(k) + b_1 u(k-1) + d_{01} u(k) u(k-1)$$

11. The generalized logistic model:
$$y(k) = a_1 y(k-1) + b_0 u(k) + e_{11} y^2(k-1)$$

12. The following model, discussed in Sec. 4.5.2:
$$y(k) = a_1 y(k-1) + b_0 u(k) + d_{01} u(k) u(k-1) + e_{11} y^2(k-1).$$

As suggested by the range of these examples, the class of polynomial NARMAX models is too broad to characterize in general terms. Almost all of the special cases listed above have been discussed previously as specific examples of other, more restrictive model classes. For example, the polynomial NARMAX class includes as proper subsets the polynomial Hammerstein models, all of the bilinear models considered in Chapter 3, and all of the classes of finite Volterra models defined in Chapter 5.

Conversely, the class of polynomial NARMAX models is not broad enough to include *all* of the other NARMAX models considered in this book. For example, none of the nonlinear homogeneous models or the positive-homogeneous models defined in Chapter 3 have polynomial NARMAX representations, a point discussed further in Chapter 7. Many of the homogeneous model examples considered in Chapter 3 are members of the larger family of rational NARMAX models defined in Sec. 4.6. Conversely, note that even when polynomial Wiener models exhibit NARMAX representations, they are generally neither polynomial nor rational NARMAX representations. As a specific example, recall that the NARMAX representation of the cubic Wiener model considered in Sec. 4.4.1 [Eq. (4.80)] involved fractional exponents. Finally, the classes of linear multimodels discussed in Chapter 6 do not belong to either the polynomial or rational NARMAX classes.

To illustrate the range of qualitative behavior typically seen in polynomial NARMAX models, the remainder of this section briefly considers two simple examples. Sec. 4.5.1 examines an autoregressive extension of a second-order Volterra model (Robinson, 1977), a polynomial member of the LARNMAX class defined in Sec. 4.2.2. Following this discussion, Sec. 4.5.2 considers a simple extension of Robinson's AR-Volterra model that includes a nonlinear autoregressive term. The subtle behavior seen in the response of this model to white noise inputs emphasizes the need for careful *a posteriori* validation of empirically determined polynomial NARMAX models.

4.5.1 Robinson's AR-Volterra model

Robinson (1977) investigated the following second-order Volterra model for non-Gaussian statistical time series:

$$y(k) = b_0 u(k) + b_1 u(k-1) + d_{01} u(k) u(k-1). \qquad (4.100)$$

Because this model belongs to the NMAX class, it exhibits finite impulse response behavior, as discussed in Sec. 4.2.2. In Robinson's analysis, $\{u(k)\}$ was taken to be a zero-mean Gaussian white noise sequence like the one described in Chapter 2. In this context, the finite impulse response character of the model (4.100) means that responses $y(k)$ and $y(k-j)$ are statistically independent if $|j| > 1$. Because this consequential restriction is extremely strong, and thus frequently a poor working assumption in practice, Robinson proposed adding linear autoregressive terms to the basic Volterra model, an idea that has been explored recently in chemical process modeling applications (Kirnbauer, 1991;

Kirnbauer and Jorgi, 1992). As a specific illustration, Robinson considered the following model, for which the resulting statistical dependence behavior is more reasonable:

$$y(k) = a_1 y(k-1) + b_0 u(k) + b_1 u(k-1) + d_{01} u(k) u(k-1). \quad (4.101)$$

In particular, by including the linear autoregressive term $a_1 y(k-1)$, responses $y(k)$ and $y(k-j)$ are correlated for all $j > 0$; this statistical characterization is closely related to the fact that the impulse response of this model decays asymptotically to zero rather than dropping to zero identically for $k > 1$, as in the original model. Also, note that the original model (4.100) is a special case of the model (4.101) for $a_1 = 0$, so the discussion presented here focuses on a comparison of the more general model with this special case.

For the parameters $b_0 = b_1 = d_{01} = 1$, the response of this model to an impulse of amplitude γ is given by:

$$\begin{aligned} y(0) &= \gamma, \\ y(1) &= (1+a_1)\gamma, \\ y(k) &= a_1^{k-1} y(1) = a_1^{k-1}(1+a_1)\gamma \quad \text{for } k > 1. \end{aligned} \quad (4.102)$$

Note that this impulse response is exactly the same as that of the linear model:

$$z(k) = a_1 z(k-1) + u(k) + u(k-1). \quad (4.103)$$

Similarly, for $k \geq 1$, the response of this model to a step of amplitude γ is the same as that of the linear model:

$$z(k) = a_1 z(k-1) + u(k), \quad (4.104)$$

for a step of amplitude $\gamma' = 2\gamma + \gamma^2$. The key point of both of these results is that, like the NMAX models discussed in Sec. 4.2, the impulse and step responses of this particular model differs slightly if at all from that of a closely related linear model.

In contrast, the sinusoidal and white noise responses differ significantly from those of linear models. This point is particularly obvious in the upper two plots in Fig. 4.13, which shows the response of the two cases $a_1 = 0$ (i.e., Robinson's unmodified Volterra model) and $a_1 = 0.8$ to the standard sinusoidal input; in both cases, the parameters b_0, b_1 and d_{01} all have the value 1.0. The left-hand plot shows the response for $a_1 = 0$ and the right-hand plot shows the response for $a_1 = 0.8$: the nonsinusoidal character of these responses immediately demonstrates the nonlinearity of both models, and the difference in both magnitude and shape of these responses clearly illustrates the influence of including the autoregressive term a_1. The responses of these models to Gaussian white noise is shown in the lower two plots; even cursory visual comparison of these responses illustrates the smoothing effects of the autoregressive term a_1, and a more careful comparison of these responses with linear model responses considered in Chapter 2 offers evidence of the model's nonlinearity.

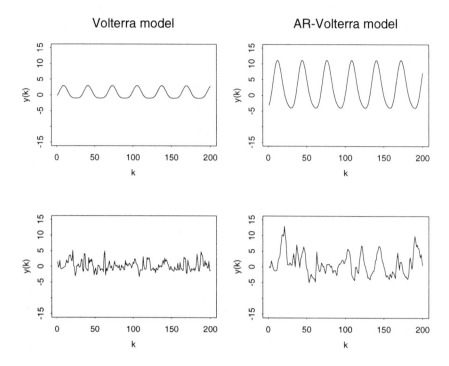

Figure 4.13: A comparison of Robinson's two models

4.5.2 A more general example

The following example includes both nonlinear autoregressive and nonlinear moving average terms and is typical of what might be obtained as the result of a stepwise regression procedure, excluding terms failing a statistical significance test from the general model defined by Eq. (4.99):

$$y(k) = a_1 y(k-1) + b_0 u(k) + b_1 u(k-1) + d_{01} u(k) u(k-1) + e_{11} y^2(k-1) \tag{4.105}$$

Despite its simplicity, this model is capable of exhibiting some relatively complex behavior, as the following discussion illustrates. In particular, consider the case $a_1 = 0.8$, $b_0 = b_1 = d_{01} = 1$, and $e_{11} = -0.02$; note that for $e_{11} = 0$, Eq. (4.105) reduces to Robinson's AR-Volterra model so the apparently small magnitude of the coefficient e_{11} considered here *might* suggest that these models should exhibit similar qualitative behavior. Fig. 4.14 shows the four standard responses introduced in Chapter 2 for the model defined by Eq. (4.105). The upper left plot in this figure shows the impulse response of this model, given by

$$y(k) = [0.8 - 0.02y(k-1)]y(k-1), \tag{4.106}$$

and driven by the initial condition $y(0) = \alpha$ for an impulse of amplitude α. For a unit impulse, it may be seen that $0 < y(k) < 1.8$ for all k, from which

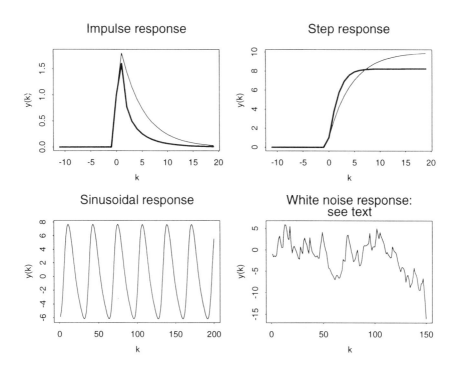

Figure 4.14: Standard responses for a general NARMAX model

it follows that $0.8 - 0.02y(k-1) \simeq 0.8$. Therefore, the impulse response looks very much like that of the linear model obtained by setting $d_{01} = e_{11} = 0$. In fact, the impulse response of this linear model is also shown in the upper left plot in Fig. 4.14 as the thin reference line. Despite the small magnitude of the nonlinear term, note the significant difference in initial decay rates exhibited by these two models. In addition, the response of the nonlinear model is unstable for sufficiently large amplitude impulses. This point may be seen by noting that

$$|0.8 - 0.2y(k-1)| > 1 \quad \Rightarrow \quad |y(k)| > |y(k-1)|, \tag{4.107}$$

implying instability if $\alpha < -10$ or $\alpha > 90$.

The step response for the polynomial NARMAX model defined in Eq. (4.105) is shown in the upper right plot in Fig. 4.14, and it also looks quite similar in general character to that of a linear model. Again, the light line in this plot shows the step response for the linear model obtained by dropping both the quadratic and bilinear terms. It is interesting to note how much faster the nonlinear model settles to its asymptotic value than the linear model does. Careful examination of the sinusoidal response shown in the lower right plot in Fig. 4.14 reveals an asymmetric, nonsinusoidal character, but nothing obviously exotic. Similarly, examining the initial portion of this model's response to the standard white-noise input sequence, nothing dramatic appears there, either. Comparing this response to that of the linear reference model, however, reveals

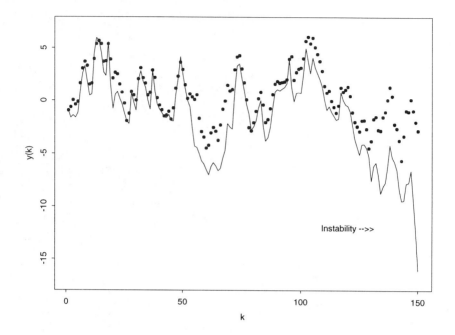

Figure 4.15: Instability in NARMAX white noise response

the onset of an amplitude-dependent instability. This comparison is shown in Fig. 4.15, where the linear model response is indicated by the solid circles and the nonlinear model response is indicated by the light line. During the first part of this data record, both model responses are quite similar, but at about $k = 140$, the difference begins to become pronounced and for $k > 150$, the nonlinear model exhibits an instability, diverging to $-\infty$. Here, the fundamental difficulty is that the autoregressive terms $0.8y(k-1) - 0.02y^2(k-1)$ have become sufficiently negative that the overall response is unstable. This possibility was noted in connection with the impulse response discussed above, but the white noise response illustrated here demonstrates how subtle this effect can be: the model *appears* to be stable with respect to this input sequence for over 100 samples but then the response diverges without warning. Given the simplicity of this particular polynomial NARMAX model, the subtle unpleasantness of this example is unsettling.

4.6 Rational NARMAX models

Although polynomials are extremely flexible, there are certain types of mathematical behavior that polynomials cannot exhibit. For example, the hard saturation phenomenon seen in the high-purity distillation column example discussed

NARMAX Models

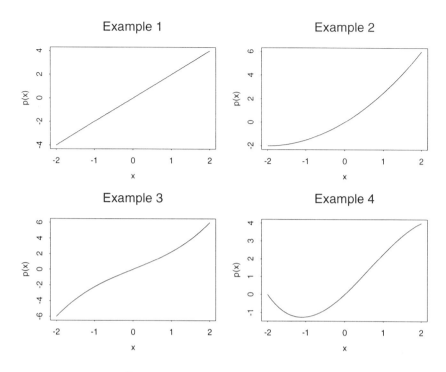

Figure 4.16: Examples of polynomials

in Chapter 1 cannot be exhibited by any polynomial since, as noted in Sec. 4.1, polynomials of order $n > 1$ are radially unbounded. Conversely, because polynomials are continuous everywhere, they are bounded on any bounded interval and so cannot exhibit singularities, either. In contrast, *rational functions* are defined as the *ratio* of two polynomials:

$$f(x) = \frac{a_0 + a_1 x + \cdots + a_n x^n}{b_0 + b_1 x + \cdots + b_m x^m}, \qquad (4.108)$$

and this class of functions can exhibit both of these types of nonpolynomial behavior with the right choices of n, m, a_i, and b_i.

The ranges of polynomial and non-polynomial behavior that may be represented by rational functions is partially illustrated the four plots in Figs. 4.16 and 4.17. Specifically, Fig. 4.16 shows plots of the following four polynomials:

1. $p(x) = 2x$
2. $p(x) = x + x^2/2$
3. $p(x) = x + x^3/4$
4. $p(x) = x + x^2/2 - x^3/4$.

All four of these polynomials are plotted on the range $-2 \leq x \leq 2$ to facilitate comparison. An important feature of polynomials in general is that $p(x)$ must approach $\pm\infty$ continuously as $x \to \infty$ or $x \to -\infty$.

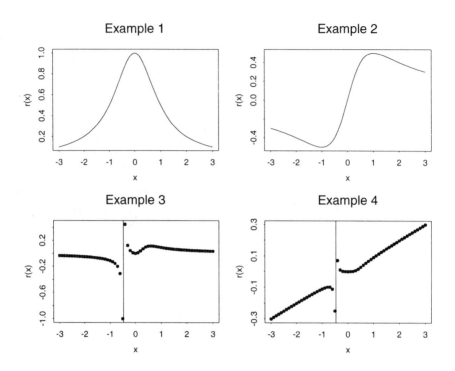

Figure 4.17: Examples of rational functions

In contrast, Fig. 4.17 shows plots of the following four rational functions:

1. $r(x) = 1/(1+x^2)$
2. $r(x) = x/(1+x^2)$
3. $r(x) = x^2/[1+10x^3]$
4. $r(x) = x^4/[1+10x^3]$.

Note that the first two of these functions are bounded for all x, decaying asymptotically to zero as $x \to \pm\infty$, a form of behavior that polynomials cannot exhibit. Similarly, the third and fourth examples both exhibit a singularity at $x = -10^{-1/3} \simeq -0.464$ where the denominator goes to zero. Asymptotically, the rational function defined in the third example goes to zero like the first two examples, whereas the fourth example exhibits the limiting behavior $r(x) \simeq x/10$ for arguments x of sufficiently large magnitude. The key point is that although the class of rational functions includes the class of polynomials as a special case, the general range of behavior the rational class can represent is substantially larger.

The class of *rational NARMAX models* is defined by permitting the function $F(\cdot)$ in either Eq. (4.1) or (4.2) to be a rational function of its arguments. That is, $F(\cdot)$ is the ratio of two multivariable polynomials of the type discussed in Sec. 4.5 for the class of polynomial NARMAX models. This class of models yields a more complex identification problem than the class of polynomials

NARMAX models because the rational models are not linear in the unknown model parameters. These problems are not severe, however, and have been discussed in the literature (Zhu and Billings, 1993, 1994). Tong (1990) discusses a subset of this class of models under the alternative name *fractional autoregressive models*. Specifically, he focuses on NARX models of the form

$$y(k) = \frac{a_0 + \sum_{i=1}^{p} a_i y^i(k-1)}{b_0 + \sum_{j=1}^{q} b_j y^j(k-1)} + u(k). \tag{4.109}$$

The motivation offered for this particular class of models is that many nonlinear functions may be represented by continuous fractions; truncating these representations leads to rational function approximations like that appearing on the right-hand side of Eq. (4.109). In particular, models of this form arise naturally as approximations of the general first-order nonlinear autoregressive model discussed in Sec. 4.4.2, suggesting that the class of models defined by Eq. (4.109) should be quite flexible. As an example of this flexibility, note that the nonlinear homogeneous model discussed in Chapter 3 [Eq. (3.34)] is a member of the rational NARMAX class, provided some limiting procedure (for example, L'Hospital's rule) is adopted to interpret the rational function $F(\cdot)$ in Eq. (4.1) when both the numerator and denominator are zero. It is shown in Chapter 7 that homogeneous models cannot exhibit polynomial representations. The following subsections briefly describe two specific rational NARMAX models and offer some useful insights into the range of behavior this class can exhibit.

4.6.1 Zhu and Billings model

A representative rational NARMAX model is that developed by Zhu and Billings (1994) for the response of a nonlinear fluid loading system to ocean waves. In fact, their paper develops two rational NARMAX models for this system, corresponding to two different sea states, and these models are denoted S_1 and S_2. Because they are very much concerned with the issues of observation noise and modeling errors, Zhu and Billings develop a model of the form (4.1) involving the modeling errors $e(k)$ for S_1, but the model for S_2 is in the simpler form (4.2) of primary interest here that does not depend on $e(k)$. Specifically, the rational function $F(\cdot)$ in Eq. (4.2) for model S_2 is

$$\begin{aligned} y(k) = &[1.74y(k-1) - 0.82y(k-2) + 0.001y^2(k-1) \\ &- 26.9u^2(k-2) + 25.9u(k-1)u(k-2)] \\ &[1 + 0.00001y(k-1) + 0.001y(k-2) \\ &- 1.70u^2(k-1) - 1.53u^2(k-2) + 3.19u(k-1)u(k-2)]^{-1}. \end{aligned} \tag{4.110}$$

The four standard responses for this model are shown in Fig. 4.18. Cursory examination might suggest these responses to be from a second-order linear model: both the impulse and step responses appear reasonably consistent with this hypothesis and the sinusoidal response gives no appreciable evidence of the model's nonlinearity. The standard white noise response exhibits evidence

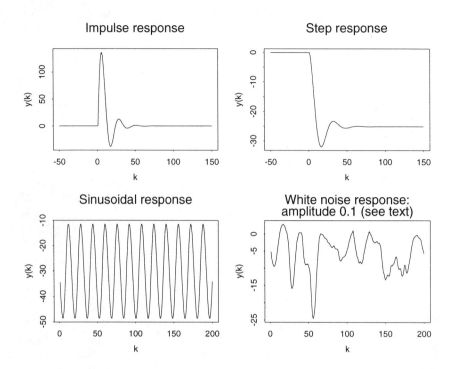

Figure 4.18: Four standard responses, Zhu and Billings model

of slow divergence, but the response to a smaller amplitude white noise input sequence (specifically, a sequence with standard deviation 0.1) shown in the lower right plot in Fig. 4.18 again appears consistent with the behavior of a lightly damped second-order linear model.

More careful examination reveals some very interesting results, however. Specifically, it is easy to show from Eq. (4.110) that for an impulse input, $y(k) = 0$ for $k < 2$ and $y(2) = -26.9u^2(0)/[1 - 1.53u^2(0)]$, reducing to $y(2) \simeq 50.8$ for a unit impulse; for impulses of sufficiently large magnitude, note that $y(2) \simeq +17.58$, regardless of the sign of the impulse. Also, note that since $y(2)$ depends on $u^2(0)$, the responses to positive and negative impulses of the same magnitude are identical. For $k > 2$, the impulse response is given by the *unforced* rational NARMAX model

$$y(k) = \frac{1.74y(k-1) - 0.82y(k-2) + 0.001y^2(k-1)}{1 + 0.00001y(k-1) + 0.001y(k-2)}. \tag{4.111}$$

It is especially interesting to consider the behavior of this impulse response as a function of the impulse magnitude α. Note that for small α (e.g., $\alpha = 0.1$), the magnitude of $y(k-j)$ remains small and the impulse response may be approximated by the of the *linear* model

$$y(k) = 1.74y(k-1) - 0.82y(k-2), \tag{4.112}$$

NARMAX Models

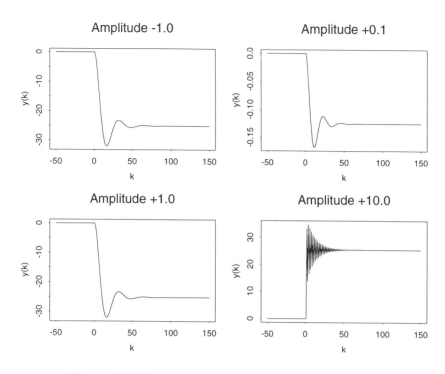

Figure 4.19: Four step responses, Zhu and Billings model

to the initial condition $y(2) = -26.9\alpha^2$ ($= -0.269$ in this example). It is not difficult to show that the poles of this second-order model occur inside the unit circle at approximately $z = 0.87 \pm i0.25$, thus implying a stable decaying oscillation for the impulse response.

Taking $u(k)$ to be a step of amplitude α, it once again follows that $y(k) = 0$ for $k < 2$, and $y(2)$ may be determined from Eq. (4.110) as $y(2) = -\alpha^2/[1 - 0.04\alpha^2]$. As before, this response depends only on the magnitude of the step and not its sign, and for sufficiently large amplitude steps, $y(2)$ approaches the limit 25.0. For $k > 2$, the step response is given by:

$$y(k) = \frac{1.74y(k-1) - 0.82y(k-2) + 0.001y^2(k-1) - \alpha^2}{1 + 0.00001y(k-1) + 0.001y(k-2) - 0.04\alpha^2}. \quad (4.113)$$

The step responses obtained from this equation are shown in Fig. 4.19 for $\alpha = -1.0, 0.1, 1.0$, and 10. The left-hand pair of plots, corresponding to $\alpha = \pm 1.0$, illustrate the point that the step response depends only on the magnitude of the step and not the sign. Also, both of these plots illustrate that the step response is negative if $|\alpha| < 5$, as does the response for $\alpha = 0.1$ shown in the upper right. The lower right plot illustrates the response for large amplitude steps ($\alpha = 10$), approaching the limiting value 25 noted previously. Finally, note that if $\alpha = \pm 5$, $y(2)$ is infinite, illustrating the phenomenon of *finite escape time*, a particularly nasty form of nonlinear instability.

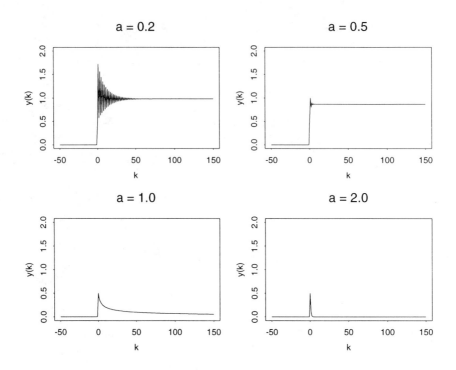

Figure 4.20: Additive rational models, impulse response

4.6.2 Another rational NARMAX example

It is easy to construct structurally additive *polynomial* NARMAX models, but it is tempting to conclude that these are the *only* structurally additive, rational NARMAX models. The following model represents a simple counterexample:

$$\begin{aligned} y(k) &= \frac{y(k-1)}{a^2 + y^2(k-1)} + \frac{u(k)}{b^2 + u^2(k)} \\ &= \frac{[b^2 + u^2(k)]y(k-1) + [a^2 + y^2(k-1)]u(k)}{[a^2 + y^2(k-1)][b^2 + u^2(k)]} \\ &= \frac{b^2 y(k-1) + a^2 u(k) + u^2(k)y(k-1) + u(k)y^2(k-1)}{a^2 b^2 + a^2 u^2(k) + b^2 y^2(k-1) + u^2(k)y^2(k-1)}. \quad (4.114) \end{aligned}$$

The first line establishes this model's structural additivity and the last line confirms its membership in the rational NARMAX class.

The responses of this model to a unit impulse input are shown in Fig. 4.20 for $a = 0.2, 0.5, 1.0$, and 2.0, respectively, with the b parameter fixed at 1 in all cases. The most striking feature of these impulse responses is that for $a = 0.2$ and 0.5, they do not decay to zero but approach a nonzero steady-state value, like the *step response* of a linear model. In contrast, for $a = 1.0$ and 2.0, the impulse response exhibits more typical behavior, decaying asymptotically to zero. It is worth noting, however, that for $a = 1$, this decay is extremely slow.

NARMAX Models

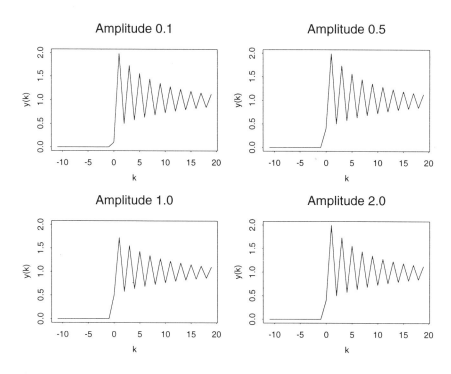

Figure 4.21: Influence of impulse intensity

To understand these responses, note that the equation for the steady-state response y_s:

$$y_s = \frac{y_s}{a^2 + y_s^2} \Rightarrow y_s^3 + (a^2 - 1)y_s = 0, \qquad (4.115)$$

has the three roots $y_s = 0, \pm(1 - a^2)^{1/2}$, which are all real if $|a| < 1$, but only the root $y_s = 0$ is real for $|a| > 1$. Hence, for $|a| > 1$, the impulse response necessarily decays to zero but for $a = 0.2$ and 0.5, the unit impulse response settles asymptotically to the second of these roots, which occur at approximately 0.98 and 0.87, respectively. The unit impulse responses do not settle to the "usual" value $y_s = 0$ because this steady-state is unstable. For $a = \pm 1$, all three roots coalesce to $y_s = 0$, so the impulse response necessarily settles to this asymptotic limit, but the multiplicity of this root results in an extremely slow (i.e., nonexponential) decay.

It is also interesting to consider the influence of the amplitude of the impulse on these results. Fig. 4.21 shows plots of the response for $a = 0.2$ to impulses of amplitude $\alpha = 0.1, 0.5, 1.0$, and 2.0. The striking feature of these results is that they are all nearly identical; this result again follows from the fact that the impulse response settles to the stable steady-state value $y_s \simeq 0.98$, which is independent of α. This behavior—a very weak dependence on the initial condition $y(0)$—represents the antithesis of the "sensitive dependence on initial

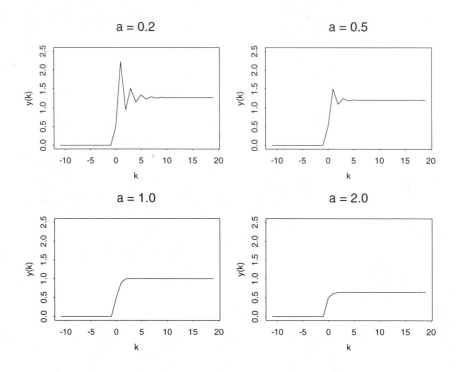

Figure 4.22: Additive rational models, step response

conditions" often given as one of the defining characteristics of deterministic chaos (Doherty and Ottino, 1988; Guckenheimer and Holmes, 1983).

The step responses of this model are shown in Fig. 4.22 for the same four values of a considered above (0.2, 0.5, 1.0, and 2.0). These responses are less surprising, approaching a constant limit that varies consistently with the model parameter a. Proceeding as before, note that the steady-state response y_s satisfies the cubic equation

$$y_s^3 - \gamma y_s^2 - (1 - a^2)y_s - a^2\gamma = 0, \qquad (4.116)$$

where γ is related to the step amplitude α by

$$\gamma = \frac{\alpha}{b^2 + \alpha^2}. \qquad (4.117)$$

For the example considered here, $\alpha = 1$ and $b = 1$, so $\gamma = 1/2$.

The sinusoidal responses of this model are shown in Fig. 4.23 for the same four values of the parameter a: 0.2, 0.5, 1.0, and 2.0. As in the case of the impulse responses, for $a = 0.2$ and $a = 0.5$, these responses are oscillatory but centered around the negative stable steady-state values -0.98 and -0.87, respectively. In both of these cases, because $u(k)$ was negative at the beginning of the simulation, these responses all lie in the domain of the negative stable

NARMAX Models

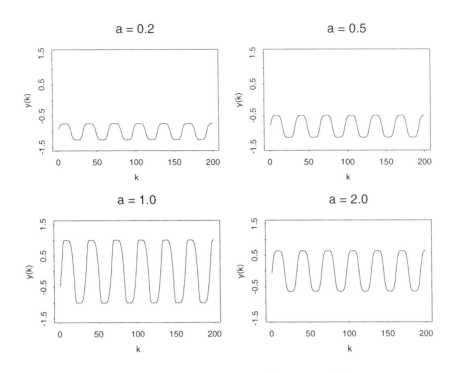

Figure 4.23: Additive rational models, sinusoidal responses

steady-state, rather than the positive stable steady-state. For $a \geq 1$, the only real steady-state solution of the model equation is $y_s = 0$, so the sinusoidal responses for $a = 1.0$ and $a = 2.0$ oscillate about zero.

White noise responses for the same four cases are shown in Fig. 4.24 and, although they are more complex, these responses are also intimately related to the steady-state analysis just presented. As with the sinusoidal results, responses for $a = 1.0$ and $a = 2.0$ (the bottom two plots) are centered around zero, the only steady state for this model. For $a = 0.5$ (upper right plot), the model response clearly lies in the domain of the positive stable steady state ($y_s \simeq 0.87$) most of the time, but exhibits a transition to the negative stable steady state toward the end of the data record. The most interesting case is $a = 0.2$, shown in the upper left plot in Fig. 4.24. There, the model response oscillates about *both* stable steady states ($y_s \simeq \pm 0.98$) and exhibits multiple transitions between them. Qualitatively, this behavior is typical of what is observed in systems like the exothermic CSTR discussed in Chapter 1 that exhibit output multiplicity.

4.7 More Complex NARMAX Models

As emphasized in Chapter 1, this book is primarily concerned with discrete-time dynamic models of *moderate complexity*. The primary exceptions to this

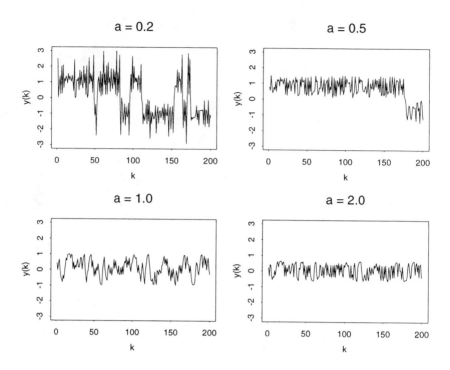

Figure 4.24: Additive rational models, white noise responses

general rule are various infinite-dimensional models, which are considered principally to illustrate the relationship between different model classes (e.g., bilinear and Volterra). In addition, certain other model classes—in particular, dynamic models based on neural networks or radial basis functions—are popular enough that it is unreasonable to omit them here, *even though they are complex enough that relatively little can be said about their behavior in general terms*. Consequently, this section briefly considers some of these models and their relation to the other NARMAX model classes introduced here. Because they provide a natural connection between structurally additive models and neural network models, this discussion begins with the class of projection-pursuit models.

4.7.1 Projection-pursuit models

Sec. 4.4 notes that a strong motivation for considering structurally additive models is that they permit the application of single-variable nonparametric model identification procedures in problems involving multiple arguments. That is, structurally additive multivariable functions are of the form

$$F(x_1, \ldots, x_m) = \sum_{i=1}^{m} f_i(x_i). \qquad (4.118)$$

NARMAX Models

The class of *projection-pursuit models* (Fan and Gijbels, 1996, p. 265) is similarly motivated, based on functions of the form

$$F(x_1,\ldots,x_m) = \sum_{i=1}^{r} f_i\left(\sum_{j=1}^{m} w_{ij}x_j\right). \tag{4.119}$$

Here, $\{w_{ij}\}$ represents a set of $r \times m$ constants that may be viewed as a collection of r vectors in R^m defining "preferred directions" onto which the vector $\mathbf{x} \in R^m$ of independent variables is to be projected. The resulting projection-pursuit model is then simply a structurally additive model based on these projections instead of the individual components of $\mathbf{x}$. Note that if $r = m$ and these vectors are taken as the unit vectors along each axis of R^m (i.e., $w_{ij} = 1$ if $i = j$, $w_{ij} = 0$ otherwise), the projection-pursuit model reduces to the structurally additive model defined in Eq. (4.118).

The class of *projection-pursuit NARMAX models* is defined by

$$y(k) = \sum_{\ell=1}^{r} f_\ell \left(\sum_{i=1}^{p} a_i^\ell y(k-i) + \sum_{j=0}^{q} b_j^\ell u(k-j) \right) \tag{4.120}$$

where $\{a_i^\ell\}$ and $\{b_j^\ell\}$ define r linear model components whose outputs are the arguments of the *scalar* static nonlinearities $f_\ell(\cdot)$ appearing in Eq. (4.120). Note that if $p = 0$, the projection-pursuit NARMAX model defined by Eq. (4.120) is a member of the NMAX model class discussed in Sec. 4.2. In fact, it is interesting to note that this class of NMAX models may be represented by the "dual Uryson" block diagram shown in Fig. 4.25. That is, whereas the r-channel Uryson model consists of r Hammerstein models connected in parallel, this model consists of r *Wiener models* connected in parallel. Each of these Wiener models consists of a linear, time-invariant subsystem with finite impulse response b_j^ℓ, followed by a static nonlinearity $f_\ell(\cdot)$. As before, if the steady-state gains of the linear models are all constrained to be 1, it follows that the steady-state behavior of the projection-pursuit model is identical to its Uryson dual obtained by reversing the order of the linear and nonlinear blocks in each of the r channels. The transient behavior, however, will be quite different, as it is between Hammerstein and Wiener models.

Conversely, if $q = 0$, the projection-pursuit model is a member of the NARX* model class but not the more common NARX model class, unless all of the functions $f_\ell(\cdot)$ are linear, thus reducing the projection-pursuit model to a linear autoregressive one. Arguably the simplest possible nonlinear projection-pursuit NARX* model is that obtained by setting $r = 1$, $f_1(x) = x^2$, $p = 1$, and $q = 0$. Explicitly, this model is given by

$$\begin{aligned} y(k) &= [a_1 y(k-1) + b_0 u(k)]^2 \\ &= a_1^2 y^2(k-1) + 2a_1 b_0 u(k) y(k-1) + b_0^2 u^2(k), \end{aligned} \tag{4.121}$$

which may be recognized as a special case of the polynomial NARMAX model defined in Eq. (4.99).

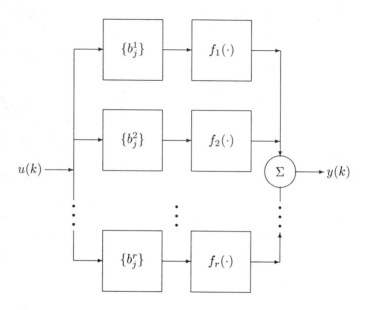

Figure 4.25: Dual Uryson structure of projection-pursuit NMAX models

Viewed as a class of NARMAX models, the projection-pursuit family appears to be new. Given the number of model structures that have Uryson representations—for example, additive NMAX models (Sec. 4.4) and diagonal Volterra models (Chapter 5)—and the fact that Wiener models seem to exhibit more flexible dynamic behavior than Hammerstein models, the class of dual Uryson models appears to be an intriguing one for future investigation. More immediately, however, the class of projection-pursuit NARMAX models provides a convenient step toward the class of NARMAX models based on neural networks.

4.7.2 Neural network models

The literature on neural networks is enormous and these models are only considered peripherally here; for a more detailed introduction to the basic ideas, see the classic reference by Rumelhart et al. (1986) or for a more application-oriented introduction, refer to Narendra and Parthasarathy (1990) or Su and McAvoy (1993). Although many different classes of artificial neural networks (ANNs) have been defined, the class of interest here is the class of feedforward networks which implement static maps $F : R^m \to R^n$ between vector spaces of arbitrary dimensions m and n. The most popular feedforward networks are three-layered structures that map an input vector $\mathbf{x} \in R^m$, first to an intermediate or *hidden layer* vector $\mathbf{v} \in R^p$ and ultimately into an output vector

NARMAX Models

$\mathbf{y} \in R^n$. Each of these transformations is formed from projection-pursuit mappings. Specifically, the input-layer mapping in a feedforward network is of the form

$$v_i = f_i\left(\sum_{j=1}^{m} w_{ij} x_j\right), \qquad (4.122)$$

where $f_i(x)$ is the *squashing function*

$$f_i(x) = s(x - \theta_i) \qquad s(x) = \frac{e^x}{e^x + 1}. \qquad (4.123)$$

The constants $\{\theta_i\}$ are called *bias weights*, and they, together with the *synaptic weights* w_{ij}, define the transformation from the input vector $\mathbf{x}$ to each component of the hidden vector $\mathbf{v}$. A second transformation of exactly the same structure maps the hidden vector $\mathbf{v}$ into the output vector $\mathbf{y}$. Combining these transformations, the relationship between the components x_ℓ of the input vector and y_i of the output vector may be expressed as

$$y_i = f_i\left(\sum_{j=1}^{p} h_{ij} g_j \left[\sum_{\ell=1}^{m} w_{j\ell} x_\ell\right]\right). \qquad (4.124)$$

It is possible to define *neural NARMAX models* based on feedforward networks by taking $n = 1$ and defining the input vector $\mathbf{x}$ as

$$\mathbf{x} = [y(k-1), \ldots, y(k-p), u(k), u(k-1), \ldots, u(k-q)]^T, \qquad (4.125)$$

where $y(k)$ is the scalar output from the network. Networks of this type have been considered by a number of authors (Narendra and Parthasarathy, 1990; Su and McAvoy, 1993) and they can be quite flexible, a point discussed further in Sec. 4.7.3. Conversely, it should be clear that mappings like those defined by Eq. (4.124) are complex enough in structure to be very difficult to understand intuitively. The general observations concerning nonlinear autoregressive behavior versus nonlinear moving average behavior discussed earlier in this chapter are applicable, and these observations may be extremely useful in deciding how to choose the input vector $\mathbf{x}$ for a particular application, but beyond general observations of this sort, it is difficult to say much about qualitative behavior.

The following *simplest possible neural NARX* model* illustrates this point. The term *recurrent network* refers to networks that take one or more components of the output vector $\mathbf{y}$ and feed them back as components of the input vector $\mathbf{x}$. In the context of neural network-based NARMAX models, *nonrecurrent networks* would correspond to neural network-based NMAX models, whereas *recurrent networks* exhibit nonlinear autoregressive dependence. The simplest possible example of this dependence would be based on a neural network mapping the input vector $\mathbf{x} = [y(k-1), u(k)]$ into the scalar output $y(k)$ through a "single unit hidden layer" [i.e., $p = 1$ in Eq. (4.124)]. To simplify the subsequent discussion slightly, consider the following *modified squashing function*

$$s^*(x) = \tanh x = 2s(2x) - 1. \qquad (4.126)$$

Note that the function $s^*(x)$ is a saturation nonlinearity with exactly the same general character as $s(x)$, but it maps R^1 into the interval $(-1, 1)$ instead of $(0, 1)$; in particular, note that $s^*(0) = 0$ whereas $s(0) = 1/2$. Adopting this modified squashing function, the neural network-based NARMAX model defined verbally above corresponds to the nonadditive, nonpolynomial, nonrational NARMAX model:

$$y(k) = \tanh\left(\tanh[a_1 y(k-1) + b_0 u(k) - \theta_1] - \theta_2\right). \quad (4.127)$$

It should be clear from this expression that, even in this almost trivially simple case (relative to the neural network architectures employed in practice), it is difficult to say much about the qualitative behavior of this model. One interesting insight that can be gained is that the condition "$u(k) = 0$ for all k implies $y(k) = 0$ for all k" reduces Eq. (4.127) to

$$\tanh\left(\tanh[-\theta_1] - \theta_2\right) = 0 \quad \Rightarrow \quad \theta_2 = -\tanh\theta_1. \quad (4.128)$$

Note that if $\theta_1 = \theta_2 = 0$, this condition is automatically satisfied, but if θ_1 and θ_2 are determined empirically, as they generally are in practice, it is unlikely that condition (4.128) will be satisfied. Analogous, but more complex, conditions will be required for more complex networks to satisfy this input-output condition.

It is important to emphasize that *typical* neural network models are much more complex than the example just described. As a specific example, Saravanan et al. (1993) describe a neural network model developed for the main engine of the space shuttle. They develop a recursive neural network model that predicts four output variables at time k from their past values at time $k-1$, along with two input values at time $k-1$. The neural network is a three-layer feedforward network like the ones described above, and it is parameterized by a total of 136 weights. Models of somewhat greater but qualitatively comparable complexity are considered by Narendra and Parthasarathy (1990), who consider six examples and use four-layer feedforward networks with 20 and 10 nodes in the two hidden layers. A typical example is a single-input, single-output network that approximates the unknown nonlinearity in a Hammerstein model with known second-order linear dynamics (Narendra and Parthasarathy, 1990, Example 1). This nonlinearity satisfies the condition $f(0) = 0$ and the network considered sets the bias weights to zero, so this condition is automatically satisfied. The nonlinearity is a sum of three sinusoids and the neural network that approximates it involves 230 weight parameters. The other five examples yield models of comparable complexity.

Overall, then, detailed quantitative analysis of neural network behavior appears to be too difficult to attempt. For this reason, little more will be said here about neural network models beyond the discussion of Cybenko's approximation results given in the next section. The one exception is the following stability result, which is revisited briefly in Sec. 4.8:

Neural network-based NARMAX models are BIBO stable.

This result follows directly from the fact either of the squashing functions $s(x)$ or $s^*(x)$ considered here map the entire real line into a bounded interval.

4.7.3 Cybenko's approximation result

One reason for the popularity of neural networks is that they are quite flexible in the range of functional behavior they can exhibit. A specific statement of this flexibility is the approximation result of Cybenko (1989), which establishes the ability of neural networks to approximate arbitrary continuous functions on a closed, bounded domain. Specifically, Cybenko considers *continuous* functions $f : I^N \to R^1$, where $I^n = [0,1]^n$ is the unit cube in R^n, and shows that they may be approximated arbitrarily well by functions of the form

$$G(\mathbf{x}) = \sum_{j=1}^{N} \alpha_j \sigma(\mathbf{c}_j^T \mathbf{x} + \theta_j). \qquad (4.129)$$

That is, he establishes sufficient conditions on the squashing function $\sigma : R^1 \to [0,1]$ such that, given $\epsilon > 0$, there exists N, $\{\alpha_j\}$, $\{\mathbf{c}_j\}$, and $\{\theta_j\}$ such that $|G(\mathbf{x}) - f(\mathbf{x})| < \epsilon$ for all $\mathbf{x} \in I^n$. One sufficient condition is that the function $\sigma(t)$ be sigmoidal, satisfying the condition

$$\sigma(t) \to \begin{cases} 1 & \text{as } t \to \infty \\ 0 & \text{as } t \to -\infty. \end{cases} \qquad (4.130)$$

Essentially, Cybenko's results establish that neural networks are as flexible as polynomials with regard to approximating continuous functions. That is, the Weierstrass theorem (Klambauer, 1975, p. 334) establishes the ability of polynomials to uniformly approximate any continuous function on a closed, bounded domain like I^n. In interpreting either of these results, it is important to keep two points in mind. First, the requirement that the domain (I^n in Cybenko's result) be compact (i.e., closed and bounded) is both crucial and restrictive. In particular, neither Cybenko's result nor the Weierstrass approximation theorem extend to the whole real line; hence, if quality of approximation over a fixed range is a reasonable measure of performance for a function $F : R^n \to R^1$, then either polynomials or neural networks may provide useful results. In dynamic modeling applications, this restriction implies that the range of both input and output variables must be known *a priori* for Cybenko's result to apply.

The second point to note here is that both Cybenko's result and the Weierstrass theorem are *approximation* results, and do *not* imply that a particular *qualitative* behavior can be exhibited. For example, it has already been noted that only linear polynomials of the form $p(x) = p_1 x$ can exhibit homogeneous behavior. Hence, although the Weierstrass approximation theorem guarantees that a homogeneous map $h : R^1 \to R^1$ can be *approximated* arbitrarily well on any fixed compact set S by a polynomial of sufficiently high degree, this approximating function cannot be homogeneous. This observation is closely related to the fact that polynomial approximations of saturation nonlinearities are typically not monotonic, since the only way a polynomial can closely approximate a constant saturation limit is to oscillate about that limit. Analogous results can be expected for neural network approximations, although the details will be different than for polynomial approximations.

4.7.4 Radial basis functions

Finally, another approach to the development of dynamic models that appears to be growing in popularity is the use of radial basis functions (RBFs). In terms of complexity, the resulting models are roughly comparable to neural networks, although their structure is somewhat different, and the model parameters appear to be somewhat easier to estimate from empirical data. Essentially, radial basis functions are maps $F: R^n \to R^1$ that have the specific form (Chen et al., 1992; Pottmann and Seborg, 1992):

$$F(\mathbf{x}) = \sum_{i=1}^{M} a_i \phi(\|\mathbf{x} - \mathbf{c}_i\|). \tag{4.131}$$

Here, the M vectors $\mathbf{c}_i \in R^n$ are called *centers* because the individual terms $\phi(\|\mathbf{x} - \mathbf{c}_i\|)$ appearing in the sum depend only on the radial distance from these points in R^n. Many different choices are possible for the scalar function $\phi(\cdot)$; two specific examples are the thin-plate spline (Chen et al., 1992)

$$\phi(r) = r^2 \ln r \tag{4.132}$$

and the reciprocal multiquadric function (Pottmann and Seborg, 1992)

$$\phi(r) = (r^2 + \beta)^{-1/2} \tag{4.133}$$

for some $\beta \geq 0$.

To apply radial basis functions to dynamic modeling problems, define the vector $\mathbf{v}(k)$:

$$\mathbf{v}(k) = [y(k-1), ..., y(k-p), u(k), ..., u(k-q)]^T \tag{4.134}$$

and consider NARMAX models of the form (Chen et al., 1992; Pottmann and Seborg, 1992)

$$y(k) = a_0 + \sum_{i=1}^{M} a_i \phi(\|\mathbf{v}(k) - \mathbf{c}_i\|). \tag{4.135}$$

As a typical example, Pottmann and Seborg (1992) consider the CSTR model discussed repeatedly throughout this book. They obtain reasonable approximations to the process dynamics with a radial basis function model based on the vector

$$\mathbf{v}(k) = [y(k-1), y(k-2), y(k-3), u(k-1), u(k-2), u(k-3)]^T \tag{4.136}$$

and the reciprocal multiquadric function defined in Eq. (4.133). They describe a number of different models, typically involving between 10 and 30 centers $\mathbf{c}_i$. Chen et al. (1992) discuss three examples in which they use the thin-plate spline function defined in Eq. (4.132). In the first example, they consider a simulated time series and build a prediction model based on the two most recent output observations, obtaining reasonable results with $M = 30$. The second example is based on experimental data, relating a pump motor voltage to the water level in a tank. Reasonable results are obtained with model based on a vector $\mathbf{v}(k)$ of dimension 8, using 40 centers $\mathbf{c}_j$. The third example involves modeling the dynamics of a heat exchanger, based on a vector $\mathbf{v}(k)$ of dimension 12 with 90 centers $\mathbf{c}_j$. The key point is that RBF models, like neural network models, are significantly more complex in structure than the models of primary interest in this book.

4.8 Summary: the nature of NARMAX models

As an aid in understanding the NARMAX class, this chapter has considered the following five structurally-defined subclasses:

1. NMAX (nonlinear moving average) models
2. NARX (nonlinear autoregressive) models
3. NAARX (structurally additive) models
4. Polynomial NARMAX models
5. Rational NARMAX models.

Probably the most significant qualitative differences arise between the NMAX models and the non-NMAX models that include nonlinear autoregressive terms. This point may be seen clearly by comparing the BIBO stability results presented in Sec. 4.2 for the NMAX model class with the (much weaker) analogous results for the NARX and NARX* model families presented in Sec. 4.3.

Interestingly, the structurally additive model class considered in Sec. 4.4 seems to represent a middle ground, yielding BIBO stability results that are strictly stronger than those for the more general NARX and NARX* models considered in Sec. 4.3, but strictly weaker than those for the NMAX models considered in Sec. 4.2. In addition, it was shown in Sec. 4.4.4 that the *steady-state behavior* of any NMAX or NARX model can be matched exactly by a structurally additive model. Further, many NARMAX models exhibit approximately linear (and therefore structurally additive) responses to inputs of sufficiently small amplitude. Hence, the general consequences of imposing a structural additivity constraint on a NARMAX model will usually only appear in moderate or high intensity transient responses. Still, these differences can be significant, as exemplified by the behavioral differences between a Hammerstein model and its dual Wiener model: the Hammerstein model is structurally additive, whereas the Wiener model is not.

The polynomial NARMAX class seems to be a popular one, but even simple members of this class typically exhibit complex (and often undesirable) input-dependent behavior. This point was illustrated in Sec. 4.5.2 with a simple quadratic model that *appeared* to be stable over the range of inputs considered but was not. To illustrate and examine this behavior further, it is useful to consider the following simple example:

$$y(k) = \alpha y^n(k-1) + u(k). \tag{4.137}$$

Here, α is any nonzero real constant, and n is any integer greater than 1; note that for $n = 1$, this model reduces to the standard first-order linear autoregressive model discussed in Chapter 2. For $n > 1$, this model is a first-order nonlinear autoregressive model belonging to the last four of the five model classes considered here, but not the first (i.e., this model has no NMAX representation).

First, note that this model exhibits output multiplicity: given a steady-state input value u_s, the possible output values are given by the real roots of the polynomial

$$\alpha y_s^n - y_s + u_s = 0. \tag{4.138}$$

Note that this polynomial has n roots, although they need not all be real or distinct. For $u_s = 0$, the general solution is $y_s = 0$ (one root), or $y_s^{n-1} = (1/\alpha)$, corresponding to $n - 1$ roots altogether, of which at most two are real. For the special case $n = 2$, the general solution is

$$y_s = \frac{1 \pm \sqrt{1 - 4\alpha u_s}}{2\alpha}. \tag{4.139}$$

Note that for $\alpha u_s < 1/4$, both of these roots are real, implying the existence of two steady-state solutions, whereas the steady-state solution is unique for $u_s = 1/4\alpha$ (i.e., $y_s = 1/2\alpha = 2u_s$), and *no* steady-state solution exists for $\alpha u_s > 1/4$.

As noted repeatedly, input-dependent qualitative behavior is typical of polynomial NARMAX models. This point is illustrated clearly in the response of the model (4.137) to an impulse of amplitude λ. It is not difficult to show by induction that this response is

$$y(k) = \alpha^{\gamma(k)} \lambda^{n^k}, \tag{4.140}$$

where $\gamma(k)$ is given by

$$\gamma(k) = \sum_{i=0}^{k-1} n^i = \frac{n^k - 1}{n - 1}. \tag{4.141}$$

Now, define $\nu(k)$ as

$$\nu(k) = \frac{\gamma(k)}{n^k} = \frac{1 - n^{-k}}{n - 1}, \tag{4.142}$$

and note that

$$0 < \nu(k) < \frac{1}{n - 1} \leq 1 \tag{4.143}$$

for all k. The impulse response may then be rewritten as

$$y(k) = [\alpha^{\nu(k)} \lambda]^{n^k}. \tag{4.144}$$

From this expression, it follows that if $|\alpha^{\nu(k)}\lambda| < 1$ for all k, the impulse response will be stable, decaying to zero *much* more rapidly than the corresponding linear model (i.e., compare the exponent n^k here with the usual linear exponent k). As a specific example, note that the following conditions are sufficient to guarantee the stability of the impulse response $y(k)$:

$$|\alpha| < 1 \qquad |\lambda| < 1. \tag{4.145}$$

A striking feature of this result—valid for all $n > 1$—is that it represents the well-known sufficient condition for the stability of the first-order linear model, *augmented by a restriction on the input amplitude*. Conversely, in the nonlinear case, it is also possible to exploit this dependence, trading off the input restriction and the model parameter restriction. For example, taking $\alpha = 2$, it is not difficult to show that the impulse response is stable provided that

$$|\lambda| < 2^{-1/(n-1)}. \tag{4.146}$$

Note that this restriction becomes *less* severe as the order n of the nonlinearity increases. Finally, if $|\alpha^{\nu(k)}\lambda| > 1$ for all k, note that the impulse response will be unstable.

More generally, it is possible to derive a sufficient condition for the stability of this model that is in the same spirit as that developed by Lee and Mathews (1994) for the bilinear model. Specifically, assume the following conditions are satisfied:

$$|u(k)| \leq M < 1 \qquad |\alpha| < 1 - M. \tag{4.147}$$

It follows from the triangle inequality (Abramowitz and Stegun, 1972) that

$$\begin{aligned} |y(k)| &= |\alpha y^n(k-1) + u(k)| \\ &\leq |\alpha||y(k-1)|^n + |u(k)| \\ &\leq (1-M)|y(k-1)|^n + M. \end{aligned} \tag{4.148}$$

From this result, it follows immediately that if $|y(k-1)| \leq 1$, then $|y(k)| \leq 1$; since $|y(0)| = |u(0)| \leq M < 1$, it follows that $|y(k)| < 1$ for all k if the conditions given in Eq. (4.147) are satisfied. Like the result of Lee and Mathews for bilinear models, this result is strictly weaker than BIBO stability: given the *fixed* bound M on the input sequence, conditions are developed under which the nonlinearity is "weak enough" to guarantee that the response remains bounded. In fact, results of this character are the best possible, since the behavior of the impulse response illustrates that polynomial NARMAX models are *not* BIBO stable, in general.

To conclude this example, it bears repeating that even if the response of a polynomial NARMAX model is bounded, it need not be "well behaved," even in the simplest of cases. This point is illustrated in Figs. 4.26 and 4.27, which show four different step responses each for the model considered here with $n = 2$ and $n = 3$, respectively, and $\alpha = -0.3$. Fig. 4.26 shows the

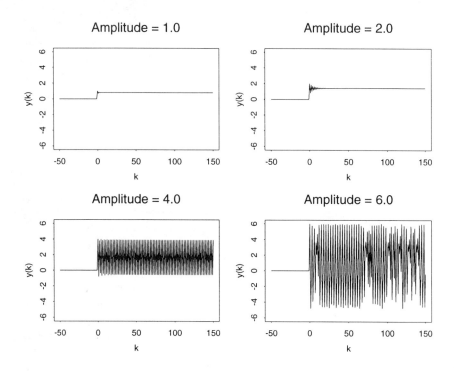

Figure 4.26: Four step responses, $n = 2$

responses to steps of amplitude $\lambda = 1.0$, 2.0, 4.0, and 6.0 for the quadratic model ($n = 2$), which exhibits the same qualitative behavior as the quadratic example considered in Sec. 4.3.2. Specifically, the step response becomes more oscillatory with increasing input amplitude, finally becoming chaotic for $\lambda = 6$. In fact, the step response is unstable for $\lambda = 7$. Fig. 4.27 shows similar plots for the cubic model ($n = 3$), for $\lambda = 1.00$, 1.40, 1.80, and 1.98. Again, the response becomes more oscillatory with increasing input amplitude, becoming unstable for $\lambda = 2$, but here it does not appear to exhibit chaotic responses.

The general conclusion drawn from this example and the others considered in Sec. 4.5 is that polynomial NARMAX models often seem to exhibit input-dependent instabilities. In addition, these examples have also illustrated that these models can exhibit other forms of exotic behavior, like persistent oscillations or deterministic chaos in their step responses. When this type of behavior is not seen in the process of interest, it is probably undesirable in an approximate model of the process dynamics. A number of alternatives may be considered to overcome these difficulties. First, recall that the BIBO stability of neural network models was briefly discussed in Sec. 4.7. This result follows directly from the fact that neural networks can only implement *bounded* maps $F(\cdots)$, so any NARMAX model based on a neural network is necessarily BIBO stable. This idea may be extended to other model classes, and it has been. For example, Hernandez and Arkun (1993) develop a modification of polynomial

NARMAX Models

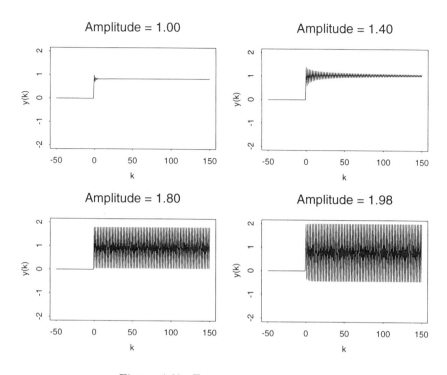

Figure 4.27: Four step responses, $n = 3$

NARMAX models in which the usual polynomial map $F(\cdots)$ is followed by a squashing function like that used in neural networks. The effect of this saturation nonlinearity is to produce a (non-polynomial) bounded map $B(\cdots)$, defined over the same domain as the original multivariate polynomial $F(\cdots)$. Another alternative is to note that whereas polynomials are necessarily unbounded, rational functions *can* be bounded. As a specific example, note that the following rational NARMAX model is BIBO stable, provided $d > 0$:

$$y(k) = \frac{ay(k-1) + cy^2(k-1)}{1 + dy^2(k-1)} + bu(k). \tag{4.149}$$

This result follows from the fact that the rational function

$$f(x) = \frac{ax + cx^2}{1 + dx^2} \tag{4.150}$$

is bounded for $d > 0$. That is, since $|f(x)| < M$ for some finite M for all x, it follows that $|y(k)| \leq M + |b||u(k)|$, thus establishing BIBO stability for this example.

The next two chapters build on the results presented here, specializing them to two other important subsets of the NARMAX family. Specifically, Chapter 5 considers the class of Volterra models, arguably the prototype of the NMAX model class. When NMAX models are appropriate (e.g., when there are no output multiplicities, conditional stability considerations, or other "complicated" steady-state or dynamic phenomena), Volterra models offer some significant advantages with respect to analysis, model identification, and control. When NMAX models are inappropriate, however (e.g., whenever the above pathologies do arise), the class of linear multimodels considered in Chapter 6 may be a reasonable alternative. These models are more subtle than they appear, however, and the results presented here for NARMAX models will be quite useful in understanding the nature of linear multimodels. In particular, one of the important features of linear multimodels is that they involve a local model selection process, and the details of this process determine whether the multimodel exhibits NMAX-like behavior, NARX-like behavior, or something more complex.

Chapter 5

Volterra Models

One of the main points of Chapter 4 is that nonlinear moving-average (NMAX) models are both inherently better-behaved and easier to analyze than more general NARMAX models. For example, it was shown in Sec. 4.2.2 that if $g(\cdots)$ is a continuous map from R^{q+1} to R^1 and if $y_s = g(u_s, \ldots, u_s)$, then $u_k \to u_s$ implies $y_k \to y_s$. Although it is not always satisfied, continuity is a relatively weak condition to impose on the map $g(\cdots)$. For example, Hammerstein or Wiener models based on moving average models and the hard saturation nonlinearity

$$g(x) = \begin{cases} -1 & x < 0 \\ 0 & x = 0 \\ +1 & x > 0, \end{cases} \tag{5.1}$$

represent discontinuous members of the class of NMAX models. This chapter considers the analytical consequences of requiring $g(\cdot)$ to be *analytic*, implying the existence of a Taylor series expansion. Although this requirement is much stronger than continuity, it often holds, and when it does, it leads to an explicit representation: Volterra models.

The principal objective of this chapter is to define the class of Volterra models and discuss various important special cases and qualitative results. Most of this discussion is concerned with the class $V_{(N,M)}$ of *finite Volterra models*, which includes the class of linear finite impulse response models as a special case, along with a number of practically important nonlinear moving average model classes. In particular, the finite Volterra model class includes Hammerstein models, Wiener models, and Uryson models, along with other more general model structures. In addition, one of the results established in this chapter is that most of the bilinear models discussed in Chapter 3 may be expressed as infinite-order Volterra models. This result is somewhat analogous to the equivalence between finite-dimensional linear autoregressive models and infinite-dimensional linear moving average models discussed in Chapter 2. The bilinear model result presented here is strictly weaker, however, since there exist classes of bilinear models that *do not* possess Volterra series representations. Specifically, it is shown in Sec. 5.6 that *completely bilinear models* do not exhibit Volterra series representations. Conversely, one of the results discussed at the end of this chapter is that the class of discrete-time *fading memory* systems may be approximated arbitrarily well by finite Volterra models (Boyd and Chua, 1985). The main point of this chapter is that the class of Volterra models is an extremely flexible and useful one, particularly in attempting to distinguish between linear behavior and *mildly nonlinear* behavior, for which Volterra models represent a useful prototype.

5.1 Definitions and basic results

Because the focus of this book is on discrete-time nonlinear models, the class of Volterra models discussed here is based on the discrete-time series representation presented below. Historically, however, the discrete-time Volterra model evolved as a discrete-time version of the continuous-time infinite series representation of the Volterra integral operator

$$V[u(t)] = \int_0^t k[t, s, u(s)] ds, \tag{5.2}$$

where $k(t, s, u(s))$ is called the *Volterra kernel*. This series representation is of the form

$$V[u(t)] = \sum_{i=1}^{\infty} \int_0^t \cdots \int_0^t k_i(t_1, \ldots, t_i) u(t - t_1) \cdots u(t - t_i) dt_1 \cdots dt_i \tag{5.3}$$

and is called the Volterra series in honor of Vito Volterra, who studied integral equations involving operators of this form in the first half of this century (Corduneanu, 1991). Hammerstein and Uryson studied similar integral equations that are closely related to the Hammerstein and Uryson models introduced in Chapter 1.

5.1.1 The class $V_{(N,M)}$ and related model classes

The class of *discrete-time Volterra models* considered here is a subset of the class of (infinite-order) nonlinear moving average models, defined by an infinite series analogous to this continuous-time Volterra series. Since *finite-order Volterra models* are also of considerable interest, the discussion of both of these model classes begins with the finite-order case and infinite-order models are treated as limiting cases of these finite-order models. More specifically, a *Volterra model of nonlinear order N and dynamic order M* is defined by the following expression:

$$y(k) = \alpha_0 + \sum_{n=1}^{N} v_M^n(k), \tag{5.4}$$

$$v_M^n(k) = \sum_{i_1=0}^{M} \sum_{i_2=0}^{M} \cdots \sum_{i_n=0}^{M} \alpha_n(i_1, i_2, \ldots, i_n) u(k - i_1) u(k - i_2) \cdots u(k - i_n).$$

For simplicity, this class of finite Volterra models is denoted by the symbol $V_{(N,M)}$ to make the nonlinear order N and dynamic order M explicit. These models belong to the following two NARMAX classes:

1. The class of polynomial NARMAX models of order N
2. The class of nonlinear moving average models of order M.

Both of these observations are useful, particularly the second, since it implies that finite Volterra models inherit all of the advantages and/or disadvantages of the nonlinear moving average class of models discussed in Chapter 4. Thus, for example, finite Volterra models cannot exhibit subharmonic generation or output multiplicities, although they can exhibit input multiplicities. Further, since Volterra models depend continuously on $u(k-i)$ for $i = 0, 1, \ldots, M$, the "well-behavedness theorem" presented in Chapter 4 (Sec. 4.2.2) for NMAX models applies here, thus establishing the BIBO stability of these models and asymptotically constant responses to asymptotically constant input sequences.

Infinite-order Volterra models may be defined in three different ways, taking the limit $N \to \infty$, taking the limit $M \to \infty$, or taking both limits simultaneously. These limiting model classes are denoted $V_{(\infty,M)}$, $V_{(N,\infty)}$, and $V_{(\infty,\infty)}$, respectively, and are discussed further in Sec. 5.5. One motivation for the consideration of infinite-order models comes from the consideration of NMAX models of the form

$$y(k) = G(u(k), u(k-1), ..., u(k-M)), \tag{5.5}$$

where $G(\cdots)$ is an *analytic* mapping from R^{M+1} to R^1. In this case, $G(\cdots)$ may be represented as a Taylor series to obtain a Volterra model of the limiting class $V_{(\infty,M)}$. Although this representation is not directly useful in developing empirical models, it can be extremely useful in establishing connections between different model classes. This point is discussed further in Sec. 5.6.

It is also useful to define the following four subclasses of $V_{(N,M)}$:

1. $V^0_{(N,M)}$, the class of Volterra models with $\alpha_0 = 0$
2. $MA(M)$, the class of linear moving average models of order M
3. $\mathcal{N}_N$, the class of Nth-order polynomial static nonlinearities
4. $\mathcal{O}_{(N,M)}$, the class of odd-symmetry Volterra models.

All of these model classes belong to the general class of *pruned Volterra models* introduced in Sec. 5.4. To motivate the definition of the class $V^0_{(N,M)}$, note that the response to the zero input sequence $u(k) = 0$ for all k is $y(k) = \alpha_0$ for all k. Control applications frequently make use of *incremental models*, in which the input and output sequences under consideration are deviations from specified reference values (typically, some specified steady state). In such cases, the zero input sequence corresponds to "no deviation from the input reference value" and the desired response is usually "no deviation from the output reference value," or $y(k) = 0$ for all k. Clearly, this situation only occurs if $\alpha_0 = 0$. Since the assumption that $u(k) \equiv 0$ implies $y(k) \equiv 0$ is often invoked in practice, it is useful to have a special notation for the class of Volterra models with $\alpha_0 = 0$, thus leading to the above definition of the model class $V^0_{(N,M)}$.

Another extremely important subset of $V_{(N,M)}$ is the class of linear moving average models. In particular, note that the class $MA(q)$ of qth-order linear moving average models defined in Chapter 2 corresponds to $V^0_{(1,q)}$. Since all time-invariant linear models may be represented as infinite-order linear moving average models—that is, they are completely characterized by their impulse responses—it follows that the class $\mathcal{L}$ of linear time-invariant models is equivalent to $MA(\infty)$, or alternatively, $V^0_{(1,\infty)}$. Conversely, any polynomial of order N may be viewed as a member of the class $V_{(N,0)}$, so the class $\mathcal{N}$ of *analytic* static nonlinearities is equivalent to $V_{(\infty,0)}$. It follows from these results that both Hammerstein and Wiener models based on analytic nonlinearities and general linear models belong to the class $V_{(\infty,\infty)}$, as shown in Sec. 5.5.

It is also useful to define the class of *odd-symmetry Volterra models*, denoted $\mathcal{O}_{(N,M)}$. These models are members of $V_{(N,M)}$ in which only terms of odd order have nonzero coefficients. An important characteristic of this class of models is that if $y(k)$ is the response to the input sequence $\{u(k)\}$ then $\{-y(k)\}$ is the response to the input sequence $\{-u(k)\}$. Similarly, it is not difficult to show that odd-order Volterra models can generate odd harmonics but not even harmonics. Also, odd-order Volterra models do not exhibit the rectification phenomenon discussed in Chapter 1 (Sec. 1.2.1). Since α_0 is an even-order term, it is necessarily zero in an odd-order Volterra model, so these models also belong to the class $V^0_{(N,M)}$. Further, since linear terms are of order 1, the class of odd-order Volterra models includes the class of linear moving average models. More generally, note that the following inclusions hold:

$$MA(M) \subset \mathcal{O}_{(N,M)} \subset V^0_{(N,M)} \subset V_{(N,M)}.$$

The class $\mathcal{N}_N$ does not lie in this chain of inclusions and may be regarded loosely as "orthogonal" to them.

Finally, an extremely useful result is the following: the class $V_{(N,M)}$ is closed under parallel combinations. In particular, consider the parallel combination of a $V_{(N_1,M_1)}$ model defined by coefficients $\{\alpha_n(i_1,\ldots,i_n)\}$ and a $V_{(N_2,M_2)}$ model defined by coefficients $\{\beta_n(i_1,\ldots,i_n)\}$. In a parallel combination, both component models are driven by the same input sequence $\{u(k)\}$ and their outputs are summed. Hence, it follows directly from Eq. (5.4) that the response of the parallel combination is given by

$$y(k) = \gamma_0 + \sum_{n=1}^{N'} v^n_{M'}(k), \tag{5.6}$$

$$v^n_{M'}(k) = \sum_{i_1=0}^{M'} \sum_{i_2=0}^{M'} \cdots \sum_{i_n=0}^{M'} \gamma_n(i_1, i_2, \ldots, i_n) u(k-i_1) u(k-i_2) \cdots u(k-i_n).$$

Volterra Models 213

where $N' = \max\{N_1, N_2\}$ and $M' = \max\{M_1, M_2\}$. The constant term in this model is simply $\gamma_0 = \alpha_0 + \beta_0$ and the other model coefficients are given by

$$\gamma_n(i_1, \ldots, i_n) = \alpha_n(i_1, \ldots, i_n) + \beta_n(i_1, \ldots, i_n), \tag{5.7}$$

where undefined terms are taken as zero (e.g., $\alpha_n(i_1, \ldots, i_n)$ for $n > N_1$). This result facilliates the discussions of Uryson and projection-pursuit models given in Sec. 5.2. Further, it is not difficult to show that the model classes $V^0_{(N,M)}$, $MA(M)$, $\mathcal{N}_N$, and $\mathcal{O}_{(N,M)}$ are also closed under parallel combinations.

5.1.2 Four simple examples

One of the big advantages of Volterra models is that their structure makes them amenable to much analysis, and one of the principal objectives of this chapter is to illustrate this point. Another big advantage of this class of models is that it includes many other interesting model classes as special cases, and another key objective of this chapter is to illustrate this point as well. As a first step toward these objectives, consider the following four members of the class $V_{(2,1)}$:

$$\begin{align}
\textbf{Model 1} \quad y(k) &= u(k) + u(k-1) \tag{5.8} \\
\textbf{Model 2} \quad y(k) &= u(k) + u(k-1) + u^2(k) + u^2(k-1) \tag{5.9} \\
\textbf{Model 3} \quad y(k) &= u(k) + u(k-1) + u^2(k) + u^2(k-1) \\
&\quad + 2u(k)u(k-1) \tag{5.10} \\
\textbf{Model 4} \quad y(k) &= u(k) + u(k-1) + 2u(k)u(k-1). \tag{5.11}
\end{align}$$

The first of these examples is a simple linear model, included for reference, whereas the other three are nonlinear. Specifically, Model 2 is a Hammerstein model, Model 3 is a Wiener model, and Model 4 is a special case of Robinson's Volterra model, discussed in Chapter 4. Comparing the responses of these models to the four standard input sequences introduced in Chapter 2 provides some useful insights into the similarities and differences between these second-order Volterra models.

Before comparing these model responses, however, it is useful to say something about the relationship between the models themselves. First, note that the Hammerstein and Wiener models considered here are duals, both based on the linear model defined in Eq. (5.8) and the quadratic nonlinearity:

$$f(x) = x + x^2. \tag{5.12}$$

In addition, all of these models may be represented as a member of the class $V_{(2,1)}$ by the appropriate choice of coefficients in Eq. (5.4). Specifically, note that $\alpha_0 = 0$ for all four of these models, and that the *linear part* of all of these models is the same, that is

$$\alpha_1(0) = 1 \quad \alpha_1(1) = 1. \tag{5.13}$$

Table 5.1: Volterra model impulse and step responses

Model	Type	$y(0)$	$h(1)$	$s(1)$
1	Linear	α	α	2α
2	Hammerstein	$\alpha + \alpha^2$	$\alpha + \alpha^2$	$2\alpha + 2\alpha^2$
3	Wiener	$\alpha + \alpha^2$	$\alpha + \alpha^2$	$2\alpha + 4\alpha^2$
4	Robinson	α	α	$2\alpha + 2\alpha^2$

The only difference, then, is in the second-order coefficients $\alpha_2(i_1, i_2)$, of which there are four. Since multiplication is commutative, however, there is an inherent over-parameterization in the Volterra model defined above. In the particular case considered here, the quadratic term $v_2^2(k)$ in the Volterra model is

$$v_2^2(k) = \alpha_2(0,0)u^2(k) + [\alpha_2(0,1) + \alpha_2(1,0)]u(k)u(k-1) + \alpha_2(1,1)u^2(k-1). \quad (5.14)$$

Clearly, both this term and the behavior of the resulting Volterra model depend only on the sum $\alpha_2(0,1) + \alpha_2(1,0)$, and not on the individual values of these two parameters. Consequently, there is no loss of generality in assuming $\alpha_2(1,0) = \alpha_2(0,1)$, thus reducing the number of independent quadratic parameters in this model to three. The four models considered here correspond to the following choices for these three parameters:

1. $\alpha_2(0,0) = 0$, $\alpha_2(0,1) = 0$, $\alpha_2(1,1) = 0$
2. $\alpha_2(0,0) = 1$, $\alpha_2(0,1) = 0$, $\alpha_2(1,1) = 1$
3. $\alpha_2(0,0) = 1$, $\alpha_2(0,1) = 1$, $\alpha_2(1,1) = 1$
4. $\alpha_2(0,0) = 0$, $\alpha_2(0,1) = 1$, $\alpha_2(1,1) = 0$.

Because these models are members of the NMAX family, both impulse and step responses achieve their asymptotic values after two steps (i.e., zero for the impulse response, and the steady-state value y_s for the step response). These values are summarized in Table 5.1. The initial values for the impulse and step responses are the same since these input sequences are identical for times $k \leq 0$; the common value of these initial responses is labelled $y(0)$ in Table 5.1. For $k \geq 1$, the impulse and step responses differ and Table 5.1 gives the values of the impulse response $h(k)$ and the step response $s(k)$ for $k = 1$. For the impulse responses, note that $h(1) = y(0)$ for all four models, although the value of this common response varies from one model to another. Also, note that the linear model and Robinson's Volterra model exhibit identical impulse responses, as do the Hammerstein and Wiener models. Conversely, note that the steady-state responses of the three nonlinear models all differ from that of the linear model. Here, these steady-state values are identical for the Hammerstein

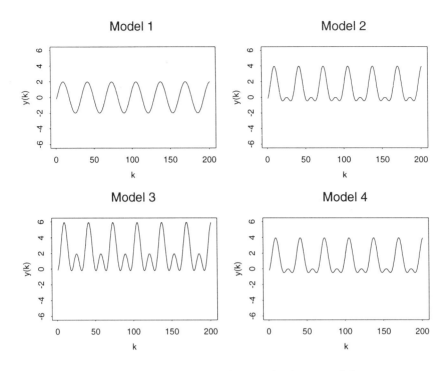

Figure 5.1: Responses to the standard sinusoidal input

model and Robinson's model, but different for the Wiener model. This result reinforces a point made earlier about the relationship between dual Wiener and Hammerstein models: their steady-state behavior is identical *provided that the steady-state gain of the linear subsystem is* 1. In this example, the steady-state gain of this subsystem (Model 1) is 2, so the Hammerstein and Wiener models exhibit different steady-state responses.

It is clear from these results that these four models cannot be distinguished on the basis of their impulse responses alone. They can be distinguished on the basis of their step responses, but plots of fixed-amplitude step responses are not immediately indicative of the pronounced nonlinearity of Models 2, 3, and 4. More direct evidence of this nonlinearity may be seen in the responses of these models to sinusoidal input sequences. Fig. 5.1 shows the responses of each of these four models to the standard sinusoidal input sequence defined in Chapter 2. The linear model response (Model 1) shown in the upper left plot in this figure is sinusoidal, as it must be for any linear model. The highly nonsinusoidal responses of the other three models gives immediate and dramatic evidence of their nonlinearity. As discussed in Chapter 4, because these models are members of the NMAX family, their response to a sinusoid of frequency f is necessarily also periodic with period $T = 1/f$. Because these nonlinear models all involve quadratic terms, however, it follows immediately from simple trigonometric identities that these responses also contain sinusoidal components of frequency $2f$, seen in the higher frequency "ripples" in all three of these responses.

In addition, all three of the nonlinear models exhibit the phenomenon of *rectification* discussed in Chapter 1: the responses all include a constant offset term that depends on the magnitude of the sinusoidal input excitation. This result also follows from simple trigonometric identities and accounts for the pronounced asymmetry of these responses: while the sinusoid exhibits equal positive and negative excursions, the nonlinear model responses exhibit much larger positive excursions than negative excursions. Finally, note that the Hammerstein model and Robinson's model exhibit *nearly* identical responses; in fact, these responses differ slightly, but these differences are not great enough to be visible in these plots.

5.1.3 Stochastic characterizations

Another advantage of Volterra models over general NARMAX models is that, as in the case of linear models, complete characterization of the response of a Volterra model to a stochastic input sequence is possible, at least in principle. In practice, stochastic input sequences have been used successfully in the identification of Volterra models (Koh and Powers, 1985; Pearson et al., 1992, 1996), especially white noise sequences. There, $\{u(k)\}$ is a sequence of statistically independent random fluctuations about a reference value $u_s = E\{u(k)\}$ (specifically, $\{u(k)\}$ is a sequence of independent, identically distributed random variables), and this assumption leads to significant algorithmic simplifications. As noted in the discussion of the class $\mathcal{O}_{(N,M)}$ in Sec. 5.1, it is common practice to develop incremental models that describe the dynamics of deviations from some consistent reference values (u_s, y_s); a popular choice of input reference value is $u_s = 0$. Further, it is also common to invoke a distributional symmetry assumption which, taken together with the zero-mean assumption $u_s = 0$ implies that $E\{u^n(k)\} = 0$ for all k and any odd integer n. The standard Gaussian white noise sequence defined in Chapter 2 is probably the best known member of this family of input sequences, but many other choices are possible, a point discussed further in Chapter 8.

Under these assumptions, the expected value of $y(k)$ may be computed explicitly as:

$$y_s = E\{y(k)\} = \alpha_0 + \sum_{n=1}^{N}\sum_{i=0}^{M} \alpha_n(i,i,\ldots,i)E\{u^n(k)\}$$

$$= \alpha_0 + \sum_{\text{even}}\sum_{i=0}^{M} \alpha_n(i,i,\ldots,i)E\{u^n(k)\}. \qquad (5.15)$$

The subscript "even" appearing in Eq. (5.15) indicates that the sum extends only over even integers, a result that follows from the even distributional symmetry assumed for $\{u(k)\}$. This result may be viewed as a quantitative characterization of the phenomenon of rectification discussed earlier in connection with

Robinson's second-order Volterra model. That is, note that $y_s \neq 0$ in general, even though $u_s = 0$. *For a particular input sequence centered about zero*, α_0 may be chosen to force $y_s = 0$, but in general, the mean response y_s to *other* zero-mean input sequences will be nonzero. This result means that the fluctuations of the input sequence are being converted to a constant offset by the even-order terms in the Volterra model.

As a corollary to this last observation, note that y_s will be identically zero for all symmetrically distributed, zero-mean white noise input sequences if and only if the following two conditions are satisfied:

$$\alpha_0 = 0 \tag{5.16}$$
$$\alpha_n(i,i,\ldots,i) = 0 \text{ for } n \text{ even and } i = 0,1,\ldots,M. \tag{5.17}$$

Recall that the class $V_{(N,M)}^0$ of Volterra models with $\alpha_0 = 0$ [Eq. (5.16)] was singled out specifically because it is the subset of $V_{(N,M)}$ for which the deterministic input sequence $u(k) \equiv 0$ yields the deterministic output sequence $y(k) \equiv 0$. The additional condition given in Eq. (5.17) here is necessary in the stochastic case to exclude the rectification phenomenon just described.

Eq. (5.17) may be satisfied in either of two ways. The first and perhaps most obvious way is to restrict consideration to the class of odd-symmetry Volterra models $\mathcal{O}_{(N,M)}$ defined and discussed earlier. Physically, this class of Volterra models is appropriate for systems whose responses to symmetrically distributed inputs are themselves symmetric. The other class of models for which Eq. (5.17) is satisfied consists of those models for which the *diagonal* terms of even order are identically zero. Robinson's Volterra model falls into this category, including the second-order term $u(k)u(k-1)$ but excluding the diagonal terms $u^2(k)$ and $u^2(k-1)$ that appear in both the Hammerstein and Wiener models considered in Sec. 5.1.2. This point is seen clearly in Fig. 5.2, which shows the responses of the four models discussed there to the standard white noise input sequence introduced in Chapter 2. In particular, both the Hammerstein and Wiener models (Models 2 and 3, respectively) clearly exhibit both rectification phenomena (i.e., nonzero means) and pronounced asymmetry in their distribution. In contrast, both the linear model (Model 1) and Robinson's Volterra model (Model 4) appear to be symmetrically distributed about a mean value of zero.

In this respect, Robinson's model is similar to the following one (Nikias and Petropulu, 1993, p. 452):

$$y(k) = u(k) + \beta u(k-1)u(k-2). \tag{5.18}$$

This model is particularly interesting because if $\{u(k)\}$ is a zero-mean Gaussian white noise sequence, $\{y(k)\}$ passes many of the usual tests for whiteness, even though it exhibits a non-trivial dependence structure (i.e., it is not white). This example was discussed in Pearson (1995) and Pearson and Ogunnaike (1997), but two points are worth noting here. First, the autocorrelation function $R_{yy}(m)$ for the response sequence $\{y(k)\}$ is identically zero for $m \neq 0$. This condition— or the equivalent one of flatness of the power spectrum—is often used to test for statistical independence, but this example illustrates that these tests are only

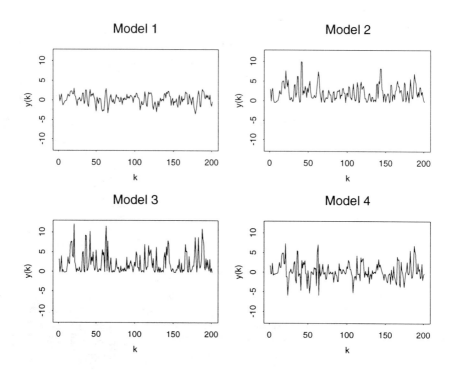

Figure 5.2: Responses to the standard white noise input

necessary conditions for independence, and not sufficient conditions. Similarly, the inherent nonlinearity of this model is not detected by the following criterion (Billings and Voon, 1986a):

$$E\{[y(k+m) - y_s][y(k) - y_s]^2\} = 0 \text{ for all } m, \tag{5.19}$$

since it may be shown with a little algebra that this expectation reduces to a sum of terms all involving odd powers of $u(k)$, and is therefore identically zero for all m, independent of the severity of the true nonlinearity (i.e., the magnitude of β).

5.2 Four important subsets of $V_{(N,M)}$

One undesirable characteristic of Volterra models is that the number of parameters required to specify an arbitrary element of the set $V_{(N,M)}$ grows combinatorially with M and N (for example, for $N = 4$ and $M = 20$, the number of required coefficients is 12,650). As a consequence, there is considerable interest in *structured Volterra models* having more efficient parameterizations. The following discussion describes four subsets of the class $V_{(N,M)}$, each corresponding to a different structural restriction.

5.2.1 The class $H_{(N,M)}$

The class $H_{(N,M)}$ is the class of finite-order Hammerstein models, obtained by combining the polynomial $g(\cdot)$ of order N

$$g(x) = \sum_{n=0}^{N} g_n x^n, \tag{5.20}$$

with the linear moving average model of order M

$$y(k) = \sum_{i=0}^{M} h(i)v(k-i). \tag{5.21}$$

For an input sequence $\{u(k)\}$, it follows that $v(k-i) = g(u(k-i))$, leading immediately to the Volterra representation

$$y(k) = \sum_{n=0}^{N} \sum_{i=0}^{M} g_n h(i) u^n(k-i). \tag{5.22}$$

The designation $H_{(N,M)}$ emphasizes that the resulting Hammerstein model is a member of the class $V_{(N,M)}$ of Volterra models.

In fact, this Volterra model corresponds to the "diagonal part" of the general Volterra series defined in Eq. (5.4). Specifically, define the class $D_{(N,M)}$ of *diagonal Volterra models* as the subset of $V_{(N,M)}$ satisfying

$$a_n(i_1, i_2, \ldots, i_n) = \begin{cases} \gamma_n(i_1) & i_1 = i_2 = \cdots = i_n, \\ 0 & \text{otherwise}, \end{cases} \tag{5.23}$$

for all $n > 0$. It follows from Eq. (5.22) that any Hammerstein model in the set $H_{(N,M)}$ is also a diagonal Volterra model with

$$\gamma_n(i) = g_n h(i) \quad \text{and} \quad \alpha_0 = \sum_{i=0}^{M} \gamma_0(i), \tag{5.24}$$

for $n = 0, 1, \ldots, N$. It is important to note that the Volterra series coefficients for a polynomial Hammerstein model are simply proportional to the impulse response coefficients $h(i)$ for the linear model, with a proportionality constant that depends on n. Consequently, the model class $H_{(N,M)}$ represents a highly constrained subset of the model class $D_{(N,M)}$.

5.2.2 The class $U^r_{(N,M)}$

The class $U^r_{(N,M)}$ consists of the r-channel Uryson models formed from N^{th}-order polynomial nonlinearities $g_j(x)$ and moving average linear models with finite impulse responses $h_j(i)$. (In interpreting this description, note that any polynomial of order $N' < N$ may also be represented as a polynomial of order N

by setting the highest-order $N - N'$ terms to zero; similarly, any linear moving average model of order $M' < M$ may also be regarded as a moving average model of order M by setting the terms of order i to zero for $i = M'+1, ..., M$.) As noted in Sec. 5.1, the parallel combination of Volterra models yields another Volterra model; together with the $H_{(N,M)}$ Volterra representation given in Eq. (5.22), this observation yields the following Volterra series representation for the Uryson model class $U^r_{(N,M)}$:

$$y(k) = \sum_{n=0}^{N} \sum_{i=0}^{M} \left[\sum_{j=1}^{r} g_n^j h_j(i) \right] u^n(k-i). \qquad (5.25)$$

For $r = 1$, the Uryson model reduces to the Hammerstein model, and this result reduces to Eq. (5.22). For $r > 1$, note that this model still exhibits diagonal structure. That is, Eq. (5.25) may be written in the form of Eq. (5.23) with

$$\gamma_n(i) = \sum_{j=1}^{r} g_n^j h_j(i) \quad \text{and} \quad \alpha_0 = \sum_{i=0}^{M} \gamma_0(i) \qquad (5.26)$$

for $n = 0, 1, ..., N$. Comparing this result with the diagonal representation for the Hammerstein model, the increased flexibility of the Uryson model is clear. Further, the following theorem establishes that any $U^r_{(N,M)}$ model with $r > N+1$ may be represented as an equivalent Uryson model with $N + 1$ channels.

Theorem:

Any model from the class $U^r_{(N,M)}$ with $r > N + 1$ is equivalent to a model from the class $U^{N+1}_{(N,M)}$.

Proof:

Define the $N + 1$-vectors

$$\mathbf{s}_i = [\gamma_0(i), \gamma_1(i), ..., \gamma_N(i)]^T \qquad \mathbf{g}_j = [g_0^j, g_1^j, ..., g_N^j]^T,$$

and note that Eq. (5.26) may be written as

$$\mathbf{s}_i = \sum_{j=1}^{r} h_j(i) \mathbf{g}_j.$$

This result expresses one vector of dimension $N + 1$ in terms of r others; if $r > N+1$, this combination cannot be linearly independent and may be simplified to one involving $N + 1$ basis vectors. This simplified combination defines an $N + 1$ channel Uryson model.

□

Volterra Models

This theorem establishes that within the class $V_{(N,M)}$, there is a class of r-channel Uryson models of maximal order $r = N + 1$. The representation of this model is not unique, but one natural choice is the monomial form shown in Fig. 5.3. This representation is particularly convenient since it establishes the equivalence between the class $U^r_{(N,M)}$ for $r = N + 1$ and the class $D_{(N,M)}$ of diagonal Volterra models. Specifically, note that any diagonal Volterra model may be written as

$$y(k) = \alpha_0 + \sum_{n=1}^{N} \sum_{i=0}^{M} \gamma_n(i) u^n(k-i). \qquad (5.27)$$

The top channel shown in Fig. 5.3 implements the constant term α_0 as the cascade connection of the nonlinearity $g_0(x) = x^0$ followed by the constant linear model $H_0(z) = \alpha_0$. Channels 1 through N each implement one of the N terms in the sum over n in Eq. (5.27), based on the static nonlinearity $g_n(x) = x^n$ and the linear dynamic model whose impulse response is $\gamma_n(i)$. This result establishes the following chain of Uryson model class inclusions:

$$H_{(N,M)} = U^1_{(N,M)} \subset U^2_{(N,M)} \cdots \subset U^{N+1}_{(N,M)} = D_{(N,M)}. \qquad (5.28)$$

5.2.3 The class $W_{(N,M)}$

The Wiener model class $W_{(N,M)}$ is the dual of the Hammerstein model class $H_{(N,M)}$, consisting of a linear $MA(M)$ dynamic model followed by a polynomial nonlinearity of order N. The response of such a Wiener model to the input sequence $\{u(k)\}$ is given by

$$y(k) = \sum_{n=0}^{N} g_n \left[\sum_{i=0}^{M} h(i) u(k-i) \right]^n$$

$$= g_0 + \sum_{n=1}^{N} g_n \sum_{i_1=0}^{M} h(i_1) u(k-i_1) \cdots \sum_{i_n=0}^{M} h(i_n) u(k-i_n). \qquad (5.29)$$

This expansion leads immediately to the $V_{(N,M)}$ representation of Eq. (5.4) with $\alpha_0 = g_0$ and

$$\alpha_n(i_1, \ldots, i_n) = g_n h(i_1) \cdots h(i_n), \qquad (5.30)$$

for $n > 0$. Like the Hammerstein model, this Volterra representation is also highly structured, but the structure is very different. In particular, note that the Wiener model not exhibit a diagonal Volterra representation if $N > 1$. This observation is a corollary of the result established in Chapter 4 that Hammerstein models are members of the additive NARMAX class but Wiener models are not, aside from the special cases of static nonlinearities and linear dynamic models.

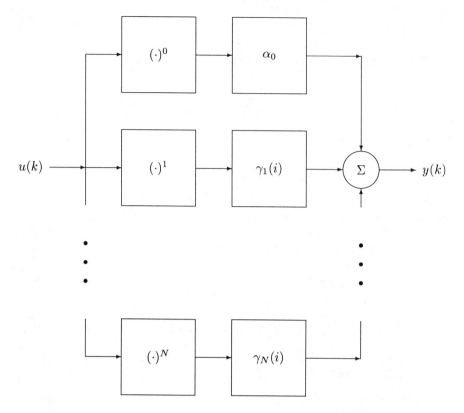

Figure 5.3: Monomial form of the Uryson model $U_{(N,M)}^{N+1}$

5.2.4 The class $P_{(N,M)}^r$

The class $P_{(N,M)}^r$ is the dual of $U_{(N,M)}^r$ and consists of the r-channel NMAX projection-pursuit models:

$$y(k) = \sum_{\ell=1}^{r} g_\ell \left(\sum_{i=0}^{M} h_\ell(i) u(k-i) \right), \qquad (5.31)$$

where $g_\ell(\cdot)$ is a polynomial of order N. As in the case of the Uryson model $U_{(N,M)}^r$, the Volterra representation for the projection-pursuit model $P_{(N,M)}^r$ follows directly from those of the r individual $W_{(N,M)}$ components connected in parallel to form this model:

$$a_n(i_1, \ldots, i_n) = \sum_{\ell=1}^{r} g_n^\ell h_\ell(i_1) \cdots h_\ell(i_n), \qquad a_0 = \sum_{\ell=1}^{r} g_0^\ell \qquad (5.32)$$

In fact, this representation is general enough that any finite Volterra model may be written as an equivalent projection-pursuit model. The key to this result lies in the following observations. First, note that the $V_{(N,M)}$ model response defined by Eq. (5.4) consists of a sum of $N+1$ terms, indexed by the degree of nonlinearity n. Since projection-pursuit models consist of r channels in parallel, it follows that the parallel combination of projection-pursuit models is again a projection-pursuit model. In particular, it follows from this observation that a $V_{(N,M)}$ model may be represented as a projection-pursuit model provided the terms α_0 and $v_M^n(k)$ may each be represented as projection-pursuit models. Since the constant term α_0 and the linear term $v_M^1(k)$ may be represented as special cases of the Wiener model, the general result follows if it is possible to represent the terms $v_M^n(k)$ for $n > 1$ as projection-pursuit models for some r.

To establish this general result, form the $(M+1)^n$-dimensional vectors

$$\mathbf{v} = [\alpha_n(0,\ldots,0), \alpha_n(1,0,\ldots,0),\ldots,\alpha_n(M,\ldots,M)]^T$$
$$\mathbf{h}_\ell = [h_\ell^n(0), h_\ell(1)h_\ell^{n-1}(0),\ldots,h_\ell^n(M)]^T. \tag{5.33}$$

Eq. (5.32) may be rewritten as

$$\mathbf{v} = \sum_{\ell=1}^{r} g_n^\ell \mathbf{h}_\ell. \tag{5.34}$$

Now, it is possible to represent the term $v_M^n(k)$ as the output of a projection-pursuit model if the model coefficients g_n^ℓ and $h_\ell(i)$ can be chosen to make $\mathbf{v}$ assume arbitrary values. This is possible if the vectors $\mathbf{h}_\ell$ can be chosen to span the $(M+1)^n$-dimensional Euclidean space in which $\mathbf{v}$ lies. In particular, if $r = (M+1)^n$, the moving average model coefficients $h_\ell(i)$ can be chosen so that the vectors $\mathbf{h}_\ell$ are mutually orthogonal (e.g., by Gram-Schmidt orthogonalization) and span this space. Given these vectors, form the square matrix $\mathbf{H}$ by taking the vectors $\mathbf{h}_\ell$ as columns. This matrix is nonsingular by construction, so the coefficients g_n^ℓ required to represent $v_M^n(k)$ as a projection-pursuit model are the elements of the vector $\mathbf{g}$ given by

$$\mathbf{g} = \mathbf{H}^{-1}\mathbf{v}. \tag{5.35}$$

This result and the previous arguments are sufficient to establish that $V_{(N,M)} \subset P_{(N,M)}^r$ for sufficiently large r. In analogy with the Uryson model inclusion chain, this result leads to the following chain of projection-pursuit model class inclusions:

$$W_{(N,M)} = P_{(N,M)}^1 \subset P_{(N,M)}^2 \cdots \subset P_{(N,M)}^r = V_{(N,M)}. \tag{5.36}$$

In the general case, r is large enough that the complexity of the resulting projection-pursuit model is unreasonable, but this result does establish the flexibility of the projection-pursuit representation. In particular, the construction procedure just described requires $(M+1)^n$ channels to implement the term $v_M^n(k)$, implying an overall projection-pursuit model of class $P_{(N,M)}^r$ with r given by

$$r = \sum_{n=0}^{N}(M+1)^n = \frac{(M+1)^N - 1}{M} \sim M^{N-1}. \tag{5.37}$$

For the example discussed at the beginning of this section with $N = 4$ and $M = 20$, this representation requires approximately 8000 $W_{(4,20)}$ models in parallel; overall, this construction requires about 200,000 total model parameters, an order of magnitude more than the original $V_{(N,M)}$ representation. Conversely, the general flexibility of the projection-pursuit model class established by this result raises the hope that, for a given application, it may be possible to find a *parametrically efficient* projection-pursuit approximation whose $P_{(N,M)}^r$ representation is substantially simpler than its $V_{(N,M)}$ representation.

5.3 Block-oriented nonlinear models

As noted in Chapter 1, the class of *block-oriented nonlinear models* consists of various series and parallel combinations of static nonlinearities and linear dynamic models. All four of the model structures discussed in Sec. 5.2 are members of this class: the Hammerstein and Wiener models consisting of series combinations of these elements, and the Uryson and projection-pursuit models consisting of parallel combinations of Hammerstein and Wiener models. It follows from the results presented in Sec. 5.2.4 that every $V_{(N,M)}$ model has a block-oriented (specifically, projection-pursuit) representation, and it will be established here that the converse also holds: every block-oriented model constructed from polynomial nonlinearities and moving average dynamic models has a $V_{(N,M)}$ representation. *Hence, it follows that the polynomial block-oriented models, the finite Volterra models, and the projection-pursuit models are all equivalent classes.* As the discussion at the end of Sec. 5.2.4 illustrates, however, not all of these representations are equally efficient. In practice, then, different applications are likely to favor different representations. The next two sections illustrate first, the range of block-oriented structures that have been investigated by various researchers and second, the equivalence between block-oriented and finite Volterra models.

5.3.1 Block-oriented model structures

Chen (1995) gives an excellent review of the block-oriented model class, discussing a representative variety of model structures, the relationship between these model structures and the structure of their Volterra series representations, and the problem of estimating model parameters from input/output data. In addition, Chen discusses multiple-input, multiple-output (MIMO) block-oriented models, dealing with the same range of issues. Chen also briefly considers applications of these models in the areas of electrical and biomedical engineering, but does not consider applications in other areas like chemical engineering. Most of Chen's results are presented in terms of continuous-time models, although his identification results assume these model responses are sampled uniformly in time, yielding a discrete-time model. The relationship between continuous-time and discrete-time block-oriented models is revisited briefly in Chapter 7.

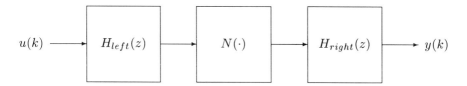

Figure 5.4: Structure of the "sandwich" or LNL model

Chen's review introduces an extremely useful notation for the class of block-oriented nonlinear models. Specifically, he uses the symbol L to designate a linear, time-invariant dynamic model and the symbol N to denote a static nonlinearity. Block-oriented nonlinear model structures are then described by combining these symbols in a way that permits reconstruction of the corresponding block diagram. As a specific example, since the Hammerstein model consists of a static nonlinearity N followed by a linear dynamic element L, Chen designates these models as NL. Note that the ordering of this notation is *opposite* that of a composition of mathematical operators: the input sequence is *first* processed by the *left-most* block and then works toward the *right*. Hence, the Wiener model is designated LN, indicating that the input sequence is processed *first* by the linear dynamic model L, and *then* by the static nonlinearity N. Similarly, the model designated LNL corresponds to the "sandwich model" shown in Fig. 5.4, consisting of a linear dynamic model (the left-most L), followed by the static nonlinearity N, and finally followed by a second linear dynamic model (the right-most L). This model has been studied by a number of authors (Billings, 1980; Brillinger, 1977; Greblicki and Pawlak, 1991), in part because it includes both the Hammerstein and Wiener models as special cases. That is, taking the right-most linear model to be the identity $[y(k) = u(k)$ or $H(z) = 1]$, the LNL model reduces the Hammerstein model NL, whereas taking the left-most linear model to be the identity yields the Wiener model LN.

For block-oriented structures like the Uryson model that involve multiple parallel paths, Chen introduces the notation PX_m to indicate that the path ($X = L, N, NL, LNL$, etc.) is repeated in parallel m times. As a specific example, since the m-channel Uryson model consists of m Hammerstein models (NL) connected in parallel, Chen's symbol for this model is PNL_m. In this way, arbitrarily complex single-input, single-output, block-oriented models may be represented as the appropriate string of symbols. A representative illustration is the PL_1NL_2 model structure shown in Fig. 5.5. Here, the left-most symbol is PL_1, indicating the input sequence first passes through a *one-channel* linear block L. The reason for including the P symbol in this designation is that the *next* part of this model—designated NL_2—consists of two Hammerstein (NL) models in parallel. Finally, an important point to note here is that the cascade connection of two nonlinearities N_1 and N_2 is simply a third nonlinearity N, so the structure NNL is equivalent to the simpler structure NL. The same observation holds for linear models, so the structure LLN simplifies to LN.

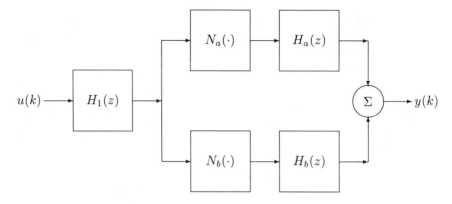

Figure 5.5: Chen's PL_1NL_2 model structure

These observations reduce the complexity of both the block diagrams for these models and their corresponding symbols in Chen's notation.

5.3.2 Equivalence with the class $V_{(N,M)}$

Since it was established in Sec. 5.2.4 that every finite Volterra model in the class $V_{(N,M)}$ may be represented as a block-oriented $P^r_{(N,M)}$ model, it is sufficient to show that every polynomial block-oriented model has a $V_{(N,M)}$ representation. This result follows from the fact that the class of finite Volterra models is closed under both series and parallel combinations. The parallel combination result was established in Sec. 5.1 and the series combination result is established in Chapter 7 where it forms the basis for the construction of an algebraic category of Volterra models (Sec. 7.4.3). Specifically, it is shown that the series combination of a model of class $V_{(N_1,M_1)}$ with a model of class $V_{(N_2,M_2)}$ is a model of class $V_{(N_1N_2,M_1+M_2)}$.

To illustrate the basic result, consider the PL_1NL_2 model shown in Fig. 5.5. If $N_a(\cdot)$ is a polynomial of order N_a and $H_a(z)$ represents a moving average model of order M_a, their series combination may be represented as a member of the $H_{(N_a,M_a)}$ model class. Similarly, the series combination of $N_b(\cdot)$ and $H_b(z)$ may be represented as a member of the $H_{(N_b,M_b)}$ model class. The parallel combination of these two models is a Uryson model belonging to the class $U^2_{(N',M')}$ where $N' = \max\{N_a, N_b\}$ and $M' = \max\{M_a, M_b\}$. If the linear model $H_1(z)$ is a moving average model of order M_1, it may be represented as a member of the Volterra model class $V_{(1,M_1)}$. Since the $U^2_{(N',M')}$ model belongs to the Volterra class $V_{(N',M')}$, it follows from the series combination result noted above that the overall PL_1NL_2 model may be represented as a member of the Volterra model class $V_{(N',M'+M_1)}$.

Clearly, this idea may be extended to any combination of moving average linear models and polynomial nonlinearities, *provided only series and parallel connections are allowed*. In particular, it is necessary to exclude *feedback connections* like those appearing in the Lur'e model discussed in Chapter 4. With this exclusion, the class of block-oriented nonlinear models is equivalent to the class of finite Volterra models. In addition, it is also possible to extend these results to the more general class of block-oriented nonlinear models based on *analytic nonlinearities* and *arbitrary linear dynamic models*. The resulting Volterra models belong to the infinite-dimensional class $V_{(\infty,\infty)}$, raising some important new issues (e.g., convergence of the Volterra series). Some of these issues and some examples of these models are discussed in Sec. 5.5.

5.4 Pruned Volterra models

One of the important practical motivations for considering restricted classes of Volterra models like the Hammerstein, Wiener, or sandwich models is that they permit efficient parameterizations. Alternatively, it is also possible to limit Volterra model complexity by explicitly restricting the coefficients in the $V_{(N,M)}$ representation. The class of *pruned Volterra models* is obtained by setting some of the model coefficients $\alpha_n(i_1,\ldots,i_n)$ to zero (Pearson et al., 1996).

5.4.1 The class of pruned Volterra models

The defining idea underlying the class of pruned Volterra models was just stated: specified model coefficients $\alpha_n(i_1, i_2, \ldots, i_n)$ are set to zero. To give a more precise and useful definition, it is necessary to introduce some notation. Specifically, let $\mathcal{M}$ denote the set of integers $0, 1, \ldots, M$ and let $\mathcal{M}^n$ be the n-fold cartesian product $\mathcal{M} \times \mathcal{M} \times \cdots \times \mathcal{M}$. Since the nth-order Volterra model coefficients $\alpha_n(i_1, i_2, \ldots, i_n)$ are specified by the indices $i_1, i_2, \ldots, i_n$, these coefficients may be associated with members of the set $\mathcal{M}^n$. However, since the functions associated with these coefficients are the polynomials $u(k-i_1)u(k-i_2)\cdots u(k-i_n)$, and since scalar multiplication is commutative, it should be clear that there is some redundancy present in this representation. In fact, this redundancy was illustrated in the second-order Wiener model discussed in Sec. 5.1.2, where it was addressed by taking $\alpha_2(0,1) = \alpha_2(1,0)$. This choice corresponds to a *symmetric representation*, one that is particularly convenient in working with second-order Volterra models (Pearson et al., 1992, 1996). For Volterra models of arbitrary order, however, it is generally more convenient to work with the following *ordered representation*.

Specifically, it follows directly from the commutativity of scalar multiplication that any Volterra model may be represented in terms of coefficients that are nonzero only if the following condition is satisfied:

$$0 \leq i_1 \leq i_2 \leq \cdots \leq i_n \leq M. \tag{5.38}$$

In the specific case of the Wiener model considered in Sec. 5.1.2, this representation would correspond to replacing the condition $\alpha(1,0) = \alpha(0,1)$ with the condition $\alpha(1,0) = 0$. To represent the same Wiener model, it would only be necessary to double the original coefficient $\alpha(0,1)$; the key point is that since the coefficient of $u(k)u(k-1)$ in the Volterra representation for this Wiener model is $\alpha(0,1) + \alpha(1,0)$, it is immaterial how this sum is partitioned between the individual components $\alpha(0,1)$ and $\alpha(1,0)$.

To define the class of pruned Volterra models, let Γ_r be the collection of elements $(i_1, i_2 \ldots, i_r)$ of $\mathcal{M}^r$ satisfying Eq. (5.38). For $r = 1, 2 \ldots, N$, let S_r be an arbitrary subset of Γ_r. The model $\mathcal{V}$ is said to be *pruned with respect to the sets* $\{S_r\}$ if $(i_1, i_2 \ldots, i_r) \in S_r$ implies $\alpha_r(i_1, i_2 \ldots, i_r) = 0$. Similarly, let S_0 be either the empty set $\emptyset$ or the singleton $\{0\}$ and define $\mathcal{V}$ to be pruned with respect to S_0 if $S_0 \neq \emptyset$ (i.e., if $S_0 = \{0\}$, implying $\alpha_0 = 0$). This definition generalizes one given earlier for symmetric second-order Volterra models (Pearson et al., 1996). It is important to note that the subsets S_r need not be *proper* subsets of Γ_r: both $S_r = \emptyset$ and $S_r = \Gamma_r$ are admissible. In fact, note that if $S_r = \emptyset$ for $r = 0, 1 \ldots, N$, then the structure of the model $\mathcal{V}$ is unconstrained: all model coefficients $\alpha_r(i_1, i_2, \ldots, i_r)$ can assume any value (including, of course, zero). At the opposite extreme, note that if $S_r = \Gamma_r$ for some r, *all* terms of order r are necessarily excluded from the Volterra model $\mathcal{V}$. As a specific example, note that if N is odd and $S_r = \Gamma_r$ for $r = 0, 2, \ldots, N-1$, then the pruned Volterra model $\mathcal{V}$ is a member of the class of odd-order Volterra models $\mathcal{O}_{(N,M)}$ introduced in Sec. 5.1.

A number of other important model classes may also be viewed as pruned Volterra models. For example, the class $\mathcal{N}_N$ of static nonlinearities defined in Sec. 5.1 may be regarded as a pruned Volterra model; there, $S_0 = \emptyset$ and, for $r = 1, 2, \ldots, N$, the sets S_r contain all elements of Γ_r *except* $(0, 0, \ldots, 0)$. Similarly, the class $V^0_{(N,M)}$ may be viewed as the class of $V_{(N,M)}$ Volterra models pruned with respect to $S_0 = \{0\}$ and $S_r = \emptyset$ for $r = 1, 2, \ldots, N$. In addition, the class $MA(M)$ of Mth-order moving average models may be regarded as the class of pruned $V_{(N,M)}$ Volterra models with $S_0 = \{0\}$, $S_1 = \emptyset$, and $S_r = \Gamma_r$ for $r = 2, 3, \ldots, N$. Finally, the class of diagonal Volterra models introduced in Sec. 5.2.1 is one of the most important members of the class of pruned Volterra models. Specifically, $\mathcal{V} \in D_{(N,M)}$ implies $\mathcal{V}$ is a $V_{(N,M)}$ model pruned with respect to the sets:

$$S_r = \Gamma_r \setminus \{(i, i, \ldots, i) | i = 0, 1, \ldots, M\}, \quad (5.39)$$

where $\setminus$ is the symbol for "set subtraction." That is, the sets S_r include all of Γ_r *except* the diagonal elements $(i, i, \ldots, i)$ for $i = 0, 1, \ldots, M$, corresponding to the fact that these elements define the only Volterra model parameters that are not constrained to be zero.

5.4.2 An example: prunings of $V_{(2,2)}$

To clarify the ideas just presented, consider the possible prunings of the $V_{(2,2)}$ class of Volterra models. Beginning with the constant term α_0, note that there are only two possibilities for *any* $V_{(N,M)}$ Volterra model: either the model is pruned by restricting α_0 to be zero, or it isn't. In terms of the set S_0, these choices correspond to the following possibilities:

$$S_0 = \emptyset \Rightarrow \alpha_0 \text{ unconstrained} \quad \text{or} \quad S_0 = \{0\} \Rightarrow \alpha_0 = 0.$$

Volterra Models

As noted earlier, the assumption $\alpha_0 = 0$ is quite common in practice. For the linear part of the $V_{(2,2)}$ model, there are eight possible prunings:

1. $S_1 = \emptyset$, implying the linear model terms are unconstrained
2. $S_1 = \{0\}$, implying $\alpha_1(0) = 0$
3. $S_1 = \{1\}$, implying $\alpha_1(1) = 0$
4. $S_1 = \{2\}$, implying $\alpha_1(2) = 0$
5. $S_1 = \{0, 1\}$, implying *only* $\alpha_1(2) \neq 0$
6. $S_1 = \{0, 2\}$, implying *only* $\alpha_1(1) \neq 0$
7. $S_1 = \{1, 2\}$, implying *only* $\alpha_1(0) \neq 0$
8. $S_1 = \mathcal{M} = \{0, 1, 2\}$, implying *all* linear model terms are zero.

Of these possble prunings, the second is probably the most common, corresponding to the usual one-sample delay for the model response. Specifically, this pruning implies $y(k)$ depends on $u(k-1)$ and $u(k-2)$, but not on the "direct feedthrough" term $u(k)$. Also, note that the fourth, sixth, and seventh prunings listed here all entail *dynamic order reduction* for the linear model, since they imply $\alpha_1(2) = 0$. Finally, note that the eighth pruning listed here corresponds to a complete elimination of linear terms from the model, in the same spirit as the class of completely bilinear models discussed in Chapter 3.

In this example, note that there are six possible quadratic terms, so the set Γ_2 is given by

$$\Gamma_2 = \{(0,0), (0,1), (0,2), (1,1), (1,2), (2,2)\}. \tag{5.40}$$

Altogether, there are 64 possible prunings of this set, enumerated as follows:

A. $S_2 = \emptyset$, 1 possible set with 0 terms
B. S_2 is a singleton, 6 possible sets with 1 term each
C. S_2 is a pair, 15 possible sets with 2 terms each
D. S_2 is a triple, 20 possible sets with 3 terms each
E. S_2 is a quadruple, 15 possible sets with 4 terms each
F. S_2 is a quintuple, 6 possible sets with 5 terms each
G. $S_2 = \Gamma_2$, 1 possible set with 6 terms.

Again, setting $S_2 = \emptyset$ corresponds to imposing no constraint on the quadratic model terms, whereas setting $S_2 = \Gamma_2$ corresponds to eliminating the quadratic terms entirely. As in the linear case, excluding those three terms involving the index 0 [i.e., the terms $(0, 0)$, $(0, 1)$, and $(0, 2)$] would correspond to the standard "one sample delay" assumption, whereas excluding those terms involving the index 2 [i.e., the terms $(0, 2)$, $(1, 2)$, and $(2, 2)$] would correspond to dynamic order reduction. Excluding the terms $(0, 1)$, $(0, 2)$, and $(1, 2)$ would restrict the model to the diagonal class $D_{(2,2)}$, whereas excluding only the term $(0, 2)$ amounts to a tridiagonal restriction discussed in Sec. 5.4.3. In the general case, enumeration of the sets of possible pruned $V_{(N,M)}$ models corresponds to the enumeration of *chains*, a classic problem in combinatorics (van Lint and Wilson, 1992). Not surprisingly, these numbers rapidly become *enormous*, further encouraging the consideration of particular subclasses.

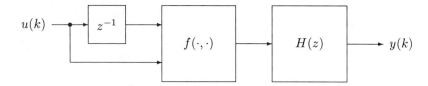

Figure 5.6: PPOD model structure

5.4.3 The PPOD model structure

Fig. 5.6 shows the block diagram for a modification of the Hammerstein model proposed by Pawlak et al. (1994) which will be subsequently abbreviated as the PPOD model. This structure replaces the Hammerstein nonlinearity $g(x)$ with the two-dimensional function $g(u(k), u(k-1))$ to obtain a member of the LARNMAX model class introduced in Chapter 4:

$$y(k) = \sum_{i=1}^{p} a_i y(k-i) + \sum_{j=0}^{q} b_j g(u(k-j), u(k-j-1)). \quad (5.41)$$

The motivation for this extension of the Hammerstein model was that it permitted the extension of nonparametric identification results for Hammerstein models (Greblicki and Pawlak, 1989) to a wider class of nonlinear dynamic models (Pawlak et al., 1994). That is, although nonparametric methods are not well suited to estimating unknown functions of many variables, these techniques are extremely practical for scalar problems, and remain practical if somewhat more complex for low-dimensional multivariable maps. Generalizing from $g(x)$ to $g(x,y)$ thus permits the exploration of a nontrivial extension of the class of representable nonlinear dynamic phenomena without abandoning the use of nonparametric methods.

Robinson's second-order Volterra model is a particularly simple member of the PPOD model family, obtained by taking $H(z)$ as the linear identity system and $g(x,y)$ as:

$$g(x,y) = x + \alpha y + \beta xy. \quad (5.42)$$

Further, note that if this linear identity model is replaced with a linear $AR(p)$ model, the result is Robinson's autoregressive extension of this second-order Volterra model. Another interesting member of the PPOD model class is the *modified Hammerstein model* (Zhu and Seborg, 1994), proposed to facillitate the solution of a nonlinear model predictive control (NMPC) problem. In particular, note that a direct replacement of the linear $MA(M)$ model normally used in the MPC problem formulation with a nonlinear $H_{(N,M)}$ model results in a requirement to solve the following equation for $u(k)$ in terms of $v(k)$:

$$v(k) = \sum_{n=0}^{N} g_n [u(k)]^n. \quad (5.43)$$

Volterra Models 231

This problem is the classical one of finding the roots of an Nth-order polynomial, but generally only *one* of these N roots represents the desired solution, with the others being "unphysical" (e.g., complex). Zhu and Seborg assume $g_0 = 0$ and replace Eq. (5.43) with

$$v(k) = g_1 u(k) + \sum_{n=2}^{N} g_n [u(k-1)]^n, \qquad (5.44)$$

noting that, given the previous MPC solution $u(k-1)$ (known at time k), the desired solution is given explicitly by

$$u(k) = g_1^{-1} \left[v(k) - \sum_{n=2}^{N} g_n [u(k-1)]^n \right]. \qquad (5.45)$$

Further, note that if the MPC solution $\{u(k)\}$ varies slowly with time, the solution given explicitly in Eq. (5.45) will not differ greatly from the desired (i.e., "physical") polynomial root obtained by inverting Eq. (5.43). Note, however, that this modification replaces the static nonlinearity on which the Hammerstein model is based with the *dynamic* nonlinearity defined in Eq. (5.44), depending on $u(k)$ and $u(k-1)$.

More generally, it can be shown that PPOD models based on polynomial nonlinearities and linear moving average models exhibit a tridiagonal Volterra model representation. To see this result, begin with the input/output relation

$$y(k) = \sum_{i=0}^{M} b_i g(u(k-i), u(k-i-1)), \qquad (5.46)$$

where the polynomial nonlinearity may be represented as the $V_{(N,1)}$ model

$$g(u(k), u(k-1)) = g_0 + \sum_{n=1}^{N} \sum_{i_1=0}^{1} \sum_{i_2=0}^{1} \cdots \sum_{i_n=0}^{1} \qquad (5.47)$$
$$\gamma_n(i_1, i_2, \ldots, i_n) u(k-i_1) u(k-i_2) \cdots u(k-i_n).$$

Since the PPOD model consists of the cascade connection of the $V_{(N,1)}$ model defined in Eq. (5.47) and a linear $V_{(1,M)}$ model, it follows that the PPOD model may be represented as the $V_{(N,M+1)}$ model

$$y(k) = \alpha_0 + \sum_{n=1}^{N} \sum_{j_1=0}^{M+1} \sum_{j_2=0}^{M+1} \cdots \sum_{j_n=0}^{M+1} \qquad (5.48)$$
$$\alpha_n(j_1, j_2, \ldots, j_n) u(k-j_1) u(k-j_2) \cdots u(k-j_n).$$

The constant term α_0 is easily seen to be

$$\alpha_0 = g_0 \sum_{i=0}^{M} b_i, \qquad (5.49)$$

a result obtained by substituting Eq. (5.47) into Eq. (5.46) and comparing the result with Eq. (5.48). Determination of the terms $\alpha_n(j_1, j_2, \ldots, j_n)$ proceeds in the same way conceptually, but the details are somewhat more complicated.

The linear terms in Eq. (5.48) are determined by those in Eq. (5.47), which are $\gamma_1(0)u(k) + \gamma_1(1)u(k-1)$. Substituting these terms into Eq. (5.46) yields the result

$$\sum_{j_1=0}^{M+1} \alpha_1(j_1)u(k-j_1) = \sum_{i=0}^{M} b_i[\gamma_1(0)u(k-i) + \gamma_1(1)u(k-i-1)]. \quad (5.50)$$

Simplifying this result, it follows that

$$\alpha_1(j) = \begin{cases} b_0\gamma_1(0) & j = 0, \\ b_j\gamma_1(0) + b_{j-1}\gamma_1(1) & 1 \leq j \leq M, \\ b_M\gamma_1(1) & j = M+1. \end{cases} \quad (5.51)$$

In general, detailed expressions for $\alpha_n(j_1, j_2, \ldots, j_n)$ become more complicated and less informative as n increases, but it is worth considering the case $n = 2$ because it illustrates the essential tri-diagonal structure of PPOD models. Specifically, note that there are three second-order model coefficients in Eq. (5.47): $\gamma_2(0,0)$, $\gamma_2(0,1)$, and $\gamma_2(1,1)$. Matching the second-order terms in both representations yields

$$\sum_{j_1=0}^{M+1} \sum_{j_2=0}^{M+1} \alpha_2(j_1, j_2)u(k-j_1)u(k-j_2) = \sum_{i=0}^{M} b_i[\gamma_2(0,0)u^2(k-i)$$
$$+ \gamma_2(0,1)u(k-i)u(k-i-1)$$
$$+ \gamma_2(1,1)u^2(k-i-1)], \quad (5.52)$$

and matching individual terms $u(k-j_1)u(k-j_2)$ on both sides of this expression ultimately gives the result

$$\alpha_2(i,j) = \begin{cases} b_0\gamma_2(0,0) & i = j = 0, \\ b_i\gamma_2(0,0) + b_{i-1}\gamma_2(1,1) & 1 \leq i = j \leq M, \\ b_M\gamma_2(1,1) & i = j = M, \\ b_i\gamma_2(0,1) & j = i+1,\ 0 \leq i \leq M, \\ 0 & \text{otherwise.} \end{cases} \quad (5.53)$$

In particular, note that the only nonzero terms in this expansion are the diagonal terms $\alpha_2(i,i)$ and the first off-diagonal terms $\alpha_2(i,i+1)$. This result is general: $\alpha_n(i_1, i_2, \ldots, i_n) = 0$ unless $|i_i - i_j| \leq 1$ for all indices i_i and i_j. This structure is a consequence of the fact that the nonlinearity $g(\cdot, \cdot)$ defining the PPOD model only involves $u(k)$ and $u(k-1)$.

5.5 Infinite-dimensional Volterra models

As noted in Sec. 5.1, three distinct infinite-dimensional extensions of the $V_{(N,M)}$ model class are possible: $V_{(\infty,M)}$, $V_{(N,\infty)}$, and $V_{(\infty,\infty)}$. This section considers four examples of infinite-dimensional Volterra models. The first three

Volterra Models

are all Hammerstein models, but each example illustrates a different infinite-dimensional model class; these models are all described in Sec. 5.5.1. The fourth example is Robinson's AR-Volterra model, described in Sec. 5.5.2. Sec. 5.6 explores the Volterra representation of bilinear models, which are generally (though not always) members of the class $V_{(\infty,\infty)}$.

5.5.1 Infinite-dimensional Hammerstein models

To illustrate some of the differences between the three Volterra model classes $V_{(\infty,M)}$, $V_{(N,\infty)}$, and $V_{(\infty,\infty)}$, the following paragraphs introduce the corresponding Hammerstein model classes, $H_{(\infty,M)}$, $H_{(N,\infty)}$, and $H_{(\infty,\infty)}$. The first of these model classes is based on an *analytic* function $g(\cdot)$ and a linear model from the class $MA(M)$. It is a standard result that analytic functions may be represented by the Taylor series expansion

$$g(x) = \sum_{n=0}^{\infty} \frac{g^{(n)}(0)}{n!} x^n. \tag{5.54}$$

Similarly, the response of the linear subsystem is of the form

$$y(k) = \sum_{i=0}^{M} h(i)v(k-i), \tag{5.55}$$

where $v(k) = g(u(k))$. Combining these results gives a $V_{(\infty,M)}$ model representation with coefficients

$$a_n(i_1,\ldots,i_n) = \begin{cases} [g^{(n)}(0)/n!]h(i_1) & i_1 = \cdots = i_n \\ 0 & \text{otherwise,} \end{cases} \tag{5.56}$$

defined for $n = 0, 1, \ldots$, and $i_j = 0, 1, \ldots, M$. An important issue in Taylor series expansions is their radius of convergence, which depends on the function $g(\cdot)$. In the case of the exponential function, the Taylor series converges for all x, but generally the radius of convergence is finite. For example, in the case of the hyperbolic tangent, the Taylor series converges for $|x| < \pi/2$. Clearly, it is possible to construct a Hammerstein model using this static nonlinearity and a linear $MA(M)$ subsystem and the response of this model is well-defined for all real-valued input sequences $\{u(k)\}$. Conversely, the $H_{(\infty,M)}$ representation defined by Eq. (5.56) for this model is not valid for $|u(k)| > \pi/2$. Generally, even if a Hammerstein model based on an analytic nonlinearity is BIBO stable, its $H_{(\infty,M)}$ representation will cease to be valid for input sequences of sufficiently large amplitude.

A more common infinite-dimensional Hammerstein model structure is the $H_{(N,\infty)}$ model, constructed from a polynomial nonlinearity of order N and a general linear model whose impulse response $h(i)$ does not decay to zero in finite time. This model has the Volterra series representation $V_{(N,\infty)}$ defined by the coefficients

$$a_n(i_1,\ldots,i_n) = \begin{cases} g_n h(i_1) & i_1 = \cdots = i_n \\ 0 & \text{otherwise,} \end{cases} \tag{5.57}$$

defined for $n = 0, 1, \ldots, N$ and $i_j = 0, 1, \ldots$. As in the previous example, convergence of the resulting infinite series is an important issue, but here the considerations are quite different. In particular, the $V_{(N,\infty)}$ representation defined by Eq. (5.57) is valid provided the linear model is stable.

Finally, the most general infinite-dimensional Hammerstein model is the $H_{(\infty,\infty)}$ model, formed from an analytic nonlinearity $g(\cdot)$ and a general linear model with impulse response $h(i)$. This model has the $V_{(\infty,\infty)}$ representation defined by the coefficients

$$a_n(i_1, \ldots, i_n) = \begin{cases} [g^{(n)}(0)/n!]h(i_1) & i_1 = \cdots = i_n \\ 0 & \text{otherwise,} \end{cases} \quad (5.58)$$

for $n = 0, 1, \ldots$ and $i_j = 0, 1, \ldots$. For this model class, *both* of the previous convergence criteria are important. In particular, the $V_{(\infty,\infty)}$ representation defined by Eq. (5.58) will be valid for stable linear subsystems, but generally only for inputs satisfying some bound $|u(k)| < \rho$.

5.5.2 Robinson's AR-Volterra model

Robinson's AR-Volterra model class was introduced in Chapter 4 and its simplest member is defined by

$$y(k) = ay(k-1) + u(k) + bu(k-1) + cu(k)u(k-1). \quad (5.59)$$

For $a = 0$, this model is a member of the $V_{(2,1)}$ class, but for $a \neq 0$, the result is an infinite-dimensional Volterra model. In fact, this model is a member of the class $V_{(2,\infty)}$, a result that is fairly easy to demonstrate and that will be useful in the discussion of bilinear models presented in the next section.

To compute the Volterra model coefficients, begin by forming the series expansion

$$y(k) = \alpha_0 + \sum_{i_1=0}^{\infty} \alpha_1(i_1) u(k - i_1)$$

$$+ \sum_{i_1=0}^{\infty} \sum_{i_2=0}^{\infty} \alpha_2(i_1, i_2) u(k - i_1) u(k - i_2) + \cdots . \quad (5.60)$$

This expansion is then substituted into Eq. (5.59) and terms of like powers in $u(k - j)$ are equated to determine the coefficients $\alpha_n(i_1, \ldots, i_n)$. Solving for the constant term yields $\alpha_0 = a\alpha_0$; since $|a| < 1$ is a necessary and sufficient condition for the stability of this model, it follows that $\alpha_0 = 0$ for stable models. The coefficients $\alpha_1(i)$ are determined by equating the linear terms in the Volterra expansion, that is

$$\sum_{i=0}^{\infty} \alpha_1(i) u(k-i) = a \left[\sum_{j=0}^{\infty} \alpha_1(j) u(k-1-j) \right] + u(k) + bu(k-1). \quad (5.61)$$

Volterra Models

To be a valid Volterra series representation for all possible input sequences, the coefficients of $u(k-i)$ must match on both sides of the equation for all i. Matching the $u(k)$ terms yields $\alpha_1(0) = 1$, matching the $u(k-1)$ terms yields $\alpha_1(1) = b + a\alpha_1(0) = b + a$ and, for $i > 1$, matching the terms $u(k-i)$ leads to the result

$$\alpha_1(i) = a\alpha_1(i-1) = (a+b)a^{i-1}. \tag{5.62}$$

Note that this result is independent of the nonlinear model coefficient c and that, for $b = 0$, it yields the impulse response for the corresponding first-order linear system.

The quadratic model coefficients are obtained by matching the second-order terms:

$$\sum_{i_1=0}^{\infty}\sum_{i_2=0}^{\infty} \alpha_2(i_1, i_2) u(k-i_1) u(k-i_2) =$$

$$a\left[\sum_{j_1=0}^{\infty}\sum_{j_2=0}^{\infty} \alpha_2(j_1, j_2) u(k-j_1-1) u(k-j_2-1)\right] + cu(k)u(k-1). \tag{5.63}$$

Since there is no term involving $u^2(k)$ on the right-hand side of this equation, it follows that $\alpha_2(0,0) = 0$. Similarly, since the only term involving $u(k)u(k-1)$ appearing on the right-hand side is the model term $cu(k)u(k-1)$, it follows that $\alpha_2(0,1) + \alpha_2(1,0) = c$. Adopting the ordered representation introduced in Sec. 5.4 implies $\alpha_2(1,0) = 0$ and thus $\alpha_2(0,1) = c$. Further, since there are no other terms involving $u(k)$ on the right-hand side of Eq. (5.63), it follows that $\alpha_2(0, i_2) = 0$ for all $i_2 > 1$. More generally, for $i_2 \geq i_1 \geq 1$, Eq. (5.63) yields the following recursion relation for the quadratic model coefficients

$$\alpha_2(i_1, i_2) = a\alpha_2(i_1-1, i_2-1). \tag{5.64}$$

Taken together, these results may be combined into the following general expression for the quadratic model coefficients

$$\alpha_2(i_1, i_2) = \begin{cases} ca^{i_1} & i_2 = i_1 + 1 \\ 0 & \text{otherwise.} \end{cases} \tag{5.65}$$

Note that this result is consistent with the tridiagonality of the finite-dimensional case $a = 0$ derived in Sec. 5.4.3. Finally, since there are no third- or higher-order terms in the original AR-Volterra model, it follows that $\alpha_n(i_1, \ldots, i_n) = 0$ for all $n > 2$, thus establishing that Robinson's AR-Volterra model belongs to the class $V_{(2,\infty)}$ for $|a| < 1$.

5.6 Bilinear models

The following discussion considers the problem of developing $V_{(\infty,\infty)}$ representations for the bilinear models introduced in Chapter 3, starting with the *matching conditions* described in Sec. 5.6.1, similar to the results presented in Sec. 5.5.2. Using these conditions, it is shown in Sec. 5.6.2 that the somewhat pathological class of *completely bilinear models* discussed in Chapter 3 *does not* exhibit Volterra realizations and it is shown in Sec. 5.6.3 that these matching conditions may be used to develop explicit Volterra representations for bilinear models that include linear terms.

5.6.1 Matching conditions

All of these results are based on the following basic idea. Suppose $y(k)$ is the response of the general bilinear model

$$y(k) = \sum_{i=1}^{p} a_i y(k-i) + \sum_{i=0}^{q} b_i u(k-i) + \sum_{i=1}^{P}\sum_{j=1}^{Q} c_{i,j} y(k-i) u(k-j). \quad (5.66)$$

Further, suppose $y(k)$ also has the $V_{(\infty,\infty)}$ Volterra series representation

$$y(k) = \alpha_0 + \sum_{n=1}^{\infty} v^n(k), \quad (5.67)$$

where $v^n(k)$ is defined as

$$v^n(k) = \sum_{i_1=0}^{\infty}\sum_{i_2=0}^{\infty} \cdots \sum_{i_n=0}^{\infty} a_n(i_1, i_2, \ldots, i_n) u(k-i_1) u(k-i_2) \cdots u(k-i_n). \quad (5.68)$$

Substituting the Volterra expansion defined by Eq. (5.67) into Eq. (5.66) yields the equation

$$\alpha_0 + \sum_{n=1}^{\infty} v^n(k) = \sum_{i=1}^{p} a_i \left[\alpha_0 + \sum_{n=1}^{\infty} v^n(k-i)\right] + \sum_{i=0}^{q} b_i u(k-i)$$
$$+ \sum_{i=1}^{P}\sum_{j=1}^{Q} c_{i,j} \left[\alpha_0 + \sum_{n=1}^{\infty} v^n(k-i)\right] u(k-j). \quad (5.69)$$

This equation may be rearranged to give

$$\alpha_0 + v^1(k) + \sum_{n=2}^{\infty} v^n(k) = \alpha_0 \left[\sum_{i=1}^{p} a_i\right]$$
$$+ \sum_{i=1}^{p} a_i v^1(k-i) + \sum_{i=0}^{q} b_i u(k-i) + \sum_{i=1}^{P}\sum_{j=1}^{Q} c_{i,j} \alpha_0 u(k-j)$$
$$+ \sum_{n=2}^{\infty} \left[\sum_{i=1}^{p} a_i v^n(k-i) + \sum_{i=1}^{P}\sum_{j=1}^{Q} c_{i,j} v^{n-1}(k-i) u(k-j)\right]. \quad (5.70)$$

Volterra Models

The key point to note here is that $v^n(k)$ is of polynomial order n in its dependence on the input sequence $\{u(k)\}$, implying among other things that it is homogeneous of order n. That is, if $\{u(k)\}$ is scaled by the arbitrary constant λ to $\{\lambda u(k)\}$, it follows that $v^n(k)$ becomes $\lambda^n v^n(k)$. Therefore, if $y(k)$ exhibits a Volterra representation, the terms of each order n must agree on both sides of Eq. (5.70). This observation provides the basis for deriving explicit Volterra series representations, or establishing that no such representation exists.

Matching the constant terms yields the condition

$$\alpha_0 = \alpha_0 \left[\sum_{i=1}^{p} a_i\right], \qquad (5.71)$$

implying $\alpha_0 = 0$, except for the special case

$$\sum_{i=1}^{p} a_i = 1. \qquad (5.72)$$

Note that for $p = 1$, this case corresponds to an unstable model. More generally, suppose that Eq. (5.72) *is* satisfied and consider Eq. (5.66) for $u(k) \equiv 0$: it follows that *any* constant solution $y(k) = y_s$ satisfies this equation.

Matching the linear terms leads to the equation

$$v^1(k) = \sum_{i=1}^{p} a_i v^1(k-i) + \sum_{i=0}^{q} b_i u(k-i) + \sum_{i=1}^{P}\sum_{j=1}^{Q} c_{ij}\alpha_0 u(k-j). \qquad (5.73)$$

If the pathological case defined by Eq. (5.72) is excluded, this expression simplifies to the linear $ARMA(p,q)$ model

$$v^1(k) = \sum_{i=1}^{p} a_i v^1(k-i) + \sum_{i=0}^{q} b_i u(k-i). \qquad (5.74)$$

Also, for $n = 1$, note that Eq. (5.68) simplifies to

$$v^1(k) = \sum_{i=0}^{\infty} \alpha_1(i) u(k-i), \qquad (5.75)$$

implying that the first-order Volterra model coefficients $\{\alpha_1(i)\}$ are simply the impulse response coefficients for the $ARMA(p,q)$ model defined in Eq. (5.74).

For $n > 1$, the matching conditions are given by:

$$v^n(k) = \sum_{i=1}^{p} a_i v^n(k-i) + \sum_{i=1}^{P}\sum_{j=1}^{Q} c_{i,j} v^{n-1}(k) u(k-j). \qquad (5.76)$$

Note that this equation is *linear* with respect to $v^n(k)$, but is driven by the nonlinear terms $v^{n-1}(k-i)u(k-j)$. Therefore, to derive a Volterra series representation for a bilinear model, proceed in order of increasing nonlinearity, starting with $v^1(k)$, then deriving the model coefficients associated with $v^2(k)$, etc. The following two examples illustrate this procedure in detail.

5.6.2 The completely bilinear case

It is instructive to consider an example for which the procedure described in Sec. 5.6.1 *cannot* be followed. Unlike the other bilinear example discussed in Sec. 5.6.3, completely bilinear models *do not* possess Volterra realizations. A short proof of this result is presented here, but the basic reason is that in the completely bilinear case, it is not possible to satisfy the matching conditions presented above. This observation—that completely bilinear models do not exhibit Volterra representations—is qualitatively consistent with the pathological dependence on initial conditions noted in Chapter 3 for this class of models.

Theorem:

> Unless Eq. (5.72) is satisfied, completely bilinear models do not exhibit Volterra realizations.

Proof:

> For the completely bilinear case, $b_i = 0$ for all i and, from Eq. (5.71), $\alpha_0 = 0$. Hence, it follows from Eq. (5.74) that $v^1(k) = 0$ for all k. Further, it follows from Eq. (5.76) that if $v^n(k)$ is not identically zero for some $n > 1$ then $v^{n-1}(k) \neq 0$ for some k. This chain of reasoning ultimately leads to the conclusion that $v^1(k)$ is not identically zero, a contradiction. Hence, it follows that $v^n(k) \equiv 0$ for all n, implying that $y(k)$ does not have a Volterra representation.

□

5.6.3 A superdiagonal example

This example illustrates that Volterra series representations generally *do* exist for the more usual class of bilinear models that include both linear and bilinear terms. Specifically, the example considered here is the superdiagonal model

$$y(k) = \phi y(k-1) + u(k) + \gamma y(k-2)u(k-1), \qquad (5.77)$$

for some $-1 < \phi < 1$ and $\gamma \neq 0$. It follows immediately from Eq. (5.71) that the constant term in the Volterra is $\alpha_0 = 0$. Similarly, it follows from Eqs. (5.74) and (5.75) that $\alpha_1(i)$ is simply the impulse response of the linear part of the model defined by Eq. (5.77), that is

$$\alpha_1(i) = \phi^i, \qquad (5.78)$$

for $i = 0, 1, \ldots$. Note that this result is independent of γ; in particular, it holds for $\gamma = 0$, as it must since it gives the correct infinite-order moving average representation for the linear model $y(k) = \phi y(k-1) + u(k)$.

Volterra Models

The second-order Volterra coefficient $\alpha_2(i_1, i_2)$ is obtained by solving the matching condition defined in Eq. (5.76) for $n = 2$. Expressed in terms of the second-order coefficients $\alpha_2(i_1, i_2)$, this condition is

$$\sum_{i_1=0}^{\infty} \sum_{i_2=i_1}^{\infty} \alpha_2(i_1, i_2) u(k - i_1) u(k - i_2) =$$

$$\sum_{i_1=0}^{\infty} \sum_{i_2=i_1}^{\infty} \phi \alpha_2(i_1, i_2) u(k - 1 - i_1) u(k - 1 - i_2)$$

$$+ \sum_{\ell=0}^{\infty} \gamma \phi^\ell u(k - 2 - \ell) u(k - 1) \quad (5.79)$$

Here, the sum over i_2 runs over $i_2 \geq i_1$, consistent with the ordered representation introduced in Sec. 5.4. Though these conditions are somewhat messy, they can be simplified by redefining the indices on the right-hand side

$$\sum_{i_1=0}^{\infty} \sum_{i_2=0}^{\infty} \alpha_2(i_1, i_2) u(k - i_1) u(k - i_2) =$$

$$\sum_{i_1=1}^{\infty} \sum_{i_2=i_1}^{\infty} \phi \alpha_2(i_1 - 1, i_2 - 1) u(k - i_1) u(k - i_2)$$

$$+ \sum_{i_2=2}^{\infty} \gamma \phi^{i_2 - 2} u(k - 1) u(k - i_2). \quad (5.80)$$

From this rearrangement, it is immediately apparent that there are no terms of the form $u(k)u(k - m)$ appearing on the right-hand side, thus implying $\alpha_2(0, m) = 0$ for all $m \geq 0$. In addition, if $i_1 > 1$, it follows that $\alpha_2(i_1, i_2)$ satisfies the simple linear recursion relation

$$\alpha_2(i_1, i_2) = \phi \alpha_2(i_1 - 1, i_2 - 1). \quad (5.81)$$

Since $\alpha_2(0, 0) = 0$, it follows immediately that $\alpha_2(i, i) = 0$ for all i. Hence, in contrast to the diagonal structure of the Volterra representation for the Hammerstein model, the second-order diagonal terms in this model are identically zero. However, it is easy to see from Eq. (5.80) that

$$\alpha_2(1, i) = \gamma \phi^{i-2}, \quad (5.82)$$

for all $i \geq 2$. Combining Eqs. (5.81) and (5.82) leads to the following general expression for the second-order Volterra model coefficients $\alpha_2(i_1, i_2)$

$$\alpha_2(i, i + j) = \gamma \phi^{i+j-2}, \quad (5.83)$$

for $i = 1, 2, \ldots$, and $j = 1, 2, \ldots$. Note that $\alpha_2(i, j) = 0$ identically for all i and j if $\gamma = 0$, as it must due to the linearity of the model in this case.

Higher-order terms are computed recursively, observing that the general condition to be satisfied is

$$v^n(k) = \phi v^n(k-1) + \gamma u(k-1)v^{n-1}(k-2). \tag{5.84}$$

As before, this equation may be converted into a recursion relation involving the Volterra coefficients $\alpha_n(i_1, i_2, \ldots, i_n)$ in terms of $\alpha_n(i_1-1, i_2-1, \ldots, i_n-1)$. Also, it may be shown that this coefficient is zero if any of the indices are zero, by the same reasoning as above. Therefore, it follows that the diagonal terms $\alpha_n(i, i, \ldots, i) = 0$ for all orders $n > 1$, generalizing the quadratic result just presented. Conversely, it is also easy to show that

$$\alpha_n(1, 2, \ldots, n) = \gamma^{n-1}, \tag{5.85}$$

thus illustrating how the coefficients $\alpha_n(\cdots)$ generally scale with the nonlinearity parameter γ. Overall, the math becomes progressively messier with increasing order, but it should be clear from this partial list of results that the Volterra representation for this bilinear model is well-defined, unlike the complete bilinear case considered above.

5.7 Summary: the nature of Volterra models

It was shown in Sec. 5.3 that finite Volterra models are equivalent to block-oriented models based on polynomial nonlinearities and moving-average linear models. For purposes of efficiently identifying model parameters from input/output data, the block-oriented representation is often preferable. This conclusion is particularly true when this structure is simple, as in the case of Hammerstein or Wiener models. For purposes of analysis, however, the Volterra structure is generally more convenient, a point that will be illustrated by several different results presented in Chapter 7. From the perspective of model structure selection, the most important question is what class of qualitative phenomena these models are capable of exhibiting. In addressing this question, the following results (Boyd and Chua, 1985) are particularly useful.

In general terms, Boyd and Chua established that any model that exhibits the *fading memory property* may be *approximated arbitrarily well* by finite Volterra models. More specifically, suppose N is a nonlinear, discrete-time dynamic model mapping bounded input sequences into bounded output sequences and let K be any subset of ℓ_∞. The dynamic model N has *fading memory on K* if the following condition holds. Given $\{u(k)\} \in K$ and $\epsilon > 0$, there exists $\delta > 0$ such that for all $\{v(k)\} \in K$:

$$\sup_{k \geq 0} |u(-k) - v(-k)|w(k) < \delta \quad \Rightarrow \quad |Nu(0) - Nv(0)| < \epsilon \tag{5.86}$$

where $\{w(k)\}$ is a sequence bounded between 0 and 1 and monotonically decreasing to 0 as $k \to \infty$. In words, this condition means that if two sequences $\{u(k)\}$ and $\{v(k)\}$ differ significantly only in the "distant past," the responses to these sequences "in the present" will be essentially the same. Mathematically, this condition may be viewed as a stronger version of the requirement that N be a continuous map from ℓ_∞ into ℓ_∞ (Boyd and Chua, 1985). Conversely, note that this condition is *weaker* than the defining conditions for the *continuous NMAX models* considered at the end of Chapter 4. In particular, recall that those models were of the form

$$y(k) = g(u(k), u(k-1), \ldots, u(k-q)), \tag{5.87}$$

where $g : R^{q+1} \to R^1$ was assumed continuous. Note that this condition is equivalent to Eq. (5.86) for the particular sequence

$$w(k) = \begin{cases} 1 & 0 \le k \le q \\ 0 & k > q, \end{cases} \tag{5.88}$$

implying these models exhibit the fading memory property on any bounded subset of ℓ_∞. The principal difference between these conditions is that the fading memory condition permits *weak* dependence on "the distant past," whereas continuous, finite-horizon NMAX models like that defined in Eq. (5.87) exhibit *no* dependence on "the distant past."

Boyd and Chua establish that if K is a bounded subset of ℓ_∞ and N is a time-invariant nonlinear dynamic model with the fading memory property on K, then given any $\epsilon > 0$, there exists a finite Volterra model $\hat{N}$ such that

$$|Nu(k) - \hat{N}u(k)| \le \epsilon, \tag{5.89}$$

for all $\{u(k)\} \in K$. It is important to note here that the fading memory condition on K is much weaker than the condition of analyticity on K. Hence, though it may not be justifiable to take an exact Volterra series expansion of N, the fading memory condition is strong enough to guarantee the existance of an approximation of arbitrary accuracy on K. In particular, note that this result applies to all of the continuous NMAX models considered in Chapter 4, by virtue of Eq. (5.88).

However, as in the discussion of Cybenko's approximation result given in Chapter 4, it is important to distinguish between *approximation* and *qualitative behavior*. For example, it might be argued that the first-order autoregressive model $y(k) = 1.01y(k-1) + u(k)$ is *almost* stable because the unstable coefficient $a_1 = 1.01$ is well approximated by the stable coefficient $a_1 = 0.99$. Clearly, however, these two models differ radically in their qualitative behavior: one is exponentially stable, whereas the other is exponentially unstable. Still, the numerical approximation issue is an important one that must be dealt with in practical problems involving the estimation of frequencies and damping ratios for systems that exhibit lightly damped oscillatory responses. In the present context, it is shown in Chapter 7 that, despite their appeal as approximations to systems that exhibit fading memory behavior, Volterra models are inherently incapable of exhibiting certain types of nonlinear behavior. Specifically, it is shown that nonlinear Volterra models cannot exhibit either homogeneous or positive-homogeneous behavior.

Another type of qualitative behavior that Volterra models cannot exhibit is output multiplicity, a point noted several times previously. In fact, if N exhibits the fading memory property on some set K, the response y_s to any constant sequence $u_s \in K$ is necessarily unique (Boyd and Chua, 1985). One approach to dealing with such output multiplicities would be to develop *local models*, each valid within the region of attraction of a particular steady state and combine these individual models into a global composite model. Fading memory models could then be considered as candidates for these local models. In fact, this approach is reasonably popular, using the simplest class of fading memory models—that is, linear models. This topic is the subject of the next chapter.

Chapter 6

Linear Multimodels

This chapter briefly discusses the class of *linear multimodels*, which are *globally nonlinear* dynamic models, obtained by "piecing together" several *local, linear* dynamic models. The motivation behind this approach to model development is the observation that an approximate model's complexity generally increases with the range of operation over which it must be valid. In particular, it has been noted repeatedly that linear models are often adequate approximations of process dynamics *over a sufficiently narrow operating range*. Thus, if the total operating range can be decomposed into small enough subsets, it is reasonable to expect that linear models will provide reasonable characterizations of process dynamics over these local regimes. The process of piecing these local models together may be approached in a number of different ways, and the details of this process are important, as subsequent examples illustrate.

The principal issues that must be addressed in developing linear multimodels are illustrated with the example of a batch fermentation reactor, described in Sec. 6.1. Three possible definitions of discrete-time linear multimodels are presented in Secs. 6.2.1 through 6.2.3. The first of these definitions is based on Johansen and Foss (1993), whereas the second definition represents an apparently slight variation on the first that can lead to fundamentally different qualitative behavior. The third definition of linear multimodels is a special case of the first, described in Tong (1990) under the name *open-loop threshold autoregressive models*. The primary difference between Tong's definition and that of Johansen and Foss is whether the regions of local model validity can overlap: they can in the models of Johansen and Foss, but they cannot in Tong's model.

An extremely important practical issue is the criterion by which local models are selected. The primary focus here is on Tong's class of linear multimodels where each local model completely describes the global model dynamics over some specified operating range. If this operating range is defined entirely in terms of the input sequence $\{u(k)\}$, these models will be designated *input-selected*, whereas if the operating range is defined entirely in terms of the output sequence $\{y(k)\}$, they will be called *output-selected*; if both inputs and outputs are involved, the term *generally selected* will be used. In practice, the generally selected case can be quite complex, involving *dynamic* combinations of inputs and outputs (e.g., state estimates obtained from a Kalman filter), but the examples considered here are fairly simple, selected to illustrate the basic ideas. In addition, another practically important multimodel strategy is to base regime selection on some external variable (e.g., some important measurable disturbance). Despite its potential importance, this last case is not considered here because such consideration requires an explicit consideration of external disturbances that lies beyond the scope of this book.

These distinctions are important because the different classes of linear multimodels behave quite differently, in general: input-selected linear multimodels behave much like NMAX models, output-selected linear multimodels behave much like NARX models, and generally selected linear multimodels can exhibit the full complexity of general NARMAX models. These different model classes are briefly introduced in Sec. 6.3, where a pair of local models is described that will serve as the basis for detailed examples considered in subsequent sections. Strictly speaking, the multimodels most widely considered in practice are *affine multimodels*, including nonzero constant terms to shift local steady-states from $y_s = u_s = 0$ to some other point in the (u_s, y_s) plane. One of the local models presented in Sec. 6.3 is strictly linear, whereas the other is affine; differences in qualitative behavior between affine models and strictly linear models are also discussed briefly in Sec. 6.3. Secs 6.4 through 6.6 then consider input-selected, output-selected, and generally selected multimodels, comparing the influence of these different selection schemes in multimodels based on the local models discussed in Sec. 6.3.

An interesting family of multimodels that is based on strictly linear local models is the class of TARMAX models introduced in Chapter 3, as a special case of the class of positive-homogeneous models. The fact that these models are linear multimodels is demonstrated in Sec. 6.7, where it is also shown that any structurally additive positive-homogeneous NMAX or NARX model can be represented as a TARMAX model, thus generalizing the linear realization results discussed at the end of Chapter 2. In addition, it follows as an immediate corollary that structurally additive, homogeneous NMAX or NARX models are necessarily linear. The case of positive-homogeneous NARX* models is somewhat more complex, and suggests the still greater complexity faced in attempting to extend these results to the general additive NARMAX class. Finally, Sec. 6.8 gives a brief summary of the main results presented here for class of linear multimodels.

6.1 A motivating example

The following example is representative of the application of linear multimodels to the control of chemical processes. Much more detailed discussions are given in the papers on which this example is based (Foss et al., 1995; Johansen and Foss, 1995, 1997), but the basic problem is the following. A stirred vessel is initially filled with a culture of the microorganism *Pseudomonas ovalis*,

Linear Multimodels

glucose, and water. The microorganism cells in this vessel consume glucose and oxygen and reproduce, generating more cells. In addition, the cells produce gluconolactone, which reacts with water to form gluconic acid, the desired end product. Assuming constant temperature operation, the kinetics of these three reactions are described by the following five-dimensional, continuous-time, nonlinear state-space model:

$$\dot{\chi} = \left[\frac{\mu_m sc}{k_s c + k_0 s + sc}\right]\chi,$$

$$\dot{p} = k_p \ell,$$

$$\dot{\ell} = \left[\frac{v_\ell s}{k_\ell + s}\right]\chi - 0.91 k_p \ell,$$

$$\dot{s} = -\left[\frac{\mu_m sc}{Y_s[k_s c + k_0 s + sc]} + \frac{1.011 v_\ell s}{k_\ell + s}\right]\chi,$$

$$\dot{c} = k_\ell a(c^* - c) - \left[\frac{\mu_m sc}{Y_0[k_s c + k_0 s + sc]} + \frac{0.09 v_\ell s}{k_\ell + s}\right]\chi. \tag{6.1}$$

The state variables in this model are the concentrations of the cells (χ), gluconic acid (p), gluconolactone (ℓ), glucose (s), and dissolved oxygen (c). The other terms in these equations are constants defined in Johansen and Foss (1995).

Based on a qualitative understanding of these reaction kinetics, it is possible to define different operating regimes. In the fermentation example, an examination of the reaction mechanisms suggests the following four regimes:

1. Initial growth regime
2. Intermediate regime A
3. Intermediate regime B
4. Final production regime.

The initial growth regime is characterized by a small initial cell contentration, resulting in little production of gluconolactone and gluconic acid. Both dissolved oxygen and glucose concentrations are high in this regime. The two intermediate regimes are characterized by rapid cell growth, accompanied by high production of gluconolactate. Both dissolved oxygen and glucose concentrations decrease with time and the difference between intermediate regimes A and B lies in whether the oxygen concentration is low enough to be rate-limiting or not. In the final production regime, both cell growth and glucolactone production are limited by a shortage of glucose and dissolved oxygen concentration again increases.

Because these different operating regimes involve different dominant kinetic mechanisms and different magnitudes of the state variables, it is possible to approximate the response of the nonlinear state-space model (6.1) by a collection of simpler local models. In Foss et al. (1995) and Johansen and Foss (1997), the four-regime decomposition described in the previous paragraph is used as a basis for multimodeling. Alternatively, it is also possible to perform this regime decomposition empirically, as in Johansen and Foss (1995), where a five-regime decomposition is obtained for the same example. The important question of how to select these operating regimes is discussed in all three of these papers.

Johansen and Foss describe a number of different model structures in their papers, and it is useful to briefly discuss two of them here, the second being a specialization of the first. For the more general structure, suppose that N operating regions are identified and that the dynamics each of these regions may be approximated by a state-space model of the form

$$\dot{x}^{(j)} = f^{(j)}(\mathbf{x}^{(j)}, u) \qquad y^{(j)} = g^{(j)}(\mathbf{x}^{(j)}). \tag{6.2}$$

These *local* models, indexed by the superscript (j), are then combined to form the following *global multimodel*:

$$\dot{\mathbf{x}} = \sum_{j=1}^{N} f^{(j)}(\mathbf{x}, u) w^{(j)}(\phi), \qquad y = \sum_{j=1}^{N} g^{(j)}(\mathbf{x}) w^{(j)}(\phi), \tag{6.3}$$

where $w^{(j)}(\phi)$ are weights assigned to the N local models, as a function of the *operating regime* ϕ. In the example considered here, $\phi(t) = [c(t), s(t)]^T$ is the dissolved oxygen/glucose concentration vector at time t. The fundamental working assumption underlying multimodeling is that such an operating regime vector can be defined. Given this vector, the reasonableness of model j at a particular operating point $\phi(t)$ is then described by a *validity function* $\rho^{(j)}(\phi)$ that assumes values in the interval $[0, 1]$. That is, if the operating regime $\phi(t)$ lies in the region of validity of model j, then $\rho^{(j)}(\phi) \simeq 1$. Usually, $\rho^{(\ell)}(\phi) \simeq 0$ for $\ell \neq j$, although it is possible to consider multimodels based on *overlapping* regimes of validity; in such cases, $\rho^{(j)}(\phi)$ may be significantly different from zero for more than one value of j. The weights $w^{(j)}(\phi)$ are simply normalized versions of these validity functions, that is

$$w^{(j)}(\phi) = \frac{\rho^{(j)}(\phi)}{\sum_{\ell=1}^{N} \rho^{(\ell)}(\phi)}. \tag{6.4}$$

Viewed as a family of functions of ϕ, these *interpolating functions* satisfy the following conditions for all ϕ:

$$w^{(j)}(\phi) \in [0, 1] \qquad \sum_{j=1}^{N} w^{(j)}(\phi) = 1. \tag{6.5}$$

Taken together, Eqs. (6.2), (6.3), and (6.4) constitute one possible definition of a multimodel, although it is important to note the following subtle point. For the global state vector $\mathbf{x}(t)$ in Eq. (6.3) to be well defined, it is necessary for all of the N local state vectors $\mathbf{x}^{(j)}(t)$ to be of the same dimension. In fact, in the best circumstances, these state vectors are *the same*. For example, in the fermentor example considered here, all models would be based on the state vector $\mathbf{x} = [\chi, p, \ell, s, c]^T$. Johansen and Foss (1993) discuss two possible approaches whenever these state vectors are *not* the same, but this particular multimodeling formulation is best suited to cases in which the local models involve the same state variables. More generally, the multimodeling approach seems best suited to situations in which the *structure* of the different local models is the same. As a specific example, the second structure considered by Johansen and Foss is the specialization of this general multimodel formulation to the case of *affine local models, all with the same structure*. They note that in this case, the resulting linear multimodel may be represented as

$$\dot{\mathbf{x}} = \mathbf{a}(\phi) + \mathbf{A}(\phi)\mathbf{x} + \mathbf{b}(\phi)u \qquad y = d(\phi) + \mathbf{c}^T(\phi)\mathbf{x}. \qquad (6.6)$$

In other words, the linear multimodel may be viewed as an affine model whose parameters depend on the operating regime vector ϕ. The global nonlinearity of this model arises from the fact that the operating regime $\phi(t)$ depends on the input $u(t)$, the output $y(t)$, or the state vector $\mathbf{x}(t)$. In the simplest case, this dependence is discontinuous: the local regions $\Omega^{(j)}$ are disjoint so at any time t, $\phi(t)$ lies in only one region j, implying $\mathbf{A}(\phi)_{m,n} = A_{m,n}^{(j)}$. Note that this choice corresponds to taking $\rho(\phi)$ as the *characteristic function* for this region, that is

$$\rho^{(j)}(\phi) = \begin{cases} 1 & \phi \in \Omega^{(j)} \\ 0 & \phi \notin \Omega^{(j)}. \end{cases} \qquad (6.7)$$

This particular choice is closely related to the discrete-time TARSO model class (Tong, 1990), discussed in some detail in Sec. 6.2.3.

It bears repeating that an extremely important feature of the model defined by Eq. (6.6) is that the local models are not strictly linear, but rather *affine*, as noted in the introduction. Specifically, the right-hand sides of the state equation and observation equation in this model both include constant terms. This inclusion is an important feature of linear multimodels, and is necessary to account for the fact that the operating points about which the N models are centered *are different*. In developing strictly linear models, it is common to work in *deviation variables*, relating deviations of input and output variables from some specified "nominal" values. Thus, it is at least conceptually possible to separate the specification of these nominal values from the development of a linear dynamic model. Linear multimodels generally combine linear models that are centered at *different* operating points. Consequently, the steady-state input, output, and state-vector values for these local linear models will generally all be nonzero. To accommodate this behavior, it is necessary to include the constant terms $\mathbf{a}(\phi)$ and $d(\phi)$ in Eq. (6.6). This inclusion introduces some slight differences in the qualitative behavior of the local models relative to the more familiar strictly linear models discussed in Chapter 2; this point is explored briefly in Sec. 6.3 for the discrete-time case.

Ultimately, Foss et al. (1995) develop affine, *discrete-time* multimodels to describe batch fermentor dynamics. These models are the logical end result of the linear multimodeling process, based on the near-equivalence of continuous-time and discrete-time *linear* models. Specifically, for the fermentor example considered here, five-dimensional linear, discrete-time state-space models were developed for each of the four operating regimes described earlier in this section.

As in the continuous-time case, it was necessary to include constant terms to these models to account for the required nonzero steady-state values of input, output, and state variables. In this particular example, the models considered were of the form

$$\mathbf{x}(k+1) = \mathbf{a}_j + \mathbf{A}_j \mathbf{x}(k) + \mathbf{b}_j u \qquad y = d_j + \mathbf{c}_j^T \mathbf{x}(k). \tag{6.8}$$

where $\mathbf{x}(k)$ represented the value of the five-dimensional state vector in the continuous-time formulation at time $t = t_k$. For the model predictive control application they considered, the authors found that the linear multimodel approach gave results that were quite comparable to those obtained using the detailed nonlinear model and much better than those obtained from a single linear model.

6.2 Three classes of multimodels

It has often been said that "the devil is in the details," a statement that is seldom more true than in the case of multimodeling. In particular, though the basic concept is simple enough—describe globally nonlinear system dynamics by piecing together local linear approximations—both the precise formulation of the model equations and the specific choice of local model selection criteria can profoundly influence qualitative model behavior. For this reason, the following discussion presents three possible definitions of multimodels and considers some of the differences between them. The first is based on Johansen and Foss (1993), the second is an apparently slight modification of this formulation, and the third is a definition proposed by Tong (1990) in the context of time-series modeling that may be viewed as a special case of the first.

6.2.1 Johansen-Foss discrete-time models

Johansen and Foss (1993) describe a linear multimodel approximation for the general NARMAX model

$$y(k) = F(\psi(k)), \tag{6.9}$$

where $\psi(k)$ is the vector of dimension $p + q + 1$:

$$\psi(k) = [y(k-1), \ldots, y(k-p), u(k), \ldots, u(k-q)]^T. \tag{6.10}$$

The function $F(\cdots)$ is assumed to be reasonably well-behaved (e.g., smooth) and the problem of interest is to approximate it on some set $\Psi \subset R^{p+q+1}$. The proposed solution is to first assume there exists a collection $\{\Psi_j\}$ of subsets of Ψ such that the following conditions hold:

$$\Psi_j \subset \Psi \text{ for all } j, \qquad \bigcup_{j=1}^{N} \Psi_j = \Psi, \tag{6.11}$$

Linear Multimodels

and that there also exists a collection local validity functions $\rho^{(j)} : \Psi \to [0,1]$ such that

$$\rho^{(j)}(\psi) \simeq 1 \text{ if } \psi \in \Psi_j, \qquad \rho^{(j)}(\psi) \simeq 0 \text{ if } \psi \notin \Psi_j. \tag{6.12}$$

Johansen and Foss note that for any function $F(\cdots)$ and any $\psi \in \Psi$, the following result is trivially true:

$$F(\psi) = \frac{\sum_{j=1}^{N} F(\psi)\rho^{(j)}(\psi)}{\sum_{j=1}^{N} \rho^{(j)}(\psi)}. \tag{6.13}$$

The utility of this result lies in the observation that, given N *local approximations* $\Phi^{(j)}(\cdot)$ such that

$$\Phi^{(j)}(\psi) \simeq F(\psi) \text{ for all } \psi \in \Psi_j, \tag{6.14}$$

the following *global approximation* may be constructed for $F(\psi)$:

$$F(\psi) \simeq \frac{\sum_{j=1}^{N} \Phi^{(j)}(\psi)\rho^{(j)}(\psi)}{\sum_{j=1}^{N} \rho^{(j)}(\psi)}. \tag{6.15}$$

In particular, taking the approximating functions $\Phi^{(j)}$ to be local affine approximations of $F(\cdot)$ on the sets Ψ_j leads to the following definition of a discrete-time linear multimodel:

$$\Phi^{(j)}(\psi) = \mathbf{g}_j^T \psi + a_0^{(j)}. \tag{6.16}$$

This result may be written in somewhat more familiar form by expressing the vectors $\mathbf{g}_j$ as

$$\mathbf{g}_j = [a_1^{(j)}, a_2^{(j)}, \ldots, a_p^{(j)}, b_0^{(j)}, b_1^{(j)}, \ldots, b_q^{(j)}]^T. \tag{6.17}$$

Define the local approximations $y^{(j)}(k)$ as

$$y^{(j)}(k) = \Phi^{(j)}(\psi) = a_0^{(j)} + \sum_{i=1}^{p} a_i^{(j)} y(k-i) + \sum_{i=0}^{q} b_i^{(j)} u(k-i), \tag{6.18}$$

and note that the original NARMAX model defined in Eq. (6.9) is approximated by the linear multimodel

$$y(k) = \sum_{j=1}^{N} y^{(j)}(k) w^{(j)}(k), \tag{6.19}$$

where the weights $\{w^{(j)}(k)\}$ are given by

$$w^{(j)}(k) = \frac{\rho^{(j)}(\psi(k))}{\sum_{j=1}^{N} \rho^{(j)}(\psi(k))}. \tag{6.20}$$

Since the vector $\psi(k)$ appearing in this expression depends on past inputs and/or outputs, it is the weights $w^{(j)}(k)$ that are responsible for the model's nonlinearity, as in the batch fermentation example discussed in Sec. 6.1.

In Johansen and Foss (1993), the linear multimodels defined by Eqs. (6.18), (6.19), and (6.20) are considered as approximations to the NARMAX model defined in Eqs. (6.9) and (6.10). The quality of this approximation depends on a number of factors, including the coefficients appearing in Eq. (6.18), the definition of the functions $\rho^{(j)}(\cdot)$, the choice of the subsets Ψ_j, and the number N of subsets chosen. The *qualitative behavior* of this linear multimodel will depend strongly on how the subsets Ψ_j are chosen, a point that is illustrated in Secs. 6.4 through 6.6.

6.2.2 A modifed Johansen-Foss model

To see the influence of an apparently small change in the multimodel problem formulation, it is useful to consider the following modification of the Johansen-Foss model. Specifically, replace Eq. (6.18) with

$$y^{(j)}(k) = a_0^{(j)} + \sum_{i=1}^{p} a_i^{(j)} y^{(j)}(k-i) + \sum_{i=0}^{q} b_i^{(j)} u(k-i), \qquad (6.21)$$

Combining this result with Eqs. (6.10), (6.19) and (6.20) as before leads to another possible definition of linear multimodels. Although this defintion *appears* quite similar to that of Johansen and Foss, there is an extremely profound, subtle difference. Specifically, note that in the class of linear multimodels defined by Johansen and Foss, the autoregressive variables appearing in Eq. (6.18) are past lags of the *global model output* $y(k)$. In Eq. (6.21), these global model outputs have been replaced with past lags of *local model outputs* $y^{(j)}(k)$. Consequently, whereas the variables $y^{(j)}(k)$ defined in Eq. (6.18) for the Johansen and Foss model are influenced by the past history of the sequence $w^{(j)}(k)$, those defined in Eq. (6.21) are not.

This apparently slight notational change can have profound consequences, as subsequent examples illustrate. In particular, because the models defined by Eq. (6.21) may be viewed as linear models with a constant offset, they may be equally well represented by their associated impulse responses $h^{(j)}(k)$, that is

$$y^{(j)}(k) = a_0^{(j)} + \sum_{i=0}^{\infty} h^{(j)}(i) u(k-i). \qquad (6.22)$$

The response of the resulting multimodel is then given by

$$y(k) = \sum_{j=1}^{N} \left[a_0^{(j)} + \sum_{i=0}^{\infty} h^{(j)}(i) u(k-i) \right] w^{(j)}(k). \qquad (6.23)$$

As it stands, this model represents an infinite-order NARMAX model, since it depends on the complete past history of the input sequence $\{u(k)\}$. In special cases, there may be simpler (i.e., finite-dimensional) representations, but it is not obvious what these special cases are. In contrast, note that in the Johansen and Foss model, combining Eqs. (6.18), (6.19), and (6.20) yields a NARMAX model with finite order parameters p and q.

Another subtle point regarding this alternative definition of linear multimodels is the precise definition of the weights $w^{(j)}(k)$. In particular, note that there are at least two options: $w^{(j)}(k)$ can be defined in terms of the validity functions $\rho^{(j)}(\psi)$ and the *global* vector $\psi(k)$ defined in Eq. (6.10) as before, or $\psi(k)$ can be replaced with a *local* vector $\psi^{(j)}(k)$ defined in terms of the local model output lags $y^{(j)}(k-i)$, that is

$$\psi(k) \rightarrow \psi^{(j)}(k) = [y^{(j)}(k-1), \ldots, y^{(j)}(k-p), u(k), \ldots, u(k-q)]^T. \quad (6.24)$$

Arguments could be advanced in support of either choice. For example, it could be argued that the weights $w^{(j)}(k)$ should provide a *global* assessment of the validity of each local model and should therefore be based on global model responses $y(k)$. Conversely, in a decentralized control scheme, Eq. (6.24) provides a basis for *local generation* of both model predictions $y^{(j)}(k)$ and the model validity function $\rho^{(j)}(\psi)$, from which the weights $w^{(j)}(k)$ could be computed via Eq. (6.20).

A practical difficulty can arise in the decentralized case of local generation of model validity functions. Specifically, in this case, there is no mechanism to prevent all local validity functions $\rho^{(j)}(\psi)$ from being zero simultaneously. In fact, this situation arises in the output-selected multimodel example considered in Sec. 6.5.3, and leads to times k for which $w^{(j)}(k)$ *cannot* be computed via Eq. (6.20). In such circumstances, it would be necessary to impose some alternative rule—implemented by the global coordinator—to guarantee that the global model output $y(k)$ is well defined at all times. Because of this difficulty, the examples considered in Secs. 6.4 through 6.6 focus primarily on the centralized modeling scheme where $\psi(k)$ is defined as in Eq. (6.10). The key point is that such choices may have significant effects on the qualitative behavior of the resulting linear multimodel. If the model validity functions $\rho^{(j)}(\cdot)$ depend only on the input variables $u(k), \ldots, u(k-q)$, however, there is no difference between the centralized and decentralized formulations.

6.2.3 Tong's TARSO model class

Though Tong does not refer to them as linear multimodels, the class of *open-loop threshold autoregressive systems* or *TARSO models* (Tong, 1990, p. 101) may be viewed as a special case of the linear multimodels of Johansen and Foss. In Tong's notation, these models are defined by the equation:

$$X_t = a_0^{(j)} + \sum_{i=1}^{m_j} a_i^{(j)} X_{t-i} + \sum_{i=0}^{m'_j} b_i^{(j)} Y_{t-i} + \epsilon_t^{(j)}. \quad (6.25)$$

In this notation, t represents the discrete time index denoted k in this book, X_t represents the observed model response, analogous to $y(k)$ in Eq. (6.10), and Y_t

represents an exogeneous input, analogous to $u(k)$ in Eq. (6.10). The constants $a_0^{(j)}$, $a_i^{(j)}$ and $b_i^{(j)}$ for $j = 1, 2, \ldots, \ell$ represent ℓ sets of linear ARMAX model coefficients, identical to those appearing in Eq. (6.18). Because Tong is concerned with nonlinear modeling of stochastic processes, the term $\epsilon_t^{(j)}$ represents a family of ℓ mutually independent zero-mean Gaussian white noise sequences.

In Tong's model, the coefficients superscripted by (j) are selected by a time-varying indicator variable J_t. That is, if $J_t = j$, then the coefficients $a_0^{(j)}$, $a_i^{(j)}$, and $b_i^{(j)}$ are selected in Eq. (6.25) to determine X_t. Tong notes that this general scheme can be extremely flexible, depending on exactly how J_t is defined. In his description of the class of open-loop threshold autoregressive systems, Tong defines J_t to be a function of the exogenous variable Y_t, although he concludes the discussion with the comment, "naturally, there are situations where other choices of J_t are more appropriate." The key point here is that J_t assumes one of ℓ discrete values at any given time t, so X_t evolves from X_{t-j} and Y_{t-j} according to the linear model Eq. (6.25) specified by this discrete value.

To establish the connection between Tong's model and that of Johansen and Foss, take $\rho^{(j)}$ in the Johansen and Foss model to be the characteristic function for the set Ψ_j:

$$\rho^{(j)}(\psi) = \begin{cases} 1 & \psi \in \Psi_j, \\ 0 & \psi \notin \Psi_j. \end{cases} \tag{6.26}$$

Under this assumption, the weights $w^{(j)}(k)$ defined by Eq. (6.20) are

$$w^{(j)}(k) = \begin{cases} 1 & \psi(k) \in \Psi_j, \\ 0 & \psi(k) \notin \Psi_j. \end{cases} \tag{6.27}$$

Substituting this expression into Eqs. (6.19) leads to the following simplified representation:

$$y(k) = y^{(j)}(k) \text{ if } \psi(k) \in \Psi_j, \tag{6.28}$$

where $y^{(j)}(k)$ is the local linear model response defined in Eq. (6.18). Therefore, making the associations between X_t and $y(k)$ and Y_t and $u(k)$ noted above and omitting the stochastic driving term $\epsilon_t^{(j)}$, Tong's model becomes a special case of the linear multimodel of Johansen and Foss. Specifically, the local model responses appearing in Eq. (6.28) are

$$y^{(j)}(k) = a_0^{(j)} + \sum_{i=1}^{p} a_i^{(j)} y(k-i) + \sum_{i=0}^{q} b_i^{(j)} u(k-i). \tag{6.29}$$

Since the class of TARSO models is a special case of the multimodel definition of Johansen and Foss, note that the local model output $y^{(j)}(k)$ exhibits autoregressive dependence on the *global* model output $y(k-i)$ rather than the *local* model output $y^{(j)}(k-i)$. Note also that for the TARSO model representation to be well-defined, it is necessary for the sets Ψ_j to be disjoint so that for any $\psi(k) \in \Psi$, it follows that $\psi(k) \in \Psi_j$ for precisely one subset index j.

6.3 Two important details

Because they are both central to the linear multimodeling problem, this section briefly considers two important details: local model selection criteria, and the differences between linear and affine models.

6.3.1 Local model selection criteria

Much of the art of linear multimodeling is inherent in the choice of the sets Ψ_j that define the "regions of validity" of the local models. In the batch fermentation example considered by Johansen and Foss, these sets were defined by the operating regime vector ϕ, which may depend on essentially any measurable or computable quantity. Conversely, one of the key points of this chapter is that the qualitative behavior of the linear multimodel is strongly dependent on the details of these choices. Consequently, it is useful to distinguish between the following special cases of the Johansen-Foss multimodel class:

1. For *input-selected* models, $\rho^{(j)}(\psi(k))$ depends only on inputs $u(k-i)$
2. For *output-selected* models, $\rho^{(j)}(\psi(k))$ depends only on outputs $y(k-i)$.

Because Tong's TARSO model class is a subset of the Johansen-Foss model class, these definitions are also applicable to TARSO models. Similarly, the definition of an input-selected multimodel given in (1) above is also applicable to the modified class of multimodels defined in Sec. 6.2.2. The definition of output-selected multimodels given in (2) is applicable without modification to the *centralized* formulation discussed in Sec. 6.2.2, but in the decentralized formulation, the global outputs $y(k-i)$ would be replaced with the local outputs $y^{(j)}(k-i)$. Multimodels based on both past inputs and past outputs will be referred to as *generally selected*.

Differences between these selection criteria can be quite profound, a point that is illustrated in Secs. 6.4 through 6.6 with examples based on the following two local models:

$$y(k) = \frac{1}{2}y(k-1) + u(k-1), \tag{6.30}$$

$$y(k) = a - \frac{1}{2}y(k-1) + u(k-1). \tag{6.31}$$

In Sec. 6.4, input-selected multimodels are developed from these local models, based on both the Johansen-Foss (J-F) formulation described in Sec. 6.2.1 and the modified formulation described in Sec. 6.2.2. Similarly, in Sec. 6.5, output-selected multimodels are developed from these same local models, again based on both the J-F and modified formulations (in particular, the centralized version of the modified formulation). Finally, in Sec. 6.6, generally selected multimodels based on these two formulations are compared. In all cases, the regions of model validity are nonoverlapping, so the J-F models considered are in fact members of Tong's TARSO family.

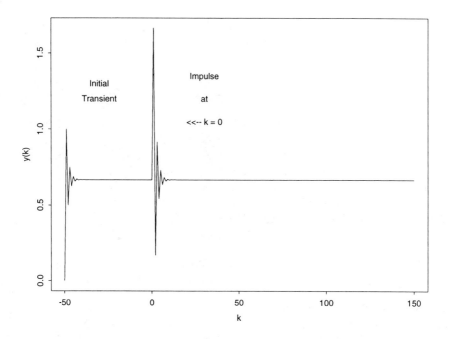

Figure 6.1: Affine model response to standard impulse input

6.3.2 Affine vs. linear models

An important distinction between the local models (6.30) and (6.31) is that the first is a strictly linear model, but the second is affine. A fundamental difference is that if $u(k) = u_s = 0$ in the linear model, $y(k) = y_s = 0$, whereas for the affine model $y(k) = y_s = (2/3)a$. In fact, this difference is precisely the motivation for basing multimodels on affine local models, as noted earlier. An important subtle point, however, is that while the standard working assumption, "the model is at steady-state for $k < 0$" implies the initial condition $y(k) = 0$ for $k < 0$, this assumption implies the nonzero initial condition $y(k) = y_s = 2a/3$ for the affine model (6.31). It is easy to overlook this point in simulations, leading to sometimes unexpected numerical results.

As a specific example, Fig. 6.1 shows the response of this model for $a = 1$ to the standard impulse input introduced in Chapter 2, adopting the inappropriate standard working assumption $y(k) = 0$ for all $k < 0$. The model's response to the impulse at $k = 0$ is exactly as expected: an oscillatory transient reflects the negative autoregressive coefficient $-1/2$ in the model, settling fairly quickly to the steady-state value $y_s = 2/3$. At $k = -50$, however, there is another oscillatory transient, reflecting the step change in steady-state value from $y(k) = 0$ imposed artificially for $k < 0$ to the natural steady-state value $y_s = 2/3$, corresponding to $u(k) = 0$ for $k = -50, -49, \ldots, -1$.

6.4 Input-selected multimodels

This section briefly considers the behavior of input-selected multimodels, beginning with the special case of local moving-average models in Sec. 6.4.1. This case is interesting because the resulting multimodel may be shown to be a member of the NMAX family, implying much about its qualitative behavior. In particular, it follows that these models can exhibit input multiplicity but not output multiplicity. In fact, this conclusion is more general, as is shown in Sec. 6.4.2: no input-selected multimodel based on affine local models can exhibit output multiplicity, even in cases where it is not equivalent to a finite-order NMAX model. In addition, there is no difference in the steady-state behavior of the Johansen-Foss and modified Johansen-Foss models, and the two formulations are equivalent for models based on input-selected local moving-average models. To see the difference between these model formulations, it is necessary to consider the dynamic behavior of multimodels with local autoregressive terms, and these differences are examined in Sec. 6.4.3 which presents a brief comparison of the qualitative dynamics of these two formulations. Finally, Sec. 6.4.4 presents a detailed discussion of two input-selected multimodel examples based on the local models defined in Eqs. (6.30) and (6.31) in Sec. 6.3.1.

6.4.1 Input-selected moving average models

Suppose the local linear models defining a linear multimodel are all $MA(q)$ models for some finite q and the local validity regions Ψ_j are defined in terms of input values alone. Under these assumptions, Eq. (6.18) reduces to

$$y^{(j)}(k) = a_0^{(j)} + \sum_{i=0}^{q} b_i^{(j)} u(k-i). \tag{6.32}$$

It follows that the global model response is

$$y(k) = \sum_{j=1}^{N} a_0^{(j)} w^{(j)}(k) + \sum_{j=1}^{N} \sum_{i=0}^{q} w^{(j)}(k) b_i^{(j)} u(k-i). \tag{6.33}$$

Since $w^{(j)}(k)$ depends only on the inputs by assumption, Eq. (6.33) may be written as a function of the input lags alone, implying that this particular model may be written as an NMAX model

$$y(k) = g(u(k), u(k-1), \ldots, u(k-q)), \tag{6.34}$$

for some (generally complicated) function $g : R^{q+1} \to R$. Consequently, the qualitative behavioral results presented in Chapter 4 for finite-order NMAX models all apply to this particular class of input-selected multimodels. In contrast, note that if the local model validity sets Ψ_j depend on past outputs $y(k-i)$, then the NARMAX representation for the linear multimodel of Johansen and Foss will involve nonlinear autoregressive behavior.

6.4.2 Steady states of input-selected models

Even if an input-selected linear multimodel model includes autoregressive terms, it is not difficult to show that such a model cannot exhibit output multiplicity. In particular, note that if $u(k) = u_s$ for all k, it follows that $w^{(j)}(k) = w_s^{(j)}$ for all k as well. Substituting this result into Eq. (6.19), it further follows that $y(k) = y_s$ for all k where y_s is given by

$$y_s = \left[1 - \sum_{j=1}^{N}\sum_{i=1}^{p} w_s^{(j)} a_i^{(j)}\right]^{-1} \sum_{j=1}^{N} w_s^{(j)} \left[a_0^{(j)} + \sum_{i=0}^{q} b_i^{(j)} u_s\right]. \qquad (6.35)$$

Note that this result reduces to the familiar steady-state characterization of a linear discrete-time model if $a_0^{(j)} = 0$, $a_i^{(j)} = a_i$, and $b_i^{(j)} = b_i$ for all j. That is, if all of the local models defining the linear multimodel are *identical*, the linear multimodel reduces to a simple linear model and Eq. (6.35) describes its steady-state behavior. Similarly, for Tong's TARSO models, note that since the regions of model validity are disjoint, it follows that if $u(k) = u_s$ for all k, $w^{(j)}(k) = 1$ for all k for some specific model index j and $w^{(\ell)}(k) = 0$ for all k if $\ell \neq j$. In that case, the steady-state response would be $y_s = y_0 + K u_s$, where

$$y_0 = \frac{a_0^{(j)}}{1 - \sum_{i=1}^{p} a_i^{(j)}} \qquad K = \frac{\sum_{i=0}^{q} b_i^{(j)}}{1 - \sum_{i=1}^{p} a_i^{(j)}} \qquad (6.36)$$

More generally, Eq. (6.35) establishes that, given a steady-state input u_s, the steady-state response y_s is unique. Depending on the details of the local model validity function $\rho^{(j)}(\psi(k))$, it may be possible for these models to exhibit *input multiplicity* but not output multiplicity.

6.4.3 J-F vs. modified J-F models

The general form of an input-selected linear multimodel in the Johansen-Foss formulation introduced in Sec. 6.2.1 is

$$y(k) = \sum_{j=1}^{N} y^{(j)}(k) w^{(j)}(k) \qquad (6.37)$$

$$y^{(j)}(k) = a_0^{(j)} + \sum_{i=1}^{p} a_i^{(j)} y(k-i) + \sum_{i=0}^{q} b_i^{(j)} u(k-i), \qquad (6.38)$$

where the weights $w^{(j)}(k)$ depend on the current and past input vector

$$\psi(k) = [u(k), \ldots, u(k-r)]^T, \qquad (6.39)$$

for some finite $r \geq 0$. In the modified Johansen-Foss formulation, Eq. (6.38) is replaced with the alternative local model definition

$$y^{(j)}(k) = a_0^{(j)} + \sum_{i=1}^{p} a_i^{(j)} y^{(j)}(k-i) + \sum_{i=0}^{q} b_i^{(j)} u(k-i). \qquad (6.40)$$

Linear Multimodels 257

It is important to emphasize that the distinction between these two model formulations lies in the autoregressive terms included in Eq. (6.38) vs. Eq. (6.40). Specifically, the J-F model formulation includes the *global* autoregressive terms $y(k - i)$ defined by Eq. (6.37), whereas the modified J-F model formulation includes the *local* autoregressive terms $y^{(j)}(k - i)$ defined by Eq. (6.40). This difference has significant behavioral consequences, a point that is illustrated here in general terms and illustrated further in Sec. 6.4.4 in connection with two specific examples.

Given these formulations, it is possible to establish a number of useful results concerning the qualitative behavior of the model's response to impulses, steps, and periodic input sequences. For the impulse and step response results, note that both of these inputs satisfy $u(k) = u_s$ for all $k > 0$. Hence, $\psi(k) = [u_s, \ldots, u_s]^T$ for all $k > r$ and it follows that $w^{(j)}(k) = w_s^{(j)}$ for all $k > r$. As a consequence, note that the impulse and step response of any input-selected linear multimodel is necessarily that of an affine model. The specific form of this affine model depends on the formulation, as the following results illustrate. For the J-F model, responses for $k > r$ are governed by the equivalent affine model

$$y(k) = \tilde{a}_0 + \sum_{i=1}^{p} \tilde{a}_i y(k - i) + \sum_{i=0}^{q} \tilde{b}_i u(k - i), \quad (6.41)$$

whose coefficients are defined by

$$\tilde{a}_0 = \sum_{j=1}^{N} a_0^{(j)} w_s^{(j)} \quad \tilde{a}_i = \sum_{j=1}^{N} a_i^{(j)} w_s^{(j)} \quad \tilde{b}_i = \sum_{j=1}^{N} b_i^{(j)} w_s^{(j)}. \quad (6.42)$$

One important difference between these responses is that $u_s = 0$ for an impulse, regardless of the amplitude, whereas u_s reflects the nonzero step amplitude. Therefore, the impulse response of an input-selected linear multimodel can exhibit amplitude dependence only in its transient behavior, whereas the step response can exhibit more general amplitude dependence.

For the modified J-F formulation, the analogous results follow from the impulse response representation given in Eq. (6.22) for the local affine models. In fact, substituting $w^{(j)}(k) = w_s^{(j)}$ into Eq. (6.23) yields, for $k > r$

$$y(k) = \tilde{a}_0 + \sum_{i=0}^{\infty} \tilde{h}(i) u(k - i), \quad (6.43)$$

where $\tilde{a}_0$ is as in Eq. (6.42) and the effective impulse response is

$$\tilde{h}(i) = \sum_{j=1}^{N} w_s^{(j)} h^{(j)}(i). \quad (6.44)$$

As in the J-F formulation, note that the amplitude dependence of the step response is potentially much more complex than that of the impulse response since $u_s = 0$ for all impulses, regardless of their intensity.

Finally, it is instructive to consider the response of these models to periodic input sequences, following the same general approach taken in Chapter 3 for the class of bilinear models. Specifically, assume that the input sequence $\{u(k)\}$ satisfies the periodicity condition $u(k+P) = u(k)$ for all k and some $P > 0$. Next, define $z(k) = y(k+P) - y(k)$ and note that if the model response $\{y(k)\}$ is also periodic with period P, then the sequence $\{z(k)\}$ is identically zero. Because the model is input-selected, it follows from Eq. (6.39) that $w^{(j)}(k+P) = w^{(j)}(k)$ for all k. Hence, in either multimodel formulation, it follows that

$$z(k) = \sum_{j=1}^{N} w^{(j)}(k) z^{(j)}(k), \qquad (6.45)$$

where $z^{(j)}(k)$ is formulation-specific. For the J-F formulation, this sequence is defined by

$$z^{(j)}(k) = y^{(j)}(k+P) - y^{(j)}(k) = \sum_{i=1}^{p} a_i^{(j)} z(k-i), \qquad (6.46)$$

from which it follows that

$$z(k) = \sum_{i=1}^{p} \alpha_i(k) z(k-i). \qquad (6.47)$$

The coefficients in this equivalent time-varying linear model are

$$\alpha_i(k) = \sum_{j=1}^{N} w^{(j)}(k) a_i^{(j)}. \qquad (6.48)$$

As in the case of the completely bilinear model considered in Chapter 3, note that one solution of Eq. (6.47) is $z(k) \equiv 0$. Further, note that if this equation exhibits a nonzero solution $\{z(k)\}$, then $\{\lambda z(k)\}$ is also a solution for any real λ. Hence, as in the bilinear model case, such nonzero solutions may be viewed as transients initiated when a periodic input sequence is applied at some time $k = 0$: so long as the time-varying model defined by Eqs. (6.47) and (6.48) is stable, this transient will decay to zero as $k \to \infty$. Consequently, subharmonic generation is generally beyond the capability of input-selected J-F models exactly as in the case of bilinear models. Similar conclusions may be drawn for the modified J-F formulation. There, the sequence $\{z^{(j)}(k)\}$ is defined by

$$z^{(j)}(k) = y^{(j)}(k+P) - y^{(j)}(k) = \sum_{i=1}^{p} a_i^{(j)} z^{(j)}(k-i), \qquad (6.49)$$

from which it follows that any nonzero transient $z(k)$ will decay to zero if all of the N local models are stable. Note how much simpler these stability criteria are than the corresponding criteria for the J-F formulation. In particular, note that the modified J-F criteria do not involve either stability assessment of time-varying linear models or the time variation of the input sequence.

Linear Multimodels 259

6.4.4 Two input-selected examples

The previous section illustrated some of the general differences between the J-F and modified J-F formulations for input-selected linear multimodels. The following section briefly compares two simple examples to provide a more detailed illustration of the similarities and differences between these model formulations. For the J-F formulation, consider the multimodel

$$y(k) = \begin{cases} \frac{1}{2}y(k-1) + u(k-1) & |u(k-1)| \le 1 \\ a - \frac{1}{2}y(k-1) + u(k-1) & |u(k-1)| > 1, \end{cases} \quad (6.50)$$

based on the local models (6.30) and (6.31) defined in Sec. 6.3.1. Similarly, for the modified J-F formulation, consider the alternative multimodel

$$y^{(1)}(k) = \frac{1}{2}y^{(1)}(k-1) + u(k-1)$$

$$y^{(2)}(k) = a - \frac{1}{2}y^{(2)}(k-1) + u(k-1)$$

$$y(k) = \begin{cases} y^{(1)}(k) & |u(k-1)| \le 1 \\ y^{(2)}(k) & |u(k-1)| > 1, \end{cases} \quad (6.51)$$

based on the same two local models. Again, it is important to emphasize the distinction between these two model formulations: in the J-F formulation, there is a single sequence $\{y(k)\}$ whose dynamics is dictated by the model selection criteria, whereas in the modified J-F formulation, *two* sequences $y^{(1)}(k)$ and $y^{(2)}(k)$ evolve at all times and the selection criteria determine which one is chosen for the overall model output. As noted in Sec. 6.2.2, since the particular model considered here is input-selected, there is no distinction between the centralized and decentralized versions of the modified J-F formulation. The following discussions briefly consider the steady-state behavior of these models, together with their responses to the four standard input sequences introduced in Chapter 2.

The steady-state behavior of the linear multimodels defined in Eqs. (6.50) and (6.51) is identical, arising from the combination of the steady-state behavior of the local linear models and the selection criterion $|u_s| \le 1$ vs. $|u_s| > 1$. These factors are illustrated in Fig. 6.2 for $a = 1$, which shows the steady-state y_s vs. u_s relationship for each of the local linear models defined in Eqs. (6.30) and (6.31). Specifically, these relationships are the lines

$$y_s = 2u_s \quad (6.52)$$

for the local model defined by Eq. (6.30) and

$$y_s = \frac{2}{3}a + \frac{2}{3}u_s \quad (6.53)$$

for the local model defined by Eq. (6.31). In addition, vertical lines at $u_s = \pm 1$ are shown that divide the (u_s, y_s)-plane into three regions of model validity,

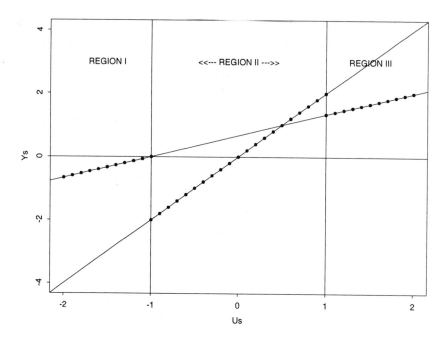

Figure 6.2: Input-selected multimodel steady states

denoted I, II, and III, respectively. Region II corresponds to the range $|u_s| \leq 1$ for which the local model defined by Eq. (6.30) is valid, whereas Regions I and III together constitute the disjoint set $|u_s| > 1$ for which the local model defined by Eq. (6.31) is valid.

The overall steady-state response function for the multimodel defined by Eq. (6.50) or (6.51) for $a = 1$ is represented by the discontinuous curve indicated with overlaid circles in Fig. 6.2. As emphasized in Sec. 6.4.2, input-selected linear multimodels cannot exhibit output multiplicities, but they can exhibit input multiplicities, and this example illustrates this point. Specifically, note that for $-2 \leq y_s < 0$, u_s may assume either of two possible values: $u_s = y_s/2$ or $u_s = (3/2)y_s - 1$. Similarly, for $4/3 < y_s \leq 2$, u_s may assume the same two possible values, relative to y_s. As in a number of examples discussed previously, note that these input multiplicities represent "intermediate-range behavior." That is, there is no input multiplicity for the "inner" range of outputs $0 \leq y_s \leq 4/3$ or for the "outer" range of outputs $|y_s| > 2$.

Turning to the dynamic behavior of these multimodels, first consider the response to an impulse of amplitude γ applied at time $k = 0$. Note that for $|\gamma| \leq 1$, the selection criterion for the local model given by Eq. (6.30) is always satisfied, so the impulse response of both the J-F model and the modified J-F model is simply this linear model's response. For $|\gamma| > 1$, slight differences between the impulse responses for these two models appear, although they are

Linear Multimodels 261

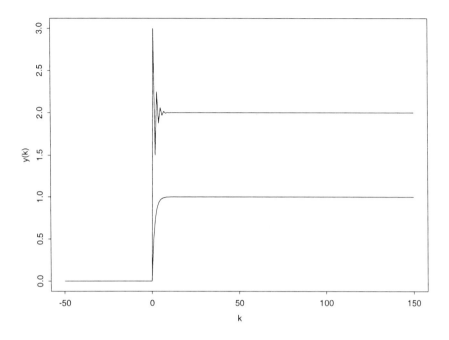

Figure 6.3: Step responses for the J-F model

not dramatic. Specifically, for $k = 1$, $y(1) = a + \gamma$ for both models since $|u(k-1)| = |\gamma| > 1$ and both multimodels select the local model defined by Eq. (6.31). Differences between these multimodel formuations appear for $k \geq 2$. For the J-F formulation, the local model defined by Eq. (6.30) is selected by the *global* initial condition $y(1) = a+\gamma$, yielding $y(2) = (a+\gamma)/2$. In contrast, for the modified J-F formulation, the local model defined by Eq. (6.30) is again selected, but with its own initial condition $y(1) = y^{(1)}(1) = \gamma$, yielding $y(2) = \gamma/2$. Since $|u(k-1)| = 0$ for all $k > 2$, the model defined by Eq. (6.30) remains selected in both multimodel formulations, so the response is $y(k) = y(2)/2^{k-2}$ for both cases, decaying at the same rate, but from different values for $y(2)$. In fact, this general behavior is precisely in line with that described in Sec. 6.4.3 for both of these models.

Plots of the responses of the J-F model to step inputs of amplitude $\gamma = 0.5$ and $\gamma = 2$ are shown in Fig. 6.3. For the smaller amplitude step, the response is monotonic, characteristic of the linear model defined by Eq. (6.30), which remains selected for all k since $|u(k-1)| \leq 1$ for all $k > 0$. In contrast, for the larger step, the linear model defined by Eq. (6.31) remains selected for all $k > 0$ and this model exhibits an oscillatory step response, consistent with its negative autoregressive coefficient $-1/2$. In fact, this behavior is exactly what would be expected for an input-selected linear multimodel: the input sequence determines which local linear model is valid, and this model remains in force at

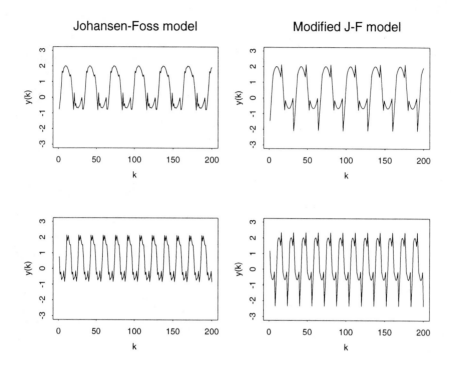

Figure 6.4: Sinusoidal model responses

all times. Note also that in this case, the response of the modified J-F model is identical to that of the original J-F model, since no transitions occur between models $y^{(1)}(k)$ and $y^{(2)}(k)$ for any $k > 0$, and it is precisely at these transitions where the differences between the two model formulations arise.

The sinusoidal responses for these models are much more interesting, clearly illustrating that the J-F and modified J-F formulations can behave quite differently. Again, for sufficiently small amplitudes, the local linear model defined by Eq. (6.30) remains valid for all k and the response to a sinusoidal input is simply the attenuated, phase-shifted sinusoidal response of this linear model. This result holds for both the J-F model and the modified J-F model, and is independent of the excitation frequency. In contrast, the response of these models to large-amplitude sinusoidal inputs exhibits highly nonsinusoidal behavior that reflects both the overall nonlinearity of the model and pronounced differences in qualitative behavior between the two multimodel formulations. This point is illustrated in Fig. 6.4, which shows the responses of both models to sinusoids of amplitude $\gamma = 2$. Plots on the left-hand side of the figure correspond to the J-F formulation, whereas plots on the right-hand side correspond to the modified formulation. Qualitative differences in the two multimodel formulations are seen clearly by comparing these two pairs of plots; in particular, note the pronounced negative "spikes" in the response of the modified model formulation that are not present in the J-F model response. The upper plots in this figure correspond

Linear Multimodels

to the frequency of the standard sinusoidal input sequence defined in Chapter 2, whereas the lower plots correspond to twice this frequency. Note that the general character of these model responses is not highly frequency dependent. Not surprisingly, the response of these multimodels to white noise excitations is complex enough that some of the differences seen clearly in the sinusoidal responses are present but not obvious from plots of these responses. For this reason, discussion of white noise responses is deferred until Sec. 6.6.1.

6.5 Output-selected multimodels

One of the principal results established in the previous section was that input-selected linear multimodels are incapable of exhibiting output multiplicity. Hence, if output multiplicity is required it follows that output-selection is also required. In fact, output-selected multimodels behave somewhat analogously to the class of NARX models introduced in Chapter 4, just as input-selected multimodels behave somewhat analogously to NMAX models. Sec. 6.5.1 illustrates this connection, establishing the dual of the result established in Sec. 6.4.1: certain output-selected autoregressive models are members of the NARX model class. Following this discussion, steady-state results are derived for the general class of output-selected multimodels in Sec. 6.5.2, again analogous to the input-selected results derived previously. Finally, Sec. 6.5.3 compares the behavior of the output-selected duals of the two simple models considered in Sec. 6.4.4.

6.5.1 Output-selected autoregressive models

Suppose the local models defining a linear multimodel are all $AR(p)$ models for some finite p and the local validity regions Ψ_j are defined in terms of output values alone. Further, assume that $b_0^{(j)} = 1$ for all j. Under these assumptions, Eq. (6.18) reduces to

$$y^{(j)}(k) = a_0^{(j)} + \sum_{i=1}^{p} a_i^{(j)} y(k-i) + u(k), \qquad (6.54)$$

It follows from the normalization condition imposed on the weights $w^{(j)}(k)$ that the global model response is

$$y(k) = \sum_{j=1}^{N} a_0^{(j)} w^{(j)}(k) + \sum_{j=1}^{N} \sum_{i=1}^{p} w^{(j)}(k) a_i^{(j)} y(k-i) + u(k). \qquad (6.55)$$

Since $w^{(j)}(k)$ depends only on the outputs by assumption, it follows that the other sums in Eq. (6.33) may be written as a function of the output lags alone, thus implying the linear multimodel has a NARX representation of the form

$$y(k) = f(y(k-1), \ldots, y(k-p)) + u(k) \qquad (6.56)$$

for some (generally complicated) function $f(\cdots)$. In contrast, note that if $b_0^{(j)}$ depends on j or if the local model validity sets Ψ_j depend on the input lags $u(k-i)$, then the NARMAX representation for the linear multimodel of Johansen and Foss will involve nonlinear moving average behavior.

6.5.2 Steady-state behavior

For an output-selected linear multimodel—with or without moving average terms—note that if $y(k) = y_s$ for all k, then the model selection weights $w^{(j)}(k) = w_s^{(j)}$ are also constant. Hence, it follows that $u(k) = u_s$ for all k, where u_s is given explicitly by

$$u_s = \left[\sum_{j=1}^{N} \sum_{i=0}^{q} w_s^{(j)} b_i^{(j)} \right]^{-1} \left\{ y_s - \sum_{j=1}^{N} w_s^{(j)} \left[a_0^{(j)} + \sum_{i=1}^{p} a_i^{(j)} y_s \right] \right\}. \qquad (6.57)$$

As before, although this equation establishes that u_s is unique given y_s—that is, an output-selected linear multimodel cannot exhibit input multiplicities—it is certainly possible for the same value of u_s to be associated with more than one value of y_s, as the example discussed in the next section illustrates.

It is important to note, however, that this result *assumes* the existence of a steady-state solution $y(k) = y_s$ for all k. It is possible, as subsequent examples illustrate, for an output-selected linear multimodel to exhibit step responses with persistent oscillations or even chaos, a point illustrated in Sec. 6.7. This observation illustrates the concept of *transition dynamics*: the global response to a transition between validity regions may be radically different in character from the dynamics of any of the local models on which the multimodel is based.

6.5.3 Two output-selected examples

In analogy with Sec. 6.4.4, the following paragraphs compare the general behavior of the output-selected J-F and modified formulations. As before, the local models are defined by Eqs. (6.30) and (6.31), but the local selection criteria are $|y(k-1)| \leq 1$ vs. $|y(k-1)| > 1$. For the J-F formulation, the resulting multimodel is

$$y(k) = \begin{cases} \frac{1}{2} y(k-1) + u(k-1) & |y(k-1)| \leq 1 \\ a - \frac{1}{2} y(k-1) + u(k-1) & |y(k-1)| > 1, \end{cases} \qquad (6.58)$$

whereas for the modified J-F formulation, the model is

$$y^{(1)}(k) = \frac{1}{2} y^{(1)}(k-1) + u(k-1)$$

$$y^{(2)}(k) = a - \frac{1}{2} y^{(2)}(k-1) + u(k-1)$$

$$y(k) = \begin{cases} y^{(1)}(k) & |y(k-1)| \leq 1 \\ y^{(2)}(k) & |y(k-1)| > 1. \end{cases} \qquad (6.59)$$

Linear Multimodels

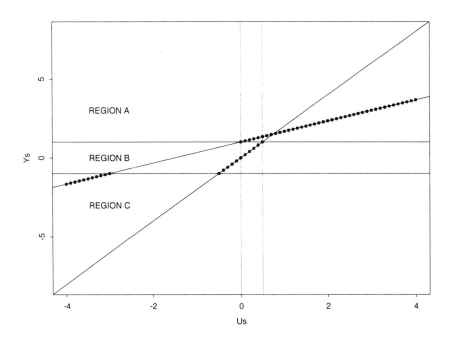

Figure 6.5: Steady-state behavior of the output-selected models

It is important to note that this modified model employs the centralized formulation discussed in Sec. 6.2.2, in which model selection is based on the *global* model output value $y(k-1)$ and *not* on the local outputs $y^{(1)}(k-1)$ and $y^{(2)}(k-1)$. The decentralized formulation was also considered, but the problem of indeterminacy (i.e., the possibility that both local validity functions $\rho^{(1)}(\psi)$ and $\rho^{(2)}(\psi)$ could be zero simultaneously) occurred in the sinusoidal responses for this model. For this reason, results are presented here only for the centralized formulation.

As in the case of the input-selected models, the steady-state behavior of the J-F model and the modified model is identical, depending on the steady-state behavior of the local models defined by Eqs. (6.30) and (6.31), along with the selection conditions $|y_s| \leq 1$ or $|y_s| > 1$. These conditions are illustrated graphically in Fig. 6.5 for the particular case $a = 3/2$ in Eq. (6.31). As before, the (u_s, y_s)-plane is partitioned into three regions, corresponding to $y_s > 1$, $|y_s| \leq 1$, and $y_s < -1$, which may be denoted regions A, B, and C, respectively. Hence, in regions A and C, the steady-state behavior is given by $y_s = (2/3)a + (2/3)u_s$, appropriate to the local model defined by Eq. (6.31), whereas in region B, the steady-state behavior is given by $y_s = 2u_s$, appropriate to the local model defined by Eq. (6.30). The overall steady-state response function for the multi-model defined by Eqs. (6.58) or (6.59) is represented by the discontinuous curve overlaid with circles in Fig. 6.5. Here, there is a region of output multiplicity,

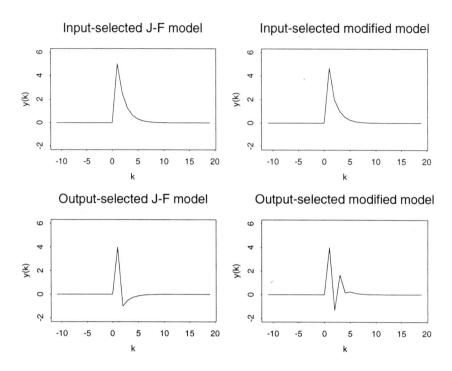

Figure 6.6: A comparison of four impulse responses

with two possible values for y_s for any u_s in the range $0 < u_s \leq 1/2$, indicated by the vertical dashed lines in the figure. In addition, note that there is also a region of negative values of u_s for which there is *no* steady-state solution y_s (specifically, the region $-5/2 - a < u_s < -1/2$). The practical implications of this observation are considered in Sec. 6.6.1 in connection with a generally selected multimodel that also exhibits a range of steady-state input values for which no corresponding steady-state output value exists.

The impulse responses of the two input-selected multimodels considered in Sec. 6.4.4 and the two output-selected multimodels considered here are shown in Fig. 6.6 for amplitude $\gamma = 4$. As noted in Sec. 6.4.4, the impulse responses for the two input-selected models are only slightly different, both essentially the same as that of the local linear model defined by Eq. (6.30). In contrast, the impulse responses for the output-selected models are both different from this input-selected response and from each other, with the modified model giving a significantly more oscillatory response. Conversely, the step responses for these models are essentially identical to those for the input-selected multimodel and are therefore not shown here. As in the case of the input-selected models, the step response for both of these output-selected models tends to remain almost entirely in the domain of one or the other of the local linear models, so these step responses are very close to what would be expected for the local models defined by Eqs. (6.30) and (6.31), depending on the magnitude of the step.

Linear Multimodels 267

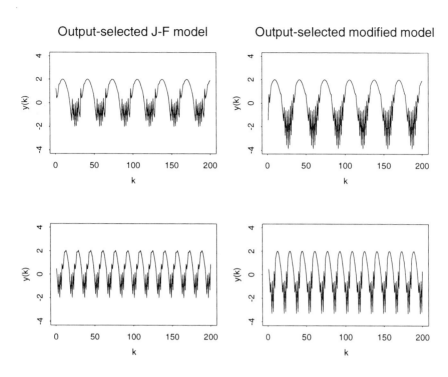

Figure 6.7: High-amplitude sinusoidal responses

As in the case of the input-selected multimodels considered in Sec. 6.4.4, the response of the output selected models to sinusoidal input sequences is extremely informative. Responses of the two output-selected models to the same two sinusoidal inputs considered in Sec. 6.4.4 are shown in Fig. 6.7. As before, the nonlinearity of the multimodel is immediately apparent from the appearance of any of these four plots, and the differences between the J-F model responses and the modified model responses are pronounced. Further, comparison of these four responses with the four shown in Fig. 6.4 illustrates the behavioral differences between the input-selected models and the output-selected models. In particular, note the high-frequency, almost noise-like "ringing" in the negative cycle of the response of the output-selected models for the lower frequency input sequence (the top two plots in Fig. 6.7). By comparison, the input-selected multimodel exhibits a response that is somewhat less oscillatory in its negative half cycles.

6.6 More general selection schemes

Clearly, both input-selection and output-selection are simply special cases of the more general model selection schemes discussed in Sec. 6.3. In the general case, model selection can depend on both inputs and outputs, and if a NARMAX representation exists, it will usually be quite complex, not belonging to the NMAX, NARX, or NARX* classes. In terms of steady-state behavior, these models can exhibit input multiplicities, output multiplicities, both, or neither. In addition, as the example discussed in Sec. 6.6.2 below illustrates, these models can even exhibit isolas (discussed in Chapters 1 and 4). Although they are by no means exhaustive, the following three subsections consider four multimodels whose selection schemes involve both inputs and outputs. The first two examples are discussed in Sec. 6.6.1 and are based on the same two local models considered in Secs. 6.4.4 and 6.5.3; as in those discussions, the two examples considered here are based on the multimodel formulations discussed in Secs. 6.2.1 and 6.2.2, and comparisons between these examples illustrate the influence of the choice of multimodel formulation. Next, Sec. 6.6.2 describes a simple Johanssen-Foss multimodel based on four first-order linear moving average models whose steady-state locus consists entirely of an isola. Finally, Sec. 6.6.3 shows that both the Wiener model and the Hammerstein model based on a simple first-order linear model and a monotonic positive-homogeneous function $\theta(\alpha, \beta; x)$ can be represented as linear multimodels that differ principally in their selection criteria. In particular, it is shown that the Hammerstein model represents an input-selected multimodel, whereas the Wiener model represents a generally selected one.

6.6.1 Johanssen-Foss and modified models

The two models considered here are again based on the local linear models defined by Eqs. (6.30) and (6.31), but now the selection criterion is based on both $u(k-1)$ and $y(k-1)$. Specifically, the first of these linear models is selected if $|u(k-1)| + |y(k-1)| \leq 1$, and the other is selected if this condition is not met. The resulting Johanssen-Foss multimodel may be represented as

$$y(k) = \begin{cases} \frac{1}{2}y(k-1) + u(k-1) & |u(k-1)| + |y(k-1)| \leq 1 \\ a - \frac{1}{2}y(k-1) + u(k-1) & |u(k-1)| + |y(k-1)| > 1, \end{cases} \quad (6.60)$$

whereas for the modified J-F formulation, the model is

$$y^{(1)}(k) = \frac{1}{2}y^{(1)}(k-1) + u(k-1)$$

$$y^{(2)}(k) = a - \frac{1}{2}y^{(2)}(k-1) + u(k-1)$$

$$y(k) = \begin{cases} y^{(1)}(k) & |u(k-1)| + |y(k-1)| \leq 1 \\ y^{(2)}(k) & |u(k-1)| + |y(k-1)| > 1. \end{cases} \quad (6.61)$$

As in the case of the output-selected multimodel discussed in Sec. 6.5.3, it would be possible to consider either the centralized or the decentralized formulation of this model. Here, only the centralized formulation is considered where selection is dependent on the global model output $y(k-1)$ and not the local model outputs $y^{(1)}(k-1)$ and $y^{(2)}(k-1)$.

In terms of steady-state behavior, once again the two multimodel formulations considered here are equivalent, depending on the steady-state behavior of

Linear Multimodels

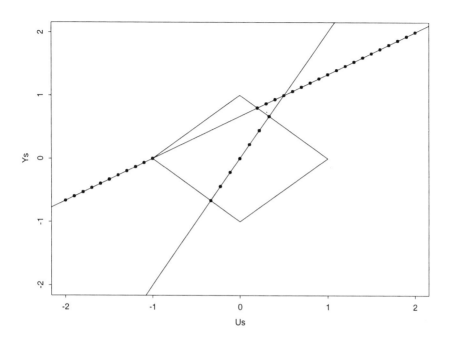

Figure 6.8: Steady-state behavior of the generally selected model

the local models, together with the local model selection criterion that $|u_s|+|y_s|$ does or does not exceed 1 in value. These conditions are shown in the y_s vs. u_s plot in Fig. 6.8 for the case $a = 1$: the local model defined by Eq. (6.30) is valid within the diamond-shaped region $|u_s| + |y_s| \leq 1$, whereas the local model defined by Eq. (6.31) is valid outside this region. The overall steady-state locus for this model is indicated by the line segments marked with overlaid circles. Note that in the range $-2/3 \leq y_s < 0$, there exist two possible values of u_s, one corresponding to the steady-state response of each of the local models. Thus, the multimodel exhibits input multiplicity over this range of output values. In contrast, note that for $y_s < -2/3$ or $y_s \geq 0$, the corresponding value of u_s is unique, so once again, this model exhibits input multiplicity only over an "intermediate" range of values. Conversely, viewing the steady-state locus shown in Fig. 6.8 as a function of increasing input values u_s, the behavior is slightly more complex. For $u_s \leq -1$, y_s corresponds to the steady-state response of the model defined by Eq. (6.31) and is unique, but for $-1 < u_s < -2/3$, no solution y_s exists. For $-2/3 \leq u_s \leq +1/5$, y_s is again unique, corresponding to the steady-state response of the model defined by Eq. (6.30). For $1/5 < u_s \leq 1/3$, two possible values of y_s exist, as either model may be selected. For $u_s > 1/3$, the steady-state response y_s is once again unique, corresponding to the model defined by Eq. (6.31).

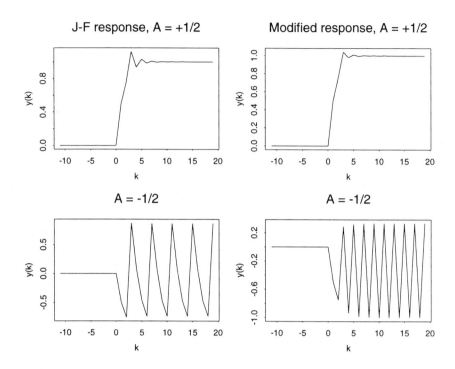

Figure 6.9: A comparison of four step responses

The fact that no steady-state response value y_s exists for $-1 < u_s < -2/3$ implies a complicated response to step inputs whose amplitude lies in this range. Step responses of amplitude $\gamma = +1/2$ and $\gamma = -1/2$ are shown in the upper and lower pairs of plots in Fig. 6.9, respectively, for both of the multimodels considered here. For $\gamma = +1/2$, the step responses for both the J-F model shown in the left-hand plot and the modified model shown in the right-hand plot are similar, settling out to the steady-state value $y_s = +1$. In contrast, both step responses for $\gamma = -1/2$ exhibit persistent oscillations, and the differences between these responses are much more pronounced. In particular, note the substantially higher frequency of the modified model's oscillatory response.

The impulse responses of these models are not terribly interesting, being essentially the same as those of the output-selected model considered in Sec. 6.5.3. Conversely, the responses of the generally selected model to high-amplitude sinusoids is quite similar to that of the input-selected multimodel. Finally, there appears to be little difference between the responses of the Johansen-Foss models and the responses of the modified Johansen-Foss models to white noise inputs. A typical example is shown in Fig. 6.10, which compares the reponses of the generally selected models for an input of amplitude 0.5. In this figure, the Johnansen-Foss model response is plotted as the solid line, and the modified model response is plotted as the solid circles. Differences between the selection criteria are more pronounced, although not dramatically so.

Linear Multimodels

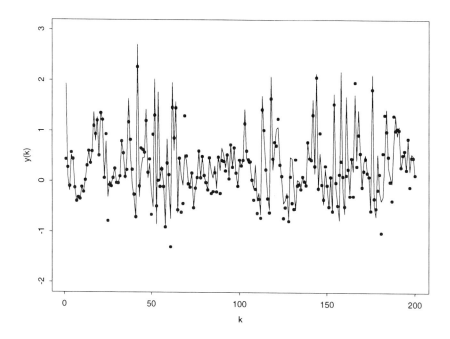

Figure 6.10: Comparison of two generally-selected models

6.6.2 The isola model

The following example is interesting because its steady-state locus is an isola:

$$y(k) = \begin{cases} -u(k-1) + 1 & y(k-1) \geq 0, u(k-1) \geq 0 \\ u(k-1) + 1 & y(k-1) \geq 0, u(k-1) < 0 \\ u(k-1) - 1 & y(k-1) < 0, u(k-1) \geq 0 \\ -u(k-1) - 1 & y(k-1) < 0, u(k-1) < 0. \end{cases} \quad (6.62)$$

The selection criterion here is defined by the four quadrants of the (u_s, y_s) plane, and the overall steady-state locus is the diamond-shaped closed curve shown in Fig. 6.11, indicated with overlaid circles on top of the four lines $y_s = \pm u_s \pm 1$ that define the steady-state behavior of the local models. The response of this model to an impulse of amplitude γ is

$$y(1) = \begin{cases} 1 - \gamma & \gamma > 0 \\ 1 + \gamma & \gamma < 0, \end{cases} \quad (6.63)$$

and, for $k > 1$,

$$y(k) = \begin{cases} 1 & y(1) \geq 0 \\ -1 & y(1) < 0. \end{cases} \quad (6.64)$$

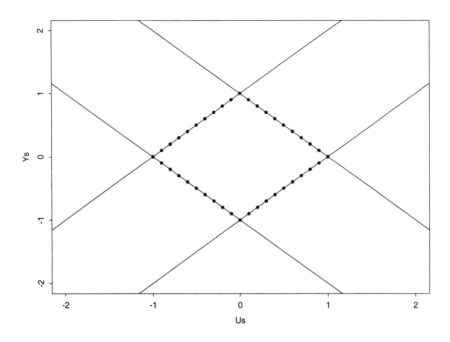

Figure 6.11: Steady-state locus for the isola model

Similarly, the repsonse of this multimodel to a step of amplitude γ must assume one of the four values $\pm\gamma \pm 1$. Depending on both the initial condition $y(0)$ and the value of γ, this response may either settle to one of these values or fluctuate between more than one value. For example, for $y(0) = 0$, the response of this model to a step of amplitude $\gamma = 1/3$ is simply a step of amplitude $2/3$, delayed by one sampling time. In contrast, the response to a step of amplitude $\gamma = 3/2$ is a persistent oscillation with period 2 between the values $\pm 1/2$.

The response of this model to sinusoidal input sequences is much more complex, as seen in Fig. 6.12. There, the upper two plots show the response of this model to two sinusoids of the standard frequency described in Chapter 2, with amplitudes 0.5 and 2.0 for the left- and right-hand plots, respectively. Both of these responses are periodic with the same period as the input sequence, but are highly non-sinusoidal, thus correctly reflecting the pronounced nonlinearity of this model. Also, note the marked change in character between the low-amplitude and high-amplitude responses. In particular, note how much smoother the low-amplitude response is than the high-amplitude response, which appears almost "noise-like." The character of the high-amplitude response becomes even more complex with increasing input frequency, as seen in the bottom two plots, which show the model responses to sinusoidal inputs of the same two amplitudes but at twice the frequency.

Linear Multimodels

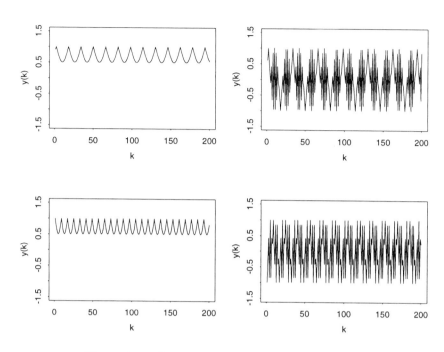

Figure 6.12: Sinusoidal responses of the isola model

The amplitude dependence of the qualitative behavior of this model is even more extreme in the case of white noise inputs. Four such responses are shown in Fig. 6.13, corresponding to the standard white noise input sequence introduced in Chapter 2, but with amplitudes (i.e., standard deviations) of 0.1, 0.5, 1.0, and 2.0. In the lowest amplitude example (the upper left plot), the model response is $y(k) = 1 - |u(k-1)| \simeq 1$ for all k; very careful examination would be required to distinguish this response from that of the affine model $y(k) = u(k-1) + 1$. In the upper right plot, the response is shown for amplitude 0.5; in this case, the response is highly reminiscent of that seen in the exothermic CSTR, which exhibits output multiplicity. In particular, note the step-like transitions between what appears to be an upper steady state at $y(k) \simeq +1$ and a lower steady state at $y(k) \simeq -1$. Increasing the input amplitude still further, the response of the multimodel to the standard white noise input sequence (standard deviation 1.0) is shown in the lower left plot. By itself, this response does not appear to be particularly informative, but its character is markedly different from that seen in the lower amplitude responses. Finally, the lower right plot shows the response of the multimodel for an input amplitude of 2.0; note the general qualitative similarity of this response to that seen in Fig. 6.10. Overall, the most important feature of these responses is their pronounced qualitative dependence on the input amplitude, clearly reflecting the nonlinearity of this linear multimodel.

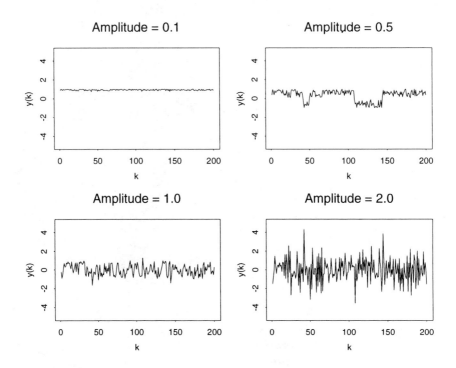

Figure 6.13: White noise responses for the isola model

6.6.3 Wiener and Hammerstein models

This section considers the Wiener and Hammerstein models constructed from the first-order linear model

$$z(k) = \phi z(k-1) + (1-\phi)v(k-1), \tag{6.65}$$

and the static nonlinearity

$$g(x) = \theta(\alpha, \beta; x). \tag{6.66}$$

Here, it is assumed that $0 < \phi < 1$, so the linear model defined by Eq. (6.65) is stable, with a steady-state gain of 1 and it is assumed that $0 \leq \beta < \alpha$ so $g(x)$ is monotonic and thus invertible, with the inverse

$$g^{-1}(x) = \theta(\gamma, \delta; x), \quad \gamma = \frac{\alpha}{\alpha^2 - \beta^2} \quad \delta = \frac{-\beta}{\alpha^2 - \beta^2}. \tag{6.67}$$

The Hammerstein model considered here is formed from the cascade connection of this static nonlinearity followed by the linear model. The NARMAX representation for this model is

$$y(k) = \begin{cases} \phi y(k-1) + (1-\phi)(\alpha+\beta)u(k-1) & u(k-1) \geq 0 \\ \phi y(k-1) + (1-\phi)(\alpha-\beta)u(k-1) & u(k-1) < 0. \end{cases} \tag{6.68}$$

Linear Multimodels

Defining $a^{(1)} = a^{(2)} = \phi$, $b^{(1)} = (1-\phi)(\alpha + \beta)$, and $b^{(2)} = (1-\phi)(\alpha - \beta)$, it follows that this Hammerstein model may be represented as an input-selected linear multimodel based on the local linear models for $j = 1, 2$:

$$y(k) = a^{(j)}y(k-1) + b^{(j)}u(k-1). \tag{6.69}$$

For comparison, the dual Wiener model consists of the the linear model defined in Eq. (6.65) followed by the static nonlinearity $g(x)$ defined in Eq. (6.66). As discussed in Chapter 4, the NARMAX representation for this model is

$$y(k) = g\left(\phi g^{-1}(y(k-1)) + (1-\phi)u(k-1)\right),$$
$$= \theta\left(\alpha, \beta; [\theta(\phi\gamma, \phi\delta; y(k-1)) + (1-\phi)u(k-1)]\right). \tag{6.70}$$

This result follows from the fact that $\phi\theta(\gamma, \delta; x) = \theta(\phi\gamma, \phi\delta; x)$. This model may be reduced to a linear multimodel in two steps, first replacing the inner nonlinearity $\theta(\phi\gamma, \phi\delta; x)$ with its piecewise linear equivalent, and then replacing the outer nonlinearity $\theta(\alpha, \beta; x)$. The first step in this process yields the intermediate result

$$y(k) = \begin{cases} \theta(\alpha, \beta; [\phi(\gamma + \delta)y(k-1) + u(k-1)]) & y(k-1) \geq 0 \\ \theta(\alpha, \beta; [\phi(\gamma - \delta)y(k-1) + u(k-1)]) & y(k-1) < 0 \end{cases} \tag{6.71}$$

Expanding the outer nonlinearity and simplifying ultimately yields the following linear multimodel:

$$y(k) = \begin{cases} \phi y(k-1) + (1-\phi)(\alpha + \beta)u(k-1) & j = 1 \\ \frac{\phi(\alpha - \beta)}{\alpha + \beta} y(k-1) + (1-\phi)(\alpha - \beta)u(k-1) & j = 2 \\ \frac{\phi(\alpha + \beta)}{\alpha - \beta} y(k-1) + (1-\phi)(\alpha + \beta)u(k-1) & j = 3 \\ \phi y(k-1) + (1-\phi)(\alpha - \beta)u(k-1) & j = 4, \end{cases} \tag{6.72}$$

where the selection index j is determined by the conditions

$$j = \begin{cases} 1 & y(k-1) \geq 0, u(k-1) \geq -\frac{\phi(1-\phi)}{\alpha + \beta}y(k-1), \\ 2 & y(k-1) \geq 0, u(k-1) < -\frac{\phi(1-\phi)}{\alpha + \beta}y(k-1), \\ 3 & y(k-1) < 0, u(k-1) \geq -\frac{\phi(1-\phi)}{\alpha - \beta}y(k-1), \\ 4 & y(k-1) < 0, u(k-1) < -\frac{\phi(1-\phi)}{\alpha - \beta}y(k-1). \end{cases} \tag{6.73}$$

The differences between these dual multimodels are striking. First, note that the Wiener model is equivalent to a linear multimodel based on *four* local models instead of two, as in the Hammerstein case. Also, note that the first and fourth of these local linear models are exactly the same as the two appearing in the multimodel representation of the Hammerstein model. Further, note that the Hammerstein model is input-selected, whereas the Wiener model is generally selected, involving both inputs and outputs. In fact, for the range of values for α, β, and ϕ assumed here, the first of these models is selected whenever $u(k-1) \geq 0$ and $y(k-1) \geq 0$, whereas the fourth is selected whenever $u(k-1) < 0$ and $y(k-1) < 0$. Note that these conditions represent refinements of the conditions under which the Hammerstein multimodel selects the same two models: $u(k-1) \geq 0$ for the first model, and $u(k-1) < 0$ for the last model. Also, note that one of these two models is selected whenever the product $y(k-1)u(k-1)$ is nonnegative. It follows, then, that if this condition is satisfied for all k, no differences will be seen between this Wiener model and its dual Hammerstein model. It is not difficult to show that this condition is satisfied by the steady-state solution $y_s = \theta(\alpha, \beta; u_s)$, consistent with the observation made several times before that the steady-state behavior of Wiener and Hammerstein models is equivalent.

For this particular pair of models, no difference is seen in the behavior of the Hammerstein and Wiener models for either impulse or step inputs. This results follows from the fact that the condition $y(k-1)u(k-1) \geq 0$ is satisfied for all k for these input sequences. In contrast, differences between the Hammerstein and Wiener models are evident in their responses to sinusoids or white noise. These differences are illustrated in Fig. 6.14, which shows the Hammerstein model responses in the plots on the left-hand side and the Wiener model responses in the plots on the right-hand side. The parameters defining these models are $\alpha = 1.1$, $\beta = 1.0$, and $\phi = 0.8$. The upper pair of plots compares their responses for the standard sinusoidal input sequence defined in Chapter 2, whereas the lower pair of plots compares the responses for the white noise input sequence defined there. In both cases, significant differences in these responses are seen, reflecting the presence of the two additional local models in the multimodel representation of the Wiener model, relative to the Hammerstein model.

6.7 TARMAX models

Although it was not originally concieved as a family of linear multimodels, the family of positive homogeneous TARMAX models introduced in Chapter 3 turns out to be a subset of the linear multimodel family of Johansen and Foss. As noted in Chapter 3, these models are positive-homogeneous members of the class of structurally additive models defined in Chapter 4; recall that they are defined by Eq. (3.67), repeated here for convenience:

$$y(k) = \sum_{i=1}^{p} \theta(a_i, c_i; y(k-i)) + \sum_{i=0}^{q} \theta(b_i, d_i; u(k-i))$$
$$= \sum_{i=1}^{p} a_i y(k-i) + \sum_{i=1}^{p} c_i |y(k-i)|$$
$$+ \sum_{i=0}^{q} b_i u(k-i) + \sum_{i=0}^{q} d_i |u(k-i)|. \qquad (6.74)$$

The remainder of this section describes this model family in some detail, starting with the simplest case: $p = 1, q = 0$, discussed in Sec. 6.7.1. Many of the results established for this model generalize to TARMAX models of arbitrary order. For example, Sec. 6.7.1 establishes that the first-order TARMAX model is a member

Linear Multimodels 277

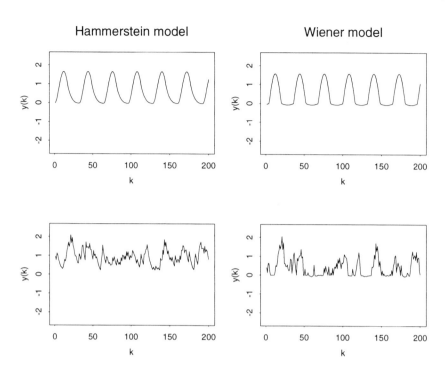

Figure 6.14: Hammerstein and Wiener model responses

of Tong's family of linear multimodels; Sec. 6.7.2 extends this result to arbitrary TARMAX models. Similarly, Sec. 6.7.1 presents an analysis of the steady-state behavior of the first-order model, and Sec. 6.7.3 extends this analysis to TARMAX models of arbitrary order. Since the two structural restrictions defining the TARMAX class—the structural additivity condition discussed in Chapter 4, representing the model in terms of scalar nonlinearities, and the restriction of these nonlinearities to the class of theta functions introduced in Chapter 3—are both rather restrictive, it is not surprising that this model class is a rather special one. What is surprising is both the great range of qualitative behavior this model class can exhibit, and the extent to which these models are amenable to detailed analysis.

6.7.1 The first-order model

Taking $p = 1$, $q = 0$, and omitting subscripts for simplicity, the TARMAX model defined by Eq. (3.67) reduces to the following first-order model:

$$\begin{aligned} y(k) &= \theta(a,c;y(k-1)) + \theta(b,d;u(k-1)) \\ &= ay(k-1) + c|y(k-1)| + bu(k) + d|u(k)|. \end{aligned} \quad (6.75)$$

As the following discussion illustrates, this model may be represented as a multimodel based on four local linear models. To proceed, first note that this model may be represented in the form of the starting equation $y(k) = F(\psi(k))$ of Johansen and Foss (Eq. (6.9)) with $\psi(k) = [y(k-1), u(k)]^T$ and $F: R^2 \to R^1$ defined by

$$F(x_1, x_2) = \theta(a, c; x_1) + \theta(b, d; x_2). \tag{6.76}$$

Next, define the following four *orthant sets* Ω_j for $j = 1, 2, 3, 4$:

$$\begin{aligned}
\Omega_1 &= \{y(k-1) \geq 0, u(k) \geq 0\}, \\
\Omega_2 &= \{y(k-1) \leq 0, u(k) \geq 0\}, \\
\Omega_3 &= \{y(k-1) \geq 0, u(k) \leq 0\}, \\
\Omega_4 &= \{y(k-1) \leq 0, u(k) \leq 0\}.
\end{aligned} \tag{6.77}$$

The key to representing the model defined in Eq. (6.75) as a linear multi-model lies in the observation that if $\psi(k) \in \Omega_j$, then the model response $y(k)$ may be written as one of the four *linear* models

$$y(k) = \alpha^{(j)} y(k-1) + \beta^{(j)} u(k), \tag{6.78}$$

for $j = 1, 2, 3, 4$. Here, the model coefficients $\alpha^{(j)}$ and $\beta^{(j)}$ are given by:

$$\begin{aligned}
\alpha^{(1)} &= a + c & \beta^{(1)} &= b + d, \\
\alpha^{(2)} &= a - c & \beta^{(2)} &= b + d, \\
\alpha^{(3)} &= a + c & \beta^{(3)} &= b - d, \\
\alpha^{(4)} &= a - c & \beta^{(4)} &= b - d.
\end{aligned} \tag{6.79}$$

Thus, the TARMAX model defined by Eq. (6.75) may be viewed as a member of Tong's class of open loop threshold autoregressive systems with the selection index J_k defined by

$$J_k = j \quad \text{when} \quad \psi(k) \in \Omega_j. \tag{6.80}$$

The only possible ambiguity in this selection scheme occurs if $\psi(k)$ can belong to more than one set Ω_j. Note that the union of these four sets consitutes the plane R^2, and that the intersection of any two of these sets corresponds to the line $y(k-1) = 0$, the line $u(k) = 0$, or the origin ($y(k-1) = u(k) = 0$). Thus, an ambiguity is possible in principle, but it does not arise because the sets Ω_i and Ω_j intersect at boundaries where the ith and j^{th} linear models are equivalent. To see this point, consider the pairwise intersections of the sets Ω_j:

$$\begin{aligned}
\psi(k) \in \Omega_1 \cap \Omega_2 &\Rightarrow y(k-1) = 0, \\
\psi(k) \in \Omega_1 \cap \Omega_3 &\Rightarrow u(k) = 0, \\
\psi(k) \in \Omega_1 \cap \Omega_4 &\Rightarrow y(k-1) = 0, u(k) = 0, \\
\psi(k) \in \Omega_2 \cap \Omega_3 &\Rightarrow y(k-1) = 0, u(k) = 0, \\
\psi(k) \in \Omega_2 \cap \Omega_4 &\Rightarrow u(k) = 0, \\
\psi(k) \in \Omega_3 \cap \Omega_4 &\Rightarrow y(k-1) = 0.
\end{aligned} \tag{6.81}$$

Note that if $\psi(k) \in \Omega_1 \cap \Omega_2$, either model 1 or model 2 from Eq. (6.78) could be applicable, but for $y(k-1) = 0$, these two models are equivalent, so there is no conflict. The same reasoning applies to the other cases of possible ambiguity listed in Eq. (6.81).

To explore the possible steady-state behavior of the first-order TARMAX model, let $u(k) = u_s$; if a constant solution $y(k) = y_s$ exists, it must satisfy

$$y_s = ay_s + c|y_s| + bu_s + d|u_s| \tag{6.82}$$

which may be rearranged to the form:

$$\theta(1-a, -c; y_s) = \theta(b, d; u_s). \tag{6.83}$$

It follows from the conditions given in Chapter 3 that $\theta^{-1}(1-a, -c; x)$ exists if and only if

$$a \neq 1 \qquad |c| < |1-a|. \tag{6.84}$$

These conditions can be related to the additive model stability conditions developed in Chapter 4 by noting that

$$\begin{aligned} |\theta(a,c;x) - \theta(a,c;y)| &= |a(x-y) + c(|x|-|y|)| \\ &\leq |a(x-y)| + |c(|x|-|y|)| \\ &= |a||x-y| + |c|(|x|-|y|)|. \end{aligned} \tag{6.85}$$

Simplification of this expression is possible since

$$\begin{aligned} |x| &= |x-y+y| \leq |x-y| + |y| \\ &\Rightarrow |x| - |y| \leq |x-y|. \end{aligned} \tag{6.86}$$

Reversing the roles of x and y yields the inequality

$$|y| - |x| \leq |y-x| = |x-y| \tag{6.87}$$

and combining the results from Eqs. (6.86) and (6.87) yields

$$|(|x|-|y|)| \leq |x-y|, \tag{6.88}$$

thereby reducing the inequality (6.85) to the Lipschitz condition

$$|\theta(a,c;x) - \theta(a,c;y)| \leq (|a|+|c|)|x-y|, \tag{6.89}$$

for all real x and y. Hence, it follows from the stability theorem for structurally additive NARMAX models presented in Chapter 4 that the model defined in Eq. (6.82) is BIBO stable if

$$|a| + |c| < 1. \tag{6.90}$$

It is not difficult to show that if this condition is satisfied, then so are both of the conditions in Eq. (6.84), and the function $\theta(a,c;x)$ is invertible.

The real point of the preceeding results is that they establish that if condition (6.90) is satisfied, then the first-order TARMAX model defined in Eq. (6.82) is BIBO stable and exhibits no output multiplicities. In particular, note that under these conditions, the unique steady-state response y_s is given by

$$y_s = \theta^{-1}\left(1-a,c;\theta(b,d;u_s)\right). \tag{6.91}$$

Similar reasoning leads to the conclusion that $\theta(b,d;x)$ is invertible if

$$b \neq 0 \qquad |d| < |b|, \tag{6.92}$$

from which it follows that the model will exhibit no input multiplicity. In this case, given a steady-state response y_s, the associated input u_s is:

$$u_s = \theta^{-1}\left(b,d;\theta(1-a,c;y_s)\right). \tag{6.93}$$

6.7.2 The multimodel representation

The multimodel representation for the first-order TARMAX model given in the previous section extends directly to TARMAX models of arbitrary order. As in the previous example, the central observation that leads to this result is that the general TARMAX model defined in Eq. (3.67) is *linear* when restricted to an *orthant set* Ω_j. Here, the argument $\psi(k)$ of the function $F(\cdots)$ in Eq. (6.9) is a vector in the space R^{p+q+1}. In this space, there are 2^{p+q+1} possible orthant sets, defined as follows. First, for any $j = 1, 2, \ldots, 2^{p+q+1}$, define the associated binary expansion coefficients as the unique numbers $s_i^{(j)}$ such that:

$$j = 1 + \sum_{i=1}^{p+q+1} 2^{(i-1)} s_i^{(j)}. \tag{6.94}$$

Thus, for example:

$$\begin{aligned}
s_i^{(1)} &= 0, \; i = 1, 2, \ldots, p+q+1, \\
s_i^{(2)} &= \begin{cases} 1 & i = 1, \\ 0 & i = 2, 3, \ldots, p+q+1, \end{cases} \\
s_i^{(3)} &= \begin{cases} 0 & i = 1, \\ 1 & i = 2, \\ 0 & i = 3, 4, \ldots, p+q+1, \end{cases} \\
s_i^{(4)} &= \begin{cases} 1 & i = 1, \\ 1 & i = 2, \\ 0 & i = 3, 4, \ldots, p+q+1, \end{cases}
\end{aligned} \tag{6.95}$$

with analogous expressions for $j = 5, 6, \ldots, 2^{p+q+1}$. Given the expansion coefficients $\{s_i^{(j)}\}$—which all have values 0 or 1—define the associated orthant sets Ω_j as follows:

$$\Omega_j = \{\mathbf{x} \in R^{p+q+1} \mid (2s_i^{(j)} - 1)x_i \geq 0\}. \tag{6.96}$$

Linear Multimodels

In particular, note that $2s_i^{(j)} - 1 = \pm 1$ for all i and j, so the defining characteristic of these sets is that, if $\mathbf{x} \in \Omega_j$, then the ith component of this vector satisfies $x_i \geq 0$ if $s_i^{(j)} = 1$ and $x_i \leq 0$ if $s_i^{(j)} = 0$.

Now, suppose $\psi(k) \in \Omega_j$. It follows by the same reasoning as in the first-order case that the response $y(k)$ of the TARMAX model defined in Eq. (3.67) may be expressed as

$$y(k) = \sum_{i=1}^{p} \alpha_i^{(j)} y(k-i) + \sum_{i=0}^{q} \beta_i^{(j)} u(k-i). \tag{6.97}$$

Here, the coefficients $\alpha^{(j)}$ and $\beta^{(j)}$ are defined by

$$\alpha_i^{(j)} = a_i + (2s_i^{(j)} - 1)c_i, \tag{6.98}$$

for $i = 1, 2, \ldots, p$, and

$$\beta_i^{(j)} = b_i + (2s_{(i+p+1)}^{(j)} - 1)d_i. \tag{6.99}$$

for $i = 0, 1, \ldots, q$. The subscripts in these expressions reflect the fact that $\psi_i(k) = y(k-i)$ for $i = 1, 2, \ldots, p$ and $\psi_{(i+p+1)}(k) = u(k-i)$ for $i = 0, 1, \ldots, q$.

Again as in the first-order case, Eq. (6.97) defines a member of Tong's open loop threshold autoregressive systems class. The selection index J_k is again defined by Eq. (6.80); as before, *potential* ambiguities arise on the common boundaries between distinct orthants Ω_i and Ω_j, but these ambiguities may be resolved arbitrarily because the associated models are equivalent on these boundaries. Finally, note that the issue of regime determination—that is, input-selection versus output-selection—is determined by the coefficients a_i, b_i, c_i, and d_i defining the TARMAX model. That is, the linear multimodel is

1. Unconditionally linear if $c_i = 0$ and $d_i = 0$ for all i
2. Input selected if $c_i = 0$ for all i
3. Output selected if $d_i = 0$ for all i
4. General (both input and output selected) if $c_i \neq 0$ and $d_j \neq 0$ for at least one i and one j.

6.7.3 Steady-state behavior

A particularly elegant feature of the class of TARMAX models is that the steady-state analysis presented in Sec. 6.7.1 for the first-order model generalizes nicely to the entire model class. Assuming as before that the response to a constant input sequence $u(k) = u_s$ is a constant output sequence $y(k) = y_s$, it follows that these constants will be related by the equation

$$y_s = \sum_{i=1}^{p} \theta(a_i, c_i; y_s) + \sum_{i=0}^{q} \theta(b_i, d_i; u_s). \tag{6.100}$$

Again, it is important to note that the response to a constant input sequence need not be constant, as an example presented later in this section illustrates. When the response to a constant input sequence *is* constant—arguably the situation of greatest practical interest—it is useful to simplify Eq. (6.100), as follows.

First, recall from Chapter 3 that

$$\sum_{i=1}^{m} \theta(\alpha_i, \beta_i; x) = \theta\left(\sum_{i=1}^{m} \alpha_i, \sum_{i=1}^{m} \beta_i; x\right). \quad (6.101)$$

Next, define the four constants

$$\bar{a} = \sum_{i=1}^{p} a_i \quad \bar{b} = \sum_{i=0}^{q} b_i \quad \bar{c} = \sum_{i=1}^{p} c_i \quad \bar{d} = \sum_{i=0}^{q} d_i. \quad (6.102)$$

It follows from Eq. (6.101) that Eq. (6.100) may be rewritten as

$$y_s = \theta(\bar{a}, \bar{c}; y_s) + \theta(\bar{b}, \bar{d}; u_s),$$
$$\Rightarrow \theta(1 - \bar{a}, -\bar{c}; y_s) = \theta(\bar{b}, \bar{d}; u_s), \quad (6.103)$$

which has exactly the same form as the steady-state equation for the first-order TARMAX model considered in Sec. 6.7.1. Therefore, the conditions for input and output multiplicities are exactly the same as in the first-order model, provided the constants a, b, c, and d in the first-order example are replaced with the aggregate constants $\bar{a}$, $\bar{b}$, $\bar{c}$ and $\bar{d}$, respectively. However, note that the condition $|\bar{a}| + |\bar{c}|$ is not sufficient to guarantee BIBO stability of the TARMAX model for $p > 1$. In particular, recall from the discussion of the NARX and NARX* stability results presented in Chapter 4 that a necessary condition for the stability of a second-order linear model is $a_2 < 1$; hence, a second-order TARMAX model with $a_1 = -3/2$, $a_2 = 2$, and $c_1 = c_2 = 0$ will satisfy the condition for steady-state uniqueness but will not be BIBO stable because it corresponds to an unstable second-order linear model.

A particularly interesting special case of the above result is that obtained when $\bar{c} = \bar{d} = 0$. In this case, Eq. (6.103) reduces to a *linear* equation, and the steady-state solution is given by

$$y_s = \left(\frac{\bar{b}}{1 - \bar{a}}\right) u_s. \quad (6.104)$$

This class of models belongs to the more general class of *static-linear* models, discussed in detail in Chapter 3. These are models whose steady-state behavior is linear even though their dynamic behavior may be highly nonlinear, as in the case considered here. As a specific example, consider the following second-order TARMAX model:

$$y(k) = 0.8y(k-1) + c|y(k-1)| - c|y(k-2)| + u(k-1). \quad (6.105)$$

Note that if $u(k) = u_s$ for all k, this equation has the steady-state solution $y_s = 5u_s$, appropriate to the linear model obtained when $c = 0$. It follows from the results of Sec. 6.7.2 that this model may be represented as an output-selected linear multimodel. Specifically, this model is a member of Tong's TARSO class, based on four local models of the form:

$$y^{(j)}(k) = a_1^{(j)} y(k-1) + a_2^{(j)} y(k-2) + u(k-1), \qquad (6.106)$$

where the local model coefficients are

$$\begin{aligned} a_1^{(1)} &= 0.8 + c & a_2^{(1)} &= -c \\ a_1^{(2)} &= 0.8 + c & a_2^{(2)} &= c \\ a_1^{(3)} &= 0.8 - c & a_2^{(3)} &= -c \\ a_1^{(4)} &= 0.8 - c & a_2^{(4)} &= c, \end{aligned} \qquad (6.107)$$

and the selection index is

$$j = \begin{cases} 1 & y(k-1) \geq 0, y(k-2) \geq 0 \\ 2 & y(k-1) \geq 0, y(k-2) < 0 \\ 3 & y(k-1) < 0, y(k-2) \geq 0 \\ 4 & y(k-1) < 0, y(k-2) < 0. \end{cases} \qquad (6.108)$$

To understand the basic nature of this model, it is instructive to consider the step responses shown in Fig. 6.15 for various values of the constant c. The plot in the upper left shows the positive and negative unit step responses for the linear model obtained by setting $c = 0$. As noted previously, the symmetry of this plot is a necessary consequence of the linearity of this model. In contrast, the same pair of step responses is shown in the upper right plot for the case $c = 0.8$, which exhibits the qualitative asymmetry seen in the exothermic CSTR example introduced in Chapter 1. In particular, note that the response to the positive step is oscillatory, characteristic of a second-order linear model, whereas the response to the negative step is monotonic, characteristic of a first-order linear model. However, the steady-state gain is $+5$ for both positive and negative steps, and both step responses ultimately settle out to this steady-state value. This behavior is a consequence of the static-linearity of the model, since such models are characterized by a single well-defined steady-state gain.

The bottom two plots in Fig. 6.15 show the response of the model defined in Eq. (6.105) for $c = -1.2$. The left-hand plot shows the positive step response, whereas the right-hand plot shows the negative step response. Both of these responses are bounded, but they exhibit persistent fluctuations and never settle out to the steady-state value $y_s = 5u_s$. In fact, both these responses appear to be chaotic in character but significantly different from each other in their general behavior. *The real point of this example is to illustrate that even if a model appears to have a constant steady-state solution, its response to asymptotically constant input sequences like steps need not approach that steady state.* Again, a key issue in dealing with models of this type is that of initial conditions: if $y(k) = y_0 \neq y_s$ for $k \leq 0$, the transient response to an input change at $k = 0$ may never decay, and the expected steady-state value may never be reached.

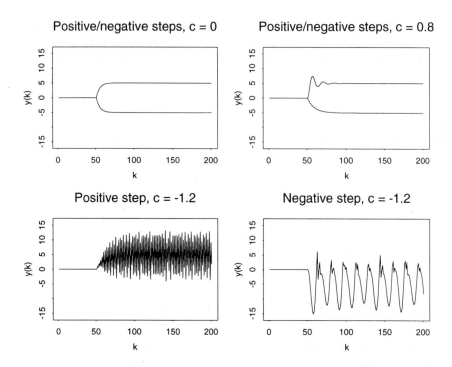

Figure 6.15: Step responses, static-linear TARMAX model

6.7.4 Some representation results

The following discussion presents four useful TARMAX model representation results. The first of these results is concerned with the special case of input-selected TARMAX models, corresponding to the restriction $c_i = 0$ and leading to the representation

$$y(k) = \sum_{i=1}^{p} a_i y(k-i) + \sum_{i=0}^{q} \theta(b_i, d_i; u(k-i)). \qquad (6.109)$$

This representation may be simplified by expressing $\theta(b_i, d_i; x)$ as

$$\theta(b_i, d_i; x) = (b_i + d_i)\theta_+(x) + (b_i - d_i)\theta_-(x), \qquad (6.110)$$

where the functions $\theta_-(x)$ and $\theta_+(x)$ were defined in Chapter 3:

$$\theta_+(x) = \theta(\frac{1}{2}, \frac{1}{2}; x) = \begin{cases} x & x \geq 0 \\ 0 & x < 0, \end{cases}$$

$$\text{and} \quad \theta_-(x) = \theta(\frac{1}{2}, -\frac{1}{2}; x) = \begin{cases} 0 & x \geq 0 \\ x & x < 0. \end{cases} \qquad (6.111)$$

Linear Multimodels

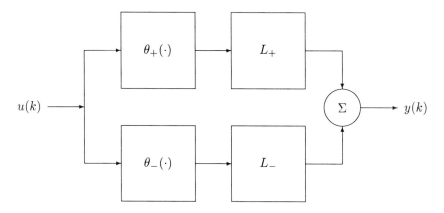

Figure 6.16: Uryson representation for $c_i = 0$

From this result, it follows that $y(k)$ may be represented as the Uryson model composed of the parallel Hammerstein models

$$y^+(k) = \sum_{i=1}^{p} a_i y^+(k-i) + \sum_{i=0}^{q} b_i^+ \theta_+(u(k-i)),$$

$$y^-(k) = \sum_{i=1}^{p} a_i y^-(k-i) + \sum_{i=0}^{q} b_i^- \theta_-(u(k-i)), \quad (6.112)$$

that is, $y(k) = y^+(k) + y^-(k)$. The structure of this Uryson model is shown in Fig. 6.16, where L_+ and L_- represent the linear models defined by the autoregressive coefficients a_i and the moving average coefficients:

$$b_i^+ = b_i + d_i \qquad b_i^- = b_i - d_i. \quad (6.113)$$

The second representation result given here is essentially the converse of the first: *any Uryson model of the form shown in Fig. 6.16 may be represented as a TARMAX model, although generally an infinite-dimensional one.* Specifically, if $h_+(k)$ represents the impulse response of the linear model L_+ and $h_-(k)$ represents the impulse response of L_-, the response of this Uryson model to an arbitrary input sequence $\{u(k)\}$ is given by:

$$\begin{aligned}
y(k) &= \sum_{i=0}^{\infty} h_+(i)\theta_+(u(k-i)) + \sum_{i=0}^{\infty} h_-(i)\theta_-(u(k-i)) \\
&= \sum_{i=0}^{\infty} h_+(i)\theta\left(\frac{1}{2},\frac{1}{2};u(k-i)\right) + \sum_{i=0}^{\infty} h_-(i)\theta\left(\frac{1}{2},-\frac{1}{2};u(k-i)\right) \\
&= \sum_{i=0}^{\infty} \theta\left(\frac{h_+(i)}{2},\frac{h_+(i)}{2};u(k-i)\right) + \sum_{i=0}^{\infty} \theta\left(\frac{h_-(i)}{2},\frac{-h_-(i)}{2};u(k-i)\right) \\
&= \sum_{i=0}^{\infty} \theta\left(\frac{h_+(i)+h_-(i)}{2},\frac{h_+(i)-h_-(i)}{2};u(k-i)\right). \quad (6.114)
\end{aligned}$$

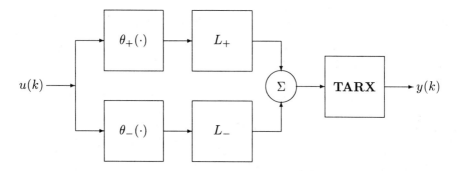

Figure 6.17: Block representation for arbitrary TARMAX models

This result is based on the following properties of the function $\theta(a, b; x)$, discussed in Chapter 3:

$$\lambda\theta(a, b; x) = \theta(\lambda a, \lambda b; x),$$
$$\theta(a, b; x) + \theta(c, d; x) = \theta(a + c, b + d; x), \qquad (6.115)$$

for any real numbers λ, a, b, c, d, and x.

The final two TARMAX representation results given here involve the following subsets of the TARMAX model class, each defined analogously to their more general NARMAX counterparts:

TMAX models: Eq. (3.67) with $p = 0$
TARX models: Eq. (3.67) with $q = 0$, $d_0 = 0$.

Since $p = 0$ implies $c_i = 0$, it follows immediately that the Uryson representation result developed in the previous paragraph applies to the TMAX model class. *Specifically, any TMAX model may be represented as the two-channel Uryson model shown in Fig. 6.16.* In this special case, note that the linear models L_+ and L_- appearing in Fig. 6.16 are both moving average models.

The last result presented here follows from the factorization for additive NARMAX models presented in Chapter 4. Specifically, note that any TARMAX model may be represented as

$$y(k) = \sum_{i=1}^{p} a_i y(k - i) + \sum_{i=1}^{p} c_i |y(k - i)| + v(k) \qquad (6.116)$$
$$v(k) = \sum_{i=0}^{q} b_i u(k - i) + \sum_{i=0}^{q} d_i |u(k - i)|. \qquad (6.117)$$

Here, Eq. (6.116) defines a TARX model and Eq. (6.117) defines a TMAX model. By the previous result, this TMAX model may be represented as the Uryson model shown in Fig. 6.16, implying that any TARMAX model may be represented as shown in Fig. 6.17.

6.7.5 Positive systems

The class of *positive systems* are defined by the discrete-time state-space model

$$\mathbf{x}(k) = \mathbf{A}\mathbf{x}(k) + \mathbf{B}\mathbf{u}(k-1), \tag{6.118}$$

subject to nonnegativity restrictions on the elements of $\mathbf{A}$, $\mathbf{B}$, $\mathbf{x}(k)$, and $\mathbf{u}(k)$ (Coxson and Shapiro, 1987). More specifically, $\mathbf{A}$ is an $n \times n$ matrix of nonnegative values, $\mathbf{x}(k)$ is a state vector with n nonnegative components, $\mathbf{B}$ is an $n \times m$ matrix of nonnegative values, and $\mathbf{u}(k)$ is a control input vector with m nonnegative components. Coxson and Shapiro consider controllability conditions for models of this class and discuss three examples, one involving a simple reversible chemical reaction, and two others from pharmacokinetics. More generally, models of this class may be reasonable approximations to the local dynamics of a wide variety of systems involving physical quantities that are necessarily nonnegative. Controllability conditions for the unrestricted case are well known (Kwakernaak and Sivan, 1972), and positive systems appear to be inherently more difficult to control.

For simplicity, restrict consideration to $m = 1$ and add a scalar observation equation to the above state-space model:

$$y(k) = \mathbf{c}^T \mathbf{x}(k), \tag{6.119}$$

where $\mathbf{c}$ is an n-vector of nonnegative components. It is not difficult to show that the input-output behavior of the combined model can be rewritten as an $ARMAX(n, r)$ model where r is at most n and the coefficients of this model are necessarily nonnegative. Further, it is possible to relax the explicit restriction on the input sequence $\{u(k)\}$ by replacing $u(k)$ with its nonnegative part:

$$v(k) = \theta(1/2, 1/2; u(k)) = \begin{cases} u(k) & u(k) \geq 0 \\ 0 & u(k) < 0 \end{cases} \tag{6.120}$$

In other words, the input restriction may be removed by replacing the linear $ARMAX(n, r)$ model with the positive-homogeneous Hammerstein model defined by this linear model in combination with the nonlinear function $g(x) = \theta(1/2, 1/2; x)$. This representation suggests that it may be interesting to view the class of positive systems a class of *nonlinear* models rather than a class of linear ones defined over a constrained input space. In particular, viewing such systems as nonlinear models makes the controllability results of Coxson and Shapiro less surprising.

6.7.6 PHADD models

This section describes the following nonlinear modeling problem and its solution:

> Given the behavioral restriction of positive-homogeneity and the structural restriction of additivity, characterize the resulting class of discrete-time dynamic models.

More specifically, define the class of PHADD (positive-homogeneous additive) models by the positive homogeneity condition

$$u(k) \to y(k) \quad \Rightarrow \quad \lambda u(k) \to \lambda y(k), \tag{6.121}$$

for all $\lambda \geq 0$, together with the structural additivity condition

$$y(k) = \sum_{i=1}^{p} f_i(y(k-i)) + \sum_{i=0}^{q} g_i(u(k-i)), \tag{6.122}$$

It follows immediately from the positive-homogeneity of the functions $\theta(a_i, c_i; x)$ and $\theta(b_i, d_i; x)$ that the class of TARMAX models discussed in Sec. 6.7 belongs to the PHADD model family. This result is intimately related to the linear multimodel structure of these models discussed in Sec. 6.7.2. In particular, note that the orthant sets Ω_j defined in Eq. (6.96) satisfy the scaling property

$$\psi(k) \in \Omega_j \quad \Rightarrow \quad \lambda \psi(k) \in \Omega_j, \tag{6.123}$$

for all $\lambda > 0$. Therefore, if the sequences $\{u(k)\}$ and $\{y(k)\}$ define a vector $\psi(k)$ that lies in Ω_j for all k, then so do the sequences $\{\lambda u(k)\}$ and $\{\lambda y(k)\}$.

The following three theorems present a partial converse to this basic result: that PHADD NMAX models, NARX models, and—under the appropriate restrictions—NARX* models are necessarily TARMAX models. These results are significant because they establish a relationship between model structure and qualitative behavior for a class of nonlinear models.

Theorem:

Any PHADD NMAX model is a TMAX model.

Proof:

Any PHADD NMAX model is defined in terms of the $q+1$ functions $g_i(\cdot)$ appearing in Eq. (6.122) for $p = 0$ and some $q \geq 0$. Further, this model must satisfy condition (6.121) for all input sequences $\{u(k)\}$. As a specific case, consider the impulse sequence

$$u(k) = \begin{cases} \alpha & k = 0, \\ 0 & k \neq 0, \end{cases}$$

where α is arbitrary. For this input, it follows that $y(0) = g_0(\alpha)$; to satisfy the scaling condition, it is then necessary that $g_0(\lambda \alpha) = \lambda g_0(\alpha)$ for all α and all $\lambda \geq 0$. Thus, $g_0(\cdot)$ is a positive homogeneous scalar function, and is therefore expressible as $g_0(x) = \theta(b_0, d_0; x)$ for some real constants b_0 and d_0. Since $y(i) = g_i(\alpha)$ for $i = 1, 2, \ldots, q$, it follows by analogous reasoning that the functions $g_i(x) = \theta(b_i, d_i; x)$ for $i = 1, 2, \ldots, q$.

□

Linear Multimodels

Corollary:

Any structurally additive, homogeneous NMAX model is linear.

Proof:

It follows from the previous theorem that the scalar functions $g_i(x)$ appearing in Eq. (6.122) are homogeneous and therefore linear.

□

Theorem:

Any PHADD NARX model is a TARX model.

Proof:

Any PHADD NARX model is defined by Eq. (6.122) for some $p \geq 1$, $q = 0$ and $b_0 \neq 0$. Further, this model must satisfy Eq. (6.121) for all $\{u(k)\}$, including

$$u(k) = \begin{cases} \alpha & k = 0, \\ 0 & k \neq 0, \end{cases}$$

for arbitrary α. For this input, $y(0) = b_0\alpha$ and $y(1) = f_1(b_0\alpha)$; to satisfy the scaling condition, $f_1(\lambda b_0 \alpha) = \lambda f_1(b_0 \alpha)$ for all α and all $\lambda \geq 0$. Hence, $f_1(x) = \theta(a_1, c_1; x)$ for some real constants a_1 and c_1. For $i = 2, 3, \ldots, p$,

$$y(i) = f_i(b_0\alpha) + \sum_{\ell=1}^{i-1} f_\ell(y(i - \ell)),$$

implying

$$f_i(b_0\alpha) = y(i) - \sum_{\ell=1}^{i-1} f_\ell(y(i - \ell)).$$

Therefore, if the functions $f_\ell(\cdot)$ are positive homogeneous for $\ell = 1, 2, \ldots, i-1$, and the scaling condition (6.121) is satisfied, it follows that

$$f_i(\lambda b_0 \alpha) = \lambda y(i) - \sum_{\ell=1}^{i-1} f_\ell(\lambda y(i - \ell)),$$

$$= \lambda y(i) - \lambda \sum_{\ell=1}^{i-1} f_\ell(y(i - \ell)) = \lambda f_i(b_0\alpha).$$

Consequently, it follows by induction that $f_i(\cdot)$ is positive homogeneous for $i = 2, 3, \ldots, p$; hence, $f_i(x) = \theta(a_i, c_i; x)$ for all i, establishing the desired result.

□

Corollary:

Any structurally additive, homogeneous NARX model is linear.

Proof:

It follows from the previous theorem that the scalar functions $f_i(x)$ appearing in Eq. (6.122) are homogeneous and therefore linear.

$\square$

In the proof that any PHADD NARX model is a TARX model, a certain controllability condition was required to establish that the function $f_1(x)$ is positive homogeneous. In particular, it was necessary to establish that $f_1(\lambda x) = \lambda f_1(x)$ for all $\lambda > 0$ *and all* x. In establishing this result, essential use was made of the fact that $x = y(0) = b_0 \alpha$ for an impulse input of amplitude α; therefore, $y(0)$ can be forced to assume arbitrary real values by choosing α appropriately. To see the importance of this condition, consider the following example:

$$y(k) = f_1(y(k-1)) + |u(k)|, \qquad (6.124)$$

where

$$f_1(x) = \begin{cases} a_1 x & x \geq 0 \\ \phi(x) & x < 0, \end{cases} \qquad (6.125)$$

and $\phi(\cdot)$ is arbitrary. With $y(k) = 0$ for all $k < 0$, it is not difficult to show that this model is positive-homogeneous for *any* choice of the function $\phi(\cdot)$. To overcome this difficulty, it is necessary to impose additional conditions that guarantee the range of $g_0(x)$ is the entire real line.

To extend these results to arbitrary PHADD NARMAX models—or even structurally additive homogeneous NARMAX models—it is necessary to impose additional conditions to eliminate "nonminimal realizations." As a specific example, consider the NARMAX model

$$y(k) = \phi(y(k-1)) + u(k-1) - \phi(u(k-2)), \qquad (6.126)$$

where the function $\phi(\cdot)$ is completely arbitrary. It is easy to see that one solution to Eq. (6.126) is $y(k) = u(k-1)$ for all k, independent of $\phi(\cdot)$, which may be regarded as a (linear) minimal realization. Exactly the same ambiguity in relating input/output behavior to model structure arises for the nonlinear, positive-homogeneous example

$$y(k) = \phi(y(k-1)) + |u(k-1)| - \phi(|u(k-2)|), \qquad (6.127)$$

which has the minimal TMAX realization $y(k) = |u(k-1)|$.

Linear Multimodels 291

6.8 Summary: the nature of multimodels

It has been stated before but bears repeating that the idea of combining local linear models to obtain a global nonlinear model can be implemented in a variety of different ways. More important, these different implementations are *not* equivalent, as the examples presented in this chapter have illustrated. Specifically, Secs. 6.2.1, 6.2.2, and 6.2.3 described three possible definitions of the class of linear multimodels that are interrelated but not equivalent. For example, if the regions of local validity are disjoint note that the Johansen-Foss formulation described in Sec. 6.2.1 may be viewed as a *single global model* whose parameters change at times k determined by the local model selection criteria. In contrast, the apparently slight modification described in Sec. 6.2.2 may be viewed as a collection of N *independent linear models*, one of which is selected at each time k according to the local model selection criteria. Many examples have been presented here to illustrate these two formulations of the linear multimodeling problem behave quite differently. Generally, it appears that the original Johansen-Foss formulation is probably more useful in practice, although exceptions may exist and the question should probably be considered in the initial stages of multimodel development.

Further, it should also be noted that the multimodel definitions considered in this chapter do not exhaust the possibilities inherent in the idea of combining local linear models: recall that Stromberg et al. (1995) developed a two-input, two-output bilinear distillation column model by linear interpolation of two local linear models. This model may be viewed as a pair of two-input, single-output models, one of which relates output $y_1(k)$ linearly to the inputs $u_1(k)$ and $u_2(k)$. The second model relates the output $y_2(k)$ to the inputs $u_1(k)$ and $u_2(k)$ and may be written as

$$y_2(k) = a_2 y_2(k-1) + \beta_1 u_1(k-4) + \beta_2 u_2(k-2) + c_2, \qquad (6.128)$$

where β_1 and β_2 are given by:

$$\beta_1 = b_{210} + b_{211} y_2(k) \qquad \beta_2 = b_{220} + b_{221} y_2(k). \qquad (6.129)$$

Represented in this form, the bilinear model defined by Eqs. (6.128) and (6.129) has the general flavor of the output-selected multimodels discussed in Sec. 6.5. That is, the terms β_1 and β_2 appearing in the "linear" model defined by Eq. (6.128) are functions of the model output $y_2(k)$. In fact, this view was precisely how Stromberg *et al.* developed this model. Because the resulting model is bilinear, it can exhibit neither input multiplicity nor output multiplicity, in contrast to the linear multimodels considered here. This observation emphasizes one difference between the end results of the local linear modeling approaches considered by Stromberg et al. and Johansen and Foss; other differences are apparent on comparing the bilinear model examples discussed in Chapter 3 with the linear multimodel examples discussed in this chapter.

Another key point about local linear modeling is that all such modeling approaches are based on some type of *validity function* that dictates either which model should be selected, as in Tong's TARSO models, or how the local linear model coefficients should vary over the operating range, as in the bilinear model example just considered. The examples presented in Secs. 6.4, 6.5, 6.6 emphasize that the details of this choice is critical in determining the overall qualitative behavior of the resulting multimodel. The same point may be seen in connection with the bilinear model of Stromberg et al. Suppose, for example, that the model coefficients β_1 and β_2 were made to depend on the model input $u_1(k)$ instead of the model output $y_2(k)$. This change would result in an input selected model something like

$$\beta_1 = b_{210} + b_{211} u_1(k-4) \qquad \beta_2 = b_{220} + b_{221} u_1(k-2). \tag{6.130}$$

Substituting these expressions into Eq. (6.128) yields the following two-input, one-output Hammerstein model:

$$y_2(k) = a_2 y_2(k-1) + g_1(u_1(k-4)) + g_2(u_2(k-2)), \tag{6.131}$$

where the static nonlinearities $g_1(\cdot)$ and $g(\cdot)$ are defined by

$$g_1(x) = \frac{c_2}{2} + b_{210} x + b_{211} x^2 \qquad g_2(x) = \frac{c_2}{2} + b_{220} x + b_{221} x^2. \tag{6.132}$$

Here, the constant term c_2 has been arbitrarily split between the quadratic polynomials $g_1(x)$ and $g_2(x)$. The point is that changing from "output selection" to "input selection" converts the model from a bilinear model with one type of qualitative behavior to a Hammerstein model with a fundamentally different type of qualitative behavior.

It is also worth noting the strong connection that exists between the linear multimodels described in this chapter and the class of *hybrid systems* (Bemporad and Morari, 1999; Blondel and Tsitsiklis, 1999; Morse, 1999). Loosely speaking, these systems combine continuous dynamics with discrete switching phenomena whose behavior may in turn depend on these or other continuous dynamics. The study of these systems is motivated by the fact that they arise frequently in practice. As a specific example, Bemporad and Morari (1999) describe a hybrid system model for the gas supply system at the Kawasaki Steel Mizushima Works. This model describes the flow of blast furnace gas, coke oven gas, and a mixture of these two gasses to a network of five boilers used to generate electricity. The resulting model belongs to the class of *mixed logical dynamical* (MLD) systems, described by time-varying linear evolution equations of the form

$$\begin{aligned} \mathbf{x}(k+1) &= \mathbf{A}_k \mathbf{x}(k) + \mathbf{B}_{1k} \mathbf{u}(k) + \mathbf{B}_{2k} \delta(k) + \mathbf{B}_{3k} \mathbf{z}(k) \\ \mathbf{y}(k) &= \mathbf{C}_k \mathbf{x}(k) + \mathbf{D}_{1k} \mathbf{u}(k) + \mathbf{D}_{2k} \delta(k) + \mathbf{D}_{3k} \mathbf{z}(k), \end{aligned} \tag{6.133}$$

subject to the inequality constraints:

$$\mathbf{E}_{2k} \delta(k) + \mathbf{E}_{3k} \mathbf{z}(k) \leq \mathbf{E}_{1k} \mathbf{u}(k) + \mathbf{E}_{4k} \mathbf{x}(k) + \mathbf{E}_{5k}. \tag{6.134}$$

Here $\mathbf{x}(k)$ is the state vector of the system at time k, consisting of both continuous-valued components and *binary components* that only assume the values 1 or

0. Similar partitionings hold for the input vector $\mathbf{u}(k)$ and the output vector $\mathbf{y}(k)$. The vector $\mathbf{z}(k)$ represents a collection of continuous-valued auxilliary variables, and the vector $\delta(k)$ represents a collection of binary auxiliary variables. Physically, binary variables arise from descriptions of fundamental limits (e.g., nonnegativity constraints on flow rates and volume limits on gas storage tanks), from configuration descriptions (e.g., boiler no. 2 is connected to the blast-furnace gas supply or it is not), or other similar phenomena.

The example just described appears in a special issue of *Automatica* devoted to hybrid systems (Morse, 1999). Another paper in this special issue notes the general behavioral complexity of the following piecewise-linear discrete-time state-space model (Blondel and Tsitsiklis, 1999):

$$\mathbf{x}_{k+1} = \begin{cases} \mathbf{A}_1 \mathbf{x}_k & \mathbf{c}^T \mathbf{x}_k \geq 0 \\ \mathbf{A}_2 \mathbf{x}_k & \mathbf{c}^T \mathbf{x}_k < 0. \end{cases} \qquad (6.135)$$

For example, it is shown that the general question of whether such a system is stable is NP-hard. Note that this system may be viewed as a *state-selected linear multimodel* in the context of the ideas discussed in this chapter. The key points here are first, that systems similar to those discussed in this chapter arise quite naturally in practice and second, that their qualitative behavior is extremely difficult to analyze, in general.

Finally, it is reasonable to conclude this chapter with a brief discussion of the class of TARMAX models introduced in Sec. 6.7. Though these models may seem somewhat artificial, it is important to emphasize that they are members of Tong's class of TARSO models, which are themselves members of the larger class of multimodels defined by Johansen and Foss. In addition, the TARMAX model family belongs to the larger family of positive-homogeneous models introduced in Chapter 3 and contains the family of linear ARMAX models discussed in Chapter 2. Further, specific examples or subsets of the TARMAX family do provide simple descriptions of phenomena seen in nature. For example, four TARMAX models have been discussed in this book as approximations of the qualitative assymetry seen in the exothermic CSTR model introduced in Chapter 1: the Uryson model discussed in Chapter 1 (Sec. 1.4.5), the two positive-homogeneous models discussed in Pearson (1995, models 2 and 3), and the static-linear TARMAX model discussed in Sec. 6.7.3. Also, it was noted in Sec. 6.7.5 that the class of positive systems (Coxson and Shapiro, 1987) may be viewed as a subset of the TARMAX model class.

Because of the structural simplicity of the TARMAX model class, it is almost as amenable to behavioral analysis as the class of linear ARMAX models it contains. For example, the Lipschitz stability conditions for structurally additive models given in Chapter 4 apply to the TARMAX class. This point was illustrated for the general first-order TARMAX model in Sec. 6.7.1, where explicit stability conditions were given in terms of the model parameters. As another example, Sec. 6.7.2 provided a multimodel interpretation for the TARMAX models by relating the local linear model selection conditions to the parameters of the TARMAX model. Sec. 6.7.3 then presented simple explicit conditions for a TARMAX model to be a member of the intriguing family of static linear models introduced in Chapter 3. Finally, Sec. 6.7.6 showed that the TARMAX model family, defined structurally, is essentially equivalent to the PHADD model family, which is defined in terms of both behavioral and structural criteria. Despite the analytical convenience of the TARMAX family, it is important to note that it is capable of exhibiting a wide variety of nonlinear behavior, including both strong asymmetry in response to symmetric inputs and chaotic step responses. Viewed as a microcosm of the world of linear multimodels, the behavior of the TARMAX models emphasizes the range of behavior—both intended and unintended—that can be described by piecing together local linear models.

Chapter 7

Relations between Model Classes

Chapter 2 has provided a brief review of some of the important characteristics of linear models, and Chapters 3 through 6 have introduced and discussed a number of specific classes of nonlinear models. The focus of this chapter is on the relationships that exist between these different model classes. Some of these relationships have already been discussed briefly in isolated places in earlier chapters, but this chapter attempts to give a much broader overview of how different model classes relate. In particular, it is useful to note that many popular model classes are included as proper subsets of other, larger classes. In more subtle cases, one class will be "almost included" in another, larger class, but a portion of the first class will fail to satisfy this inclusion. As a specific example, it was noted in Chapter 4 that the class of Hammerstein models is a proper subset of the class of additive NARMAX models. In contrast, it was also shown in Chapter 4 that Wiener models are not members of the additive NARMAX class except in the two degenerate cases where Wiener and Hammerstein models coincide: linear dynamic models and static nonlinearities. This chapter begins with a summary of these inclusion and exclusion results in Sec. 7.1, including both results assembled from previous chapters and a few new ones. Of particular interest are questions concerning the relationship between structurally-defined model classes and behaviorally-defined classes since these questions are directly related to the practical problem of initial model structure selection for empirical modeling, a topic considered further in Chapter 8.

One of the principal objectives of this chapter is to illustrate the utility of *category theory* in characterizing relations between different classes of dynamic models. Essentially, category theory is a branch of mathematics whose aim is to elucidate relations between different classes of mathematical objects in extremely general terms. More specifically, category theory deals with mathematical objects (in fact, called *objects*) and transformations between objects (called *morphisms*), requiring only that these transformations be "well behaved" with respect to successive application (called *composition of morphisms*). These basic notions are introduced in Sec. 7.2 and illustrated with simple examples involving familiar concepts like vectors and matrices. Sec. 7.3 applies these ideas to discrete-time dynamic models, introducing the category **DTDM** whose objects are input sequences $\{u(k)\}$ and whose morphisms are dynamic models mapping these input sequences into output sequences $\{y(k)\}$. Composition of morphisms then corresponds to cascade connection of dynamic models.

Given these definitions, Secs. 7.4 through 7.9 demonstrate the applicability of category theory to a number of practically important modeling issues. Specifically, Sec. 7.4 introduces the notion of a *subcategory* and illustrates its utility in defining both *structural model categories* based on explicit model structure restrictions and *behavioral model categories* based on qualitative behavior restrictions. Sec. 7.5 specializes these results further to subcategories that are closely related to important issues in empirical model identification. Sec. 7.6 introduces the notion of a *joint subcategory* and uses it to explore the relationship between model structure restrictions and qualitative behavior. One of the advantages of considering categories of nonlinear dynamic models is that category theory provides a framework for investigating relations between different categories. Sec. 7.7 introduces the notion of a *functor*, which is essentially a mapping between categories, and it is shown that this construct is useful in examining both the linearization and the inversion of nonlinear models. Next, Sec. 7.8 introduces the notion of an *isomorphism*, which is an invertible functor; a key point is that if an isomorphism exists between two categories, they are equivalent in certain important respects. This idea is used to investigate the relationship between autoregressive and moving average model structures, and between continuous-time and discrete-time dynamic models. Building on these results further, Sec. 7.9 explores the relationship between the category $\mathbf{Hom}_\phi$ of homomorphic models introduced in Sec. 7.4.3 and the categories of linear and homogeneous models, and ultimately proposes a systematic procedure for generating new categories of discrete-time dynamic models. Finally, Sec. 7.10 gives a brief synopsis of the results presented in this chapter and their practical significance.

7.1 Inclusions and exclusions

The simplest possible relations between different dynamic model classes are those based on inclusion: class $\mathcal{C}$ is included in class $\mathcal{D}$. Although some of these inclusions are fairly obvious, often they are only clear after some reflection (e.g., the class of Lur'e models is a proper subset of the class of additive NARMAX models). Conversely, relations based on *exclusion* are frequently much less obvious, as in the case of homomorphic models and NARX models discussed in Chapter 4 (Sec. 4.3.5). The following sections discuss a number of these relations, summarizing useful results from previous chapters and augmenting them with a few new ones, particularly in Sec. 7.1.2.

Relations between Model Classes 297

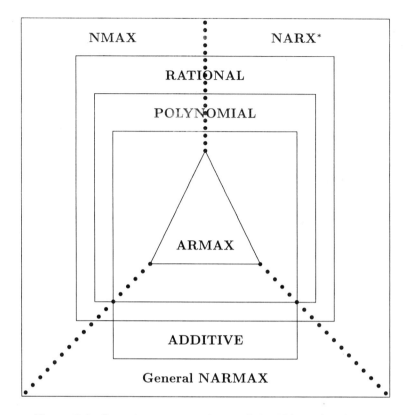

Figure 7.1: Some important subsets of the NARMAX family

7.1.1 The basic inclusions

Both structurally-defined and behaviorally-defined model classes may sometimes be ordered by inclusion and both cases are considered here. In keeping with the general organization of this book, this discussion begins with a consideration of the NARMAX model class and its principal inclusions. For convenience, Fig. 4.1 from Chapter 4 is repeated here as Fig. 7.1, illustrating the general relationships that exist between the following classes of NARMAX models:

1. NMAX models
2. NARX* models
3. Linear ARMAX models
4. Polynomial NARMAX models
5. Rational NARMAX models
6. Structurally additive NARMAX models.

All of these model classes were discussed in detail in Chapter 4; the summaries given here and in Sec. 7.1.2 emphasize inclusions, similarities, and differences within and between these model classes.

The class of NMAX models was defined in Chapter 1, discussed at length in Chapter 4, and refined in Chapter 5. This class generalizes the class $MA_{(M)}$ of order M linear moving average models and includes many popular nonlinear dynamic models as special cases. The following chain of inclusions is typical:

$$MA_{(M)} \subset H_{(N,M)} \subset D_{(N,M)} \subset V_{(N,M)} \subset NMAX_c \subset BIBO,$$

based on the Volterra representations discussed in Chapter 5. Here, $NMAX_c$ represents the class of nonlinear moving average models based on continuous functions, shown in Chapter 4 to be BIBO stable. Note that all inclusions in this chain are proper: not all BIBO stable models have NMAX representations, and not all continuous NMAX models have Volterra representations, a point illustrated further in Sec. 7.1.2. Hence, any finite-order class $V_{(N,M)}$ of Volterra models is necessarily a proper subset of the class of continuous NMAX models. Similarly, most $V_{(N,M)}$ Volterra models have nonzero off-diagonal terms—for example, all nonlinear Wiener models—thus illustrating that the class $D_{(N,M)}$ of diagonal Volterra models is a proper subset of $V_{(N,M)}$. Further, it was shown in Chapter 5 that the class of diagonal Volterra models is identical to the class of polynomial Uryson models, of which the class $H_{(N,M)}$ of polynomial Hammerstein models is a proper subset. Finally, note that the class $MA_{(M)}$ of linear moving average models is equivalent to the degenerate classes $H_{(1,M)}$, $D_{(1,M)}$, and $V_{(1,M)}$, all of which are equivalent and properly contained in the larger classes $H_{(N,M)}$, $D_{(N,M)}$, and $V_{(N,M)}$, respectively.

A similar chain of inclusions results if the Hammerstein models $H_{(N,M)}$ and the Uryson (diagonal) models $D_{(N,M)}$ are replaced with their duals

$$MA_{(M)} \subset W_{(N,M)} \subset P_{(N,M)} = V_{(N,M)} \subset NMAX_c \subset BIBO.$$

Here, $W_{(N,M)}$ represents the Wiener models that are dual to the $H_{(N,M)}$ Hammerstein models, obtained by reversing the order in which the static linearity and linear dynamic subsystems are connected. Similarly, $P_{(N,M)}$ represents the class of finite projection-pursuit NMAX models introduced in Chapter 5, obtained by reversing the order of the polynomial nonlinearities and linear dynamic subsystems appearing in the Uryson model $U_{(N,M)}$. In contrast to the previous chain of inclusions, note that the inclusion of the projection-pursuit class in the Volterra class is not proper, as it was established in Chapter 5 that these two model classes are equivalent.

Closely related to this last observation, it is interesting to note the differences between the pairs of dual model classes $[H_{(N,M)}, W_{(N,M)}]$ and $[D_{(N,M)}, P_{(N,M)}]$. That is, it was shown in Chapter 4 that the Hammerstein and Wiener model classes are almost mutually exclusive, including as common elements only linear moving average models and static nonlinearities. In contrast, because the class of projection-pursuit models is equivalent to the class of finite Volterra models, the class $P_{(N,M)}$ includes its dual class $D_{(N,M)}$ as a proper subset.

It has been emphasized repeatedly that, for linear models, equivalences exist between autoregressive and moving-average representations, a point that is discussed further in later sections of this chapter. It has also been emphasized that, for nonlinear models, these equivalences no longer hold. In fact, two different classes of nonlinear autoregressive models—the NARX* models and the NARX models—are defined and discussed in Chapter 4 and these models obey the following chain of inclusions:

$$AR(p) \subset NARX(p) \subset NARX^*(p) \subset NARMAX(p,q).$$

Here, the autoregressive model order p is specified explicitly and the designation $NARMAX(p,q)$ represents the general NARMAX class shown in Fig. 7.1. Generally, it may be said that the behavior of autoregressive models is more complex than that of moving average models. As a specific example, note that whereas the step responses of NMAX models are necessarily asymptotically constant, the step responses of BIBO stable NARX or NARX* models can exhibit either persistent oscillations or chaotic behavior. Further, although NMAX models are necessarily BIBO stable under fairly weak conditions, NARX or NARX* models can exhibit input-dependent stability behavior.

A consequence of the richer qualitative behavior exhibited by nonlinear autoregressive models is that they are also more difficult to analyze than their moving average counterparts. For example, it is noted in Chapter 5 that the class of Volterra models represents the cannonical class of NMAX models, arising naturally as the Taylor series expansion of the nonlinearity defining the NMAX model, provided this expansion exists. Further, much can be said about the class of Volterra models, as evidenced by the inclusion of an entire chapter on this model class. In contrast, the natural analog of the Volterra NMAX models would be the polynomial NARX* models, about which relatively little can be said in general terms. Among the strongest results available are the negative ones presented in Sec. 7.1.2 below: this model class does not include either nonlinear homogeneous members or nonlinear positive-homogeneous members. Certain specific NARX and NARX* models have been analyzed in considerable detail, but to date, simple but flexible *general* autoregressive model structures do not appear to have been developed. In particular, autoregressive analogs of the popular Hammerstein and Wiener models do not appear to exist at present, although the Lur'e model structure described in Chapter 4 may represent a reasonable candidate.

A particularly interesting feature of the Lur'e model is that it belongs to the class of structurally additive models introduced in Chapter 4, a class that intersects both the NMAX and NARX families. The original motivation for these models was that the additive NARMAX model structure

$$y(k) = \sum_{i=1}^{p} f_i(y(k-i)) + \sum_{i=0}^{q} g_i(u(k-i)), \qquad (7.1)$$

facilitates the use of nonparametric procedures in model identification, reducing the problem to one of determining unknown *scalar* nonlinearities. This model class was introduced here because the structural additivity restriction (7.1) has some interesting behavioral consequences. For example, it is shown in Chapter 4 that the steady-state behavior of any nonadditive NMAX or NARX model can be matched exactly by a structurally additive model. Thus, it follows that the behavioral consequences of structural additivity only appear in transient responses. As a specific example, note that Hammerstein models are structurally additive, while Wiener models are not; it was noted in Chapter 1 that these models differ only in their transient responses. Similarly, it was shown in Chapter 6 that any positive-homogeneous, structurally additive NMAX or NARX model necessarily belongs to a restricted class of linear multimodels (the TARMAX model family). Structural additivity is essential here in "reflecting" the qualitative input-output behavior (positive-homogeneity) through to structural restrictions on the model.

Overall, the structurally additive NAARX model class may be conveniently summarized by the following chain of inclusions:

$$ARMAX(p,q) \subset \left\{ \begin{array}{l} H(p,q) \\ LURE(p,q) \\ TARMAX(p,q) \end{array} \right\} \subset NAARX(p,q)$$

$$\subset PP^r(p,q) \subset NARMAX(p,q).$$

Here, $H(p,q)$, $LURE(p,q)$, and $TARMAX(p,q)$ represent the Hammerstein, Lur'e, and TARMAX model classes, respectively, each based on linear ARMAX models of autoregressive order p and moving average order q. Similarly, $NAARX(p,q)$ represents the class of structurally additive models defined by Eq. (7.1) and $PP^r(p,q)$ represents the r-channel projection-pursuit models defined in Chapter 4 and discussed further in Chapter 5.

7.1.2 Some important exclusions

In contrast to the previous section, the following discussion emphasizes the differences between model classes. In particular, this discussion is concerned with model classes that are mutually exclusive: $M \in \mathcal{C}$ implies $M \notin \mathcal{D}$. *Since almost all of the model classes considered in this book include linear models as special cases, these results will be presented "modulo linear models."* That is, statements like "model classes $\mathcal{C}$ and $\mathcal{D}$ are mutually exclusive" are to be interpreted as "any *nonlinear* model M belonging to the class $\mathcal{C}$ does not belong to the class $\mathcal{D}$ and vice versa." As a specific example, it was noted in Chapter 3 that the only homogeneous bilinear models are linear, so the classes of bilinear models and homogeneous models are mutually exclusive in this sense. In fact, this observation is a corollary of the following result.

Theorem:

> A polynomial NARMAX model is positive-homogeneous if and only if it is linear.

Relations between Model Classes

Proof:

Linear models are homogeneous, thus positive-homogeneous, so it is only necessary to prove the reverse implication. Suppose $\mathcal{M}$ is a positive-homogeneous, polynomial NARMAX model. The response of this model will be given by

$$y(k) = \sum_{n=0}^{N} \phi_n,$$

where ϕ_n is the collection of terms of order n in all arguments in the function $F(\cdots)$ defining the NARMAX model. Scaling the input sequence $\{u(k)\}$ by a positive constant λ, it follows from the positive-homogeneity of $\mathcal{M}$ that the response sequence $\{y(k)\}$ also scales by λ. Therefore, it follows that ϕ_n scales by λ^n, and positive-homogeneity requires

$$\lambda \sum_{n=0}^{N} \phi_n = \sum_{n=0}^{N} \lambda^n \phi_n \quad \Rightarrow \quad (\lambda - 1)\phi_0 + \sum_{n=2}^{N} (\lambda - \lambda^n)\phi_n = 0.$$

Since this condition must hold for all $\lambda > 0$, it follows that $\phi_0 = 0$ and $\phi_n = 0$ for $n = 2, \ldots, N$. Thus, the model $\mathcal{M}$ is linear.

□

The following corollaries of this theorem are consequences of the fact that homogeneous models are positive-homogeneous, and that Volterra models constitute a particular class of polynomial NARMAX models. These and other analogous results are discussed further in Chapter 8 in connection with model structure selection. In particular, Sec. 8.3 presents a collection of tables summarizing the relationships between various types of nonlinear qualitative behavior and most of the structural model classes discussed in this book. Included in these types of nonlinear behavior are homogeneity, positive-homogeneity, and static-linearity.

Corollary:

A polynomial NARMAX model is homogeneous if and only if it is linear.

Corollary:

A Volterra model is positive-homogeneous if and only if it is linear.

Corollary:

A Volterra model is homogeneous if and only if it is linear.

To conclude this discussion, note that mutual exclusivity relations like the previous theorem and its corollaries may also be developed between different structurally-defined model classes. These results are often not obvious and may therefore offer some useful insights into the inherent differences between different structural restrictions. For example, it was shown in Chapter 4 (Sec. 4.3.5) that the classes of $NARX(p)$ models and homomorphic systems are mutually exclusive in the sense discussed earlier (i.e., modulo linear models). Similarly, the following argument establishes that the mutual exclusivity of the $NARX$ and Hammerstein model classes. Suppose the $NARX(r)$ model

$$y(k) = F(y(k-1), \ldots, y(k-r)) + b_0 u(k) \tag{7.2}$$

also exhibited the Hammerstein representation

$$y(k) = \sum_{i=1}^{p} a_i y(k-i) + \sum_{i=0}^{q} b_i g(u(k-i)). \tag{7.3}$$

Further, suppose $y(k) = 0$ for $k < 0$ and consider the response of both models to an impulse of magnitude α at time $k = 0$. Considering $k = -1$, it follows from Eq. (7.2) that $F(0, \ldots, 0) = 0$, and matching the values of $y(0)$ from Eqs. (7.2) and (7.3) implies $g(\alpha) = \alpha$ for all real α. Hence, it follows that the only Hammerstein model with a NARX representation is linear. Very similar arguments may be used to establish the mutual exclusivity of the $NARX$ and Wiener model classes. Specifically, consider the Wiener model

$$y(k) = g(v(k)) \qquad v(k) = \sum_{i=1}^{p} a_i v(k-i) + \sum_{i=0}^{q} b_i u(k-i). \tag{7.4}$$

At $k = 0$, the response of this model to an impulse of magnitude α is $g(b_0 \alpha)$ and matching this response with that obtained from Eq. (7.2) again requires $g(x) = x$. This result is particularly interesting since it was demonstrated in Chapter 4 (Sec. 4.3.5) that Wiener models based on linear $ARX(p)$ models and invertible static nonlinearities belong to the $NARX^*(p)$ model class.

7.2 Basic notions of category theory

As noted in the introduction, category theory is concerned with very general relationships between different classes of mathematical objects (for example, linear vector spaces, groups or topological spaces). The motivation for considering category theory here is that it provides an extremely powerful framework for extending certain intuitions from linear algebra and scalar functions to specific classes of nonlinear dynamic models, as subsequent sections demonstrate. No prior exposure to category theory is assumed here, and no deep theoretical results are employed, but the basic constructs are outlined and their utility in comparing different model classes is demonstrated through a number of detailed examples. For those wishing to explore this fascinating subject further, a number of excellent introductions are available (Adamek et al., 1990; Blyth, 1986; MacLane, 1998).

7.2.1 Definition of a category

In simplest terms, a *category* is a collection of mathematical *objects* (e.g., vector spaces R^n), together with a collection of transformations or other mathematical relations between these objects called *morphisms* (e.g., matrices that map R^n into R^m). To qualify as a category, these objects and morphisms must satisfy certain compatibility conditions which are both surprisingly simple and surprisingly powerful in their consequences. More specifically, a category **C** consists of four components (Adamek et al., 1990, p. 13):

1. A class $\mathcal{O}$ of objects
2. For each pair of objects in $\mathcal{O}$, a set of morphisms relating them
3. For each object in $\mathcal{O}$, an identity morphism
4. A composition law relating morphisms.

Here, the term *class* refers to a collection that is generally larger than a set (specifically, a class is a set of sets); the distinction between classes and sets is important and is discussed in some detail in Sec. 7.2.2. If A and B represent objects in $\mathcal{O}$, the set of morphisms in the category **C** relating A and B may be written as either $\hom(A, B)$ (Adamek et al., 1990; MacLane, 1998) or $\mathrm{Mor}_\mathbf{C}(A, B)$ (Blyth, 1986). Because it is somewhat simpler, the notation $\hom(A, B)$ will generally be used here. A morphism $f \in \hom(A, B)$ is also sometimes called an *arrow* (MacLane, 1998) and is frequently denoted $f : A \to B$ for convenience. The identity morphism associated with an object A is generally denoted id_A and belongs to the morphism set $\hom(A, A)$.

The fundamental component of a category is the composition law relating morphisms, defined as follows. If $f : A \to B$ and $g : B \to C$ are morphisms in **C**, the *composition* of f and g is a morphism $g \circ f : A \to C$. For **C** to qualify as a category, the composition law must satisfy three conditions. First, the composition $g \circ f$ must be defined for any pair of morphisms $f : A \to B$ and $g : B \to C$ relating arbitrary objects A, B, and C. Second, composition must satisfy $f \circ \mathrm{id}_A = f$ and $\mathrm{id}_B \circ f = f$ for all morphisms $f \in \hom(A, B)$. Third, composition of morphisms must be *associative*, implying that if $f : A \to B$, $g : B \to C$ and $h : C \to D$ then

$$h \circ (g \circ f) = (h \circ g) \circ f. \tag{7.5}$$

The composition law imposes structure on the category **C** through its influence on the morphism sets $\hom(A, B)$, a point illustrated in Sec. 7.2.3.

One of the standard examples of a category is **Set** (Blyth, 1986, p. 2), whose class of objects is the class of all possible sets discussed in Sec. 7.2.2, and whose morphisms are all possible mappings between sets. Composition of morphisms corresponds to the usual composition of functions: if S_1, S_2, and S_3 are sets, if $f : S_2 \to S_3$, and if $g : S_1 \to S_2$, then $h = f \circ g : S_1 \to S_3$ is defined by $h(x) = f(g(x))$ for all $x \in S_1$. Further, note that for any object S in **Set**, there is an identity mapping id_S that maps S into itself, that is: $\mathrm{id}_S(x) = x$ for all $x \in S$. Similarly, it is well known that composition of functions is associative; for example, define $\phi(x) = f(g(x))$ and $\psi(x) = h(f(x))$ and note that $h(\phi(x)) = \psi(g(x))$ for all x in the domain of $f(\cdot)$.

Another standard example is **Vect**, based on the class of linear vector spaces R^n. The morphisms in this category are the $n \times m$ matrices $M : R^m \to R^n$ and composition of morphisms corresponds to ordinary matrix multiplication. That is, if $M_1 \in \text{hom}(R^n, R^m)$ and $M_2 \in \text{hom}(R^p, R^n)$, the composition of M_1 and M_2 is given by

$$M_1 \circ M_2 = M_1 M_2 \in \text{hom}(R^p, R^m). \tag{7.6}$$

Note that whereas $M_1 \circ M_2$ is well defined, as it must be for **Vect** to satisfy the defining conditions for a category, the composition $M_2 \circ M_1$ is *not* defined unless $p = m$. The identity objects id_{R^n} in the category **Vect** are the $n \times n$ identity matrices I_n. The requirement for that $id_Y \circ f = f$ and $g \circ id_Y = g$ follows from the well-known fact that $M_1 I_n = M_1$ and $I_n M_2 = M_2$ for the $m \times n$ and $n \times p$ matrices M_1 and M_2. Similarly, the associativity requirement for composition of morphisms follows from the associativity of matrix multiplication: if M_1 and M_2 are $n \times m$ and $p \times n$ matrices as before and M_3 is a $q \times p$ matrix, it is a standard result that

$$M_1[M_2 M_3] = [M_1 M_2]M_3. \tag{7.7}$$

In addition to being relatively simple and familiar, this example is important because many constructs from linear algebra can be extended to a more general setting in terms of categories. Since linear algebra is closely related to linear system theory, this observation raises the possibility that category theory may provide a useful vehicle for transferring linear systems intuition to classes of nonlinear systems.

7.2.2 Classes vs. sets

The definition of a category given in the previous section notes first, that the collection of objects on which a category is based is a *class* and second, that a class is distinct from the more familiar notion of a *set*. The following discussion briefly examines this distinction and its importance. Loosely speaking, a set is a "small collection of objects" whose elements may be characterized explicitly; examples include the finite set $\{a, b, c\}$ of real variables, the set Z of integers, and the real line R which may be viewed as the set of all possible real numbers. In contrast, *classes* are "large collections of objects" like the collection of all sets or the collection of all vector spaces. More specifically, *the members of a class are sets and any collection of sets defines a class*. Consequently, any set may be viewed as a class, but not all classes are sets; as a specific example, the class of all sets is *not* a set. Classes that may be viewed as sets are called *small classes*, whereas classes that may not be viewed as sets are called *proper classes* or *large classes*. The following discussion briefly describes some of the differences between classes and sets; for a more detailed disucssion of these differences, refer to Adamek et al. (1990, ch. 2).

The reason the distinction between sets and classes is important is that certain fundamental difficulties may arise if this distinction is not made. One specific example is *Russell's paradox* (Blyth, 1986, p. 1): consider the class $\mathcal{U}$ of all sets and the class $\mathcal{V}$ of sets that do not contain themselves:

$$\mathcal{V} = \{X \in \mathcal{U} | X \notin X\}. \tag{7.8}$$

If the class $\mathcal{U}$ were a set, then $\mathcal{V}$ would be a subset of $\mathcal{U}$ and therefore also a set. It would then be possible to ask the question, "does $\mathcal{V}$ contain itself?" Suppose not: by the definition (7.8), $\mathcal{V} \notin \mathcal{V}$ implies $\mathcal{V}$ is contained in this collection of sets that do not contain themselves; that is, $\mathcal{V} \notin \mathcal{V}$ implies $\mathcal{V} \in \mathcal{V}$. The basis of this difficulty is that $\mathcal{U}$ is not a set, but rather a proper class. Hence, the subcollection $\mathcal{V}$ is also a proper class and therefore *not* a member of the class $\mathcal{U}$ of all sets.

Again, the primary distinction between classes and sets is that the members of a class are sets, but the members of a set cannot be classes. Further, if X is an element of the set $\mathcal{S}$ and the symbol $P(X)$ is interpreted to mean "X has mathematical property P" (e.g., X is a real number, an interval, or a differentiable function), then the collection

$$\mathcal{A} = \{X \in \mathcal{S} | P(X)\} \tag{7.9}$$

forms a *set* which is a *subset* of $\mathcal{S}$. Conversely, if $\mathcal{C}$ is a class, the corresponding collection

$$\mathcal{B} = \{X \in \mathcal{C} | P(X)\} \tag{7.10}$$

forms a *class*, which may or may not be a set. Other common set operations may or may not extend to proper classes. For example, the operations of union and intersection do extend to classes: if $\mathcal{A}$ and $\mathcal{B}$ are classes, the union $\mathcal{A} \cup \mathcal{B}$ is the class whose elements are those sets belonging to either $\mathcal{A}$ or $\mathcal{B}$, or both; similarly, $\mathcal{A} \cap \mathcal{B}$ is the class whose elements are those sets belonging to both $\mathcal{A}$ and $\mathcal{B}$. Conversely, if $\mathcal{A}$ and $\mathcal{B}$ are both sets, then the collection $\{\mathcal{A}, \mathcal{B}\}$ is a set, but if $\mathcal{A}$ and $\mathcal{B}$ are both proper classes, this construction does *not* define a class since the members of a class must be sets.

The following result illustrates the fundamental difference between sets and classes. If $\mathcal{I}$ is any set, a *family of sets* $(X_i)_{i \in \mathcal{I}}$ is defined as a function mapping the set $\mathcal{I}$ (called the *index set*) into the class $\mathcal{U}$ of all sets. The *image* of this family is the class

$$\mathcal{Y} = \{X_i | i \in \mathcal{I}\}, \tag{7.11}$$

and an important result (Adamek et al., 1990, p. 6) is that this class is in fact a *set*. As a consequence, no *proper class* can be represented as an indexed family of sets X_i for any index set $\mathcal{I}$. This observation is called the *Axiom of Replacement* (Adamek et al., 1990, p. 7) and implies that proper classes are in some important sense "larger" than sets, thus motivating the terminology *small classes* for sets and *large classes* for proper classes. *Essentially, this result* means that it is not possible even in principle to explicitly enumerate the sets that belong to a proper class.

Many of the categories of interest in this chapter are *large categories*, defined as categories whose objects form a large class. For example, the category **Set** introduced in the previous section is a large category since the class of all possible sets is a large class. Similarly, the category **DTDM** of discrete-time dynamic models defined in Sec. 7.3 is also a large category. Conversely, the category **Vect** introduced in the previous section is a *small category*, defined as categories whose objects form a small class (that is, categories whose objects form a set). To see that **Vect** is a small category, first note that any vector space R^n may be viewed as the set of all possible n-dimensional vectors. Hence, the object on which **Vect** is based is the image of the family of sets R^n indexed by the positive integers n. Of more interest here, the IO subcategories defined in Sec. 7.5 also define small categories. The key point is the care required in dealing with the class of objects on which large categories are based (again, is worth emphasizing that these objects cannot be indexed by any set, however uncountably infinite its membership).

7.2.3 Some illuminating examples

To illustrate both the range of choices that are generally possible in defining category and the influence of those choices, the following discussion consider some simple variations of the two category examples introduced in Sec. 7.2 These examples are offered to help bring the concept of a category into sharp focus and also to illustrate the range of variation possible through slight modifications of the definition of a specific category. In particular, given a category **C**, other categories may be obtained by modifying the objects, the morphism or the composition rule on which it is based; generally, these choices dicta the definition of the identity morphism, as the category **Had** discussed bel illustrates. Some of these ideas are developed more systematically in connecti with the concept of subcategories in Secs. 7.4 through 7.6.

The first set of examples includes three variations of the category **Set**, based on the class of real numbers as objects. Specifically, this class is real line R^1, which constitutes an uncountably infinite *set* of objects; her these three categories are all small. Further, the morphisms on which the categories are based are different types of real-valued functions and comp tion of morphisms corresponds to composition of mappings, as in the categ **Set**. Similarly, the identity morphisms associated with each object are ba on the identity mapping $f(x) = x$. The first example is **Rfun**, the categor arbitrary real-valued functions; it is easy to see that this collection of functi constitutes a category because it is large enough to include the identity m ping and it is closed under composition. The second category considered he **Mfun**, for which the set $\hom(x, y)$ consists of all *monotone* (that is, increa or decreasing) mappings $f : x \rightarrow y$. Note that the identity mapping $f(x)$ is an increasing function and that monotonicity is preserved under functi composition. For example, if $f(x)$ is an increasing function and $g(x)$ is a creasing function, the composition $h(x) = [g \circ f](x) = g(f(x))$ is decreasing: if $x < y$, then $f(x) \leq f(y)$, implying $g(f(x)) \geq g(f(y))$.

The third variation of **Set** considered here is **Poly**, the category of polynomials. Morphisms in this category correspond to the real-valued functions

$$f_n(x) = \sum_{i=0}^{n} a_i x^i \qquad (7.12)$$

where n is any positive integer and $\{a_i\}$ is any set of n real coefficients. Taking $n = 1$, $a_0 = 0$ and $a_1 = 1$ yields the identity mapping $f_1(x) = x$, and the composition of two polynomials $f_n(x)$ and $f_m(x)$ leads to a third polynomial of order nm:

$$[f_n \circ f_m](x) = \sum_{i=0}^{n} a_i \left[\sum_{j=0}^{m} b_j x^j\right]^i \equiv \sum_{k=0}^{nm} c_k x^k = f_{nm}(x). \qquad (7.13)$$

Before leaving this example, note the following difference between the categories **Mfun** and **Poly**: the first is *behaviorally defined*, obtained by specifying the qualitative behavior of the morphisms, whereas the second is *structurally defined*, obtained by specifying the explicit mathematical structure of the morphisms. Analogous differences are seen in the categories of dynamic models of principal interest in this chapter.

The other set of examples considered here consists of five variations of the category **Vect** of finite-dimensional vector spaces. The first of these variations is the category **Square** of square matrices. As in **Vect**, composition of morphims corresponds to ordinary matrix multiplication and the identity morphisms are the $n \times n$ identity matrices. Since the product of two $n \times n$ matrices is a third $n \times n$ matrix it follows that **Square** satisfies the defining conditions for a category. Note, however, that most of the morphism sets in this category are empty; in particular, $\hom(R^n, R^m) = \emptyset$ if $n \neq m$. The second variation on **Vect** considered here is **Orthog**, the category of *orthogonal* matrices (Horn and Johnson, 1985, p. 71). These $n \times n$ matrices O are nonsingular and satisfy the condition

$$O^{-1} = O^T. \qquad (7.14)$$

Geometrically, orthogonal matrices represent rotations of a vector in R^n about the origin. To define the category **Orthog**, the objects are the linear vector spaces R^n as before, the morphisms are the orthogonal matrices, and composition of morphisms again corresponds to matrix multiplication. To see that this collection is closed under composition of morphisms, note that if O_1 and O_2 are orthogonal matrices, then

$$[O_1 O_2]^{-1} = O_2^{-1} O_1^{-1} = O_2^T O_1^T = [O_1 O_2]^T. \qquad (7.15)$$

Hence, the product of two orthogonal matrices is again orthogonal, thus establishing that the composition of two morphisms in **Orthog** defined by matrix multiplication does indeed yield a third morphism. It is also clear that the $n \times n$ identity matrices satisfy Eq. (7.14) and are thus morphisms in **Orthog**, as they must be for this collection of objects and morphisms to constitute a category.

The third variation of the category **Vect** considered here is the category **Had** of square matrices and Hadamard products. Here, the objects and the morphisms are the same as they are in the category **Square**—that is, the objects are the real vector spaces R^n and the morphisms correspond to $n \times n$ matrices—but a different composition rule is adopted. As a consequence, the categories **Square** and **Had** are very different, illustrating the point made in Sec. 7.2.1 about the influence of the composition rule on the nature of a category. Specifically, composition of morphisms in the category **Had** is defined by the *Hadamard product* of two square matrices, which is an *elementwise* matrix product, given by (Horn and Johnson, 1985, p. 321)

$$[AB]_{ij} = A_{ij}B_{ij}. \tag{7.16}$$

For any two $n \times n$ matrices, the Hadamard product yields a third $n \times n$ matrix, implying the square matrices are closed under this definition of composition. In addition, if A, B, and C are all $n \times n$ matrices, note that the ij element of the composition sequence $[A \circ B] \circ C$ is $A_{ij}B_{ij}C_{ij}$, the same as the ij element of the composition $A \circ [B \circ C]$, thus establishing the associativity of this composition rule. Finally, note that the identity morphisms in the category **Had** are *not* the usual $n \times n$ identity matrices I_n, but rather the $n \times n$ matrices E_n composed of all 1's (i.e., $E_n^{ij} = 1$ for all i, j). This difference illustrates the point noted earlier that the composition law imposes structure on the category through its influence on the morphism sets.

The fourth variation on the category **Vect** considered here is denoted **Vect**$^+$ and defined as follows. The objects of this category are the *positive orthants* R_+^n of the linear vector space R^n, corresponding to all vectors in R^n with nonnegative components. Similarly, the morphisms of this category are the *nonnegative matrices* (Berman and Plemmons, 1979), the $n \times m$ matrices whose elements are all nonnegative. It is easily verified that any $n \times m$ nonnegative matrix maps R_+^m into R_+^n. Composition of morphisms in this category again corresponds to ordinary matrix multiplication, and it is not difficult to show that if M_1 is a nonnegative $n \times m$ matrix and M_2 is a nonnegative $m \times p$ matrix, then the matrix product $M_1 M_2$ is a nonnegative $n \times p$ matrix. Similarly, it is immediate that the $n \times n$ identity matrix I_n corresponds to the identity object $id_{R_+^n}$ for R_+^n. Thus, **Vect**$^+$ is a category; similarities and differences between the categories **Vect** and **Vect**$^+$ will be used in subsequent discussions to illustrate a number of important category-theoretic notions.

The fifth and final variation of **Vect** considered here is a generalization of **Vect**$^+$, denoted **Sub**$_n$ and defined as follows. Take n as a fixed integer greater than 1 and consider as objects all subspaces S of R^n, including R^n itself (the unique n-dimensional subspace), the origin **0** (the unique 0-dimensional subspace), and all subspaces of intermediate dimension $1 \leq j \leq n-1$. The morphisms of **Sub**$_n$ are all $n \times n$ matrices **M** that leave a given subspace S

Relations between Model Classes

invariant, that is $MS \subseteq S$. Composition of morphisms again corresponds to ordinary matrix multiplication, and if $\mathbf{M}_1, \mathbf{M}_2 \in \hom(S,S)$, then

$$\mathbf{M}_1[\mathbf{M}_2 S] \subseteq \mathbf{M}_1 S \subseteq S. \tag{7.17}$$

That is, the product of two matrices that leaves the subspace S invariant also leaves this subspace invariant. Also, note that the identity morphism id_S associated with any subspace S is simply the $n \times n$ identity matrix, which leaves all subspaces of R^n invariant. Finally, note that in contrast to the category $\mathbf{Vect}^+$, the morphisms in the category $\mathbf{Sub}_n$ are specified *behaviorally* rather than structurally. That is, an element-wise constructive prescription can be given for the morphisms in $\mathbf{Vect}^+$ but the morphisms in $\mathbf{Sub}_n$ are specified by their qualitative behavior: they map a specified subspace S into itself.

7.2.4 Some simple "non-examples"

The previous section provided some specific illustrations of categories to clarify the notions of objects, morphisms, and composition of morphisms. The following examples illustrate the point that not all collections of objects and transformations relating objects satisfy the defining conditions for a category. The key point is that it is the relationships that exists between candidate objects, morphisms, and composition rules that determines whether a given collection defines a category or not. In particular, note that *any* collection of mathematical objects defines a category called the *discrete category*, obtained by simply taking, for each object X, the identity morphism id_X as the only element in the morphism set $\hom(X,X)$ (Blyth, 1986, p. 80). Consequently, a collection of objects, possible morphisms, and a possible composition rule generally fails to define a category either because the set of possible morphisms does not include everything that it must (e.g., identity morphisms) or because it is not closed under the proposed composition rule.

For example, the collection of all elementary trigonometric functions of the form $A_n \sin nx$ or $B_n \cos nx$ does not form a category: first, it does not include identity mappings, and second, it is not closed under composition (e.g., $f(x) = A_n \sin[B_m \cos mx]$ does not belong to this set of functions). Similarly, the collection of $n \times n$ antisymmetric matrices does not form a category under the usual definition of matrix multiplication. Specifically, note that antisymmetric matrices must satisfy $A^T = -A$, whereas the product of two antisymmetric matrices is

$$(AB)^T = B^T A^T = BA, \tag{7.18}$$

which is not the same as $-AB$, in general. For example, consider the 2×2 antisymmetric matrices

$$A = \begin{bmatrix} 0 & -1 \\ 1 & 0 \end{bmatrix}, B = \begin{bmatrix} 0 & 3 \\ -3 & 0 \end{bmatrix} \Rightarrow AB = \begin{bmatrix} 3 & 0 \\ 0 & 3 \end{bmatrix}. \tag{7.19}$$

It is clear that this product is symmetric and therefore is not a member of the class of antisymmetric matrices.

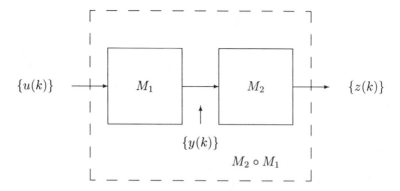

Figure 7.2: Cascade connection of discrete-time models

7.3 The discrete-time dynamic model category

The primary category of interest in this chapter is the category **DTDM** of discrete-time dynamic models, defined loosely as follows. The objects on which this category is based are input sequences $\{u(k)\}$ and the morphisms are discrete-time dynamic models that map these input sequences into output sequences $\{y(k)\}$. Composition of morphisms then naturally corresponds to the cascade connection of dynamic models, and identity morphisms correspond to the identity system $y(k) = u(k)$. The following discussions make these notions more precise, in the following order. First, Sec. 7.3.1 presents a detailed discussion of the cascade connection of discrete-time dynamic models since this construct provides the basis for all of the model categories considered here. Next, Sec. 7.3.2 gives a precise formulation of the category **DTDM** and discusses the class of objects on which this category is based. Finally, Sec. 7.3.3 explores some of the morphism sets in the category **DTDM**, emphasizing the system-theoretic interpretations of these sets.

7.3.1 Composition of morphisms

Fig. 7.2 illustrates the cascade connection of two discrete-time dynamic models, M_1 and M_2. There, $\{u(k)\}$ represents the input sequence to model M_1, $\{y(k)\}$ represents the corresponding output sequence, and $\{z(k)\}$ represents the response of the model M_2 to its input sequence, $\{y(k)\}$. Alternatively, $\{z(k)\}$ may be regarded as the response sequence of the composite model $M_2 \circ M_1$ to the input sequence $\{u(k)\}$. Note that for any two discrete-time systems defined over comparable classes of input sequences, this combination of two models unambiguously defines a third model. Further, cascade connection is associative since the discrete-time dynamic models considered here may be viewed as mappings between sequences; hence, cascade connection is equivalent to composition of mappings and associativitiy follows from the fact that composition of mappings is associative. For these reasons, cascade connection provides a basis for a well-defined composition rule for morphisms in categories of dynamic models and all

Relations between Model Classes

of the model categories considered in this chapter are based on this composition rule. *Consequently, for a collection of discrete-time dynamic models to define a category, it must be closed under cascade connection.* The following discussion briefly considers several of the discrete-time dynamic model classes introduced in previous chapters and examines their behavior under cascade connection.

A simple, illuminating, and important example of a model class that is closed under cascade connection is the class of linear moving average models. Specifically, the cascade connection of the $MA(q)$ model M_1 and the $MA(q')$ model M_2 defines a $MA(q'')$ model M_3 for some finite q''. To establish this result, suppose the impulse responses of M_1 and M_2 are $h_1(k)$ and $h_2(k)$, respectively, so the sequences $\{y(k)\}$ and $\{z(k)\}$ appearing in Fig. 7.2 may be expressed as

$$y(k) = \sum_{i=0}^{q} h_1(i)u(k-i) \qquad z(k) = \sum_{i=0}^{q'} h_2(i)y(k-i). \tag{7.20}$$

Substituting the first of these equations into the second yields

$$z(k) = \sum_{i=0}^{q'} h_2(i) \sum_{j=0}^{q} h_1(j)u(k-i-j). \tag{7.21}$$

To simplify this result, define the "zero-padded" impulse response

$$\tilde{h}_1(j) = \begin{cases} h_1(j) & 0 \leq j \leq q \\ 0 & \text{otherwise}. \end{cases} \tag{7.22}$$

Writing $\ell = i + j$ and interchanging the sums converts Eq. (7.21) to

$$z(k) = \sum_{\ell=0}^{q+q'} \left\{ \sum_{i=0}^{q'} h_2(i)\tilde{h}_1(\ell-i) \right\} u(k-\ell) \equiv \sum_{\ell=0}^{q''} h_3(\ell)u(k-\ell), \tag{7.23}$$

thus establishing the desired result: $\{z(k)\}$ is related to $\{u(k)\}$ by a transformation defined by a $MA(q'')$ linear model for $q'' = q + q'$. The last part of this equation follows from the definition

$$h_3(\ell) = \sum_{i=0}^{q'} h_2(i)\tilde{h}_1(\ell-i), \tag{7.24}$$

for $\ell = 0, 1, \ldots, q''$. This expression represents the well-known result that the impulse response of the cascade connection of two linear systems is given by the convolution of the original two impulse responses. It is also important to note that the dynamic order of the cascade connection of two models of order $q > 0$ and $q' > 0$ is strictly greater than that of either of the original models.

As a direct extension of this result, it was shown in Chapter 4 (Sec. 4.2.2) that the cascade connection of NMAX models is another NMAX model of higher dynamic order. Similarly, it was also shown in Chapter 4 (Sec. 4.3.5) that the cascade connection of NARX models is also another NARX model, again of higher dynamic order. These results are used in Sec. 7.4.3 to show that these model classes both define categories. Conversely, it follows from the fact that the dynamic order increases on cascade connection that *fixed-order* NMAX or NARX model classes [e.g., the classes $NMAX(q)$ and $NARX(q)$] do *not* define categories because they are not closed under cascade connection. Similarly, it was also shown in Chapter 4 (Sec. 4.3) that the larger class of NARX* models is not closed under cascade connection so it is not possible to define a category of NARX* models, either. Finally, note that the additive NARMAX model class is also not closed under cascade connection: recall from Chapter 4 (Sec. 4.4) that static nonlinearities and linear dynamic models are both additive, but the Wiener model is not. Since the Wiener model represents a cascade connection of these two model components, it follows that the additive model class is not closed under cascade connection. Unfortunately, these difficulties also extend to the class of general NARMAX models: these models are not closed under cascade connection, either. This conclusion follows from the fact (Sec. 4.4.1) that the Wiener model based on the first-order linear dynamic model and the hard saturation nonlinearity has no NARMAX representation.

The previous examples illustrate the necessity of imposing structural restrictions on the class of NARMAX models to satisfy the defining conditions for a category. The following examples consider two specific restrictions, one of which is sufficient to define a category of models and the other is not. It was shown in Chapter 5 that the parallel connection of two Volterra models yields a third Volterra model and the following discussion extends this result to cascade connections: the combination of a $V_{(N,M)}$ Volterra model with a $V_{(N',M')}$ Volterra model is a Volterra model of the type $V_{(NN',M+M')}$. This result combines both the behavior of polynomials—recall that the composition of polynomials of order N and N' is a third polynomial of order NN'—and linear moving average models—recall that the cascade connection of an $MA(M)$ model and an $MA(M')$ model is an $MA(M + M')$ model. To establish this result, consider the cascade connection $V_{(N,M)} \circ V_{(N',M')}$ of the models:

$$z(k) = z_0 + \sum_{n=1}^{N} w_n(k), \qquad (7.25)$$

$$w_n(k) = \sum_{i_1=0}^{M} \cdots \sum_{i_n=0}^{M} \alpha(i_1, \ldots, i_n) y(k - i_1) \cdots y(k - i_n),$$

and

$$y(k) = y_0 + \sum_{m=1}^{N'} v_m(k), \qquad (7.26)$$

$$v_m(k) = \sum_{j_1=0}^{M'} \cdots \sum_{j_m=1}^{M'} \beta(j_1, \ldots, j_m) u(k - j_1) \cdots u(k - j_m).$$

Relations between Model Classes

Note that the product $y(k-i_1)\cdots y(k-i_n)$ appearing in Eq. (7.25) may be expanded as

$$y(k-i_1)\cdots y(k-i_n) = y_0^n + y_0^{n-1}\sum_{\ell_1=1}^{n}\sum_{m_1=1}^{M'} v_{m_1}(k-i_{\ell_1}) \qquad (7.27)$$

$$+\cdots + \sum_{m_1=1}^{M'}\cdots\sum_{m_n=1}^{M'} v_{m_1}(k-i_1)\cdots v_{m_n}(k-i_m).$$

Further, note that $v_{m_j}(k-i_j)$ is a multivariable polynomial of order m_j in the variables $u(k-i_j)$ through $u(k-i_j-M')$. Hence, the highest order term appearing on the right-hand side of Eq. (7.27) arises when $m_1 = m_2 = \cdots = m_n = N'$. Consequently, this term is of order nN' in the variables $u(k-i_{(1)})$ through $u(k-i_{(n)}-M')$, where $i_{(1)}$ is the smallest index in the sequence $\{i_1,\ldots,i_n\}$, and $i_{(n)}$ is the largest index in this sequence. Since n varies from 1 to N, the highest order term appearing in the expansion for $z(k)$ will be of order NN'; similarly, since $i_{(1)}$ can be as small as 0 and $i_{(n)}$ can be as large as M, it follows that the expansion for $z(k)$ will involve the terms $u(k)$ through $u(k-M-M')$. Overall, then, $z(k)$ may be represented as a Volterra model of nonlinear order NN' and dynamic order $M+M'$, as claimed.

Finally, the next example illustrates that, in contrast to the polynomial NMAX models $V_{(N,M)}$, the class of polynomial NARMAX models is not closed under cascade connection. Specifically, consider the composition of the linear model M_1

$$y(k) = \alpha y(k-1) + \beta u(k) \qquad (7.28)$$

followed by the bilinear model M_2

$$z(k) = \delta z(k-1) + \epsilon y(k) + \nu y(k)z(k-1), \qquad (7.29)$$

and note that both of these models belong to the polynomial NARMAX class. Combining these equations yields

$$z(k) = \delta z(k-1) + [\epsilon + \nu z(k-1)][\alpha y(k-1) + \beta u(k)]. \qquad (7.30)$$

To obtain an input/output model relating $u(k)$ to $z(k)$, it is necessary to eliminate $y(k-1)$ from this equation. Solving Eq. (7.29) for $y(k)$ gives

$$y(k) = \frac{z(k) - \delta z(k-1)}{\epsilon + \nu z(k-1)} \Rightarrow y(k-1) = \frac{z(k-1) - \delta z(k-2)}{\epsilon + \nu z(k-2)}. \qquad (7.31)$$

Substituting this expression into Eq. (7.30) and simplifying ultimately yields the following $NARX^*(2)$ model:

$$z(k) = [\epsilon + \nu z(k-2)]^{-1}[(\delta + \alpha)z(k-1) - \alpha\delta z(k-2)$$
$$+ \alpha\nu z^2(k-1) + (1-\alpha\nu)\delta z(k-1)z(k-2)$$
$$+ \epsilon\beta u(k) + \nu\beta z(k-2)u(k)]. \qquad (7.32)$$

Note that this model belongs to the rational NARMAX family but not to the polynomial NARMAX family, thus establishing that the polynomial NARMAX family is not closed under cascade connection.

7.3.2 The category DTDM and its objects

As noted previously, the category **DTDM** is based on the following general constructs:

1. Objects are collections of input sequences
2. Morphisms are defined by discrete-time dynamic models
3. Composition of morphisms corresponds to cascade connection
4. Identity morphisms correspond to the identity system $y(k) = u(k)$.

The basic idea and some important details of cascade connection were discussed in the preceeding section and the nature of the identity morphisms is clear enough that no further discussion is required. It is both important and highly instructive to give careful consideration to the objects and morphisms on which this category is based, however, and these tasks are taken up in this section and the next.

The class of objects on which the category **DTDM** is based is a large class that will be denoted $\mathcal{O}$ in subsequent discussions. The objects in this class include all sets of real-valued sequences $\{x(k)\}$. Specific objects in the class $\mathcal{O}$ include:

1. Arbitrary individual sequences $\{x(k)\}$
2. The space ℓ_∞ of essentially bounded sequences
3. The spaces ℓ_p of p-power absolute summable sequences
4. The spaces S^P of period P sequences
5. The space S^1 of constant sequences
6. The space S^∞ of arbitrary sequences
7. The space $\mathcal{A}$ of bounded, asymptotically constant sequences
8. The space ℓ_∞^+ of essentially bounded nonnegative sequences
9. The set $\mathcal{M}$ of monotone sequences
10. The set $\mathcal{B}$ of binary sequences.

Recall from Sec. 7.2.2 that it is not possible to index a large class, so it is not even possible in principle to completely describe the class $\mathcal{O}$, but it is instructive to briefly discuss some of the specific members of $\mathcal{O}$ listed here because they illuminate some important points that will be useful in subsequent discussions.

First, note that any individual real-valued input sequence $\{u(k)\}$ is included in the class $\mathcal{O}$. In fact, taking this subcollection alone—that is, the *set* of all possible input sequences $\{u(k)\}$—leads to the definition of the *IO subcategories* of **DTDM** introduced in Sec. 7.5. The diferences between IO subcategories of models based on individual sequences alone and the category **DTDM** are important as will be seen in subsequent discussions, beginning with the preliminary exploration of the morphism sets in **DTDM** presented in the next section.

Relations between Model Classes 315

The class $\mathcal{O}$ also includes all possible *linear vector spaces* $\mathcal{S}$ of sequences, defined by the following requirement: if $\{u(k)\} \in \mathcal{S}$ and $\{v(k)\} \in \mathcal{S}$, then $\alpha u(k) + \beta v(k) \in \mathcal{S}$ for all real numbers α and β. These spaces may be viewed as *sets* of vectors, but endowed with this additional linearity requirement and possibly others as well. For example, the linear vector space ℓ_∞ also imposes the condition of *essential boundedness*: if $\{x(k)\} \in \ell_\infty$, the supremum of $|x(k)|$ is finite. Similarly, the spaces ℓ_p are defined by

$$\sum_{k=-\infty}^{\infty} |x(k)|^p < \infty. \tag{7.33}$$

The set of all periodic sequences $\{x(k)\}$ with fixed period P also defines a vector space, for any positive integer P: if $u(k+P) = u(k)$ and $v(k+P) = v(k)$ for all k, it follows that $z(k) = \alpha u(k) + \beta v(k)$ also satisfies $z(k+P) = z(k)$ for all k. In addition, it is interesting to note that if $\{x(k)\} \in S^P$ for some P, then $\{x(k)\} \in S^{nP}$ for all $n > 1$: in words, a periodic sequence with period P is also necessarily periodic with period nP. As a corollary, note that S^1 is the linear vector space of constant sequences and it is a proper subspace of all periodic sequence spaces S^P for $P > 1$. Further, note that taking the limit as $P \to \infty$ yields the linear vector space of all "period ∞ sequences," containing all possible real-valued sequences $\{x(k)\}$. Overall, this collection of periodic spaces is highly structured, including chains of subspace inclusions like the following:

$$S^1 \subset S^P \subset S^{2P} \subset S^{4P} \subset S^{8P} \subset \cdots \subset S^\infty$$
$$S^1 \subset S^P \subset S^{3P} \subset S^{6P} \subset S^{12P} \subset \cdots \subset S^\infty$$
$$S^1 \subset S^P \subset S^{3P} \subset S^{9P} \subset S^{18P} \subset \cdots \subset S^\infty. \tag{7.34}$$

The key point is that the collection $\mathcal{O}$ includes some extremely highly structured sequence sets and examination of the morphisms between these objects can be quite illuminating.

Another extremely interesting linear vector space of sequences is the space $\mathcal{A}$ of bounded, asymptotically constant sequences, based on the concept introduced in Chapter 4 (Sec. 4.2.2) in connection with the behavior of NMAX models. Specifically, recall that a sequence $\{x(k)\}$ is asymptotically constant if there exists some constant c and some integer N such that, given $\epsilon > 0$, $|x(k) - c| < \epsilon$ for all $k > N$. Note that if $\{x(k)\}$ is asymptotically constant, then so is $\{\lambda x(k)\}$ for any real λ. Hence, it is only necessary to establish that the sum of any two asymptotically constant sequences is asymptotically constant to establish linearity. Consider two asymptotically constant sequences $\{x_1(k)\}$ and $\{x_2(k)\}$: given $\epsilon > 0$, there exist constants c_1 and c_2 and integers N_1 and N_2 such that $|x_1(k) - c_1| < \epsilon/2$ and $|x_2(k) - c_2| < \epsilon/2$. Hence, it follows from the triangle inequality that if $k > \max\{N_1, N_2\}$ then

$$|[x_1(k) + x_2(k)] - [c_1 + c_2]| \leq |x_1(k) - c_1| + |x_2(k) - c_2| < \epsilon, \tag{7.35}$$

implying that $\{x_1(k) + x_2(k)\}$ is an asymptotically constant sequence.

Not all of the interesting objects in $\mathcal{O}$ exhibit the full richness of this linear vector space structures. One example is ℓ_∞^+, the space of essentially bounded *nonnegative* sequences, which may be considered a linear vector space but only if the scaling parameters α and β are restricted to be nonnegative. The set $\mathcal{M}$ of monotone sequences cannot be regarded as a linear vector space even under this restriction since the sum of an increasing sequence and a decreasing sequence is generally nonmonotonic. Conversely, restricting consideration to either increasing or decreasing sequences does provide the basis for defining linear vector spaces but again, only if the scaling parameters are restricted to be nonnegative. Finally, another interesting sequence set is $\mathcal{B}$ of binary sequences $\{x(k)\}$ that can assume only two values, either $x(k) = x_1$ or $x(k) = x_2$ for all k. Despite the fact that neither $\mathcal{M}$ nor $\mathcal{B}$ define linear vector spaces, both of these sets are highly structured. In fact, these sets are much more highly structured than the set S^∞ defined previously, which is a linear vector space. The key point here is that the class $\mathcal{O}$ of objects in the category **DTDM** includes some extremely interesting sets and certain useful insights may be obtained by examining the morphisms between some of these sets. This idea is made more precise in the next section.

7.3.3 The morphism sets in DTDM

In the category **DTDM**, the morphism sets $\hom(A, B)$ define classes of discrete-time dynamic models mapping the set A of input sequences into the set B of output sequences. The fact that $\hom(A, B)$ is a *set* follows from the fact that, given any two sets A and B, the class of all functions $f : A \to B$ is a set (Adamek et al., 1990, p. 6). Consequently, although the class $\mathcal{O}$ of objects in **DTDM** is not a set (that is, $\mathcal{O}$ is a proper class), given any two objects A and B (each of which *are* sets, since $\mathcal{O}$ is a class) the class of morphisms between A and B is a set, as it is in all categories. The key point here is that by choosing interesting input and output sets, the resulting morphism sets can define some interesting types of systems. The following discussion illustrates this point with a number of examples and subsequent sections of this chapter pursue this idea further, introducing some additional notions from category theory.

First, consider the simple case where the objects A and B are the individual sequences $\{u(k)\}$ and $\{y(k)\}$. In this case, the set $\hom(A, B)$ is the set of all discrete-time dynamic models mapping the input sequence $\{u(k)\}$ into the output sequence $\{y(k)\}$. Characterization of these morphism sets is closely related to the problem of empirical model identification, and this connection is discussed in detail in Sec. 7.5. Other important system-theoretic questions may be examined by considering other choices of objects in $\mathcal{O}$. For example, the morphism set $\hom(\ell_\infty, \ell_\infty)$ defines the class of BIBO stable dynamic models. Similarly, replacing ℓ_∞ with the sequence space ℓ_p for some finite p yields a more restrictive definition of stability. In particular, the resulting morphism sets contain models that exhibit *asymptotic stability* (Elaydi, 1996, p. 166) since any sequence $\{x(k)\} \in \ell_p$ necessarily decays to zero as $k \to \infty$. The even stronger notion of *exponential stability* discussed in Chapter 2 results if ℓ_p is replaced with the space $\mathcal{E}$ of exponentially bounded sequences: $\{x(k)\} \in \mathcal{E}$ if and only if there exist constants $A > 0$ and $0 < \alpha < 1$ such that

$$|x(k)| \leq A\alpha^k \tag{7.36}$$

for all k. To see that $\mathcal{E}$ is a linear vector space, first note that if $\{x(k)\}$ belongs to $\mathcal{E}$ then so does $\{\lambda x(k)\}$ for any real λ; this result follows by replacing A with $|\lambda|A$ in Eq. (7.36). Next, consider two sequences $\{x_1(k)\}$ and $\{x_2(k)\}$, each satisfying Eq. (7.36) with constants A_i and α_i for $i = 1, 2$. Define α_* as the larger of α_1 and α_2 and note that

$$|x_1(k)| \leq A_1 \alpha_1^k \leq A_1 \alpha_*^k \qquad |x_2(k)| \leq A_2 \alpha_2^k \leq A_2 \alpha_*^k. \tag{7.37}$$

Hence, by the triangle inequality

$$|x_1(k) + x_2(k)| \leq |x_1(k)| + |x_2(k)| \leq [A_1 + A_2]\alpha_*^k, \tag{7.38}$$

thus establishing that $\mathcal{E}$ is a linear vector space. Note that the morphism set hom$(\mathcal{E}, \mathcal{E})$ consists of all exponentially stable systems, including all stable, finite-dimensional linear models. Conversely, this morphism set *does not include* the stable, infinite-dimensional slow-decay models considered in Chapter 2. In addition, it is interesting to ask what can be said about the nonlinear models included in these morphism sets.

Another interesting system-theoretic characterization is that of zero dynamics, introduced in Chapter 4 (Sec. 4.3.5). Recall that the zero dynamics of a model M are those sequences $\{z(k)\}$ such that $M[\{z(k)\}] = \mathbf{0}$, where $\mathbf{0}$ is the zero sequence $x(k) \equiv 0$. Given a set $\mathcal{Z}$ of input sequences, the morphism set hom$(\mathcal{Z}, \mathbf{0})$ consists of the set of discrete-time dynamic models with zero dynamics $\mathcal{Z}$. Arguably, the two extremes of this characterization are the sets hom$(\mathbf{0}, \mathbf{0})$, consisting of all dynamic models with trivial zero dynamics and hom$(S^\infty, \mathbf{0})$, consisting only of the zero model $y(k) \equiv 0$ that maps all possible input sequences into $\mathbf{0}$. A more interesting intermediate case is hom$(S^1, \mathbf{0})$, consisting of all models whose steady-state gain is identically zero and including various linear and nonlinear differentiation filters. Analogously, the set hom$(S^P, \mathbf{0})$ includes all dynamic models whose response to any periodic input sequences of period P is identically zero.

Closely related to this last example is the morphism set hom(S^P, S^P), defining those dynamic models that preserve the periodicity of period P sequences. This set of models includes all linear models, all static nonlinearities, and all nonlinear moving average models, as discussed in Chapter 4 (Sec. 4.2.2). In addition, since the zero sequence $\mathbf{0}$ is an element of S^P for any integer P, it follows that both of the morphism sets hom$(\mathbf{0}, \mathbf{0})$ and hom$(S^P, \mathbf{0})$ are subsets of hom(S^P, S^P). Similarly, the set hom(S^P, S^P) also includes the set hom(S^1, S^P), which contains all *oscillators* that map a constant input sequence into an output sequence of period P. A particularly interesting class of systems included in this set are the homogeneous models, corresponding to *voltage-controlled oscillators* whose amplitude depends linearly on an applied constant input voltage. Finally, note that the morphism set hom(S^P, S^{nP}) includes nonlinear dynamic models that generate subharmonics of order $1/n$ in response to an input of period P. Conversely, since S^P is a subspace of S^{nP} for all $n > 1$, it follows that all of the morphisms in hom(S^P, S^P) also belong to hom(S^P, S^{nP}). In any case, the characterization of nonlinear systems according to their responses to periodic input sequences appears to be both an interesting issue and one that may be approached from the perspective of category theory, a point that is discussed further in Sec. 7.10.

Two other interesting morphism sets are hom($\mathcal{A}, \mathcal{A}$) and hom($\mathcal{M}, \mathcal{M}$). The first of these morphism sets defines those dynamic models that preserve asymptotic constancy. It was shown in Chapter 4 (Sec. 4.2.2) that this set includes all NMAX models based on continuous functions $g : R^{q+1} \to R^1$. In addition, this model class also includes all asymptotically stable linear models since if $u(k) \to u_s$ as $k \to \infty$, it follows that $y(k) \to K u_s$ for some finite steady-state gain K. (Note that if $K = 0$, these linear models are also included in the morphism set hom($S^1, \mathbf{0}$) discussed in the preceeding paragraph.) Conversely, not all BIBO stable NARX models with well-defined steady-states are included in this set, as examples presented in Chapter 4 illustrate. The set hom($\mathcal{M}, \mathcal{M}$) includes all discrete-time dynamic models that exhibit monotonic responses to monotonic input sequences. Examination of these sets leads to some intereting insights, a point that is disucssed further in Sec. 7.6. Similarly, the morphism set hom($\mathcal{B}, \mathcal{B}$) consists of all discrete-time dynamic models that transform one binary sequence into another. Aside from the identity morphism $id_\mathcal{B}$, note that this system class includes very few linear models, but it includes all static nonlinearities and all *digital logic systems* that implement mappings between sequences of 1's and 0's.

To conclude this discussion, note that the category **DTDM** is flexible enough to deal with systems that exhibit nonunique responses. In particular, note that if a nonlinear dynamic model exhibits output multiplicity, its response to a constant input sequence $u(k) \equiv u_s$ is not well-defined. In most cases of practical interest, however, the number of possible responses to a given steady-state input u_s is finite, consisting of the distinct real values $\{y_s^i\}_{i=1}^m$. Nonlinear models of this type are included in the morphism set hom($u_s, \{y_s^i\}_{i=1}^m$) where u_s and y_s^i represent the constant sequences $u(k) \equiv u_s$ and $y(k) \equiv y_s^i$. This approach to output multiplicity does introduce some unpleasant complications, but the key point here is that the category **DTDM** is flexible enough to address issues like output multiplicity. Conversely, note that input multiplicity poses no particular difficulty since the response to a constant input sequence $u(k) \equiv u_s$ is well-defined: it is the value of u_s corresponding to a particular value of this response that is not unique.

7.4 Restricted model categories

It should be clear from the preceeding discussions of the objects and morphisms in the category **DTDM** that this category is too large to be directly useful in understanding nonlinear systems. Conversely, these discussions have also illustrated that many questions of system-theoretic interest arise quite naturally as a consequence of examining some of the specific objects from this category and their associated morphism sets. This observation leads directly to the concept of a *subcategory*, defined and discussed at some length in the following section. Specifically, the definition of a subcategory is given in Sec. 7.4.1, after which it is specialized to the problem of dynamic model characterization in Secs. 7.4.2 through 7.4.4. That is, Sec. 7.4.2 defines four different categories of linear models, all of which are subcategories of **DTDM**. This discussion illustrates the notion of a *structural model category*, which is extended to nonlinear dynamic models in Sec. 7.4.3. Alternatively, it is also possible to define *behavioral model categories*, which are again subcategories of **DTDM**, and this notion is described in Sec. 7.4.4. Both structural model categories and behavioral model categories represent constructs that have much greater direct utility in systems analysis than the parent category **DTDM**, despite the fact that they are also large categories.

7.4.1 Subcategories

A category **S** is a *subcategory* of **C** if the following four conditions are satisfied (Blyth, 1986, p. 7):

1. Every object of **S** is also an object of **C**
2. For all objects X, Y of **S**, $\text{Mor}_{\mathbf{S}}(X, Y) \subseteq \text{Mor}_{\mathbf{C}}(X, Y)$;
3. Composition morphisms in **S** and **C** is the same
4. For all objects X in **S**, id_X is the same as in **C**.

Further, if $\text{Mor}_{\mathbf{S}}(X, Y) = \text{Mor}_{\mathbf{C}}(X, Y)$, **S** is called a *full subcategory* of **C** (Blyth, 1986, p. 7). Examples of subcategories include the following. The categories **Mfun** of monotonic functions and **Poly** of polynomials are both subcategories of **Rfun**, the category of real-valued functions. In both cases, condition (1) is satisfied since all three of the categories considered here are based on the same class of objects (i.e., the set of real numbers). Similarly, composition of morphisms corresponds to the composition of functions and the identity morphisms correspond to the identity mapping $f(x) = x$ in all three categories so conditions (3) and (4) are also satisfied. Hence, both **Mfun** and **Poly** are subcategories of **Rfun** because their morphism sets are both subsets of those for **Rfun**, as required by condition (2). Note, however, that neither **Mfun** nor **Poly** are full subcategories of **Rfun** because these inclusions are proper.

Two additional examples of subcategories are **Orthog** and **Square**, both of which are subcategories of **Vect**. As in the previous examples, all three of these categories are based on the same collection of objects (specifically, the vector spaces R^n), the same definition of composition of morphisms (ordinary matrix multiplication), and the same identity morphisms (the $n \times n$ identity matrices I_n). Consequently, conditions (1), (3), and (4) are satisfied for both **Orthog** and **Square**. The difference between these categories lies in their morphism sets. In **Square**, the morphism sets $\hom(R^n, R^n)$ are exactly the same as in

Vect, but the morphism sets hom(R^n, R^m) are empty for all $m \neq n$. In the category **Orthog**, the sets hom(R^n, R^m) are again empty for $m \neq n$ and the sets hom(R^n, R^n) are proper subsets of the corresponding morphism sets in **Vect**, containing only orthogonal $n \times n$ matrices. In both cases, condition (2) is satisfied, establishing that **Orthog** and **Square** are subcategories of **Vect**. As in the previous pair of examples, note that neither of these categories are full subcategories of **Vect** since not all morphisms in **Vect** are included in these subcategories. Further, since the morphism sets in **Orthog** are subsets of those in **Square**, it follows that **Orthog** is a subcategory of **Square** but again, not a full subcategory.

It is also instructive to consider some examples that are *not* subcategories. Specifically, note that neither of the categories **Vect**$^+$ and **Sub**$_n$ defined in Sec. 7.2.3 are subcategories of **Vect**. That is, since both **Vect**$^+$ and **Sub**$_n$ involve objects that are not objects in **Vect** (that is, positive orthants and subspaces of R^n, respectively), defining condition (1) for a subcategory is not satisfied in either case. Similarly, note that **Had** is not a subcategory of **Vect**. Here, both conditions (1) and (2) are satisfied since the objects and morphisms are the same in both categories. However, because both composition of morphisms and identity morphisms are different in **Vect** and **Had**, subcategory conditions (3) and (4) are not satisfied.

Again, it is worth noting that none of the subcategory examples just described are full subcategories since not all morphisms from the original category are retained in the subcategory. This behavior is characteristic of the structural model subcategories defined and discussed in Sec. 7.4.3. To obtain a full subcategory, it is necessary to restrict the objects in the original category. For example, **Rfun** is a full subcategory of **Set**, the category of set-valued mappings: all morphisms in **Set** that relate sets of real numbers appear as morphisms in **Rfun** and those morphisms in **Set** that do not appear in **Rfun** involve objects that are not present in **Rfun**. This behavior is seen in some of the behavioral model subcategories considered in Sec. 7.4.4 (specifically, the categories **BIBO**, **ASYMP** and **MONO**).

7.4.2 Linear model categories

Chapter 2 defined and briefly discussed four linear model realizations:

1. Linear moving average models, denoted $MA(q)$
2. Linear autoregressive models, denoted $AR(p)$
3. Linear autoregressive moving average models, denoted $ARMA(p,q)$
4. Linear state space models.

In fact, all four of these model structures provide the basis for well-defined subcategories of **DTDM**, denoted **MA**, **AR**, **ARMA**, and **State**, respectively. In all cases, the class of objects on which these model categories are based is the class $\mathcal{O}$ of objects on which the category **DTDM** is based. Consequently, the first of the four subcategory conditions given in Sec. 7.4.1 is automatically satisfied by all of these linear model categories. Similarly, since all of these

Relations between Model Classes

model classes are subsets of the set of discrete-time dynamic models on which the category **DTDM** is based, it follows that the second subcategory condition is also satisfied for all of these categories. Further, composition of morphisms is defined by cascade connection in all of these linear model categories, just as it is in **DTDM**, so the third of the four subcategory conditions is also satisfied. Therefore, to verify that these model classes define subcategories of **DTDM**, two things are necessary: first, it must be verified that these model classes define categories and second, that the fourth subcategory condition is satisfied. These requirements are satisfied if each model class is closed under cascade connection and contains the identity system $y(k) = u(k)$.

For the category **MA**, each morphism is defined by a linear moving average model of the form

$$y(k) = \sum_{i=0}^{q} h(i)u(k-i) \equiv h(k) * u(k) \tag{7.39}$$

for some finite q, where the symbol $*$ denotes the operation of discrete convolution. The identity mapping $y(k) = u(k)$ may be obtained from Eq. (7.39) by taking $q = 0$ and $h(0) = 1$, and it was shown explicitly in Sec. 7.3.1 that this model class is closed under cascade connection. Hence, the linear moving average model class defines a subcategory of **DTDM**.

For the category **AR**, each morphism is defined by a linear autoregressive model of the form

$$y(k) = \sum_{i=1}^{p} a_i y(k-i) + b_0 u(k). \tag{7.40}$$

Analogously to the previous example, taking $p = 0$ and $b_0 = 1$ in Eq. (7.40) yields the identity mapping $y(k) = u(k)$. To establish that this model class is closed under cascade connection, it is easiest to first transform to the z-domain description

$$Y(z) = \left[\frac{b_0}{A(z)}\right] U(z), \tag{7.41}$$

where

$$A(z) = 1 - \sum_{i=1}^{p} a_i z^{-i}. \tag{7.42}$$

Consider the cascade connection of this model with another of order p', defined by the transfer function

$$V(z) = \left[\frac{d_0}{C(z)}\right] Y(z). \tag{7.43}$$

Since cascade connection corresponds to multiplication of transfer functions, it follows that the composite model is characterized by

$$V(z) = \left[\frac{b_0 d_0}{A(z)C(z)}\right] U(z). \tag{7.44}$$

To obtain the desired result, the denominator of this composite transfer function may be written as

$$A(z)C(z) = \left[1 - \sum_{i=1}^{p} a_i z^{-i}\right]\left[1 - \sum_{i=1}^{p'} c_i z^{-i}\right] \equiv 1 - \sum_{i=1}^{p+p'} g_i z^{-i}, \qquad (7.45)$$

from which it follows that

$$v(k) = \sum_{i=1}^{p+p'} g_i v(k-i) + b_0 d_0 u(k). \qquad (7.46)$$

This result establishes that the cascade connection of two linear autoregressive models is indeed another linear autoregressive model.

Essentially the same construction leads to the category **ARMA** of linear autoregressive moving average models. Specifically, the morphisms in this category are defined by the linear models

$$y(k) = \sum_{i=1}^{p} a_i y(k-i) + \sum_{j=0}^{q} b_j u(k-j). \qquad (7.47)$$

Again, taking $p = 0$, $q = 0$, and $b_0 = 1$ defines the **DTDM** identity morphisms, and the closure of this class of models under cascade connection follows from the same basic arguments as in the previous example. Specifically, note that the cascade connection of an $ARMA(p,q)$ model and an $ARMA(p',q')$ model is an $ARMA(p'',q'')$ model where $p'' = p + p'$ and $q'' = q + q'$. The category **State** is based on linear state-space models of the form

$$x_k = Ax_{k-1} + Bu_{k-1} \qquad y_k = Cx_k + Du_k. \qquad (7.48)$$

Here, x_k is a state vector of dimension n and part of the model specification is that $x_k = 0$ for all $k \leq 0$. Note that the direct ffedthrough term Du_k is necessary here to guarantee that this model class includes the identity morphism $y_k = u_k$. To see that this model class is closed under cascade connection, consider the connection of the model defined in Eq. (7.48) with the second state-space model

$$z_k = Fz_{k-1} + Gy_{k-1} \qquad w_k = Hz_k + My_k. \qquad (7.49)$$

Closure requires the existence of a linear state-space model of this form that directly relates the input u_k to the composite output w_k. To establish the existence of this model, first combine Eqs. (7.48) and (7.49) to obtain

$$\begin{aligned} z_k &= Fz_{k-1} + G[Cx_{k-1} + Du_{k-1}] \\ w_k &= Hz_k + M[Cx_k + Du_k]. \end{aligned} \qquad (7.50)$$

The desired result follows on defining the augmented state vector s_k, which evolves according to the linear state-space model

$$s_k = \begin{bmatrix} x_k \\ z_k \end{bmatrix} = \begin{bmatrix} A & 0 \\ GC & F \end{bmatrix} s_{k-1} + \begin{bmatrix} B \\ GD \end{bmatrix} u_{k-1}$$

$$w_k = [MC \;\vdots\; H]s_k + MDu_k. \qquad (7.51)$$

Hence, the collection of linear state-space models defined by Eq. (7.48) forms the basis for the category **State** of linear dynamic models and this category satisfies all of the defining requirements for a subcategory of **DTDM**.

All four of these categories of linear dynamic models are extremely useful in constructing categories of nonlinear dynamic models. This point is illustrated in the next section, which describes a number of structurally-defined subcategories of nonlinear models, many of which are based on these linear model categories. Behaviorally-defined model categories—both linear and nonlinear—are then discussed in Sec. 7.4.4.

7.4.3 Structural model categories

The constructions used in the previous section to define the categories **MA**, **AR**, and **State** may be extended to the nonlinear case, leading to the model categories **NMAX**, **NARX**, and **NState**. The following discussion presents these definitions, along with definitions of three other structurally-defined nonlinear model categories: the category **Volt** of Volterra models, the category **Hom**$_\phi$ of homomorphic models, and the category **LN**$_\infty$ of block-oriented cascades. All of these nonlinear model categories are subcategories of **DTDM**, obtained by restricting the morphism sets. In particular, these categories are all based on the class $\mathcal{O}$ of objects underlying the category **DTDM**, composition of morphisms is defined by cascade connection of dynamic models, and the identity morphisms are defined by the identity system $y(k) = u(k)$. Hence, it again follows that the model classes considered here define subcategories of **DTDM** if the following two conditions are satisfied: the class includes the identity model, and it is closed under cascade connection.

The category **NMAX** is based on essentially the same construction as its linear counterpart, **MA**; in fact, **MA** is a subcategory of **NMAX**. The morphisms in **NMAX** are defined by the NMAX models

$$y(k) = G(u(k), \ldots, u(k-q)) \qquad (7.52)$$

for some finite $q \geq 0$. As shown in Chapter 4 (Sec. 4.2.2), this model class is closed under cascade connection, and the identity mapping may be obtained by taking $q = 0$ and $G(x) = x$. Similarly, the linear model category **AR** is a subcategory of the nonlinear model category **NARX**, defined as follows. The morphisms on which this category is based are the nonlinear autoregressive models

$$y(k) = F(y(k-1), \ldots, y(k-p)) + b_0 u(k) \qquad (7.53)$$

for some finite $p \geq 0$. As in the category **AR**, the identity system is obtained by taking $p = 0$ and $b_0 = 1$ and the closure of this model class under cascade connection was demonstrated in Chapter 4 (Sec. 4.3). Hence, both the model classes NMAX and NARX define subcategories of **DTDM**.

In contrast to the linear ARMA model class, the class of NARMAX models *does not* define a category. This result follows from the fact that the NARMAX model class is not closed under cascade connection, as demonstrated in

Sec. 7.3.1. Similarly, neither the NARX* model class nor the structurally additive NAARX model class define categories since they are not closed under cascade connection, either. In contrast, it is possible to define a category **NState** of nonlinear state-space models. Morphisms in this category are defined by the nonlinear state-space models

$$x_k = F(x_{k-1}, u_{k-1}) \qquad y_k = G(x_k, u_k). \tag{7.54}$$

As in the linear case, x_k is a state vector of dimension n and the initial condition $x_k = 0$ for $k \leq 0$ is included as part of the model specification. Also, for consistency with the state and input restrictions, it is necessary that the nonlinear mappings defining this model satisfy $F(0,0) = 0$ and $G(0,0) = 0$. Taking $G(x, u) = u$, it follows that this class of models includes the identity model $y_k = u_k$. To see that this model class is closed under cascade connection, proceed as in the linear case. Specifically, consider the composite model

$$z_k = \Phi(z_{k-1}, y_{k-1}) \qquad w_k = \Gamma(z_k, y_k). \tag{7.55}$$

Combining this equation with Eq. (7.54) then gives

$$z_k = \Phi(z_{k-1}, G(x_{k-1}, u_{k-1})) \qquad w_k = \Gamma(z_k, G(x_k, u_k)). \tag{7.56}$$

Augmenting the state vector as in the linear case leads to the nonlinear state-space model

$$s_k = \Omega(s_{k-1}, u_{k-1}) \qquad w_k = \Lambda(s_k, u_k). \tag{7.57}$$

This result establishes that nonlinear state-space models are closed under cascade connection.

A category that will be considered frequently in subsequent discussions is **Volt**, the category of Volterra models. The morphisms in this category are the $V_{(N,M)}$ models discussed at length in Chapter 5. Note that the identity model $y(k) = u(k)$ may be represented as a member of the $V_{(1,0)}$ model class. Similarly, it was established in Sec. 7.3.1 that the class of finite Volterra models is closed under cascade connection, thus establishing that **Volt** satisfies the defining conditions for a category. This category is interesting both because it is highly structured (e.g., it has many interesting subcategories, as discussed in Sec. 7.4) and because it exhibits some interesting relationships with other model categories.

Another extremely interesting subcategory of **DTDM** is the category **Hom**$_\phi$, based on the homomorphic systems defined in Chapter 3, that is

$$y(k) = \phi^{-1}\left[\sum_{i=0}^{\infty} h(i)\phi[u(k-i)]\right] = \phi^{-1}[h(k) * \phi[u(k)]]. \tag{7.58}$$

Here, $\phi[x]$ is an invertible scalar function and $\{h(i)\}$ represents the impulse response of the linear $ARMA(p, q)$ model defined by Eq. (7.47) for some finite

Relations between Model Classes

p and q. Taking $p = q = 0$ and $b_0 = 1$ in this linear model, it follows that $y(k) = u(k)$ for any invertible function $\phi[\cdot]$. To show that this model class is closed under cascade connection, suppose $\mathcal{M}_1$ and $\mathcal{M}_2$ are homomorphic systems based on the same nonlinearity $\phi[\cdot]$ and consider the response of the cascade model $\mathcal{M}_1 \circ \mathcal{M}_2$ to the input sequence $\{u(k)\}$. Now, if $y(k) = \mathcal{M}_1[u(k)]$ and $z(k) = \mathcal{M}_2[y(k)]$, it follows that

$$z(k) = \phi^{-1}[h_2(k) * \phi[y(k)]] = \phi^{-1}[h_2(k) * (h_1(k) * \phi[u(k)])]$$
$$\equiv \phi^{-1}[h_3(k) * \phi[u(k)]], \qquad (7.59)$$

where $h_3(k) = h_2(k) * h_1(k)$ is simply the impulse response of the cascade of the two linear systems on which the homomorphic models $\mathcal{M}_1$ and $\mathcal{M}_2$ are based. In other words, the output nonlinearity $\phi^{-1}[\cdot]$ of model $\mathcal{M}_1$ is cancelled by the input nonlinearity $\phi[\cdot]$ of model $\mathcal{M}_2$ in the cascade connection. Since the class of linear ARMA models is closed under cascade connection, this result extends to the class of homomorphic systems for any fixed $\phi[\cdot]$.

Finally, it is useful to observe that, although many practically important dynamic model classes *do not* define categories, they can generally be imbedded in larger categories and this process sometimes leads to interesting new model classes. For example, since the class of Hammerstein models is not closed under cascade connection, it cannot define a category. Despite this limitation, it is possible to define several categories of nonlinear models that includes Hammerstein models as morphisms. One of these categories is **LN**$_\infty$, defined as follows. Morphisms are taken as series block-oriented models of the form $L \circ N \circ \cdots \circ L \circ N$; in more colorful terms, these morphisms may be viewed as "polymeric Hammerstein models," formed from the series connection of $m \geq 1$ individual Hammerstein models, each based on an arbitrary static nonlinearity $g_j(x)$ and a linear $ARMA(p,q)$ model. Overall, the input-output relationship for each of these Hammerstein model is of the form

$$y(k) = \sum_{i=1}^{p} a_i^j y(k-i) + \sum_{i=0}^{q} b_i^j g_j(u(k-i)). \qquad (7.60)$$

Clearly, this structure is closed under cascade connection: composition of a model built from m components with one built from m' components yields a model of the same structure built from $m + m'$ components. Similarly, the identity system $y(k) = u(k)$ may be viewed as the degenerate Hammerstein model defined by Eq. (7.60) with $g(x) = x$, $p = 0$, $q = 0$, and $b_0 = 1$.

It is clear that any Hammerstein model of the type $H_{(N,M)}$ introduced in Chapter 5 belongs to the category **LN**$_\infty$. It is perhaps less obvious that the Wiener models $W_{(N,M)}$ are also members of this category. To see this point, note that the $W_{(N,M)}$ Wiener model structure may also be written as the composition $M_2 \circ M_1$ of the degenerate Hammerstein models $M_1 = L \circ 1 \in H_{(1,M)}$ and $M_2 = 1 \circ N \in H_{(N,0)}$. Also, note that this structure is somewhat reminiscent of the multi-layer feedforward neural network structure. Since multilayer networks are inherently much more flexible than single-layer networks, it is interesting to speculate about the relative flexibility of these "m-layer Hammerstein models" relative to the $H_{(N,M)}$ model class. In fact, the case $m = 2$ has been considered in modeling the behavior of neurons in bullfrogs (Segal and Outerbridge, 1982).

7.4.4 Behavioral model categories

The subcategories of **DTDM** introduced in the preceeding section were all based on *structural* restrictions of the dynamic models that define morphisms. As illustrated in Chapter 3, it is also possible to define model classes via *behavioral* restrictions. The following discussion defines four categories on the basis of these behavioral restrictions:

1. The category **L** of linear models
2. The category **H** of homogeneous models
3. The category **PH** of positive-homogeneous models
4. The category **SL** of static-linear models.

As in the structurally-defined model categories introduced previously, all of these model categories are based on the object class $\mathcal{O}$ from the category **DTDM**, composition of morphisms is defined by cascade connection, and the morphism sets considered are subsets of those from **DTDM**. Hence, to establish that each of these model classes defines a subcategory of **DTDM**, it is again enough to show that they are closed under cascade connection and include the identity system $y(k) = u(k)$. In addition, it is also possible to define behavioral subcategories by suitably restricting the class of objects on which they are based; this approach is illustrated with the three categories **BIBO**, **ASYMP** and **MONO** defined at the end of this section.

It was noted at the end of Chapter 2 that linearity may be defined in terms of the the superposition principle. That is, a model $\mathcal{M}$ is linear if it satisfies

$$\mathcal{M}[\alpha u(k) + \beta v(k)] = \alpha \mathcal{M}[u(k)] + \beta \mathcal{M}[v(k)] \qquad (7.61)$$

for all real α and β and all input sequences $\{u(k)\}$ and $\{v(k)\}$. In contrast to the four linear model categories introduced in Sec. 7.4.2, this description is entirely behavioral and says nothing directly about the inherent structure of the model $\mathcal{M}$. Despite the implicit nature of this description, it is easy to demonstrate that this class of models forms the basis for a category. For example, it is clear that the identity system $\mathcal{M}[u(k)] = u(k)$ belongs to this class and defines the identity morphisms in **L**. To see that this class of systems is closed under cascade connection, consider the composition $\mathcal{M}_1 \circ \mathcal{M}_2$, noting that

$$\begin{aligned}\mathcal{M}_1 \circ \mathcal{M}_2[ax(k) + by(k)] &= \mathcal{M}_1[\mathcal{M}_2[ax(k) + by(k)]] \\&= \mathcal{M}_1[a\mathcal{M}_2[x(k)] + b\mathcal{M}_2[y(k)]] \\&= a\mathcal{M}_1[\mathcal{M}_2[x(k)]] + b\mathcal{M}_1[\mathcal{M}_2[y(k)]] \\&= a\mathcal{M}_1 \circ \mathcal{M}_2[x(k)] + b\mathcal{M}_1 \circ \mathcal{M}_2[y(k)].\end{aligned} \qquad (7.62)$$

Relations between Model Classes

It was also demonstrated in Chapter 2 that any linear, time-invariant discrete-time dynamic model may be characterized completely in terms of its unit impulse response $\{h(k)\}$. This observation is important because it means that the morphisms in the category **L** may be represented by the set of all possible impulse response sequences.

As discussed in Chapter 3, relaxing the superposition principle leads to three classes of nonlinear models: the homogeneous class, the positive-homogeneous class, and the static-linear class. In fact, all of these relaxations lead to categories as the following discussions illustrate. First, note that since all three of these model classes relax the condition (7.61), it follows that all of the models included in the linear class defined by this condition are also members of these nonlinear model classes. In particular, it follows that the identity system belongs to all of these model classes, so it is enough to show that they are closed under cascade connection.

The category **H** is defined by relaxing the superposition principle (7.61) to the homogeneity condition

$$\mathcal{M}[\lambda x(k)] = \lambda \mathcal{M}[x(k)] \tag{7.63}$$

for all real λ. To see that this model class is closed under cascade connection, note that

$$\mathcal{M}_1 \circ \mathcal{M}_2[\lambda x(k)] = \mathcal{M}_1[\mathcal{M}_2[\lambda x(k)]] = \mathcal{M}_1[\lambda \mathcal{M}_2[x(k)]]$$
$$= \lambda \mathcal{M}_1[\mathcal{M}_2[x(k)]] = \lambda \mathcal{M}_1 \circ \mathcal{M}_2[x(k)] \tag{7.64}$$

for all real λ. Similarly, the category **PH** of positive-homogeneous models is defined by further relaxing the superposition principle, requiring Eq. (7.63) to hold only for $\lambda > 0$. It follows from Eq. (7.64) that this model class is also closed under cascade connection.

Finally, the category **SL** is defined by taking a restricted subset of the static-linear dynamic models defined in Chapter 3 as morphisms. Here, a little more care is required, since the class of static-linear models includes some peculiar systems like the time-varying linear model

$$y(k) = u(k)(-1)^k, \tag{7.65}$$

whose response to the input $u(k) = u_s$ is not constant. The cascade connection of this system with the static-linear model

$$z(k) = y(k) + y^2(k) - y(k)y(k-1) \tag{7.66}$$

yields the following response to the input sequence $u(k) = u_s$:

$$z(k) = (-1)^k u_s + 2u_s^2. \tag{7.67}$$

The nonlinearity of this response establishes that the general class of static-linear dynamic models defined in Chapter 3 is not closed under cascade connection.

Conversely, if consideration is restricted to static-linear models whose response to constant inputs is a constant, like that defined in Eq. (7.66), then static-linearity is preserved under cascade connection. In particular, consider the cascade connection of static-linear models $\mathcal{M}_1$ and $\mathcal{M}_2$ to the constant input sequence $u_s = \alpha u_1 + \beta u_2$. This response is given by:

$$\begin{aligned} y_s &= \mathcal{M}_2[\mathcal{M}_1[\alpha u_1 + \beta u_2]] \\ &= \mathcal{M}_2[\alpha \mathcal{M}_1[u_1] + \beta \mathcal{M}_1[u_2]] \\ &= \alpha \mathcal{M}_2[\mathcal{M}_1[u_1]] + \beta \mathcal{M}_2[\mathcal{M}_1[u_2]]. \end{aligned} \qquad (7.68)$$

Thus, it follows that the composite model $\mathcal{M}_2 \circ \mathcal{M}_1$ defined by cascade connection is again static linear, generating a constant response to constant input sequences. *Restricting consideration to this subset of the class of static-linear models provides the basis for a consistent definition of the category* **SL** *of static-linear models.*

As noted, it is also possible to define behavioral model subcategories by restricting the class of objects on which it is based. The following discussion illustrates this idea, introducing three additional behavioral categories:

1. The category **BIBO**, based on BIBO stability
2. The category **ASYMP**, based on asymptotic constancy
3. The category **MONO**, based on monotonicity.

The essential basis for defining each of these categories is some set $\mathcal{S}$ of sequences $\{x(k)\}$ that has some important behavioral interpretation. The class of objects on which the category is based is the *power set* of $\mathcal{S}$, denoted $\mathcal{P}(\mathcal{S})$ and defined as the *set* of all subsets of $\mathcal{S}$ (Adamek et al., 1990, p. 5). More specifically, given the set $\mathcal{S}$, the behavioral subcategories considered here are the full subcategories of **DTDM** based on the object class $\mathcal{O}_S = \mathcal{P}(\mathcal{S})$. Note that since $\mathcal{O}_S$ is a set, the resulting behavioral subcategory is a small category.

The category **BIBO** is defined by choosing $\mathcal{S} = \ell_\infty$, the linear vector space of essentially bounded sequences. This choice is based on the observation made in Sec. 7.3.3 that the morphism set $\hom(\ell_\infty, \ell_\infty)$ in **DTDM** defines the behavioral class of BIBO stable models. Note that the object set $\mathcal{O}_{BIBO}$ on which the category **BIBO** is based includes the following subsets:

1. All spaces ℓ_p of p-power summable sequences $\{x(k)\}$
2. The space ℓ_∞^+ of nonnegative bounded $\{x(k)\}$
3. The set $\mathcal{M}_\infty$ of monotone bounded $\{x(k)\}$
4. The space $\mathcal{A}$ of bounded, asymptotically constant $\{x(k)\}$
5. The set $\mathcal{B}$ of binary sequences
6. The space S^1 of constant sequences
7. The space S_∞^P of bounded period P sequences
8. The space $\mathcal{E}$ of exponentially bounded sequences
9. Arbitrary individual sequences from any of these sets
10. All finite sets of these sequences.

Relations between Model Classes 329

The key point is, although **BIBO** is a small category, its morphism sets define many interesting model types, including some with extremely special qualitative behavior.

The category **ASYMP** is the full subcategory of **DTDM** obtained by restricting objects to the set $\mathcal{S} = \mathcal{A}$, the linear vector space of bounded, asymptotically constant sequences. It follows from the results presented in Chapter 4 that the morphism sets in this category contain all NMAX models, including the Volterra models and all of their special cases. The question of what other model structures define morphisms in **ASYMP** is investigated in Sec. 7.6. Finally, the category **MONO** is obtained by restricting the objects from **DTDM** to the power set of $\mathcal{M}$, the set of all monotone sequences. Note that the objects in this set include the space S^1 of all constant sequences, along with the restricted vector spaces of increasing and decreasing sequences. Qualitatively, the morphisms in this category correspond to discrete-time dynamic models whose response to monotone input sequences is also monotone. Characterization of the models in this class also yields some interesting structural restrictions, another topic examined in Sec. 7.6.

7.5 Empirical modeling and IO subcategories

Essentially, the problem of empirical modeling is one of determining which model(s) from a particular model class $\mathcal{C}$ best conform to a given input/output pair $(\{u(k)\}, \{y(k)\})$. In the category-theoretic framework considered here, this problem may be viewed as one of characterizing sets like $\text{hom}(\{u(k)\}, \{y(k)\})$ from some structurally-defined model category **X**. In the terminology of set-theoretic parameter estimation, this set consists of all models that are *not falsified* by this input/output pair (Willems, 1991). If the set $\text{hom}(\{u(k)\}, \{y(k)\})$ contains more than one morphism, it follows that the pair $(\{u(k)\}, \{y(k)\})$ does not provide a basis for distinguishing these models. Conversely, if the set $\text{hom}(\{u(k)\}, \{y(k)\})$ is empty this input/output pair may be viewed as inconsistent with the model class defining the category **X**. Arguably, the best case occurs when all morphism sets contain precisely one member, since each model in the class may then be identified uniquely with a specific input/output pair. To discuss this and other related problems in more detail, it is useful to introduce the concept of an *IO subcategory*, along with a few more standard constructs from category theory.

7.5.1 IO subcategories

Most of the subcategories of **DTDM** defined in Sec. 7.4 were obtained by restricting the morphism sets but retaining all of the objects in the class $\mathcal{O}$ on which **DTDM** is based. Consequently, none of these subcategories are full subcategories of **DTDM**. The exceptions to this statement are the three categories **BIBO**, **ASYMP** and **MONO**, defined by restricting consideration to a subclass of $\mathcal{O}$. The following discussion specializes this construction by restricting the class on which the subcategory is based to a set $\mathcal{O}^{IO}$ whose members are individual real-valued input/output sequences $\{x(k)\}$. Because the resulting subcategories are intimately related to the input/output behavior of discrete-time dynamic models, they are designated *IO subcategories*. Different IO subcategories of **DTDM** are obtained by restricting consideration to different sets of input sequences (e.g., bounded sequences, monotone sequences, binary sequences, etc.). The essential difference between these specialized IO subcategories and the categories like **BIBO** introduced at the end of Sec. 7.4.4 is that the object set $\mathcal{O}^{IO}$ does not contain *sets* of sequences like ℓ_∞^+, but only *individual* sequences $\{x(k)\}$.

More generally, if **X** is any subcategory of **DTDM**, the corresponding IO subcategory $\mathbf{X}^{IO}$ is obtained as follows. First, define the class of objects $\mathcal{O}^{IO}$ as the set of all individual input sequences $\{x(k)\}$ contained in the object class on which **X** is based. The category $\mathbf{X}^{IO}$ is then defined as the full subcategory of **X** specified by the set $\mathcal{O}^{IO}$. As a specific example, the IO subcategory $\mathbf{NARX}^{IO}$ is based on the object set $\mathcal{O}^{IO}$ consisting of all real-valued sequences $\{x(k)\}$; the morphism sets in $\mathbf{NARX}^{IO}$ are the sets $\hom(\{u(k)\}, \{y(k)\})$ containing those NARX models that map the input sequence $\{u(k)\}$ into the output sequence $\{y(k)\}$. Similarly, the IO subcategory $\mathbf{MONO}^{IO}$ is based on the set $\mathcal{O}^{IO}$ of all *monotone* sequences and the morphism sets $\hom(\{u(k)\}, \{y(k)\})$ consist of all dynamic models mapping the specific monotone input sequence $\{u(k)\}$ into the specific monotone output sequence $\{y(k)\}$. Finally, note that if the object set for **X** does not contain individual sequences but only sets of sequences, the subcategory $\mathbf{X}^{IO}$ is the *empty category* whose only object is the empty set $\emptyset$ and whose only morphism is the associated identity morphism $id_\emptyset$ (Adamek et al., 1990, p. 16).

In order to establish connections with the empirical modeling issues noted in the introduction to this section, it is useful to introduce the following additional constructs from category theory. A category **C** is called *thin* if each morphism set $\hom(A, B)$ contains *at most* one element (Adamek et al., 1990, p. 24). Consequently, if an IO subcategory $\mathbf{X}^{IO}$ is thin, then each input/output pair $(\{u(k)\}, \{y(k)\})$ is either inconsistent with all models in the category, or it is consistent with precisely one model. Further, suppose **C** is a thin category and consider the set $\hom(A, A)$. Since this set can contain at most one element and since it must contain the identity morphism id_A, it follows that $\hom(A, A) = \{id_A\}$ for all objects A in **C**. In the context of IO subcategories, this observation is quite interesting since it rules out the possibility of *root sequences* as objects in thin categories, defined as input sequences that are invariant under the action of the dynamic model. The consequences of this observation are discussed further in Sec. 7.5.4 in connection with nonlinear digital filters and revisited in Chapter 8 in connection with input sequence design for empirical modeling.

An object A in a category **C** is called *initial* if the set $\hom(A, X)$ contains precisely one element for all objects X in **C** and *terminal* if the set $\hom(Y, A)$ contains precisely one element for all objects Y in **C** (Blyth, 1986, p. 16). Again in the context of IO subcategories, an initial object corresponds to a "good input sequence" $\{u(k)\}$ in the sense that precisely one model can be determined from the response to this input sequence, for all possible model responses $\{y(k)\}$. Conversely, a terminal object represents a "good output sequence" $\{y(k)\}$ in the sense that precisely one model can be determined from this response in combination with any possible input sequence $\{u(k)\}$. An object Z in a category **C** that is both initial and terminal is called a *zero object* (Blyth, 1986, p. 18); in connection with IO categories, zero objects would appear to be highly desirable, but their presence has some significant consequences, as the following discussion illustrates.

A particularly important example of an initial object is the following. Consider the category $\mathbf{L}^{IO}$ defined as the full subcategory of the linear model category **L** restricted to the class $\mathcal{O}^{IO}$ of all real-valued sequences $\{x(k)\}$. Note first, that this object class includes the unit impulse sequence $\{\delta(k)\}$ and second, that all morphisms in $\mathbf{L}^{IO}$ may be completely characterized by their response $\{h(k)\}$ to this input sequence. Consequently, it follows that for each object $\{x(k)\}$ in $\mathbf{L}^{IO}$, the set $\hom(\{\delta(k)\}, \{x(k)\})$ contains the single element correspond to the linear model whose unit impulse response is $h(k) = x(k)$. Therefore, the object $\{\delta(k)\}$ is initial in the category $\mathbf{L}^{IO}$. This observation illustrates the good input sequence interpretation of initial objects just suggested. Conversely, note that the unit impulse object is not terminal in $\mathbf{L}^{IO}$ since the morphism set $\hom(\{0\}, \{\delta(k)\})$ is empty: there is no linear, time-invariant dynamic model whose response to the zero sequence is the unit impulse sequence. Consequently, the unit impulse object is not a zero object in $\mathbf{L}^{IO}$. In fact, no zero object exists in this category, as the following argument demonstrates.

A category **C** is called *connected* if the set $\hom(A, B)$ is not empty for all objects A and B in **C** (Blyth, 1986, p. 19). As a specific example, note that any category based on only one object A is necessarily connected, since the single morphism set $\hom(A, A)$ in this category must contain at least the identity morphism id_A and therefore cannot be empty. In the context of IO subcategories, connectedness implies that the system class on which the category is based is large enough to include at least one model relating every possible input sequence to every possible output sequence. In connection with the previous results, it can be shown that every category containing a zero object is connected and, conversely, that every connected category with an initial object and a terminal object has a zero object (Blyth, 1986, p. 20). Consequently, it follows from these observations that the category $\mathbf{L}^{IO}$ contains no zero object: since the morphism set $\hom(\{0\}, \{\delta(k)\})$ is empty, $\mathbf{L}^{IO}$ is not connected.

Returning to the empirical modeling problem, recall the best case suggested in the introduction to this section: every input/output pair $(\{u(k)\}, \{y(k)\})$ corresponds to precisely one model in some category $\mathbf{X}^{IO}$. In category-theoretic terms, this requirement implies that $\mathbf{X}^{IO}$ is both thin and connected. To see the consequences of this requirement, suppose the category **C** is both thin and connected and first consider any single object A: since **C** is both thin and connected, it follows that the sets $\hom(A, X)$ and $\hom(Y, A)$ contain precisely one element for all objects X and Y in **C**. Hence, every object of **C** is both initial and terminal, and therefore a zero object. Conversely, suppose every object of some category **C** is a zero object: it follows from the existence of *any* zero object that **C** is connected. Further, consider any object A: since it is a zero object, the sets $\hom(A, X)$ and $\hom(Y, A)$ contain precisely one member for all objects X and Y in **C**. Hence, *a category is thin and connected if and only if all of its objects are zero objects.*

Next, suppose **C** is thin and connected and consider any two distinct objects A and B. Since **C** is connected, there exist morphisms $M_1 \in \hom(A, B)$ and $M_2 \in \hom(B, A)$, and since **C** is thin, these morphisms are unique. Further, $M_1 \circ M_2 \in \hom(B, B)$, implying $M_1 \circ M_2 = id_B$ since **C** is thin. A morphism $f : A \to B$ such that $g \circ f = id_B$ for some $g : B \to A$ is called a *section* and the corresponding morphism g is called a *retraction* (Blyth, 1986, p. 14). Hence, it follows that M_1 is a retraction and M_2 is a section. Conversely, $M_2 \circ M_1$ is also well-defined and belongs to the set $\hom(A, A)$, implying $M_2 \circ M_1 = id_A$; therefore, M_1 is also a section and M_2 is also a retraction. A morphism that is both a section and a retraction is called an *isomorphism* (Blyth, 1986, p. 15). Consequently, every morphism in **C** is an isomorphism. This requirement is extremely restrictive since isomorphisms are necessarily invertible: in the example just considered, note that $\mathcal{M}_2 = \mathcal{M}_1^{-1}$. Overall, these results show *that the only thin, connected categories are extremely special ones in which every object is a zero object and every morphism is an isomorphism.*

As a specific example, consider the category **Mfun*** of strictly monotone sequences: the class of objects on which this category is based is the set of all strictly increasing or strictly decreasing sequences, and the morphisms in this category are all transformations that preserve strict monotonicity. To see the structure of this category, consider the sequence $z(k) = k$ and note that it is strictly increasing. Further, this sequence represents an initial object in **Mfun*** since any strictly monotone sequence $\{x(k)\}$ is uniquely defined by some function f mapping k into $x(k)$; in particular, this unique function defines the only morphism in $\hom(\{z(k)\}, \{x(k)\})$. Since $\{x(k)\}$ is strictly monotonic, the function f is also strictly monotonic and therefore invertible, defining an inverse morphism $f^{-1} \in \hom(\{x(k)\}, \{z(k)\})$. In fact, this function is also unique, establishing that $\{z(k)\}$ is also terminal and hence a zero object in **Mfun***. It follows as an immediate consequence that this category is connected. Next, consider any morphism $h \in \hom(\{x(k)\}, \{y(k)\})$ for any strictly monotonic sequences $\{x(k)\}$ and $\{y(k)\}$. Taking $f : \{x(k)\} \to \{z(k)\}$ and $g : \{z(k)\} \to \{y(k)\}$, note that $h \circ f$ belongs to the set $\hom(\{z(k)\}, \{y(k)\})$, implying $h \circ f = g$ since $\{z(k)\}$ is an initial object. Further, since f is invertible, it follows that $h = g \circ f^{-1}$, implying h is unique. Hence, the category **Mfun*** is both thin and connected.

The consequences of these requirements in the context of IO subcategories are illustrated by the following three examples. The first is the category **Aff**^{ss} which is connected but not thin and illustrates the general limitations inherent in attempting to characterize dynamic models from their responses to a class of input sequences that are not sufficiently rich to excite the full character of the model (here, their steady-state responses). The second example is the category **Gauss** which is both thin and connected and illustrates that, although rare, such categories do exist. The third example is the category **Median**^{IO} which is

Relations between Model Classes 333

neither thin nor connected. This example is typical of the situation encountered in nonlinear empirical modeling: certain input/output pairs are inconsistent with all models in the class under consideration, and other input/output pairs are insufficient to distinguish between several different members of this model class. In addition, this example also provides a nice illustration of the notion of a root sequence mentioned earlier in this section. The practical implications of some of these results for input sequence design in empirical modeling are discussed in Chapter 8.

7.5.2 Example 1: the category $\mathbf{Aff}^{ss}$

The category $\mathbf{Aff}^{ss}$ is the IO subcategory of steady-state responses of *causal, affine dynamic models*, defined by the input/output relation

$$y(k) = a + \sum_{i=0}^{\infty} h(i)u(k-i). \qquad (7.69)$$

The parent category $\mathbf{Aff}$ is defined by restricting the morphisms in $\mathbf{DTDM}$ to this model class, and the subcategory $\mathbf{Aff}^{ss}$ is the full subcategory obtained by restricting the object class to the set $\mathcal{O}^{ss}$ of constant, real-valued sequences $x(k) \equiv x_s$. Note that each model in this class may be characterized by the steady-state relationship $y_s = a + Ku_s$ where K is given by

$$K = \sum_{i=0}^{\infty} h(i). \qquad (7.70)$$

In particular, note that the morphism set $\hom(u_s, y_s)$ consists of all affine models satisfying this steady-state relationship for some finite K and a. It is not difficult to see that this category is connected (that is, none of these morphism sets are empty). For example, the set $\hom(0, y_s)$ consists of all affine models with constant term $a = y_s$. More generally, for $u_s \neq 0$, the equation $y_s = a + Ku_s$ has an uncountably infinite number of solutions for any value of a. Consequently, all morphism sets in the category $\mathbf{Aff}^{ss}$ contain more than one element, thus establishing that this category is not thin.

This situation is typical for IO subcategories whose object class $\mathcal{O}^{IO}$ is "too small." As an extreme example, recall that any category $\mathbf{C}$ based on a single object is necessarily connected, but it is thin if and only if it is discrete (that is, the only morphism in the category is the identity morphism for the only element in the category). In the context of empirical modeling, IO subcategories that are connected but not thin arise when the input sequences in the object class $\mathcal{O}^{IO}$ are not sufficiently rich in character to exhibit consistently distinguishable model responses. Consequently, to improve the identifiability of an IO model category $\mathbf{X}^{IO}$, it is necessary to either enlarge the class of objects $\mathcal{O}^{IO}$ or restrict the class of dynamic models defining morphisms in $\mathbf{X}$.

7.5.3 Example 2: the category Gauss

The category **Gauss** is defined as follows. The objects on which this category is based are the zero-mean, stationary, discrete-time Gaussian stochastic processes $\{x(k)\}$ with nonzero variance. These sequences belong to the class of *purely nondeterministic sequences* (Priestley, 1981, p. 758), which have the following unique $MA(\infty)$ representation:

$$x(k) = \sum_{i=0}^{\infty} g(i)\epsilon(k-i) \equiv g(k) * \epsilon(k). \tag{7.71}$$

Here, $\{\epsilon(k)\}$ is a zero-mean Gaussian white noise process and $g(i) \in \ell_2$. This result is known as *Wold's decomposition*, and it is often normalized so that $g(0) = 1$ (Priestley, 1981, p. 756); alternatively, $g(0)$ may be determined from $x(k)$, requiring instead that $\{\epsilon(k)\}$ be the standard (i.e., unit variance) Gaussian white noise sequence. The following discussion adopts this latter choice because it fixes $\epsilon(k)$, independent of the variance of $\{x(k)\}$.

Taking the class of Gaussian stochastic processes just described as objects, the morphisms on which the category **Gauss** is based are the linear $MA(\infty)$ models of the form (7.71). It follows from the results presented in defining the category **MA** that the cascade connection of two $MA(\infty)$ models is another $MA(\infty)$ model whose coefficients are given by the discrete convolution of the two original model coefficients. Therefore, as in the category **MA**, cascade connection provides a well-defined composition of morphisms. For any object in **Gauss**, the morphism defined by taking $g(0) = 1$ and $g(i) = 0$ for all $i > 0$ in Eq. (7.71) defines the identity morphism.

To establish the claim that **Gauss** is a connected category, first note that the standard white noise process $\{\epsilon(k)\}$ represents an initial object. This observation follows immediately from the uniqueness of Wold's decomposition: for any object $\{x(k)\}$, the set $\hom(\{\epsilon(k)\}, \{x(k)\})$ consists of the unique $MA(\infty)$ model defined by Eq. (7.71). It is also possible to show that the standard white noise object $\{\epsilon(k)\}$ is terminal in **Gauss**. Specifically, recall from Chapter 2 (Sec. 2.2.2) that an explicit construction was given for converting $MA(\infty)$ models [defined by their unit impulse response $\{g(i)\}$] into infinite-order autoregressive models. Hence, it is possible to represent $x(k)$ as:

$$x(k) = \sum_{i=1}^{\infty} \gamma(i) x(k-i) + \epsilon(k). \tag{7.72}$$

This model may be re-arranged to obtain the following $MA(\infty)$ model relating the *input sequence* $\{x(k)\}$ to the *response* $\{\epsilon(k)\}$:

$$\epsilon(k) = x(k) - \sum_{i=1}^{\infty} \gamma(i) x(k-i) \equiv \lambda(k) * x(k), \tag{7.73}$$

where $\lambda(0) = 1$ and $\lambda(k) = -\gamma(k)$ for all $k > 0$. Since the coefficients $\gamma(k)$ are unique, the only morphism in the category **Gauss** relating $\{x(k)\}$ to $\{\epsilon(k)\}$ is

Relations between Model Classes 335

the infinite-order moving average model defined in Eq. (7.73). Consequently, the object $\{\epsilon(k)\}$ is terminal and, since this object is also initial, it is a zero object and it follows that **Gauss** is a connected category.

Next, consider the morphism set $\hom(\{x(k)\}, \{y(k)\})$ for any pair of objects $\{x(k)\}$ and $\{y(k)\}$ in **Gauss**. This set contains all $MA(\infty)$ models relating these two sequences; the following construction shows that there is precisely one such model for any pair of objects. Specifically, note that $\{y(k)\}$ may be expressed as in Eq. (7.71) for some coefficient sequence $\{g(i)\}$, from which it follows that

$$y(k) = g(k) * \epsilon(k) = g(k) * [\lambda(k) * x(k)] = [g(k) * \lambda(k)] * x(k) \equiv \phi(k) * x(k), \quad (7.74)$$

where $\phi(k)$ is unique since $g(k)$ and $\lambda(k)$ are unique. Consequently, $\phi(k)$ defines the only element of the set $\hom(\{x(k)\}, \{y(k)\})$ and, since $\{x(k)\}$ and $\{y(k)\}$ are arbitrary elements of **Gauss**, it follows that **Gauss** is a thin category.

Since thin, connected categories are rare, it follows that **Gauss** is an extremely special IO subcategory. In particular, recall from Sec. 7.5.1 that every object in a thin, connected category is a zero object and every morphism is an isomorphism. In the context of IO subcategories, this object condition means that every sequence $\{x(k)\}$ in **Gauss** is both a good input sequence in the sense of generating a distinctive response from every linear model and a good output sequence in the sense of being a distinctive response to every possible input sequence. Similarly, the fact that every morphism is an isomorphism reflects the fact that every model is invertible and the inverse model also belongs to the category. As the next example and other subsequent examples demonstrate, these conditions are rarely satisfied in IO subcategories.

7.5.4 Example 3: the category Median

The IO subcategory **Median** defined here is interesting both because it contains a useful collection of nonlinear digital filters and because it offers some unexpected insights into the problem of input sequence design for empirical model identification. The object set $\mathcal{O}^{IO}$ assumed here is the set of all real-valued sequences $\{x(k)\}$, as in the **NARX**^{IO} subcategory defined in Sec. 7.5.1. This category is based on the family of *noncausal* median filter of order q, denoted M_q and defined by the input/output relation

$$y(k) = \mathrm{median}\{u(k-q), \ldots, u(k), \ldots, u(k+q)\} \quad (7.75)$$

for some $q \geq 0$. Note that taking $q = 0$ yields the identity system $y(k) = u(k)$ and thus defines the identity morphisms in **Median**. By itself, this family of median filters is *not* closed under cascade connection, so it does not define a model category. As in the category $\mathbf{LN}_\infty$, however, it is possible to imbed this family of nonlinear filters into a larger category. Specifically, the morphisms on which the category **Median** is based are the median filters defined in Eq. (7.75), together with arbitrary cascades of the form $M_q \circ \cdots \circ M_{q'}$.

The reason this category is based on noncausal median filters is that much is known about the *root sequences* for these filters, defined as those sequences $\{x(k)\}$ that are invariant under the action of the median filter (Arce and Gallagher, 1988; Astola et al., 1987; Bangham, 1993; Brandt, 1987; Gallagher and Wise, 1981; Wendt, 1990). Specifically, $\{x(k)\}$ is a root sequence for a nonlinear filter $\mathcal{F}$ if $\mathcal{F}[\{x(k)\}] = \{x(k)\}$. The class of root sequences for median filters includes all constant and monotonic sequences, together with sequences composed of piecewise concatenations of sufficiently long constant or monotonic subsequences (Gallagher and Wise, 1981). In addition, median filters have also been shown to exhibit rapidly oscillating binary root sequences (Astola et al., 1987; Brandt, 1987). In the context of the category **Median**, it follows that if $\{x(k)\}$ is a root sequence for some median filter M_q, then M_q belongs to the set $\hom(\{x(k)\}, \{x(k)\})$.

The presence of root sequences in an IO subcategory have some interesting consequences. First, note that if an object $\{x(k)\}$ in an IO subcategory $\mathbf{X}^{IO}$ is a root sequence for any morphism in $\mathbf{X}^{IO}$, then it cannot be either initial or terminal. That is, note that if A is an initial or terminal object in some category $\mathbf{C}$, then the set $\hom(A, A)$ necessarily contains only the identity morphism id_A. This observation suggests that root sequences for any class $\mathcal{C}$ of dynamic models are undesirable input sequences for empirical identification of models in that class, an idea that is explored further in Chapter 8. Also, it follows by the same argument that any category containing root sequences is not thin.

Having established that the category **Median** is not thin, next consider the question of connectedness. This question may also be addressed through an examination of root sequences, although in a slightly more general context. Let $\mathcal{R}_q$ denote the set of root sequences for the median filter M_q defined in Eq. (7.75); it is known (Gallagher and Wise, 1981) that these sets are nested as

$$\mathcal{R}_\infty \subset \cdots \subset \mathcal{R}_q \subset \mathcal{R}_{q-1} \subset \cdots \subset \mathcal{R}_0 = S^\infty. \tag{7.76}$$

That is, since M_0 is simply the identity mapping, the root set $\mathcal{R}_0$ is simply the set S^∞ of all real-valued sequences. As q increases, the requirements for a locally constant or monotonic subsequence become harder to satisfy and the root sets become smaller. Conversely, sequences that are monotone or constant for all k and not just locally constant or locally monotone are invariant to median filters of all orders. Hence, the set $\mathcal{R}_\infty$ consists of those *universal root sequences* that are invariant to all median filters (and consequently, arbitrary cascades). In particular, the set $\mathcal{R}_\infty$ is nonempty and contains sequences that are invariant under all of the filter structures defined by the category **Median**. Therefore, if $\{x(k)\} \in \mathcal{R}_\infty$ and $\{y(k)\}$ is *any other sequence*, the morphism set $\hom(\{x(k)\}, \{y(k)\})$ is empty, thus establishing that **Median** is not a connected category.

Finally, it is worth noting that many of the morphisms included in the category **Median** have been discussed in the signal processing literature. First and foremost, the median filter itself has been studied extensively and it has been shown that any sequence of finite length will be mapped into a root sequence by a finite number of compositions of a median filter M_q of fixed width q (Gallagher and Wise, 1981). This finite composition of median filters has been called a *root median filter* and forms the basis for another cascade structure composed of median filters of increasing widths called the *data sieve* (Bangham, 1993).

7.6 Structure-behavior relations

One of the primary points illustrated in Sec. 7.4 is that discrete-time dynamic model categories may be defined either structurally or behaviorally. Since the relationship between nonlinear model structure and qualitative behavior is one of great practical importance, it is interesting to consider these relationships from the perspective of category theory. To pursue this investigation, Sec. 7.6.1 introduces the notion of a *joint subcategory* that is common to two compatible categories **C** and **D**. Secs. 7.6.2 through 7.6.4 then apply this notion to the characterization of linear models (in Sec. 7.6.3), of Volterra models (in Sec. 7.6.3) and of homomorphic systems (in Sec. 7.6.4).

7.6.1 Joint subcategories

Consider two categories **C** and **D**, both based on the same composition rule for morphisms, with object classes $\mathcal{O}_C$ and $\mathcal{O}_D$ and morphism sets $\text{Mor}_C(X,Y)$ and $\text{Mor}_D(X,Y)$, respectively. Also, assume that if an object X belongs to both $\mathcal{O}_C$ and $\mathcal{O}_D$, the identity morphism id_X is the same in categories **C** and **D**. Define the *joint subcategory* of **C** and **D** as follows, denoting it **Joint(C, D)**. First, define the class $\mathcal{O}_J$ of objects for the joint subcategory as the intersection

$$\mathcal{O}_J = \mathcal{O}_C \cap \mathcal{O}_D. \tag{7.77}$$

Recall from the discussion in Sec. 7.2.2 that the intersection of two classes is well-defined, regardless of whether the classes $\mathcal{O}_A$ and $\mathcal{O}_B$ are large or small (Adamek et al., 1990, p. 7). Next, define the morphism set $\text{Mor}_J(X,Y)$ as

$$\text{Mor}_J(X,Y) = \text{Mor}_C(X,Y) \cap \text{Mor}_D(X,Y) \tag{7.78}$$

for all object pairs $X, Y \in \mathcal{O}_J$. Note that since the identity morphism id_X is the same in categories **C** and **D**, it follows that this same identity morphism also belongs to the set $\text{Mor}_J(X, X)$.

This construction has a number of interesting consequences. First, it follows that **Joint(C, D)** satisfies the subcategory requirements for both parent categories **C** and **D**. Also, note that if **C** or **D** is a small category, then so is their joint subcategory; this result follows from the fact that $\mathcal{O}_J$ is contained in both $\mathcal{O}_C$ and $\mathcal{O}_D$ and is therefore a set if either of these parent object classes is a set. Similarly, note that if either **C** or **D** is a thin category, it follows that **Joint(C, D)** is also a thin category. Conversely, there is no relationship between the connectedness of these categories. For example, if $\mathcal{O}_J$ contains a single object, the joint category is connected regardless of the nature of the parent categories. At the other extreme, if the object classes $\mathcal{O}_C$ and $\mathcal{O}_D$ are identical but the only common morphisms between the two categories are the identity morphisms, the resulting joint category is the discrete category based on these common objects, which is never connected if it contains more than one object. Finally, note that if **C** is a subcategory of **D**, the joint subcategory is simply **C**.

Joint subcategories provide a useful tool for examining structure/ behavior relations. Specifically, consider a structural model category like **NMAX**, **NARX**, or **Hom**$_\phi$ and a behavioral category like **H** or **SL**; the morphism sets in the resulting joint category define models satisfying both the structural and the behavioral restrictions on which the parent categories are based. The following sections illustrate the utility of this idea with a number of specific examples. In particular, each of these sections considers the joint category defined by one of the structural model categories defined in Sec. 7.4.2 or 7.4.3 with each of the behavioral model categories defined in Sec. 7.4.4. In all cases, both the structural and the behavioral model categories are IO subcategories whose class of objects is a set of (possibly restricted) input sequences.

7.6.2 Linear model characterizations

The following discussion considers the joint subcategories formed from each of the four linear model categories defined in Sec. 7.4.2 and the seven behavioral model categories defined in Sec. 7.4.4. In fact, considerable simplification is possible since the behavioral linear model category **L** is a subcategory of the homogeneous category **H**, the positive-homogeneous category **PH**, and the static-linear category **SL**, and all of the structural categories considered here are subcategories of **L**. Hence, of the seven behavioral categories introduced in Sec. 7.4.4, it is only necessary to consider joint categories defined by the following three:

1. The category **BIBO** of BIBO stable models
2. The category **ASYMP** of models preserving asymptotic constancy
3. The category **MONO** of models preserving monotonicity.

Each of these joint subcategories leads to some useful insights. Since all of the structural model categories considered here are subcategories of **L**, it is useful to first consider the joint subcategories of **L** with **BIBO**, **ASYMP**, and **MONO**. Further, note that each of the three behavioral subcategories listed above are defined as full subcategories of **DTDM** based on an object restriction of the form $\mathcal{O}^{sub} = \mathcal{P}(\mathcal{S})$ where $\mathcal{S}$ is a set of real-valued sequences exhibiting some important behavioral characteristic and $\mathcal{P}(\mathcal{S})$ denotes the set of all subsets of $\mathcal{S}$. Also, recall that the linear model categories considered here are all based on the full object set $\mathcal{O}$ on which the category **DTDM** is based. Hence, it follows from the definition of a joint subcategory that the object classes $\mathcal{O}^J$ for all of the joint subcategories considered here are simply the object sets $\mathcal{O}^{sub}$ on which the behavioral categories listed above are based. Complete characterization of

Relations between Model Classes

these joint subcategories then corresponds to a complete characterization of the morphism sets.

To characterize the morphisms in **Joint(L, BIBO)**, first recall that the morphisms in **L** are defined by their impulse responses $\{h(k)\}$. Hence, it follows that

$$|y(k)| \leq \sum_{i=0}^{\infty} |h(i)| \cdot |u(k-i)| \tag{7.79}$$

and a sufficient condition for BIBO stability is

$$\sum_{i=0}^{\infty} |h(i)| < \infty. \tag{7.80}$$

In fact, this condition is also necessary as the following argument shows. Suppose Eq. (7.80) does not hold, thus implying that the sequence of partial sums

$$S_M = \sum_{i=0}^{M} |h(i)| \tag{7.81}$$

increases monotonically without bound. Now, consider the response $y_M(k)$ of this model to the input sequence

$$u_M(k) = \begin{cases} 0 & k < 0 \\ \text{sign}[h(M-k)] & 0 \leq k \leq M \\ 0 & k > M. \end{cases} \tag{7.82}$$

This response is given by

$$y_M(k) = \sum_{i=0}^{k} h(i)\,\text{sign}[h(M-k+i)]. \tag{7.83}$$

In particular, note that $y_M(M) = S_M$, meaning that $|y(k)|$ can be made as large as desired by choosing M large enough, even though $|u(k)| \leq 1$ for all k. Consequently, Eq. (7.80) represents a necessary and sufficient condition for BIBO stability of linear, time-invariant discrete-time dynamic models. Equivalently, this condition characterizes all of the morphisms in the category **Joint(L, BIBO)**.

Since this condition is automatically satisfied if $h(i)$ is nonzero for only a finite number of indices i, it follows that all models defining morphisms in **MA** also define morphisms in **Joint(MA, BIBO)**. Hence, this joint subcategory is a full subcategory of **MA** but not of **BIBO**. Conversely, it was noted in Chapter 2 that stability restrictions must be imposed on the autoregressive coefficients in AR and ARMA models, so the categories **Joint(AR, BIBO)** and **Joint(ARMA, BIBO)** are not full subcategories of **AR** and **ARMA**, respectively. Similarly, it was also noted in Chapter 2 that eigenvalue restrictions must be imposed on the **A** matrix defining a stable linear state-space model, thus implying that the category **Joint(State, BIBO)** is not a full subcategory of **State**. Finally, two additional points concerning linear model stability are noteworthy. First, it is a standard result that any finite-dimensional, linear, time-invariant system is asymptotically stable if and only if it is exponentially stable (Elaydi, 1996, p. 166). This condition implies the impulse response decays exponentially: there exist constants $A > 0$ and $0 < \alpha < 1$ such that $|h(i)| \leq A\alpha^i$ for all $i \geq 0$. Note that this condition is more restrictive than Eq. (7.80) since

$$\sum_{i=0}^{\infty} |h(i)| \leq A \sum_{i=0}^{\infty} \alpha^i = \frac{A}{1-\alpha} < \infty. \tag{7.84}$$

Second, note that because the morphisms in the category **L** are not restricted to finite-dimensional models, this category includes systems like the slow decay models considered in Chapter 2, defined by impulse responses of the form

$$h(k) = \frac{1}{(k+1)^\alpha}. \tag{7.85}$$

It was noted in Chapter 2 that this model is not exponentially stable but it is BIBO stable for $\alpha > 1$; indeed, this impulse response satisfies Eq. (7.80) if and only if $\alpha > 1$.

Next, consider the joint subcategories between the linear model categories and the **ASYMP** category. Recall that the objects on which this category is based are bounded sequences $\{x(k)\}$ that approach a constant limit x_s as $k \to \infty$, and the morphisms in these categories are discrete-time dynamic models that preserve this qualitative behavior. In fact, for linear systems, this behavior is very closely related to BIBO stability, as the following discussion illustrates. To establish these connections, the following observations will be useful for any sequence $\{u(k)\}$ in the set $\mathcal{O}^{ASYMP}$:

1. $|u(k)| \leq M$ for all k and some finite M
2. Given $\epsilon > 0$, there exists N such that $|u(k) - u_s| < \epsilon$ for all $k > N$
3. $|u(k) - u_s| \leq 2M$ for all k.

Next, consider the response of any linear time-invariant model to this input sequence and note that

$$|y(k) - Ku_s| = |\sum_{i=0}^{\infty} h(i)[u(k-i) - u_s]| \leq \sum_{i=0}^{\infty} |h(i)| \cdot |u(k-i) - u_s| \equiv S(k). \tag{7.86}$$

Choose $\epsilon > 0$ and write $S(k) = S_1(k) + S_2(k)$ where

$$S_1(k) = \sum_{i=0}^{k-N-1} |h(i)| \cdot |u(k-i) - u_s|$$

$$S_2(k) = \sum_{i=k-N}^{\infty} |h(i)| \cdot |u(k-i) - u_s| \tag{7.87}$$

Relations between Model Classes

where N is as defined in condition 2 listed above and assume $k > N$. Since $k - i > N$ for $i = 0, 1, \ldots, k - N - 1$, it follows that

$$S_1(k) \leq \epsilon \sum_{i=0}^{k-N-1} |h(i)| \leq \epsilon \sum_{i=0}^{\infty} |h(i)|. \tag{7.88}$$

For a given BIBO stable linear model, this final sum is finite as noted in the preceeding discussion and $S_1(k)$ may be made arbitrarily small by choosing ϵ appropriately. For $S_2(k)$, note from (3) above that

$$S_2(k) \leq 2M \sum_{i=k-N}^{\infty} |h(i)|. \tag{7.89}$$

Again, if the linear model under consideration is BIBO stable note that $S_2(N)$ is simply the infinite sum in Eq. (7.80) and if this sum is finite, $S_2(k)$ decreases monotonically to zero with increasing k. In particular, there exists N' such that $k - N > N'$ implies $S_2(k) < \epsilon$ for any $\epsilon > 0$. Hence, any BIBO stable linear model preserves asymptotic constancy. Further, it follows from the results presented in Chapter 4 that the categories **Joint(MA, BIBO)** and **Joint(MA, ASYMP)** are the same; since the object set $\mathcal{O}^{ASYMP}$ is a subset of $\mathcal{O}^{BIBO}$, it follows that **Joint(MA, ASYMP)** is a full subcategory of **Joint(MA, BIBO)**. Conversely, it is not obvious whether this result extends to the infinite-dimensional case. For example, the infinite-dimensional linear model defined by $h(k) = (-1)^k/(k+1)$ exhibits the finite steady-state gain $K = \ln 2$ (Gradshteyn and Ryzhik, 1965, no. 0.232) and thus an asymptotically constant step response, but is not BIBO stable since it does not satisfy Eq. (7.80). In particular, note that the input sequence $u(k) = (-1)^k$ generates an unstable response for this model. It is not clear whether this model exhibits asymptotically constant responses to all asymptotically constant input sequences, or whether there exist asymptotically constant but oscillatory input sequences that decay sufficiently slowly to generate an unbounded response.

Finally, consider the joint subcategories between the linear model categories **MONO**. The fundamental objects on which the category **MONO** is based are monotone sequences $\{x(k)\}$, which may be characterized by their first difference $d(k) = x(k) - x(k-1)$. In particular, the sequence $\{x(k)\}$ is increasing if and only if $d(k) \geq 0$ and it is decreasing if and only if $d(k) \leq 0$. The morphisms in the category **Joint(L, MONO)** may be characterized by first noting that

$$y(k) - y(k-1) = \sum_{i=0}^{\infty} h(i)[u(k-i) - u(k-i-1)]. \tag{7.90}$$

Hence, a linear model is monotone if $h(i) \geq 0$ for all i. Conversely, consider the response of this system to the unit step input: the corresponding difference sequence $u(k) - u(k-1)$ is then the unit impulse and it follows from Eq. (7.90) that $y(k) - y(k-1) = h(k)$. Therefore, if $h(k) < 0$ for any k, it follows that the step response is not monotone, thus establishing nonnegativity of the impulse response as a necessary and sufficient condition for linear model to preserve monotonicity.

It is interesting to note the connection between this last result and the previous two. In particular, the steady-state gain of a positive linear system is

$$K = \sum_{i=0}^{\infty} h(i) = \sum_{i=0}^{\infty} |h(i)|, \qquad (7.91)$$

implying BIBO stability if and only if $K < \infty$. Similarly, the response of a step of amplitude u_s settles to a constant value $y_s = Ku_s$ if and only if K is finite, so positive systems that are not BIBO stable do not preserve asymptotic constancy, even in the infinite-dimensional case.

7.6.3 Volterra model characterizations

This section and the next one each consider the joint subcategories obtained from a structurally-defined category of nonlinear models and the seven behavioral model categories defined in Sec. 7.4.4. The structural model category considered here is the IO subcategory of Volterra models $\mathbf{Volt}^{IO}$, whose object class is the set of all real-valued sequences $\{x(k)\}$ and whose morphism sets are defined by the $V_{(N,M)}$ models discussed at length in Chapter 5. Since the linear model category $\mathbf{MA}^{IO}$ is a subcategory of $\mathbf{Volt}^{IO}$ and $\mathbf{Volt}^{IO}$ contains no other linear model structures, it follows immediately that the joint subcategory with $\mathbf{L}$ is simply $\mathbf{MA}^{IO}$. Further, it also follows from the results presented in Sec. 7.1.2 that the joint subcategories of $\mathbf{Volt}^{IO}$ with $\mathbf{H}$ and $\mathbf{PH}$ are also $\mathbf{MA}^{IO}$. Similarly, since Volterra models are continuous NMAX models, it follows from the results presented in Chapter 4 for NMAX models (Sec. 4.2.2) that all $V_{(N,M)}$ models are BIBO stable and preserve asymptotic constancy. Hence, the joint subcategories of $\mathbf{Volt}^{IO}$ with $\mathbf{BIBO}$ and $\mathbf{ASYMP}$ are both simply $\mathbf{Volt}^{IO}$. Conversely, the joint subcategories of $\mathbf{Volt}^{IO}$ with $\mathbf{SL}$ and $\mathbf{MONO}$ define classes of structurally restricted nonlinear Volterra models.

For the category $\mathbf{Joint}(\mathbf{Volt}^{IO}, \mathbf{MONO})$, again consider the first difference $y(k+1) - y(k)$ in response to a monotone input sequence $\{u(k)\}$. It follows from the defining equations for the $V_{(N,M)}$ model that this difference is

$$y(k+1) - y(k) = \sum_{i_1=0}^{M} \alpha_1(i_1)[u(k+1-i_1) - u(k-i_1)]$$

$$+ \cdots + \sum_{n=2}^{N} \sum_{i_1=0}^{M} \cdots \sum_{i_n=0}^{M} \alpha_n(i_1,\ldots,i_n) \Delta_n(k) \qquad (7.92)$$

where the term $\Delta_n(k)$ is defined as

$$\Delta_n(k) = u(k+1-i_1) \cdots u(k+1-i_n) - u(k-i_1) \cdots u(k-i_n). \qquad (7.93)$$

Two observations follow immediately from this result: first, that the monotonicity condition does not constrain the constant term y_0 and second, that a necessary condition for any Volterra model to exhibit monotonic responses to monotonic inputs is $\alpha_1(i) \geq 0$ for all i. The implications for the nonlinear terms $\alpha_n(i_1, \ldots, i_n)$ are more complicated.

As a specific example, consider the case $n = 2$. The general expression for $\Delta_2(k)$ is simplified by defining the dummy indices $m = k - i_1$ and $n = k - i_2$:

$$\Delta_2(k) = u(m+1)u(n+1) - u(m)u(n). \tag{7.94}$$

If $\{u(k)\}$ is a *nonnegative increasing sequence*, it follows that $u(k+1) \geq u(k) \geq 0$ for all k and it is easily seen that $\Delta_2(k) \geq 0$ for all k. Hence, $\alpha_2(i_1, i_2) \geq 0$ for all i_1, i_2 is a necessary condition for monotonicity since otherwise a nonnegative increasing sequence could be chosen to force $y(k+1) - y(k) < 0$ for some k. In fact, restricting consideration to nonnegative monotone input sequences, it is not difficult to show that necessary and sufficient conditions for the response to also be nonnegative and monotone is the nonnegativity of all coefficients $\alpha_n(i_1, \ldots, i_n)$. This result is interesting because many physical systems preserve both positivity and monotonicity.

Alternatively, consider the increasing sequence: $u(1) = -2$, $u(2) = -1$, $u(3) = 1$, $u(4) = 3$. Taking $m = 3$ and $n = 1$ in Eq. (7.94) then yields $\Delta_2(k) = u(4)u(2) - u(3)u(1) = -1$. It follows from this example that preservation of monotonicity either requires further constraints on the input sequences considered (e.g., positivity) or further constraints on the model parameters (e.g., $\alpha_2(i_1, i_2) = 0$). To see that the category **Joint(Volt**^{IO}, **MONO)** does contain nonlinear models, note that Hammerstein models based on the static nonlinearity $g(x) = g_1 x + g_3 x^3$ with positive coefficients and positive linear dynamic models preserve monotonicity. If this positive linear model is a moving average model of order M, note that this Hammerstein model represents a member of the restricted class $\mathcal{O}_{(N,M)}$ of odd-order Volterra models introduced in Chapter 5. In fact, it is not difficult to show that the class of odd-order *nonnegative* Volterra models is closed under cascade composition and defines a structural model category, denoted $\mathbf{O}^+$ in subsequent discussions.

Next, consider the morphisms in the category **Joint(Volt**^{IO}, **SL)**. The response of any $V_{(N,M)}$ model to the constant input sequence $u(k) \equiv u_s$ is

$$y_s = y_0 + \sum_{n=1}^{N} \gamma_n u_s^n,$$

$$\gamma_n = \sum_{i_1=0}^{M} \cdots \sum_{i_n=0}^{M} \alpha_n(i_1, \ldots, i_n). \tag{7.95}$$

Consequently, this Volterra model is static-linear if and only if the following conditions are satisfied

$$y_0 = 0$$

$$\sum_{i_1=0}^{M} \cdots \sum_{i_n=0}^{M} \alpha_n(i_1, \ldots, i_n) = 0, \; n = 2, 3, \ldots, N. \tag{7.96}$$

Under these conditions, the steady-state relationship for the Volterra model is $y_s = K u_s$ where K is given by

$$K = \sum_{i=0}^{M} \alpha_1(i). \tag{7.97}$$

Finally, returning to the category of odd-order, nonnegative Volterra models, consider the joint subcategory **Joint**$(\mathbf{O^+}, \mathbf{SL})$. Since $\alpha_n(i_1, \ldots, i_n)$ is identically zero for all even n for morphisms in the category $\mathbf{O^+}$, Eq. (7.96) is automatically satisfied for n even. Conversely, since the coefficients of the odd-order terms are necessarily nonnegative, it follows from Eq. (7.96) that the nonlinear model coefficients must also vanish for n odd. Hence, the only morphisms **Joint**$(\mathbf{O^+}, \mathbf{SL})$ correspond to finite-order linear moving average models with nonnegative coefficients.

7.6.4 Homomorphic system characterizations

Recall that the category $\mathbf{Hom}_\phi$ is defined by homomorphic systems based on a specified invertible function $\phi[\cdot]$ and arbitrary linear $ARMA(p, q)$ models for some p and q. Since $\mathbf{Hom}_\phi$ reduces to the linear model category **ARMA** if $\phi[\cdot]$ is linear, the following discussion assumes $\phi[\cdot]$ is nonlinear. As in the previous discussions, characterizations are obtained by examining the morphism sets in the joint subcategories formed between $\mathbf{Hom}_\phi^{IO}$ and the seven qualitative model categories defined in Sec. 7.4.4.

It is instructive to first consider the category **Joint**$(\mathbf{Hom}_\phi^{IO}, \mathbf{SL})$. Specifically, note that the response to the steady-state input sequence $u(k) \equiv u_s$ of any morphism in $\mathbf{Hom}_\phi^{IO}$ is

$$y_s = \phi^{-1}[K\phi[u_s]] \tag{7.98}$$

where K is the steady-state gain of the linear model on which this morphism is based. For this morphism to exhibit static linearity, it must be possible to express this steady-state relationship as $y_s = G u_s$ for some real constant G. Combining these requirements leads to the following functional equation that must be satisfied by any static-linear morphism in $\mathbf{Hom}_\phi^{IO}$, for all u_s:

$$\phi^{-1}[K\phi[u_s]] = G u_s \Rightarrow \phi[G u_s] = K\phi[u_s]. \tag{7.99}$$

If $K = 1$, this equation is satisfied by $G = 1$ for any invertible function $\phi[\cdot]$. For $K \neq 1$, this equation becomes much more restrictive, as the following result demonstrates. Taking $u_s = 1$ leads to the result $G = \phi^{-1}[K\phi[1]]$, which may be rewritten as $K = \phi[G]/\phi[1]$. This observation reduces Eq. (7.99) to

$$\phi[G u_s]\phi[1] = \phi[G]\phi[u_s], \tag{7.100}$$

and defining $\psi[x] = \phi[x]/\phi[1]$ further reduces this equation to

$$\psi[G u_s] = \psi[G]\psi[u_s]. \tag{7.101}$$

Relations between Model Classes 345

This equation is known as *Cauchy's power equation* (Aczel and Dhombres, 1989, p. 29) and has the following three solutions:

$$\psi[x] \equiv 0 \quad \psi[x] = |x|^\nu \quad \psi[x] = |x|^\nu \operatorname{sign} x, \tag{7.102}$$

where ν is any real constant. As in the case of Cauchy's equation discussed at the end of Chapter 2, other solutions exist, but these three are the only nonpathological cases (e.g., solutions exhibiting continuity at any point). Note that the first two of these solutions are not invertible whereas the third one is invertible if $\nu \neq 0$. Overall, these results provide a complete characterization of the morphisms in **Joint**($\mathbf{Hom}_\phi^{IO}$, **SL**). For any invertible nonlinear function $\phi[\cdot]$, these morphsisms include all homomorphic systems based on linear models with steady-state gain $K = 1$. In addition, if $\phi[x] = |x|^\nu \operatorname{sign} x$ for any real constant $\nu \neq 0$, it follows that *all* morphisms in $\mathbf{Hom}_\phi^{IO}$ are static-linear.

Next, note that since **L** and **H** are both subcategories of **SL**, the joint subcategories of $\mathbf{Hom}_\phi^{IO}$ formed with these categories are necessarily subcategories of **Joint**($\mathbf{Hom}_\phi^{IO}$, **SL**). In fact, since **L** is a subcategory of **H**, it is reasonable to first consider **Joint**($\mathbf{Hom}_\phi^{IO}$, **H**) and then examine **Joint**($\mathbf{Hom}_\phi^{IO}$, **L**). Initially, suppose $\phi[x] = |x|^\nu \operatorname{sign} x$ for some $\nu \neq 0$ and consider the effect of this nonlinearity on an arbitrarily scaled input sequence $\{\lambda u(k)\}$. It follows from Cauchy's power equation that

$$\phi[\lambda u(k)] = \phi[\lambda]\phi[u(k)]. \tag{7.103}$$

Since this sequence represents the input to the linear dynamic component of any model in $\mathbf{Hom}_\phi^{IO}$, it follows that the output of this linear model is

$$z_\lambda(k) = \phi[\lambda] z_1(k) \tag{7.104}$$

where $z_1(k)$ is the linear model output for $\lambda = 1$. The overall homomorphic system output is

$$y_\lambda(k) = \phi^{-1}[z_\lambda(k)] = \phi^{-1}[\phi[\lambda]z_1(k)]. \tag{7.105}$$

Note that Cauchy's power equation may be rearranged to give

$$\phi[xy] = \phi[x]\phi[y] \quad \Rightarrow \quad \phi^{-1}[\phi[x]z] = x\phi^{-1}[z], \tag{7.106}$$

a result obtained by defining $z = \phi[y]$. Applying this result to Eq. (7.105) finally yields

$$y_\lambda(k) = \lambda \phi^{-1}(z_1(k)) = \lambda y_1(k) \tag{7.107}$$

where $y_1(k)$ is the response of the homomorphic system to the unscaled input sequence $\{u(k)\}$. In other words, this result establishes that for $\phi[x] = |x|^\nu \operatorname{sign} x$

$$\mathbf{Joint}(\mathbf{Hom}_\phi^{IO}, \mathbf{H}) = \mathbf{Joint}(\mathbf{Hom}_\phi^{IO}, \mathbf{SL}) = \mathbf{Hom}_\phi^{IO}. \tag{7.108}$$

The question of whether there exist nonlinear homogeneous models for other choices of $\phi[\cdot]$ leads to an interesting generalization of Cauchy's power equation. Note that the response of a general homomorphic system may be expressed as

$$y(k) = \phi^{-1}\left[\sum_{i=0}^{\infty} h(i)\phi[u(k-i)]\right] \quad (7.109)$$

where $\{h(i)\}$ is the impulse response of the linear ARMA model on which the overall nonlinear system is based. Now, suppose this nonlinear model is homogeneous and consider its response to an impulse of amplitude $\lambda\alpha$. In particular, consider the response $y_\lambda(0)$ to this input at time $k = 0$. If $\phi[0] = 0$, it follows from Eq. (7.109) that

$$\phi[y_\lambda(0)] = \phi[\lambda y_1(0)] = h(0)\phi[\lambda\alpha]. \quad (7.110)$$

Further, taking the special case $\lambda = 1$, it follows that

$$h(0) = \frac{\phi[y_1(0)]}{\phi[\alpha]}. \quad (7.111)$$

Defining $x = \alpha$, $y = y_1(0)$ and $z = \lambda$, Eqs. (7.110) and (7.111) may be combined to obtain

$$\phi[x]\phi[yz] = \phi[xz]\phi[y]. \quad (7.112)$$

Note that if $\phi[\cdot]$ satisfies Cauchy's power equation, both sides of this equation reduce to $\phi[x]\phi[y]\phi[z]$. It is not obvious whether other solutions exist or, if such solutions do exist, whether they yield other homogeneous, homomorphic nonlinear models. Finally, it follows from Eq. (7.109) that homomorphic models are linear if and only if they satisfy the generalized Pexider equation discussed in Chapter 4 (Sec. 4.4.1), thus implying that the function $\phi[\cdot]$ is linear.

Characterization of the morphisms in the category $\mathbf{Joint}(\mathbf{Hom}_\phi^{IO}, \mathbf{PH})$ proceeds in essentially the same way as for homogeneous models. In particular, strictly positive-homogeneous models (that is, models that are positive-homogeneous but not homogeneous) are obtained by considering homomorphic models based on nonlinear functions of the form

$$\phi[x] = \begin{cases} x^\nu & x > 0 \\ 0 & x = 0 \\ -|x|^\mu & x < 0, \end{cases} \quad (7.113)$$

where ν and μ are two distinct real constants. Note that this function is both continuous and invertible but since $\phi[-x] \neq -\phi[x]$, the homomorphic system obtained from any linear model via this nonlinearity is strictly positive-homogeneous unless the steady-state gain of the linear model is 1.

If the function $\phi[\cdot]$ is assumed everywhere continuous, its inverse is also everywhere continuous. Consequently, both of these functions are bounded on compact sets and strictly monotonic. For continuous functions, these observations reduce the characterization of the joint subcategories of $\mathbf{Hom}_\phi^{IO}$ with **BIBO**, **ASYMP**, and **MONO** to that of the corresponding joint subcategories of $\mathbf{ARMA}^{IO}$. Specifically, note that $\phi[\cdot]$ maps bounded input sequences $\{u(k)\}$ for the homomorphic model into bounded input sequences $\{\phi[u(k)]\}$ for the linear ARMA model on which the system is based. Further, the output of the homomorphic system is bounded if and only if the output of this linear model is bounded. Hence, the category of homomorphic BIBO stable models simply contains the homomorphic models based on BIBO stable ARMA models. The same reasoning applies to the category **ASYMP**: since $\phi[\cdot]$ and $\phi^{-1}[\cdot]$ both preserve asymptotically constant sequences, the category **Joint**($\mathbf{Hom}_\phi^{IO}$, **ASYMP**) contains all homomorphic systems based on linear ARMA models that preserve asymptotic constancy. Finally, the same reasoning also applies to the joint subcategory with **MONO** since both $\phi[\cdot]$ and $\phi^{-1}[\cdot]$ are strictly monotonic.

7.7 Functors, linearization and inversion

One of the principal advantages of category theory is that it provides considerable machinery for discussing relationships between different categories. One fundamental construct is that of a *functor*, which may be regarded loosely as a mapping between categories that is sufficiently well behaved to preserve the basic category structure. This concept is defined precisely and illustrated with a number of simple examples in Sec. 7.7.1. In the context of dynamic models, functors provide a useful mechanism for discussing linearization and two possible approaches to this problem are presented in Sec. 7.7.2. The first of these approaches defines a functor but the second does not and this contrast provides some useful insights into both functors and linearization. Sec. 7.7.3 demonstrates that functors are also natural constructs for discussing model inversion, illustrating this idea with the case of inverse NARX models.

7.7.1 Basic notion of a functor

Functors come in two basic flavors: *covariant* and *contravariant*; unmodified, the term generally refers to a covariant functor, which is a prescription taking objects and morphisms from one category **C** to another category **D**. Specifically, a covariant functor $\mathcal{F}$ (Blyth, 1986, p. 73):

1. Assigns to every object A in **C** an object X in **D**
2. Assigns to every morphism α in **C** a morphism ϕ in **D**
3. Preserves identities: $\mathcal{F}id_A = id_{\mathcal{F}A}$ for all objects A in **C**
4. Preserves composition of morphisms: $\mathcal{F}(\alpha \circ \beta) = [\mathcal{F}\alpha] \circ [\mathcal{F}\beta]$.

It is frequently easy to construct mappings that satisfy the first three of these conditions, but the fourth condition imposes significant restrictions on those mappings that qualify as functors. This point is demonstrated explicitly in Sec. 7.7.2.

Arguably the simplest covariant functor is the *identity functor*, which maps a category **C** into itself by simply preserving all objects and all morphisms. For an arbitrary category **C**, this identity functor will be denoted 1_C. While this example may seem trivial, identity functors play an important role in category theory, analogous to the role of identity matrices in linear algebra. Another simple example of a functor is the *inclusion functor* obtained by mapping any subcategory **S** of a category **C** into the larger category. That is, the inclusion functor $\mathcal{F} : \mathbf{S} \to \mathbf{C}$ takes all objects and morphisms in **S** into themselves, but as objects and morphisms of the larger category **C**. As a specific example, note that the inclusion mapping $\mathcal{F} : \mathbf{Hom}_\phi \to \mathbf{LN}_\infty$ relating the category of homomorphic models based on a fixed nonlinearity $\phi[\cdot]$ to the category of block-oriented cascades is a functor. Another example of an inclusion functor is that taking the subcategory **MA** into the larger category **NMAX**. Both of these examples illustrate the concept of a *forgetful functor*, which maps a more highly structured category into a less highly structured one, thus forgetting some or all of the details of the original category. In this second example, the result of the inclusion mapping is to forget the linear structure of the original model, burying it in the larger category of nonlinear moving average models.

A possibly more interesting example is the functor $\mathcal{F}_X : \mathbf{Square} \to \mathbf{Square}$, defined as follows. The objects R^n in **Square** are each mapped into themselves, and every matrix A in the set $\hom(R^n, R^n)$ is mapped into the similar matrix $X_n A X_n^{-1}$ where $\{X_n\}$ is a family of nonsingular $n \times n$ matrices, defined for all n. Taking A and B as any two $n \times n$ matrices, it follows that

$$\mathcal{F}_X[A \circ B] = X_n A B X_n^{-1} = [X_n A X_n^{-1}][X_n B X_n^{-1}] = \mathcal{F}_X[A] \circ \mathcal{F}_X[B]. \quad (7.114)$$

Clearly, the $n \times n$ identity matrices are also preserved under this transformation, establishing that it defines a functor from **Square** into itself.

To define a *contravariant functor*, it is necessary to first define the notion of *duality*. Given a category **C**, its *dual* or *opposite* category $\mathbf{C}^{op}$ is defined as follows (Blyth, 1986, p. 11). Both categories are based on the same set of objects and the same set of morphisms, but if $\mathbf{D} = \mathbf{C}^{op}$ then $\mathrm{Mor}_D(A, B) = \mathrm{Mor}_C(B, A)$. As a consequence, note that composition of morphisms in $\mathbf{C}^{op}$ is *reversed* relative to **C**. That is, if $f : A \to B$ and $g : B \to C$ are morphisms in **C**, then the composition $g \circ f$ defines a morphism from A to C but the morphism $f \circ g$ does not exist, in general. Conversely, in the dual category $\mathbf{C}^{op}$, if $f \in \hom(A, B)$ and $g \in \hom(B, C)$ then $f : B \to A$ and $g : C \to B$ and the composition $f \circ g : C \to A$ belongs to $\hom(A, C)$, but the composition $g \circ f$ generally does not exist. Also, note that the dual $[\mathbf{C}^{op}]^{op}$ of the dual category $\mathbf{C}^{op}$ is simply the original category **C**. As a specific illustration, consider the dual category of **Vect**. In both **Vect** and $\mathbf{Vect}^{op}$, the objects are the n-dimensional linear vector spaces R^n. In **Vect**, the morphism set $\hom(R^n, R^m)$ consist of the the $m \times n$ matrices A, whereas in $\mathbf{Vect}^{op}$ the morphism set $\hom(R^n, R^m)$ consists of the $n \times m$ matrices A^T obtained by transposing the morphisms A in the original category. Composition of morphisms in **Vect** corresponds to ordinary matrix multiplication, whereas composition of morphisms in $\mathbf{Vect}^{op}$ corresponds to the transposed matrix product $(AB)^T = B^T A^T$.

Relations between Model Classes 349

A *contravariant functor* $\mathcal{G} : \mathbf{C} \to \mathbf{D}$ is defined as a covariant functor from the category $\mathbf{C}$ to the dual category $\mathbf{D}^{op}$. Hence, a contravariant functor $\mathcal{G}$:

1'. Assigns to every object A in $\mathbf{C}$ an object X in $\mathbf{D}^{op}$
2'. Assigns to every morphism α in $\mathbf{C}$ a morphism ϕ in $\mathbf{D}^{op}$
3'. Preserves identities: $\mathcal{G} id_A = id_{\mathcal{G}A}$ for all objects A of $\mathbf{C}$
4'. Preserves composition of morphisms: $\mathcal{G}(\alpha \circ \beta) = [\mathcal{G}\beta] \circ [\mathcal{G}\alpha]$.

The most significant difference between the defining conditions for a covariant functor and those for a contravariant functor is the reversal of the order that occurs in the composition of morphisms (4').

To illustrate the concept of a contravariant functor with a simple example, consider the category **Nsing** of nonsingular $n \times n$ matrices with the vector spaces R^n taken as objects. Note that this construction defines a category since the product of two nonsingular $n \times n$ matrices A and B is a third $n \times n$ matrix whose inverse is given explicitly as $(AB)^{-1} = B^{-1}A^{-1}$. The dual category **Nsing**^{op} again has the vector spaces R^n as objects and the transposed $n \times n$ matrices A^T as morphisms. Now, consider the matrix inverse functor $\mathcal{G} : \mathbf{Nsing} \to \mathbf{Nsing}$. This functor is contravariant since matrix inversion reverses the order of matrix multiplication, as was just noted. Consequently, $\mathcal{G}$ may be represented as the equivalent *covariant* functor $\mathcal{F} : \mathbf{Nsing} \to \mathbf{Nsing}^{op}$. Specifically, this functor maps the $n \times n$ matrix A in **Nsing** into the $n \times n$ matrix $(A^{-1})^T$ in **Nsing**^{op}. Consequently, it follows that

$$\mathcal{F}[AB] = ([AB]^{-1})^T = (B^{-1}A^{-1})^T = (A^{-1})^T(B^{-1})^T = \mathcal{F}[A] \circ \mathcal{F}[B]. \quad (7.115)$$

In the context of categories of discrete-time dynamic models, duality and contravariant functors arise naturally in connection with model inverses, a point illustrated in Sec. 7.7.3.

7.7.2 Linearization functors

In the context of discrete-time dynamic models, a particularly interesting class of functors relating linear and nonlinear model categories is the class of *linearization functors*, defined as any functor $\mathcal{F}$ from an arbitrary discrete-time dynamic model category $\mathbf{X}$ to the category $\mathbf{L}$ of linear dynamic models. As the name implies, such a functor associates a linear model from the category $\mathbf{L}$ with every model in the original category $\mathbf{X}$. Clearly, if $\mathbf{X}$ is a linear model category like **AR** or **MA**, the most natural linearization functor is simply the inclusion functor. If $\mathbf{X}$ is a nonlinear model category, however, the requirement that functors preserve composition of morphisms turns out to be quite restrictive and provides some useful insights into the differences between linear and nonlinear dynamic models.

In particular, note that if M_1 and M_2 are two nonlinear models in $\mathbf{X}$ and $\mathcal{L}_X$ is a linearization functor, this requirement implies that:

$$\mathcal{L}_X(M_1 \circ M_2) = [\mathcal{L}_X M_1] \circ [\mathcal{L}_X M_2]. \quad (7.116)$$

Since the morphisms on the right-hand side of this equation represent linear models in the category **L**, it follows that composition is commutative. That is, since order is immaterial in cascade connections of single-input, single-output linear dynamic models, $A \circ B = B \circ A$ for any morphisms A and B in **L**. Hence, it follows from Eq. (7.116) that

$$\mathcal{L}_X(M_1 \circ M_2) = \mathcal{L}_X(M_2 \circ M_1). \tag{7.117}$$

This observation has some profound consequences, as the following two examples illustrate.

First, consider the standard linearization of Volterra models: for any model in the class $V_{(N,M)}$, associate the linear $V_{(1,M)}$ model obtained by simply retaining the linear terms in the Volterra expansion. This procedure defines a mapping of morphisms from the category **Volt** of Volterra models to the category **L** of linear models. In addition, note that this mapping preserves identity morphsism since these morphisms are defined by the linear identity model $y(k) = u(k)$ for both categories. Similarly, both categories are defined on the same set of objects. Hence, this linearization procedure defines a functor if it satisfies Eq. (7.117). It follows directly from the results presented in Sec. 7.3.1 for the cascade connection of Volterra models [specifically, Eqs. (7.25) and (7.26)] that the only terms contributing to the linear term in the composition $M_1 \circ M_2$ of two models are the linear terms in M_1 and M_2 individually. Hence, it follows that the linear part of this composite Volterra model $M_1 \circ M_2$ is simply the cascade connection of the linear parts of the models M_1 and M_2. Consequently, the standard linearization of Volterra models defines a functor $\mathcal{L}_V : \textbf{Volt} \to \textbf{L}$.

As an interesting consequence of this result, note that both the Hammerstein model $H_{(N,M)}$ and its dual Wiener model $W_{(N,M)}$ represent morphisms in the category **Volt**. Further, the Hammerstein model may be represented as the composition $L \circ N$ for some linear moving average model L and some polynomial N, whereas its dual Wiener model may be represented as the composition $N \circ L$. It follows that $\mathcal{L} : L \to L$ (i.e., Volterra linearization preserves the linear dynamic model L) but $\mathcal{L} : N \to G$ where G is the effective gain for the polynomial static nonlinearity N, equal to the coefficient of the linear term. Consequently, it follows form the preceeding discussion that the Volterra linearization of the Hammerstein model and its dual Wiener model are identical, simply given by $L \circ G = G \circ L$. This result provides an interesting addition to the Hammerstein and Wiener model comparisons presented in previous chapters. In particular, recall that the steady-state behavior of a Hammerstein model and its dual Wiener model are identical, provided the steady-state gain of the linear subsystem is 1. Conversely, examples discussed in earlier chapters have illustrated clearly that the transient responses of these two models can be dramatically different. However, the Volterra linearization result just presented implies that moderate to large amplitude transient responses must be examined to see these differences, since the linearized approximations of both models are identical.

The second example considered here is a linearization procedure that does unambiguously define a mapping from morphisms in a nonlinear model category **X** to the linear model category **L**, but which *does not* define a functor

Relations between Model Classes 351

between these categories. This example is instructive both for what it says about linearization and for what it says about functors. Specifically, note that the unit impulse response for any model M in $\mathbf{X}$ is well-defined, and denote this impulse response h_M. Since every linear model in $\mathbf{L}$ is uniquely defined by its impulse response, taking the linear model whose impulse response is h_M defines a mapping $\mathcal{H}$ from the category $\mathbf{X}$ to the category $\mathbf{L}$. Although this map is well-defined, it generally fails to satisfy the condition (7.117) and is therefore not a functor.

To see this last point, suppose $\mathbf{X}$ is the category $\mathbf{LN}_\infty$ and consider the Hammerstein and Wiener models M_H and M_W, respectively, constructed from the following components:

L: $y(k) = \alpha y(k-1) + (1-\alpha)u(k)$
N: $f(x) = x + \gamma x^3$.

Since the cubic nonlinearity N maps the unit impulse response into an impulse of amplitude $1 + \gamma$, the unit impulse response of the Hammerstein model is simply $1 + \gamma$ times the unit impulse response of the linear model L. This linear model response is the exponential decay $h(k) = (1-\alpha)\alpha^k$, so the impulse response mapping $\mathcal{H}$ takes the Hammerstein model M_H into the linear model whose impulse response is

$$h_M(k) = (1+\gamma)(1-\alpha)\alpha^k. \tag{7.118}$$

In contrast, the impulse response of the Wiener model is

$$h_W(k) = h(k) + \gamma h^3(k) = (1-\alpha)\alpha^k + \gamma(1-\alpha)^3 \beta^k, \tag{7.119}$$

where $\beta = \alpha^3$. Hence, matching the impulse response of the Hammerstein model leads to a first-order linear model, whereas matching the impulse response of the Wiener model leads to a *second-order* linear model whose impulse response is the sum of two distinct exponential decays: one proportional to α^k and one proportional to β^k. Consequently, impulse response matching does not satisfy the condition (7.117), so the map $\mathcal{H}$ does not constitute a functor. Also, note that this example illustrates the difficulty inherent in attempting to discuss the "dynamic order" of linear approximations of nonlinear dynamic models: this Wiener model exhibits a second-order impulse response even though it is built from a first-order linear dynamic model.

7.7.3 Inverse NARX models

One of the advantages of the NARX model structure noted in Chapter 4 is that NARX models are easily inverted. Specifically, recall that if M is the NARX model

$$y(k) = F(y(k-1), \ldots, y(k-p)) + b_0 u(k) \tag{7.120}$$

and $b_0 \neq 0$, the inverse model is given by

$$u(k) = b_0^{-1}[y(k) - F(y(k-1), \ldots, y(k-p))]. \tag{7.121}$$

In fact, this class of inverse NARX models defines an interesting structural IO subcategory of **DTDM**. Specifically, the objects on which this category is based are those in the set $\mathcal{O}^{IO}$ of all real-valued sequences, morphisms are based on the models defined by Eq. (7.121), composition of morphisms is based on cascade connection of models, and the identity morphisms are based on the identity system $y(k) = u(k)$. It follows that these models form a category if they are closed under cascade connection.

To establish this result, consider the cascade connection of the two inverse NARX models

$$v(k) = b_1^{-1}[y(k) - F_1(y(k-1),\ldots,y(k-p))]$$
$$u(k) = b_2^{-1}[v(k) - F_2(v(k-1),\ldots,v(k-r))]. \quad (7.122)$$

Substituting the first equation into the second, it follows that

$$u(k) = b_2^{-1}\{b_1^{-1}[y(k) - F_1(y(k-1),\ldots,y(k-p))]$$
$$\quad - H(y(k-1),\ldots,y(k-p-r))\}$$
$$\equiv (b_1 b_2)^{-1}[y(k) - G(y(k-1),\ldots,y(k-p-r))]. \quad (7.123)$$

Here, the function $H(\cdot)$ is defined as

$$H(y(k-1),\ldots,y(k-p-r)) =$$
$$F_2(b_1^{-1}[y(k-1) - F_1(y(k-2),\ldots,y(k-p-1))],\ldots,$$
$$b_1^{-1}[y(k-r) - F_1(y(k-r-1),\ldots,y(k-r-p))]), \quad (7.124)$$

exactly as in the forward NARX model cascade connection results presented in Chapter 4 [specifically, Eq. (4.21) in Sec. 4.3]. It follows from the last line of Eq. (7.123) that the cascade connection of the inverse NARX models of orders p and r defined in Eq. (7.122) is an inverse NARX model of order $p+r$. Hence, the inverse NARX models form a category, which will be denoted **INARX** in subsequent discussions. Since this inverse model is a nonlinear moving average model of order p, it follows that **INARX** is a subcategory of **NMAX**.

Not surprisingly, the inversion procedure for NARX models defines a functor $\mathcal{I} : \mathbf{NARX}^\dagger \to \mathbf{INARX}$ where $\mathbf{NARX}^\dagger$ is the IO subcategory of NARX models based on the object class $\mathcal{O}^{IO}$ with $b_0 \neq 0$. The functor $\mathcal{I}$ preserves objects, identity morphisms, and the composition rule in going from one category to the other and maps every morphism in $\mathbf{NARX}^\dagger$ into the corresponding inverse NARX model in **INARX**. To establish that this mapping is a functor, it is only necessary to show that it preserves composition of morphisms. This conclusion follows from a direct comparison with the results presented in Chapter 4. Specifically, Eqs. (4.20) through (4.22) in Sec. 4.3 apply to the cascade connection of the NARX models corresponding to the inverse models defined in Eq. (7.122). The cascade NARX model defined by Eq. (4.22) is

$$y(k) = G(y(k-1),\ldots,y(k-p-r)) + b_1 b_2 u(k) \quad (7.125)$$

Relations between Model Classes 353

and the inverse of this model is easily seen to be the composite inverse model defined by the last line of Eq. (7.123). Hence, model inversion defines a functor from the **NARX**† category to the **INARX** category.

It is important to note that, like the matrix inversion functor considered in Sec. 7.7.1, the model inversion functor $\mathcal{I}$ defined here is a *contravariant functor*. Specifically, this functor takes the NARX model $M : \{u(k)\} \to \{y(k)\}$ into the inverse model $M^{-1} : \{y(k)\} \to \{u(k)\}$. This reversal of the role of the inputs and outputs between the categories **NARX**† and **INARX** means that the functor $\mathcal{I}$ is necessarily contravariant.

7.8 Isomorphic model categories

If $\mathcal{F} : \mathbf{C} \to \mathbf{D}$ and $\mathcal{G} : \mathbf{D} \to \mathbf{E}$ are covariant functors, it follows from the defining conditions given in Sec. 7.7.1 that the composition $\mathcal{G} \circ \mathcal{F}$ defines a third covariant functor from the category $\mathbf{C}$ to the category $\mathbf{E}$. It is also possible to define compositions involving contravariant functors, although this is probably most easily done by reducing the problem to one involving the equivalent covariant functors. For example, if $\mathcal{F} : \mathbf{C} \to \mathbf{D}$ is a covariant functor but $\mathcal{G} : \mathbf{D} \to \mathbf{E}$ is a contravariant functor, the composition $\mathcal{G} \circ \mathcal{F}$ may be constructed in terms of the equivalent covariant functor $\mathcal{G}' : \mathbf{D} \to \mathbf{E}^{op}$ involving the dual category of $\mathbf{E}$. It follows immediately that the composition $\mathcal{G} \circ \mathcal{F}$ may be represented in terms of the covariant functor $\mathcal{G}' \circ \mathcal{F} : \mathbf{C} \to \mathbf{E}^{op}$. Note that since this composition takes the category $\mathbf{C}$ into the dual category $\mathbf{E}^{op}$, it is contravariant when viewed as a functor from $\mathbf{C}$ to $\mathbf{E}$.

An *isomoprhism* $\mathcal{F}$ is a functor between categories $\mathbf{C}$ and $\mathbf{D}$ satisfying the following condition: there exists a second functor $\mathcal{G}$ such that $\mathcal{G} \circ \mathcal{F} = 1_C$ and $\mathcal{F} \circ \mathcal{G} = 1_D$ where 1_C and 1_D are the identity functors for the categories $\mathbf{C}$ and $\mathbf{D}$. It follows immediately that if $\mathcal{F}$ is an isomorphism, so is its inverse functor $\mathcal{G}$; also, $\mathcal{G}$ is covariant or contravariant the same as $\mathcal{F}$. The categories $\mathbf{C}$ and $\mathbf{D}$ are then said to be *isomorphic* and may be regarded as "equivalent" in many useful respects. The following two sections illustrate this point with two sets of examples: the relationship between autoregressive and moving average models, and the relationship between continuous-time and discrete-time models. As noted earlier, in favorable cases it is possible to show that categories of these models are isomorphic (e.g., the categories **AR** and **MA**) but in less favorable cases, they are not (e.g., the categories **NARX** and **NMAX**). Following these discussions, Sec. 7.9 uses these ideas to obtain some useful characterizations and extensions of the category **Hom**$_\phi$ of homomorphic models.

7.8.1 Autoregressive vs. moving average models

It has been noted repeatedly throughout this book that, in the linear case, autoregressive and moving average representations are equivalent. In fact, this equivalence provides a nice illustration of the notion of an isomorphism between categories. In addition, a closer examination of this result reveals some interesting and subtle differences between these model classes, even in the linear case. In the nonlinear case, these differences become more pronounced and, not surprisingly, the isomorphism that holds in the linear case fails in the nonlinear case.

More specifically, suppose **C** and **D** are any two categories of models defined over the same class of input sequences. If every model X in the category **C** has a unique realization as a model Y in the category **D**, identifying these realizations defines a *realization functor* $\mathbf{F} : \mathbf{C} \to \mathbf{D}$. As a specific example, define the category **AR** as the IO subcategory of all linear, infinite-order autoregressive models of the form

$$y(k) = \sum_{i=1}^{\infty} a_i y(k-i) + b_0 u(k) \tag{7.126}$$

where $y(k) \equiv 0$ for $k < 0$ and $b_0 \neq 0$. The object class on which this category is based is the set $\mathcal{O}^{IO}$ of all real-valued input sequences defined for $k \geq 0$. Note that this model is causal and its impulse response is defined by the recursion relation

$$h(k) = \sum_{i=1}^{k} a_i h(k-i) \qquad h(0) = b_0. \tag{7.127}$$

Similarly, define the category **MA** as the IO subcategory of all linear, infinite-order moving average models of the form

$$y(k) = \sum_{i=0}^{\infty} g_i u(k-i) \tag{7.128}$$

based on the same object class $\mathcal{O}^{IO}$ as **AR** and again with the restriction $g_0 \neq 0$. Note that Eq. (7.128) defines a causal model whose impulse response is given by $h(k) = g_k$ for all $k \geq 0$. Define the realization functor $\mathcal{R} : \mathbf{AR} \to \mathbf{MA}$ by identifying each model in **AR** with the model having the same impulse response in **MA**. In contrast to the linearization of $\mathbf{LN}_\infty$ models considered in Sec. 7.7.2, note that here, impulse response matching does define a functor. In fact, this functor is an isomorphism since, given any impulse response $h(k) = g_k$ defining a model in **MA**, it is possible to invert the recursion relation (7.127) to obtain the corresponding autoregressive model coefficients, as discussed in Chapter 2. Consequently, this realization defines the inverse functor $\mathcal{R}^{-1}$, establishing that the categories **AR** and **MA** are isomorphic.

It is interesting to note how apparently minor changes in the definitions of these categories can destroy this isomorphism. Specifically, consider the category $\mathbf{AR}^0$ taking as morphisms all models defined by Eq. (7.126) and restricting the object class to the set $\mathcal{O}^0$ of all real-valued sequences *except* the zero sequence $x(k) \equiv 0$. Note that this object restriction automatically imposes the requirement $b_0 \neq 0$ on the morphism sets. Similarly, define the category $\mathbf{MA}^0$ as the category of linear moving average models over this same object set $\mathcal{O}^0$.

Relations between Model Classes 355

Here, note that the effect of this object restriction is much more pronounced, excluding all linear moving average models with nontrivial zero dynamics. In particular, note that this restriction excludes all linear moving average models with real zeros. As a specific example, note that the first-order moving average model $y(k) = g_0 u(k) + g_1 u(k-1)$ exhibits an identically zero response for any input sequence $\{u(k)\}$ satisfying the recursion relation

$$u(k) = (-g_1/g_0) u(k-1) \tag{7.129}$$

for any initial value $u(0)$. Hence, this model does not define a morphism in the category $\mathbf{MA}^0$ even though it does have a unique infinite-order autoregressive representation. As a consequence, the categories $\mathbf{MA}^0$ and $\mathbf{AR}^0$ are not isomorphic despite the apparently minor differences between the categories $\mathbf{AR}$ and $\mathbf{AR}^0$.

Another interesting isomorphism is the inversion functor $\mathcal{I} : \mathbf{AR} \to \mathbf{MA}$, although it is important to note that $\mathcal{I}$ is a contravariant functor. This functor may be viewed as a restriction of the contravariant functor $\mathcal{I} : \mathbf{NARX}^\dagger \to \mathbf{INARX}$ to the linear model categories $\mathbf{AR}$ and $\mathbf{MA}$. Specifically, note that any model defined by Eq. (7.126) with $b_0 \neq 0$ exhibits the unique inverse model

$$u(k) = b_0^{-1} \left[y(k) - \sum_{i=1}^{\infty} a_i y(k-i) \right] \equiv \sum_{i=0}^{\infty} g_i y(k-i). \tag{7.130}$$

Conversely, given any infinite-order moving average model defined by Eq. (7.128) with $g_0 \neq 0$, the corresponding inverse model is the infinite-order autoregressive model defined by $b_0 = 1/g_0$ and $a_i = -g_i/g_0$ for $i > 0$. Consequently, it is also possible to define the inverse functor $\mathcal{I}^{-1} : \mathbf{MA} \to \mathbf{AR}$, which is also contravariant.

An important property of isomorphisms is that the composition of isomorphisms is another isomorphism. Consequently, it follows from the previous two results that the composition $\mathcal{R} \circ \mathcal{I}^{-1}$ defines a *contravariant* isomorphism from $\mathbf{AR}$ into itself. Equivalently, this composition may be viewed as a covariant isomorphism from $\mathbf{AR}$ into the dual category $\mathbf{AR}^{op}$. Analogous results for the category $\mathbf{MA}$ may be obtained by considering the isomorphism $\mathcal{I} \circ \mathcal{R}^{-1}$. Since most categories are not isomorphic to their dual categories, this result provides further evidence of the extremely special nature of the categories $\mathbf{AR}$ and $\mathbf{MA}$ relative to their nonlinear extensions $\mathbf{NARX}$ and $\mathbf{NMAX}$. For example, note that although the model inversion functor $\mathcal{I} : \mathbf{NARX}^\dagger \to \mathbf{INARX}$ is easily seen to be an isomorphism, the category $\mathbf{INARX}$ is only a subcategory of $\mathbf{NMAX}$ due to the special structure of inverse NARX models. As a specific example, consider the following generalized Pythagorean model

$$y(k) = [c_0 u^3(k) + c_1 u^3(k-1)]^{1/3} \tag{7.131}$$

and its inverse

$$u(k) = [-(c_1/c_0) u^3(k-1) + (1/c_0) y^3(k)]^{1/3}. \tag{7.132}$$

Although this inverse model belongs to the NARX* model class defined in Chapter 4, it does not belong to the NARX class. In general, note that if an inverse NMAX model exists, it necessarily belongs to the NARX* class, that is

$$y(k) = G[u(k), \ldots, u(k-q)] \quad \Rightarrow \quad u(k) = F[u(k-1), \ldots, u(k-q), y(k)] \tag{7.133}$$

for some function $F : R^{q+1} \to R^1$. Conversely, since the model class NARX* does not define a category, this observation is much less useful than the corresponding linear result.

7.8.2 Discrete- vs. continuous-time

An extremely important practical question is the nature of the relationship between continuous-time dynamic models and discrete-time dynamic models. As the following discussion illustrates, some useful insights into this question may also be obtained by considering it in terms of categories. For example, consider the following question:

> Given a category **C** of continuous-time dynamic models, does there exist a category **D** of discrete-time dynamic models that is isomorphic to **C**?

As subsequent discussions demonstrate, the answer to this question is generally "no," but useful insights into *why* this answer is "no" results from the attempt to formulate even apparently trivial versions of this question in terms of categories.

To illustrate this point, let **C** be the *discrete* category defined by taking as objects all possible bounded, continuous-time input signals $u(t)$. That is, the objects defining this category are functions $u(t)$ defined on L_∞ (i.e., the Lebesgue space of essentially bounded, real-valued functions) and the morphisms are the associated identity morphisms, $y(t) = u(t)$. Similarly, let **D** be the corresponding category of bounded, discrete-time input sequences $\{u(k)\}$. While these categories appear to be analogous, they are more interesting than they first appear, precisely because they are *not* isomorphic. To establish this result, first define the *uniform sampling functor* $\mathcal{U}_T : \mathbf{C} \to \mathbf{D}$, mapping continuous-time signals $u(t)$ into discrete-time input sequences $u(k)$ by the following prescription:

$$\mathcal{U}_T u(t) = \{u(kT)\}. \tag{7.134}$$

Here, T represents a fixed sampling interval in continuous-time. Because the only morphisms in the categories considered here are identity morphisms, $\mathcal{U}_T$ simply maps the identity associated with each object in **C** into the identity associated with the corresponding object in **D**. Physically, this functor describes one of most popular computer-based sampling schemes employed in practice, and is often assumed when the relationship between continuous-time and discrete-time dynamics is discussed.

The difficulty here is that the uniform sampling functor is *not* invertible for these objects. To overcome this difficulty, note that the Shannon sampling theorem (Oppenheim and Schafer, 1975, pp. 26-30) gives conditions under which a continuous-time signal may be reconstructed from its discrete-time samples. In particular, a signal $u(t)$ is said to be *band-limited* (or more precisely, W-bandlimited) if its Fourier transform vanishes outside a specified frequency band, that is if

$$\int_{-\infty}^{\infty} u(t)e^{-i\omega t}dt = 0 \tag{7.135}$$

for $|\omega| > W$. The usual statement of the Shannon sampling theorem is that the function $u(t)$ may be reconstructed from the uniformly sampled sequence $\{u(k)\}$, provided $T < \pi/W$. Under these conditions, $u(t)$ is given explicitly as:

$$u(t) = \sum_{k=-\infty}^{\infty} u(k) \frac{\sin[(\pi/T)(t-kT)]}{(\pi/T)(t-kT)}. \tag{7.136}$$

Note that when $t = kT$, this expression may be evaluated by L'Hospital's rule to obtain the interpolation result that $u(t) = u(kT)$ at $t = kT$.

Motivated by this result, define the discrete category $\mathbf{C}^W$ of continuous-time signals $u(t)$ that satisfy the band limit condition (7.135) and define $\mathbf{D}^T$ as the discrete category of sequences $\{u(k)\}$ such that the infinite sum in (7.136) converges. Note that these categories are subcategories of $\mathbf{C}$ and $\mathbf{D}$, respectively, and that $\mathcal{U}_T$ defines a functor between these subcategories. Further, since Eq. (7.135) defines the unique W-limited interpolation of the samples $\{u(k)\}$ if $T < \pi/W$, it follows that the inverse functor $\mathcal{U}_T^{-1} : \mathbf{D} \to \mathbf{C}$ also exists. Hence, the Shannon sampling theorem provides necessary and sufficient conditions for the discrete categories $\mathbf{C}^W$ and $\mathbf{D}^T$ to be isomorphic.

This result is important for a number of reasons. First and foremost, the restriction that $u(t)$ be band-limited is fairly severe. In particular, it is a standard result that a function cannot be both band-limited and time-limited; in addition, many of the "standard" signals commonly considered in practice that are not time-limited (e.g., steps, exponential decays, etc.) are not band-limited, either. Hence, even in categories that include no nontrivial models, isomorphism imposes some significant restrictions on the objects on which the categories are based. Next, consider the question of what other models could be included in useful extensions of these categories. In particular, note that since linear models are incapable of generating superharmonics, any linear time-invariant model will transform a band-limited input signal $u(t)$ into a band-limited output signal $y(t)$. Conversely, nonlinear models generally do generate superharmonics, implying that the objects $\mathcal{O}^W$ on which the discrete category $\mathbf{C}^W$ is based do *not* constitute a basis for a category of continuous-time nonlinear models. As a specific example, consider the continuous-time analog of the category $\mathbf{LN}_\infty$ of block-oriented discrete-time dynamic models introduced in Sec. 7.4.3. Defined over the class of arbitrary real-valued signals $x(t)$, these models define a category, but they do not define a category over the class $\mathcal{O}^W$: if $u(t) \in \mathcal{O}^W$, it follows only that $u^n(t) \in \mathcal{O}^{nW}$. Since even the discrete category over arbitrary real-valued signals is not isomorphic to its discrete-time counterpart, it follows that the continuous-time and discrete-time versions of the category $\mathbf{LN}_\infty$ are not isomorphic. This observation illustrates the fact that nonlinearity introduces some additional complications into the relationship between continuous-time and discrete-time models, a point discussed further in both Chapters 1 and 8.

7.9 Homomorphic systems

The categories $\mathbf{Hom}_\phi$ form an interesting family that has been examined from various perspectives throughout this book. The following discussions present two additional characterizations of this family of categories and one extremely useful extension that provides the basis for constructing some interesting new nonlinear model categories. Specifically, Sec. 7.9.1 shows that each of these categories is isomorphic to the linear model category $\mathbf{ARMA}$ and briefly considers some of the consequences of this isomorphism. Next, Sec. 7.9.2 shows that the family of these models is not large enough to include the category $\mathbf{H}$ of homogeneous models. Finally, Sec. 7.9.3 extends the isomorphism between $\mathbf{Hom}_\phi$ and $\mathbf{ARMA}$ to define a new family of isomorphisms that take any subcategory of $\mathbf{DTDM}$ into another, possibly new, subcategory of $\mathbf{DTDM}$.

7.9.1 Relation to linear models

To see that $\mathbf{Hom}_\phi$ is isomorphic to the category $\mathbf{ARMA}$ for any fixed choice of ϕ, consider the functor $\mathcal{H}_\phi : \mathbf{Hom}_\phi \to \mathbf{ARMA}$ defined as follows. Each object X in $\mathbf{Hom}_\phi$ is mapped into the object $\phi[X]$ in $\mathbf{ARMA}$ and each morphism f in the set $\mathrm{hom}(X, Y)$ in $\mathbf{Hom}_\phi$ is mapped into the ARMA model L on which it is based. Written in terms of the cascade composition operator, this transformation may be represented as

$$f = \phi^{-1} \circ L \circ \phi \quad \Rightarrow \quad L = \phi \circ f \circ \phi^{-1}. \tag{7.137}$$

Conversely, the inverse functor $\mathcal{H}^{-1} : \mathbf{ARMA} \to \mathbf{Hom}_\phi$ is obtained by mapping each object A in $\mathbf{ARMA}$ into the object $\phi^{-1}[A]$ in $\mathbf{Hom}_\phi$ and each morphism in $\mathbf{ARMA}$ into the corresponding homomorphic system $\phi^{-1} \circ L \circ \phi$ in $\mathbf{Hom}_\phi$.

It was noted in Sec. 7.7.2 that any functor from a category $\mathbf{X}$ of nonlinear models to a category of linear models must preserve the commutativity of composition of morphisms in the linear category. To see that this requirement holds here, consider any two morphisms f_1 and f_2 in $\mathbf{Hom}_\phi$ and define $L_1 = \mathcal{H}_\phi f_1$ and $L_2 = \mathcal{H}_\phi f_2$. It follows that

$$\begin{aligned} f_1 \circ f_2 = \mathcal{H}_\phi^{-1}[L_1] \circ \mathcal{H}_\phi^{-1}[L_2] &= \mathcal{H}_\phi^{-1}[L_1 \circ L_2] \\ &= \mathcal{H}_\phi^{-1}[L_2 \circ L_1] = \mathcal{H}_\phi^{-1}[L_2] \circ \mathcal{H}_\phi^{-1}[L_1] = f_2 \circ f_1. \end{aligned} \tag{7.138}$$

In other words, since $\mathbf{Hom}_\phi$ is isomorphic to $\mathbf{ARMA}$ and composition of morphisms is commutative in $\mathbf{ARMA}$, it is necessarily also commutative in $\mathbf{Hom}_\phi$. The mechanics of this commutativity may be seen by noting that in the cascade connection of any two homomorphic models based on the function $\phi[\cdot]$, the output nonlinearity $\phi^{-1}[\cdot]$ of the first system is cancelled by the input nonlinearity of the second system. Note the marked contrast in this respect between the category $\mathbf{Hom}_\phi$ and the larger category $\mathbf{LN}_\infty$: the inequivalence of Hammerstein and Wiener models demonstrates the non-commutativity of $\mathbf{LN}_\infty$, which therefore cannot be isomorphic to any linear model category, nor to any category like $\mathbf{Hom}_\phi$ that is isomorphic to a linear category.

7.9.2 Relation to homogeneous models

The use of homomorphic models to construct homogeneous models was demonstrated in Chapter 3 and the joint subcategory of $\mathbf{Hom}_\phi$ and $\mathbf{H}$ was considered in some detail in Sec. 7.6.4. The following discussion proves the assertion made earlier that not all homogeneous models have homomorphic representations. In fact, two results are demonstrated by counterexample: first, that the category $\mathbf{H}$ of homogeneous models is not isomorphic to any category $\mathbf{Hom}_\phi$ for *fixed* $\phi[\cdot]$ and second, the stronger result that there exist homogeneous models that have no $\mathbf{Hom}_\phi$ representation for *any* $\phi[\cdot]$.

To establish the first result, it is enough to demonstrate that the category $\mathbf{H}$ is not commutative with respect to composition of morphisms. A simple example that illustrates this point is the cascade connection of the homogeneous models H_1

$$y(k) = \frac{u(k)u(k-1)}{u(k) + u(k-1)} \tag{7.139}$$

and H_2

$$y(k) = [u^3(k) + u^3(k-1)]^{1/3}. \tag{7.140}$$

Direct algebraic manipulation leads to the composite models $H_1 \circ H_2$

$$z(k) = \left[\left(\frac{u(k)u(k-1)}{u(k) + u(k-1)}\right)^3 + \left(\frac{u(k-1)u(k-2)}{u(k-1) + u(k-2)}\right)^3\right]^{1/3} \tag{7.141}$$

and $H_2 \circ H_1$

$$z(k) = \frac{[u^3(k) + u^3(k-1)]^{1/3}[u^3(k-1) + u^3(k-2)]^{1/3}}{[u^3(k) + u^3(k-1)]^{1/3} + [u^3(k-1) + u^3(k-2)]^{1/3}}. \tag{7.142}$$

Since these composite models are different, it follows that the category $\mathbf{H}$ is not commutative with respect to composition and therefore not isomorphic to $\mathbf{Hom}_\phi$ for any fixed $\phi[\cdot]$.

Conversely, note that model H_1 represents a morphism in $\mathbf{Hom}_\phi$ for $\phi[x] = 1/x$ whereas H_2 represents a homomorphic model for $\phi[x] = x^3$. Consequently, both of these models and their cascade connections represent morphisms in the category $\mathbf{LN}_\infty$ although this is not obvious on inspection, particularly for the composite models. It follows that neither of the cascade models 7.141 or 7.142 belong to $\mathbf{Hom}_\phi$ for any $\phi[\cdot]$, establishing the stronger result that, not only is $\mathbf{H}$ not isomorphic to any member of the family $\mathbf{Hom}_\phi$, the class of morphisms from the entire family is not large enough to contain all of the morphisms in $\mathbf{H}$. In particular, note that both of these cascade models exhibit the general structure $M = N_1 \circ L \circ N_2 \circ L \circ N_3$ where L is the simple moving average linear model $y(k) = u(k) + u(k-1)$ and N_1, N_2 and N_3 are static nonlinearities. For the model $H_1 \circ H_2$, these nonlinearities are $N_1[x] = x^3$, $N_2[x] = x^{-1/3}$, and $N_3[x] = x^{-1}$, whereas for the model $H_2 \circ H_1$, these nonlinearities are $N_1[x] = x^{-1}$, $N_2[x] = x^{-3}$ and $N_3[x] = x^{1/3}$.

Finally, it is interesting to consider one more example to establish the still stronger result that not all homogeneous models exhibit the $\mathbf{LN}_\infty$ structure of these last two examples. Specifically, recall that the median filter is a homogeneous model whose impulse response is identically zero and consider the category $\mathbf{LN}_\infty^0$ constructed as follows. The objects on which this category is based are arbitrary real-valued sequences and the morphisms are arbitrary cascade connections of linear dynamic models L and static nonlinearities N that satisfy two additional conditions: they are invertible, and $N[0] = 0$. Note that this category is a subcategory of $\mathbf{LN}_\infty$ and it includes $\mathbf{Hom}_\phi$ as a subcategory for any invertible function satisfying $\phi[0] = 0$. Now, consider any morphism f in $\mathbf{LN}_\infty^0$ that exhibits an identically zero impulse response. Specifically, note that this morphism may be written as

$$f = L_n \circ N_n \circ \cdots \circ L_1 \circ N_1. \tag{7.143}$$

Since $y(k) \equiv 0$ for this model, it follows that the input sequence to the linear model component L_n is identically zero, further implying the input to the nonlinear model component N_n is also identically zero. This chain of reasoning extends all the way to the input of the nonlinearity N_1, implying the $u(k) \equiv 0$. Hence, it follows that no morphism in the category $\mathbf{LN}_\infty^0$ can exhibit the impulse response behavior of the median filter. Since the median filter is homogeneous, it follows that not all homogeneous models exhibit $\mathbf{LN}_\infty^0$ representations.

7.9.3 Constructing new model categories

The isomorphism $\mathcal{H}_\phi$ between $\mathbf{Hom}_\phi$ and $\mathbf{ARMA}$ is a natural consequence of the homomorphic system structure on which the category $\mathbf{Hom}_\phi$ is based. The following discussion extends this isomorphism to a very general procedure for generating new categories of nonlinear discrete-time dynamic models and briefly discusses one of the important consequences of this procedure.

Suppose $\mathbf{C}$ is any subcategory of the category $\mathbf{DTDM}$ defined in Sec. 7.3, based on the class $\mathcal{O}_C$ of objects. Next, define $\mathbf{S}_C$ as the full subcategory of $\mathbf{DTDM}$ defined by $\mathcal{O}_C$. Note that either $\mathbf{C} = \mathbf{S}_C$ (that is, these categories are the same if $\mathbf{C}$ is a full subcategory of $\mathbf{DTDM}$) or $\mathbf{C}$ is a subcategory of $\mathbf{S}_C$. The primary points are first, that if ψ is a morphism in $\mathbf{S}_C$, then $\psi[X]$ is well-defined for all objects X in $\mathbf{C}$ and second, that the most interesting results arise when

Relations between Model Classes 361

the categories **C** and $\mathbf{S}_C$ are different. Now, suppose the morphism $\psi : A \to B$ is invertible: there exists a morphism ψ^{-1} in $\mathbf{S}_C$ such that $\psi^{-1} \circ \psi = id_A$ and $\psi \circ \psi^{-1} = id_B$.

From **C**, construct an isomorphic category **D** as follows. Objects X in **D** are the inverse image of objects A in **C** under $\psi[\cdot]$, that is

$$\mathcal{O}_D = \{\psi^{-1}[A] \mid A \in \mathcal{O}_C\}. \tag{7.144}$$

Similarly, morphisms g in **D** are defined by the similarity transformation induced by ψ on the morphisms f of **C**, that is

$$g = \psi^{-1} \circ f \circ \psi. \tag{7.145}$$

As in the original category **C**, composition of morphisms in the category **D** is defined by cascade connection, and identity morphisms are defined by the identity system $y(k) = u(k)$. For convenience, denote the isomorphism from **C** to **D** as $\mathcal{F}_\psi$ and the inverse by $\mathcal{F}_\psi^{-1}$.

The simplest example of this isomorphism is the functor $\mathcal{H}_\phi^{-1} : \mathbf{ARMA} \to \mathbf{Hom}_\phi$ introduced in Sec. 7.9.1, but innumberable other possibilities exist, some more interesting than others. For example, suppose ψ is any invertible static nonlinearity and consider the influence of $\mathcal{F}_\psi$ on the category $\mathbf{LN}_\infty$. Since any morphism in $\mathbf{LN}_\infty$ may be represented as in Eq. (7.143), it follows immediately that morphisms in the category $\mathbf{D} = \mathcal{F}_\psi \mathbf{LN}_\infty$ are necessarily of the form

$$\begin{aligned} g &= \psi^{-1} \circ L_n \circ N_n \circ \cdots \circ L_1 \circ N_1 \circ \psi \\ &= id_Y \circ \psi^{-1} \circ \cdots \circ L_1 \circ [N_1 \circ \psi]. \end{aligned} \tag{7.146}$$

In other words, the resulting morphism is again simply a cascade of linear and nonlinear models, implying $\mathbf{D} = \mathbf{LN}_\infty$. In fact, this result illustrates the point made earlier that the most interesting results are generally obtained when the categories **C** and $\mathbf{S}_C$ are not the same. Specifically, note that if ψ is a morphism in **C**, then the new morphism g is necessarily also a morphism in **C** and the isomorphism $\mathcal{F}_\psi$ simply takes **C** into itself, as in this example. Conversely, the results of this procedure can be both much more interesting than this example and somewhat counterintuitive at first glance. The next three examples illustrate these points.

For the first example, begin by defining the category **N** of static nonlinearities: the objects on which this category is based is the class $\mathcal{O}$ on which the **DTDM** category is based, and the morphisms consist of static nonlinearities $y(k) = \eta(u(k))$. Next, let ψ be any stable, minimum phase $ARMA(p,q)$ model in **DTDM**, from which it follows that ψ^{-1} is also a stable, minimum phase $ARMA(q,p)$ model. The functor $\mathcal{F}_\psi$ then takes the category **N** of static nonlinearities into the category $\mathbf{Sand}_\psi$ of *sandwich models* defined in Chapter 5, based on the cascade connection of first, the linear dynamic model ψ, then the static nonlinearity $\eta(\cdot)$ and finally, the inverse linear dynamic model ψ^{-1}. Note that this model structure is precisely the dual of the homomorphic model structure on which the category $\mathbf{Hom}_\phi$ is based.

The second example is more unusual and is based on the following construction. For the starting category **C** take the category **Median** introduced in Sec. 7.5.4 and take ψ the same as in the previous example. The resulting category **D** is isomorphic to **Median** but exhibits behavior that *appears* to be quite different, illustrating an important point. Specifically, consider the following special case. The morphism ψ is the first-order linear autoregressive model

$$y(k) = ay(k-1) + bu(k) \tag{7.147}$$

and the specific morphism f from **Median** considered here is the symmetric median filter with $q = 1$, generating the response

$$z(k) = \text{median}\{y(k-1), y(k), y(k+1)\}. \tag{7.148}$$

The corresponding morphism in **D** is the composition $\psi^{-1} \circ f \circ \psi$ whose output is given by

$$w(k) = \frac{z(k) - az(k-1)}{b}. \tag{7.149}$$

Overall, Eqs. (7.147) through (7.149) define a nonlinear digital filter that is isomorphic to the median filter but, as the following results demonstrate, not "equivalent" in terms of its apparent input/output behavior.

To see the profound differences between these isomorphic filters, consider the impulse response. It was noted several times that one of the characteristic features of the median filter is that its impulse response is identically zero, regardless of the amplitude of the impulse input. Further, the median filter is homogeneous and, since linear dynamic models are also homogeneous, the category **D** defined here is a subcategory of **H**. Hence, it is sufficient to consider the unit impulse response of this new filter to begin to see how it differs from the median filter. First, assume $b > 0$ and $0 < a < 1$ and note that the impulse response of the linear filter ψ is $y(k) = 0$ for $k \leq 0$ and $y(k) = a^k b$ for $k \geq 0$. From this result, it is not difficult to show that $z(k) = 0$ for $k < 0$, $z(0) = ab$, and $z(k) = y(k) = a^k b$ for $k > 0$, reflecting the fact that all monotone sequences are root sequences of the median filter. Since $y(k) = a^k b$ is the natural response of the forward filter ψ, it also corresponds to the zero dynamics of the inverse filter ψ^{-1}, implying $w(k) = 0$ for $k > 1$; also, note that $w(k) = 0$ for $k < 0$ by causality. Hence, the impulse response of the isomorphic median filter is identically zero except for $w(0) = a$ and $w(1) = a(1-a)$.

If $-1 < a < 0$, the character of the isomorphic median filter exhibits an interesting qualitative change that reflects the strong interaction between the individual components on which this model is based. Specifically, since $a < 0$, the impulse response of the linear filter is no longer monotonic. As a consequence, the output of the median filter at each time k turns out to be the value closest to zero in magnitude; specifically, $z(k) = 0$ for $k < 0$ and $z(k) = y(k+1) = a^{k+1}b$ for $k \geq 0$. Here again, this response corresponds to the zero dynamics of the linear inverse filter ψ^{-1} for $k \geq 1$, so the overall filter response is $w(k) = a$ for

$k = 0$ and $w(k) = 0$ otherwise. In summary, this example has considered three "equivalent" nonlinear digital filters: the median filter, the isomorphic filter for $a > 0$ and the isomorphic filter for $a < 0$. The impulse responses of these filters differ significantly, from identically zero for the median filter, to a response of duration 2 for $a > 0$ to a response of duration 1 for $a < 0$. Also, note that the qualitative nature of the step responses of these models illustrates other differences. In particular, for $a > 0$, the step response of the linear model ψ is monotonic and thus invariant to the action of the median filter. Consequently, the effect of the inverse linear model ψ^{-1} is to recreate the original step; in fact, since the linear model is a positive system for $a, b > 0$, it preserves the monotonicity of input sequences, which are unmodified by the median filter and subsequently restored by the inverse linear filter. In contrast, monotonicity is not preserved by the linear filter if $a < 0$, so the median filter modifies the transformed sequence and the inverse filter does not restore the original input sequence.

This example emphasizes the point that while isomorphisms preserve certain characteristics of objects and morphisms in the original category **C**, they generally do not preserve all "important" characteristics because importance is a context-sensitive concept. In this particular example, the isomorphism $\mathcal{F}_\psi$ preserves the homogneity of the models in the category **Median**, preserves monotonicity of response if $a > 0$, but does not preserve the median filter's characteristic insensitivity to impulsive noise for any $a \neq 0$. On reflection, none of these observations are really surprising, but neither are they entirely obvious from the outset.

Finally, as an even less obvious case consider the category **D** defined as follows. For the original category **C** take the category **NMAX** of nonlinear moving average models and for the morphism ψ take any fixed NARX model. Applying the functor $\mathcal{F}_\psi$ then yields the category **D** whose morphisms are nonlinear dynamic models consisting of the cascade connection of the NARX model, an arbitrary NMAX model, and finally the inverse of the original NARX model. Since this inverse NARX model is simply another NMAX model, it follows that this overall structure may be reduced to the cascade connection of a fixed NARX model followed by a single NMAX model that is not fixed, but which is not completely arbitrary, either. In cases where the dynamic flexibility of NMAX models is inadequate (e.g., input-dependent stability or subharmonic generation), composite structures of this type might offer significant modeling advantges since the overall input-output behavior of the composite model is partially determined by the qualitative behavior of the NARX model. Conversely, it should also be noted that although the input/output behavior of the NMAX model class may be fairly well understood, the nature of the transformation introduced by the NARX model may be fairly difficult to understand in detail. For example, conditions for preservation of monotonicity are easily determined for the first-order linear model considered in the previous example; similar conditions for a particular NARX model may be substantially more difficult to obtain.

7.10 Summary: the utility of category theory

Although category theory has sometimes been disparaged as "the abstraction of all abstractions" (Blyth, 1986, preface), in fact it is an extremely powerful, flexible framework for considering general mathematical questions. One of the primary objectives of this chapter has been to demonstrate the applicability of category theory to the problem of characterizing discrete-time dynamic models. In fact, many of the topics considered in Chapters 1 through 6 may be expressed simply in terms of objects and morphisms, and viewing them in these terms leads immediately to a number of interesting and important questions. The following discussion illustrates this point with a few selected examples.

Chapter 1 introduced six types of inherently nonlinear behavior, and all of these types of behavior may be described in terms of morphism sets in the category **DTDM** introduced in Sec. 7.3. For example, it was noted in Sec. 7.3.3 that both subharmonic generation and nonperiodic (e.g., chaotic) responses to period P input sequences correspond to morphisms in the sets $\hom(S^P, S^{nP})$ and $\hom(S^P, S^\infty)$, respectively. Both linear models and nonlinear models that generate superharmonics belong to the set $\hom(S^P, S^P)$, so a different choice of objects is required to distinguish these model classes. To approach this problem, define the category $\mathbf{P}^\omega$ of periodic responses of angular frequency ω. Specifically, the objects in this category are the sets $\mathcal{F}_r^\omega$ of real-valued sequences $\{x(k)\}$ with r-term Fourier series representations:

$$x(k) = x_0 + \sum_{n=1}^{r}[A_n \sin nk\omega + B_n \cos nk\omega]. \tag{7.150}$$

Morphisms in this category are arbitrary discrete-time dynamic models, and the simple question "what do these morphism sets look like?" leads immediately to the following classification:

$\hom(\mathcal{F}_0^\omega, \mathcal{F}_0^\omega)$: systems with constant steady-state responses
$\hom(\mathcal{F}_0^\omega, \mathcal{F}_1^\omega)$: sinusoidal oscillators
$\hom(\mathcal{F}_0^\omega, \mathcal{F}_r^\omega)$: complex waveform generators ($r > 1$)
$\hom(\mathcal{F}_r^\omega, \mathcal{F}_1^\omega)$: perfect rectifiers ($r \geq 1$)
$\hom(\mathcal{F}_r^\omega, \mathcal{F}_r^\omega)$: linear models plus what else?
$\hom(\mathcal{F}_r^\omega, \mathcal{F}_{r'}^\omega)$: nonlinear models generating superharmonics for $r' > r$
$\hom(\mathcal{F}_r^\omega, \mathcal{F}_{r'}^\omega)$: ideal linear filters plus what other models for $r' < r$?

Given this classification, it is interesting to ask what types of nonlinear models are possible in sets like $\hom(\mathcal{F}_r^\omega, \mathcal{F}_r^\omega)$, particularly for $r = 1$, where these models generate neither subharmonics nor superharmonics in response to arbitrary sinusoids of frequency ω. It is clear that these models are not morphisms in the category $\mathbf{LN}_\infty$ since all nonlinear models in this category generate superharmonics, at least for some excitation amplitude. Conversely, this behavior is possible in generalized Lur'e type structures consisting of a linear dynamic element in the forward path of the model and a nonlinear *dynamic* feedback element (e.g., an NMAX model) that exhibits zero response to sinusoids of angular frequency ω. Given their nonlinear autoregressive character, such models are likely to exhibit exotic behavior like input-dependent instability or subharmonic generation at some other frequency ω'. It is important to note, however, that since these models belong to $\hom(\mathcal{F}_r^\omega, \mathcal{F}_r^\omega)$, they cannot be distinguished from linear models on the basis of sinusoidal inputs of frequency ω.

Within a restricted model class, these same questions can lead to very different answers. For example, consider the question, "what do the morphism sets look like in **Volt**$^\omega$?" This category is defined as the full subcategory of **Volt** based on the objects $\mathcal{F}_r^\omega$ defined in Eq. (7.150). These morphism sets yield the following classification:

$\hom(\mathcal{F}_0^\omega, \mathcal{F}_0^\omega)$: includes all $V_{(N,M)}$ models
$\hom(\mathcal{F}_0^\omega, \mathcal{F}_r^\omega)$: empty for $r \geq 1$
$\hom(\mathcal{F}_r^\omega, \mathcal{F}_0^\omega)$: perfect rectifiers for $r \geq 1$
$\hom(\mathcal{F}_r^\omega, \mathcal{F}_r^\omega)$: linear $V_{(1,M)}$ models only
$\hom(\mathcal{F}_r^\omega, \mathcal{F}_{r'}^\omega)$: nonlinear $V_{(N,M)}$ models only, $r' = Nr$
$\hom(\mathcal{F}_r^\omega, \mathcal{F}_{r'}^\omega)$: ideal linear filters and nonlinear extensions $r' < r$.

In this last morphism set, note that the unweighted average of a sinusoid over a complete period $P = 2\pi/\omega$ yields an identically zero response and can therefore reduce the number of harmonics present in a signal. Similarly, note that arbitrary $W_{(N,M)}$ models built from this linear filter also exhibit zero response to any period P input, so long as the constant term in the polynomial is zero. Hammerstein models based on odd-order polynomial nonlinearities and these unweighted linear moving averages will also exhibit zero response. In fact, this observation may be used to develop perfect rectifiers, as follows. Suppose the period P is twice the window width and consider the response of the $H_{(2,P/2)}$ Hammerstein model constructed from the linear unweighted average and the quadratic nonlinearity $f(x) = x^2$. This nonlinearity converts any sequence in $\mathcal{F}_1^\omega$ into the sum of a constant term and a second harmonic term with period $P/2$; the unweighted average filter eliminates this harmonic term completely, yielding a constant output. The point of this example is that simple questions like "what morphisms are in the set $\hom(A, B)$?" can lead quickly to a variety of specific model characterizations.

A particular advantage of category theory as a framework for discussing discrete-time dynamic models is that it forces the simultaneous consideration of both the models and their associated input sequences. Consequently, many important issues in empirical modeling seem particularly well suited to the category-theoretic formulation. For example, it was noted that initial objects in IO subcategories represent good input sequences for model identification in the sense that they uniquely characterize all models in the category. Important examples in linear model categories include both impulse and step responses, the first two of the four standard input sequences introduced in Chapter 2. Even better, the Gaussian white noise sequence—the fourth of these standard input sequences—represents a zero object in the category **Gauss** of linear models introduced in Sec. 7.5.3, implying it is both a good input sequence and a good response sequence (i.e., observing this response to any input sequence also uniquely defines the linear model relating these sequences). In fact, this category is rare because it is both thin and connected, implying first that *every* object in the category is a zero object and second, that every model defining a morphism in this category is invertible. These observations immediately suggest some related questions. For example, are there other thin, connected categories like **Gauss** based on stochastic input sequences? It follows immediately from the results presented in Sec. 7.9 that the answer to this question is yes: any isomorphism $\mathcal{F}_\phi$ based on an invertible static nonlinearity $\phi[\cdot]$ yields a second category that is, like **Gauss**, both thin and connected. In fact, the result is a category of nonlinear homomorphic systems and non-Gaussian input sequences; choosing certain functions $\phi[\cdot]$ leads to classes of non-Gaussian random variables that have been studied in the statistics literature. As a specific example, taking $\phi[x] = e^x$ leads to a category of *lognormal* random variables (Johnson et al., 1994, ch. 14), although with a dependence structure that is, unlike the Gaussian case, not completely determined by the autocorrelation function. Another interesting question posed by the category **Gauss** is whether there exist thin, connected categories based on *deterministic* input sequences. In fact, this question appears to be more interesting than the first one since, if such a category exists, it requires a precarious balance between objects and morphisms. For example, note that any thin, connected category of *linear* models must exclude sinusoidal input sequences of different frequencies since no linear model can alter the frequency of a sinusoid. Conversely, any category whose object class includes the set of sinusoidal sequences of frequency ω must necessarily exclude nontrivial linear models if it is thin since this set is invariant under linear models.

A less ambitious but still daunting variation on the previous question is the following: given a nonlinear model category $\mathbf{X}^{IO}$ based on real-valued sequences $\{x(k)\}$, what can be said about the existence of initial objects. In categories like $\mathbf{Hom}_\phi$ that are isomorphic to a linear model category, these input sequences are simply the image of the unit impulse sequence under the isomorphism, but this observation leads immediately to two related points. First, note that for very general isomorphisms like the NARX-based example considered in Sec. 7.9.3, the character of this transformed initial object may not be at all obvious. In particular, note that the input sequence that plays the role of the unit impulse in such a category corresponds to the impulse response of the inverse of the NARX model on which the homomorphism is based. Second, it would be even more interesting to understand what restrictions on morphisms (i.e., models) or objects (i.e., input sequences) are necessary for initial objects to exist in categories that are *not* isomorphic to any linear model category. One interesting result presented in this chapter is that initial objects cannot be root sequences of any of the models defining morphisms in this category.

Another useful feature of category theory is that it provides a common framework for defining both structural and behavioral model categories. In particular, note that all three of the behaviorally defined model classes introduced in Chapter 3 define behavioral model categories, as do the BIBO stability, asymptotic constancy, and monotonicity characterizations considered in Sec. 7.4.4. Sim-

Relations between Model Classes 367

ilarly, it is possible to define a category **O** of odd symmetry models $\mathcal{M}$ for which $\mathcal{M}[\{-u(k)\}] = -\mathcal{M}[\{u(k)\}]$. It is easy to show that this behavior is preserved under cascade connection and in fact, corresponds to a relaxation of homogeneity. Consequently, it follows that **H** is a subcategory of **O** and asymmetric response models are morphisms in **DTDM** that are not morphisms in the subcategory **O**. Important structural model categories defined here include the category **NMAX** of nonlinear moving average models, the category **NARX** of nonlinear autoregressive models, the category **Volt** of Volterra models, and the category **Hom**$_\phi$ of homomorphic models. In addition, the results presented in this chapter have led to explicit characterizations of certain model classes (e.g., the class of static-linear homomorphic models considered in Sec. 7.6.4) and to general construction procedures for new categories of discrete-time dynamic models, as discussed in Sec. 7.9.3.

It has been emphasized repeatedly that not all interesting nonlinear model classes define categories. Specific examples include the classes of bilinear models, polynomial NARMAX models, structurally additive NARMAX models, and NARX* models. Conversely, any *specific* model may be imbedded in a category by considering arbitrary cascades, as in the category **LN**$_\infty$ of "polymeric Hammerstein models" introduced in this chapter. In fact, defining such categories can lead to some interesting new model classes that generalize both the original model class and others that were not previously recognized as related. For example, note that the class **LN**$_\infty$ includes both the motivating Hammerstein model, its dual Wiener structure, and the categories **Hom**$_\phi$ of homomorphic systems for arbitrary invertible nonlinearities $\phi[\cdot]$. It would be interesting to apply this approach to larger model classes like the structurally additive NARMAX models or the NARX* models. In fact, note that the factorization results presented in Chapter 4 for structurally additive models provides an immediate extension of the category LN_∞ that includes all structurally additive NARMAX models as morphisms and includes **LN**$_\infty$ as a subcategory. Specifically, recall that any structurally additive NARMAX model may be represented as the cascade connection of a structurally additive NMAX model followed by a structurally additive NARX model. Extending this construction to arbitrary cascades leads to a model class that is closed under cascade connection by construction, exactly as in the category **LN**$_\infty$. Further, since both linear ARMA models and static nonlinearities belong to the class of structurally additive NARMAX models, it follows that **LN**$_\infty$ is a subcategory of this new model category, designated **Add**$_\infty$ to emphasize the analogy with **LN**$_\infty$. Since the structurally additive NARMAX class was originally proposed on the basis of empirical model identification considerations, it is logical to return to these considerations and pose questions like, "are there initial objects in this new model category?" Since both impulses and steps represent initial objects in linear model categories, it follows that at least two initial objects exist in the isomorphic category **Hom**$_\phi$. Further, because **ARMA** and **Hom**$_\phi$ are both subcategories of **Add**$_\infty$, these results make it clear that initial objects exist at least in some useful subcategories of this additive model category. This observation leads in turn to the question of what other interesting subcategories exist within **Add**$_\infty$. One interesting subcategory of **Add**$_\infty$ is the category **TMAX**$_\infty$, based on arbitrary cascades of the TMAX model introduced in Chapter 3:

$$y(k) = \sum_{i=0}^{q} b_i u(k-i) + \sum_{i=0}^{q} d_i |u(k-i)|. \qquad (7.151)$$

The cascade connection of two TMAX models may be written as

$$y(k) = \sum_{i=0}^{q}\sum_{j=0}^{p} a_j b_i u(k-i-j) + \sum_{i=0}^{q}\sum_{j=0}^{p} b_i c_j |u(k-i-j)|$$
$$+ \sum_{i=0}^{q} d_i \left| \sum_{j=0}^{p} a_j u(k-i-j) + \sum_{j=0}^{p} c_j |u(k-i-j)| \right|. \qquad (7.152)$$

It follows from this result that the class of TMAX models is not closed under cascade composition, but the following observations are extremely interesting. First, note that like the individual TMAX model components on which it is based, this cascade model belongs to the class of linear multimodels discussed in Chapter 6. Further, note that the resulting linear multimodel is input-selected like its TMAX components, from which it follows that this model can exhibit input multiplicity but not output multiplicity and, because it belongs to the NMAX class, this model is necessarily BIBO stable and preserves asymptotic constancy. In addition, since TMAX models are positive-homogeneous, it follows that the cascade connection is also positive-homogeneous, thus establishing that **TMAX**$_\infty$ is a subcategory of **PH**. Another interesting subcategory of **Add**$_\infty$ is the category **TARX**$_\infty$ of cascade TARX models. Like **TMAX**$_\infty$, the morphisms in this category correspond to positive-homogeneous linear multimodels. Here, however, these models are output-selected so they can exhibit output multiplicities, input-dependent stability, chaotic impulse and step responses, subharmonic generation, and various other forms of exotic behavior. Returning to the question of empirical modeling, it would be particularly interesting to explore the existence of initial objects in these categories to facilitate the design of input sequences for the identification of linear multimodels from input/output data.

In summary, the results and examples presented here have attempted to illustrate both the broad applicability of category theory and the ease with which important system-theoretic questions may be posed in category-theoretic terms. Specific topics considered in this chapter include discretization of continuous-time models, linearization of nonlinear models, empirical model identification, model inverses, zero dynamics, equivalent realizations, and procedures for constructing new model categories from known examples. Given the general difficulty of analyzing the qualitative behavior of nonlinear dynamic models, any analytical tool this flexible would seem to merit further examination.

Chapter 8

The Art of Model Development

The primary objective of this book has been to present a reasonably broad overview of the different classes of discrete-time dynamic models that have been proposed for empirical modeling, particularly in the process control literature. In its simplest form, the empirical modeling process consists of the following four steps:

1. Select a class $\mathcal{C}$ of model structures
2. Generate input/output data from the physical process $\mathcal{P}$
3. Determine the model $\mathcal{M} \in \mathcal{C}$ that best fits this dataset
4. Assess the general validity of the model $\mathcal{M}$.

The objective of this final chapter is to briefly examine these four modeling steps, with particular emphasis on the first since the choice of the model class $\mathcal{C}$ ultimately determines the utility of the empirical model, both with respect to the application (e.g., the difficulty of solving the resulting model-based control problem) and with respect to fidelity of approximation. Some of the basic issues of model structure selection are introduced in Sec. 8.1 and a more detailed treatment is given in Sec. 8.3, emphasizing connections with results presented in earlier chapters; in addition, the problem of model structure selection is an important component of the case studies presented in Secs. 8.2 and 8.5. The second step in this procedure—input sequence design—is discussed in some detail in Sec. 8.4 and is an important component of the second case study (Sec. 8.5). The literature associated with the parameter estimation problem—the third step in the empirical modeling process—is much too large to attempt to survey here, but a brief summary of some representative results is given in Sec. 8.1.1. Finally, the task of model validation often depends strongly on the details of the physical system being modelled and the ultimate application intended for the model. Consequently, detailed treatment of this topic also lies beyond the scope of this book but again, some representative results are discussed briefly in

Sec. 8.1.3 and illustrated in the first case study (Sec. 8.2). Finally, Sec. 8.6 concludes both the chapter and the book with some philosophical observations on the problem of developing moderate-complexity, discrete-time dynamic models to approximate the behavior of high-complexity, continuous-time physical systems.

8.1 The model development process

In practice, the empirical model development process is not as simple as the four-step procedure presented in the previous section might suggest. Because it is an art, there is no single "correct" way to approach model development, but it is important to be both systematic and thorough and it is often useful to approach the problem iteratively. For example, one possibility is to begin with a preliminary exploration phase that attempts to capture gross qualitative features and then to focus on one or a few model classes that appear promising enough for further investigation. These model classes may then be investigated more thoroughly, obtaining representative examples from each class that best fit input/output data. This distinction between preliminary exploration and model refinement is useful because the analytical procedures used in these two phases are typically different. For example, the preliminary modeling phase might consist of the following sequence of steps:

P1. Select a set C_i of *candidate model classes*
P2. Select a set $\mathcal{I}_k$ of *screening inputs*
P3. Generate input/output datasets for the process $\mathcal{P}$
P4. Assess the compatibility of these datasets with each model class.

Initially, C_i might represent a few very large classes like the NMAX and the NARX classes, and the screening inputs might be a small set of fixed amplitude steps or fixed frequency sinusoids. Preliminary model structure selection could then be based on the general nature of the responses observed: subharmonic or nonperiodic responses to sinusoids would dictate the need for a NARX model, whereas observation of only (super)harmonic responses would suggest that the simpler NMAX class could be investigated further. Similarly, if the screening inputs were a sequence of step inputs observed in historical data records, the candidate model classes might be the Hammerstein, Wiener, and Lur'e models. Observation of amplitude-dependent apparent time constants in these step responses would then argue against the Hammerstein and Wiener models and in favor of the Lur'e model or some other structure involving nonlinear autoregressive terms. In either case, the objective of this preliminary analysis would be to take whatever information was either available *a priori* or could be obtained with relative ease (e.g., analysis of historical data records) and use it to select one or more model classes that might be suitable for more careful examination.

Given such an initial model class $\mathcal{C}$, the model refinement procedure typcially proceeds along the following lines. First, input sequences are chosen on the basis of their effectiveness in distinguishing different models within the class

The Art of Model Development 371

$\mathcal{C}$. These requirements are generally not the same as those for a good screening input sequence, a point illustrated in Sec. 8.4.1. One or more input/output datasets are then collected from the physical process $\mathcal{P}$ on the basis of these input sequences; this task is often somewhat invasive, requiring disruption of normal operation for the purposes of obtaining responses that are sufficiently rich to provide good model parameter estimates. These process responses are then used to select the model from class $\mathcal{C}$ that best match them according to some quantitative goodness-of-fit criterion. As a practical matter, data pretreatment is extremely important whenever direct empirical modeling is undertaken, since physical systems are prone to *outliers, unmeasurable disturbances*, or other *artifacts* that can completely invalidate empirical modeling results; Sec. 8.1.2 introduces this topic, illustrating its practical importance and providing references to more detailed treatments. Once a model has been obtained that best fits the available data, it is important to assess its validity in terms of other measures, such as agreement with other datasets (cross-validation) or agreement with important behavioral criteria (for example, open-loop stability or monotonicity of step responses); these ideas are discussed further in Sec. 8.1.3.

8.1.1 Parameter estimation

As noted in the introduction, the literature on parameter estimation is much too extensive to attempt to survey here. Instead, the following discussion presents a brief overview of some key ideas and representative results, citing some useful references for more detailed discussions. The traditional empirical model development process is summarized graphically in Fig. 8.1, which shows the following three components:

1. The physical process $\mathcal{P}$
2. A measured input/output dataset $\mathcal{D}$
3. A mathematical process model $\mathcal{M}$.

In a strictly empirical model development, the process $\mathcal{P}$ is represented entirely in terms of the measured dataset $\mathcal{D}$. The mathematical process model $\mathcal{M}$ is obtained by optimizing some *goodness-of-fit measure* between the input/output responses predicted by the model $\mathcal{M}$ and those actually observed in the dataset $\mathcal{D}$. That is, given a dataset $\mathcal{D}$ of input/output data pairs $(u(k), y(k))$ and a candidate model $\mathcal{M}$, a second dataset $\hat{\mathcal{D}}$ of predicted input/output data pairs $(u(k), \hat{y}(k))$ is generated. From these two datasets, a sequence of prediction errors $e(k) = \hat{y}(k) - y(k)$ is computed, and some measure of the "size" of this sequence is taken as a quantitative measure of the goodness-of-fit between the model $\mathcal{M}$ and the dataset $\mathcal{D}$.

One of the most popular criteria is the mean square prediction error

$$MSE = \frac{1}{N} \sum_{k=1}^{N} e^2(k). \tag{8.1}$$

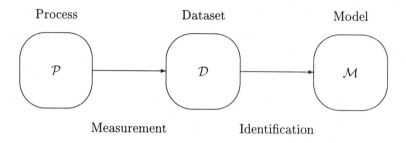

Figure 8.1: The traditional empirical modeling sequence

This quantity gives a measure of the overall magnitude of the error sequence $\{e(k)\}$ that is nonnegative for all sequences and is zero if and only if $e(k)$ is identically zero for all k. As noted in Chapter 4, empirical dynamic models are most often developed using parametric procedures; there, individual members of the model class $\mathcal{C}$ are uniquely defined by a vector $\theta \in R^p$ of p model parameters. Consequently, in parametric modeling, the error sequence $\{e(k)\}$ associated with each model is determined by both the input sequence $\{u(k)\}$ from which the process and model responses are generated, and the parameter vector θ. A "best fit" model obtained by minimizing the MSE criterion is generally called a *least squares* model, an approach that dates back to the work of Gauss and Legendre at the beginning of the nineteenth century (Rousseeuw and Leroy, 1987).

The least squares approach is popular in part because if $e(k)$ depends *linearly* on the model parameters θ_i, minimizing the mean square prediction error leads to a set of simultaneous linear equations to be solved for the least squares model parameters. Another possible approach is the method of *least absolute deviations*, obtained by minimizing the mean (absolute) deviation

$$MD = \frac{1}{N} \sum_{k=1}^{N} |e(k)|. \tag{8.2}$$

In fact, this idea is even older than least squares, dating back to Gallileo in the first half of the seventeenth century, and the basic procedure was studied by Boscovitch around the middle of the eighteenth century and by Laplace near the end of that century (Dodge, 1987). In contrast to the simultaneous linear equations obtained for least squares problems, if $e(k)$ depends linearly on the model parameter vectors θ_i, the method of least absolute deviations leads to simultaneous linear inequalities that may be solved by linear programming techniques (Bloomfield and Steiger, 1983; Dodge, 1987). In the terminology of functional analysis, these criteria are the ℓ_p-norms of the error sequence $\{e(k)\}$: MSE is the 2-norm $||e||_2$ and MD is the 1-norm $||e||_1$ (Wheeden and Zygmund, 1977, p. 130). Two important practical issues are first, that minimizing the ℓ_2 norm often leads to problems with unique solutions, whereas minimizing the ℓ_1 norm generally does not (Bloomfield and Steiger, 1983) and second, that the ℓ_1

norm is generally less sensitive to outliers than the ℓ_2 norm. The problem of outliers is important enough that it is revisited in Secs. 8.1.2 and 8.2.

Historically, the method of least squares has been the most popular basis for parametric model identification procedures, for two reasons. First, the computational and analytical advantages of least squares procedures are overwhelming relative to any other alternative. As a consequence, most of the parameter estimation algorithms described in the literature are based on minimization of some MSE criterion. Second, the limitations of least squares procedures with respect to outliers have not been widely recognized until fairly recently, although they were partially understood by the developers of the methods almost two centuries ago (Rousseeuw and Leroy, 1987). The following paragraphs present brief summaries of a few representative results, primarily based on least squares criteria. Following this discussion, Sec. 8.1.2 considers the question of how outliers can influence the results of these procedures and what can be done to overcome these and other similar problems in practice.

A useful summary of the basic issues, model parameter estimation algorithms, and various important practical aspects of *linear* model identification is given by Ljung (1987). This book provides a well-balanced coverage of both important theoretical issues like estimator bias and consistency and important practical issues like input sequence design and model validation. Further, many of the basic ideas presented in this book provide the basis for *nonlinear* empirical modeling procedures for specific model structures. Conversely, certain useful aspects of the linear model identification problem *do not* extend to most classes of nonlinear models, so such extensions should not be applied blindly. As a specific example, Ljung presents useful criteria for optimal input sequence design that reduce to the specification of an optimal power spectrum. However, specification of the spectrum does not define the *distribution* of the input sequence and the class of *pseudorandom binary sequences* (PRBS) is introduced as a way of satisfying practical restrictions on the input sequence (e.g., range limits and simplicity of generation) while still achieving the desired optimal spectral characteristics. In fact, more detailed treatments of the PRBS design problem are available (Godfrey, 1993), but it has been noted by various authors in conjunction with different model classes that PRBS sequences are *not* generally adequate for nonlinear model identification (Billings and Fakhouri, 1982; Leontaritis and Billings, 1987; Nowak and van Veen, 1994; Pearson et al., 1996).

Probably because of its structural simplicity, the Hammerstein model has been quite popular in practice and many different approaches to identifying these models have been described in the literature. One of the most popular (Narendra and Gallman, 1966) is based on the following two-step sequence: first, the static nonlinearity is fixed and the linear model parameters are estimated and second, the linear model parameters are fixed and the static nonlinearity is estimated; this procedure is repeated until convergence is achieved. It has been shown that this method can fail to converge, although it often seems to work well in practice (Stoica, 1981). Indeed, the Narendra-Gallman algorithm was one of three parameter estimation methods considered by Eskinat et al.

(1991), who found it gave the best results. Although most applications of the Narendra-Gallman algorithm assume the static nonlinearity is polynomial, the basic approach extends readily to any functional expansion of the form

$$g(x) = \sum_{i=1}^{m} \alpha_i \phi_i(x) \tag{8.3}$$

for arbitrary sets of known basis functions $\{\phi_i(x)\}$ (the polynomial case simply corresponds to $\phi_i(x) = x^i$). Alternatively, it has been noted that if the steady-state gain of the linear dynamic component of the Hammerstein model is constrained to be 1, the steady-state gain of the system is simply the static nonlinearity. There are applications in which the steady-state gain of the physical system is known *a priori*, and in these applications the Hammerstein model identification problem may be reduced to a constrained linear model identification problem (Pearson and Pottmann, 1999) and solved by the method of constrained least squares (Draper and Smith, 1998, p. 229). Conversely, nonparametric methods have also been developed for Hammerstein models (Greblicki and Pawlak, 1989), appropriate to cases where very little is known about the static nonlinearity and it is desirable to make the least restrictive assumptions possible.

Wiener model identification appears to be generally more difficult than Hammerstein model identification. For example, the Narendra-Gallman algorithm applied to Hammerstein model identification leads to a pair of linear least squares problems, but for Wiener model identification one of these problems becomes nonlinear in the parameters. In cases where the static nonlinearity is invertible (i.e., strictly monotonic) and the linear dynamics are minimum phase, note that the inverse of the Wiener model is a stable Hammerstein model. This observation has been used as the basis for identifying Wiener models, essentially applying Hammerstein model identification procedures to the "output-input" data $(y(k), u(k))$ (Pajunen, 1984, 1992; Greblicki, 1992). Alternatively, recursive prediction error methods have been developed for applications where these invertibility assumptions are too restrictive, both for the case where the static nonlinearity is unknown (Wigren, 1993) and the case where this nonlinearity is known (Wigren, 1994). If the static nonlinearity $g[\cdot]$ is both known and invertible, note that $u(k)$ and $g^{-1}[y(k)]$ are related by a linear model; a constrained weighted least squares algorithm for solving this model identification problem has also been proposed (Pearson and Pottmann, 1999). As in the case of Hammerstein models, nonparametric parameter estimation procedures have also been developed for Wiener models (Greblicki, 1992).

A particularly interesting algorithm has been proposed for the additive NARMAX model class (Bai, 1998)

$$y(k) = \sum_{i=1}^{p} \sum_{j=1}^{q} a_i d_j g_j[y(k-i)] + \sum_{i=1}^{n} \sum_{j=1}^{m} b_i c_j f_j[u(k-i)] \tag{8.4}$$

where the functions $f_j[\cdot]$ and $g_j[\cdot]$ are known. In fact, this assumption is not restrictive since it is a generalization of the Hammerstein model condition (8.3):

it simply means that consideration is restricted to parametric classes of nonlinearities that are linear in the parameters $\{c_j\}$ and $\{d_j\}$. Further, note that Bai's model class includes the Hammerstein model as a special case, obtained by taking $q = 1$ and $g_1(x) = x$. In addition, this model class also includes the first-order NARX model class discussed in Chapter 4 as a proper subset. Bai (1998) describes a detailed two-stage algorithm for estimating the unknown model parameters $\{a_i\}$, $\{b_i\}$, $\{c_j\}$, and $\{d_j\}$ and presents a detailed convergence analysis. Unfortunately, the block diagram presented in this paper is incorrect, suggesting both that the model class includes the non-additive Wiener model as a special case and that the model described by Eq. (8.4) shares the qualitative behavioral restrictions of the feedforward block-oriented models.

It was noted earlier that PRBS sequences are unsuitable as identification inputs for many classes of nonlinear models. One class of nonlinear models for which these sequences *are* adequate is the class of bilinear models. In particular, a noniterative correlation-based algorithm for estimating the parameters of a bilinear model from its responses to PRBS input sequences has been proposed and shown to be effective (Baheti et al., 1980). A more complex recursive least squares algorithm for bilinear model identification has also been proposed (Dai and Sinha, 1989); an advantage of this algorithm is that it incorporates some ideas from robust statistics to overcome the outlier-sensitivity of unmodified least squares procedures. Both of these algorithms initially assume the bilinear model is represented in the state-space form discussed in Chapter 3, but these algorithms ultimately exploit an input-output representation for these models and they should extend without difficulty to the $BL(p, q, P, Q)$ model class described in Chapter 3.

In contrast, one of the nonlinear model classes for which PRBS sequences are *not* suitable identification inputs is the class of Volterra models. Specifically, it has been shown (Nowak and van Veen, 1994) that the identification of Volterra model coefficients of nonlinear degree n generally requires input sequences that assume at least $n + 1$ distinct values. For second-order Volterra models, it is possible to exploit the fact that the dynamic fourth moments of a Gaussian sequence may be expressed in terms of products of second moments (i.e., autocorrelations) to obtain an explicit identification algorithm based on Gaussian input sequences (Koh and Powers, 1985). An extension of this result to the class of elliptically distributed random variables has also been given (Pearson et al., 1996). Alternatively, symmetrically-distributed non-Gaussian white noise sequences are also effective identification inputs, with two particular advantages: first, they lead to extremely simple expressions for parameter estimates and second, it is possible to say something about how the distribution of these sequences influences the parameter estimates (Pearson et al., 1996). Additional discussions of Volterra model identification with white noise inputs are available for both the second-order case (Cho and Powers, 1994) and the third-order case (Tseng and Powers, 1995).

One important practical disadvantage of Volterra models is their parametric complexity, as noted in Chapter 5. In fact, generally less severe but still significant difficulties of this kind arise in the development of polynomial NARMAX models. For example, taking $p = q = 2$ in the general polynomial NARMAX model requires 21 parameters to specify a quadratic model and 56 parameters to specify a cubic model, and the numbers increase rapidly as p, q or the polynomial model order n increase further. One solution to this problem is to identify restricted models, based on the observation that many of these parameters contribute little to the goodness-of-fit for the model and may therefore be set to zero with little degradation of fit and substantial simplification of the resulting model. The method of *stepwise regression* (Draper and Smith, 1998) is a practical regression method based on this idea, and one of the most popular methods of fitting polynomial NARMAX models is based on these ideas (Billings and Voon, 1986b).

8.1.2 Outliers, disturbances and data pretreatment

Despite their popularity, relative computational simplicity and analytical advantages over other methods, least squares procedures are generally quite sensitive to the presence of *outliers* or *anomalous data points* in the available data. Modeling procedures that are relatively insensitive to outliers are often called *robust* or *resistant* (Rousseeuw and Leroy, 1987) and are the sujbect of considerable research interest. The following discussion gives a very brief introduction to the topics of outliers and robust estimation, with three objectives: first, to emphasize the fact that outliers do occur in real data sequences, second to illustrate that these outliers can profoundly influence the results of an otherwise reasonable analysis and third, to give some idea of the practical precautions required in dealing with outlier-contaminated data. In addition, other types of artifacts also arise in real datasets and this section briefly considers one of these examples. *The key point of this section is that results computed from real-world measurement data should never be accepted uncritically.* More detailed introductions to the subject of outliers in data analysis and parameter estimation are available (Hampel, 1985; Huber, 1981; Rousseeuw and Leroy, 1987), as are treatments specific to *dynamic* data characterization in the presence of outliers (Martin and Thomson, 1982; Pearson, 1999; Poljak and Tsypkin, 1980).

Fig. 8.2 shows plots of four different sequences of industrial measurement data. The upper left plot shows 1024 samples of a pressure data sequence: all of the points lie in the range $71.7 \leq p(k) \leq 92.2$ *except* the single value $p(122) = 104.1$. This single value is visually inconsistent with the rest of the data sequence, lying approximately six standard deviations from the mean pressure value. Since data sequences are often *assumed* to be "nominally Gaussian," points lying further than about three standard deviations from the mean are often regarded as anomalies. In fact, this criterion is not effective in general—a point discussed further in the following paragraphs—but in this case, it does yield the correct result: the point in question is a "pressure spike" that, although valid, is not representative of the nominal pressure variations in the process. The lower left plot in Fig. 8.2 shows a sequence of about 2500 recycle flow rate measurements, approximately 80% of which lie in the interval 340 to 440 pounds per hour, corresponding to the nominal operating range of this pro-

The Art of Model Development

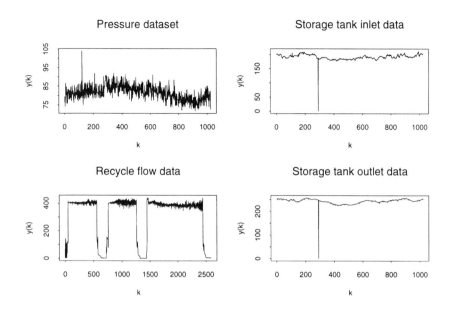

Figure 8.2: Four industrial process datasets

cess. The remaining 20% of the data corresponds to a sequence of partial process shut-downs; the point of this example is that anomalous data values—of whatever origin—can occur in clusters and need not constitute a "small" fraction of the total dataset. Finally, the two right-hand plots in Fig. 8.2 show physical property measurements obtained from on-line measurements at the inlet and outlet of a product storage tank. Both plots show a total of 1024 data points; for the inlet data, all but one data value lies in the range $177 \leq I(k) \leq 213$ and for the outlet data, all but one data value lies in the range $224 \leq O(k) \leq 257$. In both cases, the single anomalous value 0 occurs at $k = 291$, lying about 30 standard deviations below the mean. Here, a temporary measurement system failure occurred at time $k = 291$, causing a spurious value of 0 to be substituted for both missing data values. The key point of this example is that outliers need not occur independently in different variables, but they may be due to *common mode failures* or external disturbances that exert a simultaneous influence on several measurable responses.

This last example provides a nice illustration of the potentially devastating influence of outliers in dynamic data characterization (e.g., empirical modeling). One useful approach to characterizing the mean residence time and the degree of mixing in a storage tank is correlation analysis, based on the following observations. First, consider the simplified situation in which short-term physical property variations in the inlet stream are highly irregular and no mixing occurs in the storage tank. This nonmixing assumption is called *plug flow* and implies that $O(k) = I(k-d)$ where d is the transport delay through the storage tank. If the physical property variations are sufficiently irregular, it may be

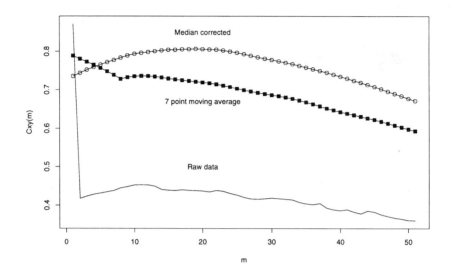

Figure 8.3: Storage tank cross-correlations

reasonable to approximate them by a white noise sequence, implying that $I(k)$ and $I(j)$ are uncorrelated for $k \neq j$. Under these assumptions, the normalized cross-correlation function

$$C_{IO}(m) = \frac{E\{[I(k) - \bar{I}][O(k) - \bar{O}]\}}{\sigma_I \sigma_O} \quad (8.5)$$

will exhibit a value of 1 for $m = d$ and 0 for $m \neq d$. Here, $\bar{I}$ and $\bar{O}$ are the average inlet and outlet values and σ_I and σ_O are their standard deviations. Mixing in the tank would result in broadening of this peak, so correlation analysis represents one potentially useful way of characterizing the performance of the storage tank. Specifically, correlation analysis yields both an estimate of the mean residence time $\bar{d}$ (the location of the maximum in $C_{IO}(m)$) and some measure of the mixing in the tank (the width of this peak).

Fig. 8.3 shows three different estimates of the normalized cross-correlation function computed from the inlet and outlet datasets shown in Fig. 8.2. The top curve in Fig. 8.3 (open circles) shows the estimate of the cross-correlation function obtained when the zero values in the inlet and outlet data sequences are replaced with the median values for these data sequences. This estimated correlation function exhibits a broad maximum at $k \simeq 18$, suggesting reasonable mixing in the tank and a residence time of about 18 hours. In this particular application, these results were consistent with expectations on the basis of other considerations. In contrast, the light line at the bottom of Fig. 8.3 represents the correlation function estimate obtained from the raw data. This estimated correlation function also has a clear physical interpretation: the storage tank

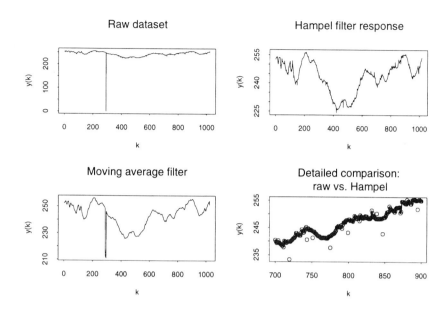

Figure 8.4: Comparison of data cleaning strategies

exhibits a very short residence time and essentially no mixing occurs, suggesting severe fouling and channeling in the tank. The key point here is that the difference in these interpretations derives from the single pair of simultaneous outliers in the original data at $k = 291$.

The third curve shown in Fig. 8.3 (solid squares) shows the results of applying a 7-point linear moving average filter to the inlet and outlet data sequences. Linear filtering is popular as a method of removing high-frequency noise from a data sequence, but the point of this example is to illustrate that linear filtering is generally ineffective in dealing with outliers in data sequences. In particular, suppose the data sequence $y(k) = x(k) + o(k)$ is observed where $x(k)$ is the sequence of interest and $o(k)$ is a single isolated outlier, of the form $o(k) = \gamma\delta(k - k_0)$. Applying a linear filter whose impulse response is $h(k)$ then yields the filtered response

$$f(k) = h(k) * x(k) + \gamma h(k - k_0). \tag{8.6}$$

To minimize the influence of the outlier, it is thus desirable to make $h(k)$ as small as possible for all k, but attempting to do this will result in distortion in the filtered signal term $h(k) * x(k)$. *Consequently, removal of outliers generally requires the use of nonlinear digital filters.*

This point is illustrated in Fig. 8.4, which presents four views of the storage tank outlet dataset. The upper left plot shows the raw data with the prominent outlier at $k = 291$ and the lower left plot shows the results of applying the 7-point linear unweighted average filter to this data sequence. Although this linear filter does attenuate the outlier, it also broadens it from a single anomalous data value to a patch of 7 successive anomalous values. This effect is responsible for the broadening of the spurious peak at $m = 0$ in the linearly smoothed autocorrelation estimate shown in Fig. 8.3. For comparison, the upper right plot in Fig. 8.4 shows the response of a simple nonlinear data cleaning filter (Pearson, 1999), called the *Hampel filter* because it is based on an outlier detection procedure that has been called the *Hampel identifier* (Davies and Gather, 1993). The lower right plot shows a detailed comparison of the response of this filter with a portion of the original data sequence that does not include the single missing data value at $k = 291$. Two points are evident from this plot: first, that most of the data values are left unmodified by this data cleaning procedure and second, that other outliers appear to be present in this data sequence (i.e., the open circles that appear relatively far from the curve defined by the majority of the 200 data points in this segment).

The basis of the Hampel filter is a robust extension of the "3σ edit rule" mentioned in connection with the pressure data sequence (the upper left plot in Fig. 8.2). Specifically, the 3σ edit rule is a popular but frequently ineffective procedure, consisting of the following steps:

1. Estimate the mean $\bar{x}$ of $\{x(k)\}$
2. Estimate the standard deviation $\hat{\sigma}$ of $\{x(k)\}$
3. Declare $x(k)$ anomalous if $|x(k) - \bar{x}| > 3\sigma$.

The basic difficulty with this procedure is that neither $\bar{x}$ nor $\hat{\sigma}$ are robust with respect to outliers in the dataset. As a specific example, if this procedure is applied to the recycle flow rate data shown in the bottom left plot in Fig. 8.2, no outliers are detected. This point is illustrated in Fig. 8.5, which shows the recycle flow rate data together with solid horizontal lines indicating the mean value ($\bar{x} \simeq 315$), the upper 3σ limit ($\bar{x} + 3\hat{\sigma} \simeq 781$) and the lower 3σ limit ($\bar{x} - 3\hat{\sigma} \simeq -150$). It is clear from this figure that all of the data values from nominal operation and the shutdown episodes fall within these limits; in particular, note that the negative lower limit corresponds to an unphysical flow reversal. The unreasonably wide range of these data validation limits is due to the fact that the standard variance estimate $\hat{\sigma}$ is badly inflated by the presence of the outliers in the dataset. In addition, note that the mean value is also shifted toward the outliers.

Fig. 8.5 also shows, as dashed lines, the model validation limits obtained by the Hampel identifier mentioned in the preceeding discussion. There, the basic idea is to replace the mean $\bar{x}$ and the standard deviation estimate $\hat{\sigma}$ in the 3σ edit rule with more robust measures of *location* and *scale*, respectively. The alternative location estimator used in the Hampel identifier is the median, defined as follows: first, the data sequence $\{x(k)\}$ is rank-ordered to obtain the sequence $\{x_{(j)}\}$ satisfying:

$$x_{(1)} \leq x_{(2)} \leq \cdots \leq x_{(N-1)} \leq x_{(N)}. \quad (8.7)$$

The median is simply the middle element of this list ($x^\dagger = x_{([N-1]/2)}$) if the sequence length N is odd, or the average of the middle two elements ($x^\dagger =$

The Art of Model Development   381

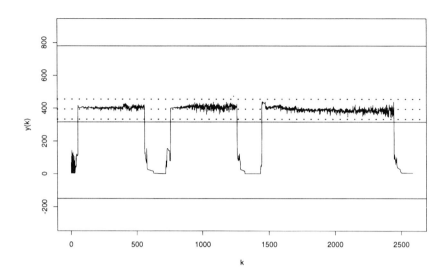

Figure 8.5: Ineffectiveness of the 3σ edit rule

$[x_{(N/2)} + x_{([N/2]+1)}]/2)$ if N is even. The alternative scale estimator used in the Hampel identifier is the MAD (median absolute deviation) scale estimate, also based on rank-ordered data. Specifically, given the median $x^\dagger$, the absolute deviations $d(k) = |x(k) - x^\dagger|$ are first computed. From these values, the raw scale estimate S_0 is then computed, defined as the median of the sequence $\{d(k)\}$. The MAD scale estimate S^* is finally defined as $S^* = 1.4826 S_0$, where the normalization constant is chosen to make the expected value of S^* equal to the standard deviation σ for Gaussian data sequences.

The Hampel identifier considered here simply replaces the mean $\bar{x}$ with the median $x^\dagger$ and the standard deviation estimate $\hat{\sigma}$ with the MAD scale estimate S^* in the 3σ edit rule described at the beginning of this discussion. The results obtained with the Hampel identifier for the recycle flow rate data are shown as dotted lines in Fig. 8.5; note that these data validation limits clearly separate the nominal operating data values from the shutdown episodes in the dataset. The Hampel filter is a moving-window version of the Hampel identifier just described, followed by median replacement for any data point $x(k)$ that is identified as an outlier. More specifically, the Hampel filter generates the cleaned data sequence $\{y(k)\}$ defined by

$$y(k) = \begin{cases} x(k) & |x(k) - x_k^\dagger| \leq \lambda S_k^* \\ x_k^\dagger & |x(k) - x_k^\dagger| > \lambda S_k^*, \end{cases} \quad (8.8)$$

where $x_k^\dagger$ is the median value in the moving data window

$$\mathbf{w}_k = \{x_{k-K}, \ldots, x_k, \ldots, x_{k+K}\}, \quad (8.9)$$

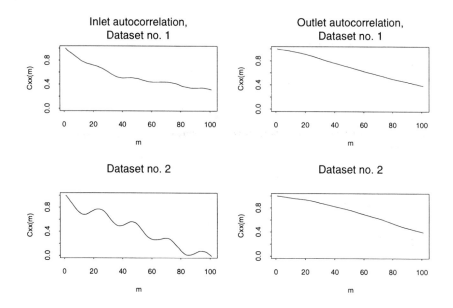

Figure 8.6: Second storage tank dataset autocorrelations

S_k^* is the MAD scale estimate for this data window, and λ is a threhsold parameter, taken as 3 in the example discussed here. A more complete discussion of the resulting nonlinear digital filter, originally proposed in the image processing literature, is available (Astola and Kuosmanen, 1997) under the name *decision-based filter*. Applications of this filter to both linear and Volterra model parameter estimation are also available (Pearson, 1999).

Despite the simplicity of the storage tank example considered here, its consequences are extremely far-reaching since many of the parameter estimation algorithms discussed in Sec. 8.1.1 are based either explicitly or implicitly on correlation estimates. Hence, the presence of outliers can badly degrade parameter estimates, a point illustrated in Sec. 8.2 for the case of bilinear models.

Fig. 8.6 illustrates a different type of artifact that can also cause erroneous results in dynamic analysis of related data sequences. Specifically, the upper plots show the estimated autocorrelation functions for the storage tank data sequences considered previously where the outlying zero values have been replaced by the median values for each dataset. In both of these upper plots, the estimated autocorrelation function decays approximately monotonically and the most obvious difference between them is that the outlet autocorrelation function decays more slowly than the inlet autocorrelation function. This behavior is consistent with both the smoother visual apperance of the outlet data sequence and the expectation that mixing occurs in the storage tank. The bottom two plots show the corresponding autocorrelation functions estimated from a second pair of datasets, obtained for the same storage tank but over a different time interval. This second dataset contains no gross outliers, but the autocorrela-

The Art of Model Development

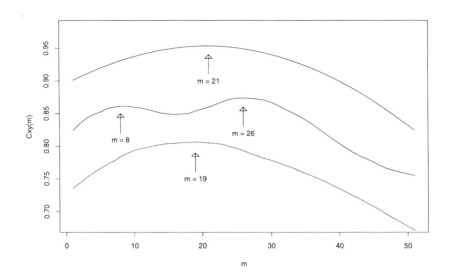

Figure 8.7: Cross-correlation estimates, dataset no. 2

tion function estimated from the second inlet dataset clearly exhibits a different character from the other three autocorrelations shown in this figure. In particular, note the pronounced ripples in this autocorrelation function: physically, these ripples correspond to oscillations in the measured data sequence with a period of 24 hours, arising from diurnal temperature variations. The influence of these periodic fluctuations on the estimated inlet-outlet cross-correlations is shown in Fig. 8.6 as the middle curve. Contrary to expectations, this estimated correlation function exhibits *two* peaks, one at $k = 8$ and the other at $k = 26$. In fact, these peaks are both spurious as the following results illustrate.

One effective method for removing periodic interference components of known period P is to apply a linear unweighted moving average filter of width P. This point is illustrated in the upper curve in Fig. 8.7, which shows the correlation estimate obtained from the second storage tank dataset after the application of a 24-hour linear unweighted average filter. The normalized cross-correlation function computed from the first dataset (after outlier removal) is shown as the bottom curve in Fig. 8.7 and it is clear that the results obtained from the second dataset after linear filtering are in reasonable agreement with these first results. The basis for this filtering is the following observation: given the data sequence $\{y(k)\}$, define the averaged data sequence

$$f(k) = \frac{1}{P} \sum_{i=0}^{P-1} y(k-i). \tag{8.10}$$

and note that if the observed data may be decomposed as $y(k) = x(k) + v(k)$

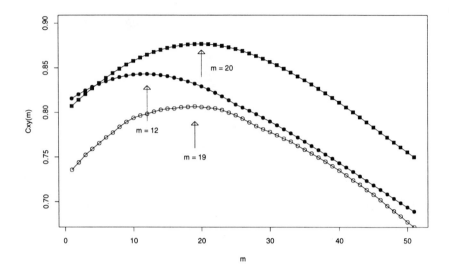

Figure 8.8: Cross-correlation estimates, dataset no. 1

where $v(k + P) = v(k)$ represents the periodic interference, the averaged sequence is given by

$$f(k) = \frac{1}{P} \sum_{i=0}^{P-1} x(k-i) + \bar{v}, \qquad (8.11)$$

where $\bar{v}$ is simply the average of the periodic component $v(k)$ over its period. If the smoothing of the fluctuations $\{x(k)\}$ of primary interest is not too severe, filtering the data sequences may improve the estimated correlations significantly, as seen in this example.

Finally, note that on careful examination of the inlet autocorrelations for the first storage tank dataset (the upper left plot Fig. 8.6), it appears that this data sequence may also be contaminated by periodic fluctuations, although much less intense fluctuations than those seen in the second dataset. Applying the 24-hour linear unweighted average filter to this dataset (after median replacement) and recomputing the cross-correlations yields the top curve in Fig. 8.8. This correlation estimate exhibits a broad maximum at $m = 20$, even closer in both location and magnitude to that ultimately obtained for the second dataset ($m = 21$) than the unfiltered result for the first dataset. The primary point of this comparison is twofold: first, once an artifact is detected and clearly identified in one part of a large data collection, it may be reasonable to look for manifestations of the same or related artifacts in other parts of the collection. In this example, there was little reason to suspect 24-hour temperature variations in the analysis of the first storage tank dataset, but their obvious presence in the second inlet dataset provides support for the tenuous evidence seen in the first inlet autocorrelation function on closer examination.

The second point of this comparison is that various forms of data pretreatment may often be combined advantageously. In this example, applying outlier detection/data cleaning strategies *first*, followed by linear filtering appears to have been effective in overcoming the effects of both outliers and periodic components. More generally, linear filters are often effective in suppressing high-frequency noise or eliminating periodic components, but they are completely ineffective against outliers, which require nonlinear filtering; consequently, it may be possible to obtain reasonable protection against both of these artifacts by adopting a "Hammerstein strategy" in which nonlinear outlier filtering is performed first (e.g., the Hampel filter described earlier in this section), *followed by* linear filtering for noise suppression or periodic component removal. The consequence of reversing this order—that is, the inadvisability of adopting a "Wiener strategy"—is also illustrated in Fig. 8.8. Specifically, the middle curve (marked with solid circles) shows the correlation estimates obtained by *first* applying the 24-hour unweighted average filter and *then* applying the Hampel filter. As in the previous example, the effect of the linear averaging filter is to broaden the single large outlier in each original dataset into a smaller but broader *patch* of succesive outliers. This computational approach also yields autocorrelation estimates with a fairly broad maximum, but centered at $m = 12$, far from the estimates obtained from either the first dataset after median replacement for outlier removal or the second dataset after linear filtering for periodic component removal.

Although a more detailed discussion of data pretreatment is beyond the scope of this book, two points should be noted. First, some form of data pretreatment is almost always necessary in applications of direct empirical modeling. At the very least, all variables should be examined graphically, their ranges of variation should be assessed, and simple consistency checks should be performed with respect to known ranges before *any* model parameters are estimated. Second, it is also important to note that any particular data pretreatment procedure can be helpful in some situations and harmful in others, as the examples considered here have demonstrated. Consequently, it is often prudent to perform *several* analyses of any given dataset and compare the results. In particular, if a specific anomaly is suspected, it may be very informative to compare the results obtained with and without corrections applied (e.g., averaged vs. unaveraged or with and without suspected outliers, etc.). Large differences in the results are often indicative of the presence of anomalies that should be investigated further.

8.1.3 Goodness-of-fit is not enough

Once a model has been estimated from a given dataset, it is important to assess its performance in terms of one or more criteria *other than* goodness-of-fit with respect to the original dataset. In particular, it is important to emphasize that all of the goodness-of-fit criteria considered in Sec. 8.1.1 represent quantitative measures of discrepancy between the model $\mathcal{M}$ being assessed and *one* dataset $\mathcal{D}$. Inherent in the use of any of these criteria, then, is the assumption that the dataset $\mathcal{D}$ is fully representative of the physical data generation process of interest. As the discussion in Sec. 8.1.2 emphasized, datasets collected from physical processes generally reflect the influence of various nonrepresentative phenomena like measurement noise, outliers, or other artifacts. Even in the case of indirect empirical modeling from simulateded responses of a detailed fundamental model, no single finite dataset is generally adequate to fully characterize the process dynamics because no finite-duration input sequence can excite all of the significant responses of the original system equations. Consequently, an exact match between the model $\mathcal{M}$ and the dataset $\mathcal{D}$ is generally undesirable, as discussed in Chapter 1 (Sec. 1.1.4). Conversely, since this exact match represents the best possible goodness-of-fit, it follows that *other criteria* should also be satisfied by a good empirical model.

One of the most popular of these "auxillary goodness criteria" is *simplicity*, often invoked as *Ockham's razor*: the fourteenth century philosopher William of Ockham advised that, when faced with the choice of two equally plausible alternatives, the simpler one is to be preferred. In empirical modeling, this advice is often interpreted in terms of the number of free parameters θ_i required to specify the model. Conversely, it is important to note that excessive adherence to Ockham's razor can also lead to the adoption of poor models. That is, in the pursuit of simplicity, it is important not to lose sight of Einstein's advice:

Everything should be as simple as possible, but no simpler.

In particular, although simple models are generally desirable, they must be complex enough to capture the essential behavior of interest. The remainder of this section discusses three different ways of combining goodness-of-fit criteria with additional constraints to increase the likelihood of obtaining reasonable models. The first two directly constrain model complexity, whereas the third imposes behavioral constraints and may be combined with either of the other two.

One way of trading off goodness-of-fit for model simplicity is to augment the goodness-of-fit measure with an explicit penalty on complexity. Probably the most popular embodiment of this idea is the Akaike Information Criterion, commonly abbreviated AIC (Tong, 1990; Ljung, 1987). This criterion applies to a family of model classes $\mathcal{M}_p$, each characterized by p free parameters; hence, if $k > j$, any model of class $\mathcal{M}_k$ may be regarded as more complex than any model of class $\mathcal{M}_j$. The AIC represents a penalized likelihood function of the form

$$AIC = -2\ln M + 2p \quad (8.12)$$

so that minimizing the AIC over the classes $\mathcal{M}_p$ trades off goodness-of-fit (expressed in terms of the likelihood M) for model complexity (expressed as the number of parameters p). In particular, note that since M necessarily increases with increasing p, the first term in this expression decreases monotonically with

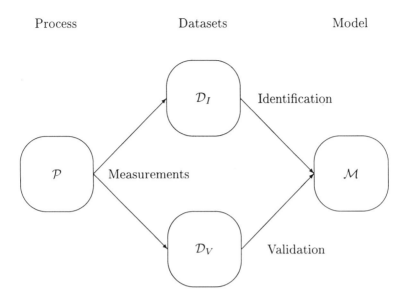

Figure 8.9: The simplest version of cross-validation

increasing p, but the second term increases linearly. Hence, AIC exhibits a minimum at some value p^* and the maximum likelihood model with p^* parameters is interpreted as the optimum choice, maximizing goodness-of-fit while minimizing complexity. In practice, it is frequently assumed that the prediction errors $e(k) = \hat{y}(k) - y(k)$ may be approximated by an independent, identically distributed sequence of zero-mean, Gaussian random variables. Under this assumption, the AIC is given explicitly by the expression

$$AIC = -N \ln \hat{\sigma}^2 + 2p \qquad (8.13)$$

where $\hat{\sigma}^2$ is the mean square error (MSE) defined in Eq. (8.1).

The AIC constrains complexity by explicitly penalizing the number of terms in the model, but it relies on statistical assumptions about the prediction error sequence $\{e(k)\}$. If these assumptions are not met—a common situation in practice—the model obtained by minimizing the AIC may represent a poor compromise between complexity and goodness-of-fit. An alternative is to replace these assumptions with a second dataset—not used in the parameter estimation step—to *validate* the models obtained by fitting the first dataset. This idea is illustrated graphically in Fig. 8.9, in which the dataset $\mathcal{D}$ appearing in Fig. 8.1 is replaced by *two* datasets:

1. An *identification dataset* $\mathcal{D}_I$
2. A *validation dataset* $\mathcal{D}_V$.

As in the discussion of the AIC, consider a family $\mathcal{M}_p$ of models, indexed by the number p of free parameters. Cross-validation proceeds in two stages: first, estimate the models $\mathcal{M}_k^*$ that best fit the identification dataset $\mathcal{D}_I$ for $k = 1$, 2,...,K. Note that if k is taken too large, $\mathcal{M}_k^*$ will approximate minor details of the dataset $\mathcal{D}_I$ at the expense of the general behavior of the process $\mathcal{P}$ from which this dataset was obtained. To overcome this problem, the performance of the model sequence $\mathcal{M}_k^*$ is assessed with respect to goodness-of-fit for the second dataset $\mathcal{D}_V$. For small k, $\mathcal{M}_k^*$ can be expected to fit $\mathcal{D}_V$ better and better with increasing k, but once k becomes large enough that $\mathcal{M}_k^*$ begins to fit the details of the dataset $\mathcal{D}_I$, goodness-of-fit with respect the dataset $\mathcal{D}_V$ will decrease. In the cross-validation approach, the optimum model is taken as that which best predicts the validation dataset $\mathcal{D}_V$, among all of the best fit models $\mathcal{M}_k^*$ computed from the identification dataset $\mathcal{D}_I$.

If only one dataset is available, a number of variations on this theme are possible. For example, *K-fold cross validation* (Efron and Tibshirani, 1993, ch. 17) splits the dataset $\mathcal{D}$ into K roughly equal subsets and one of the most popular versions of this idea is the "leave-one-out" approach that results when $K = N$. There, denote the subset of $\mathcal{D}$ obtained by *omitting* the ith data point as $\mathcal{D}_{(i)}$ and begin by estimating the models $\mathcal{M}_{(i)}$ for $i = 1, 2, ..., N$ that best fit these N subsets. Next, compute the predictions $y^{(i)}(i)$ of each of these omitted values, obtained from the model $\mathcal{M}_{(i)}$ and the dataset $\mathcal{D}_{(i)}$. The cross-validation error estimate CV is the mean square error

$$CV = \frac{1}{N} \sum_{i=1}^{N} [y^{(i)}(i) - y(i)]^2, \qquad (8.14)$$

which summarizes the overall ability of these N models to predict their omitted data points. The basic mechanics and limitations of cross-validation are examined further in Sec. 8.2.

One of the reasons that goodness-of-fit is an inadequate measure of model fidelity is that it is essentially a measure of numerical approximation with respect to the identification dataset. To see that such approximation accuracy is not sufficient to determine qualitative behavior, consider the first-order linear model

$$y(k) = \phi y(k-1) + u(k-1). \qquad (8.15)$$

If the single unknown parameter is estimated as $\hat{\phi} = 0.99$, the resulting model is stable, whereas the parameter estimate $\hat{\phi} = 1.01$ defines an unstable model. Hence, there is a substantial difference between the qualitative behavior of these two models even though the estimated parameters only differ by 2%. Further, note that $\phi = 0.99$ corresponds to a marginally stable model whose response decays slowly, whereas $\phi = 1.01$ corresponds to a marginally unstable model whose response grows rather slowly. Consequently, it is likely that the difference in goodness-of-fit with respect to a short sequence of input/output data values would not be great for these two models. In particular, it is not unreasonable to expect that, due to the effects of observation noise, unmodelled higher-order linear dynamics, or neglected nonlinear dynamics, the unstable model might actually exhibit better goodness-of-fit.

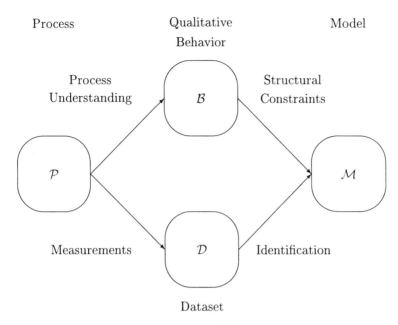

Figure 8.10: Qualitative model structure selection

This is exactly the situation described by Tulleken (1993), who compared a number of empirical linear models fit to the responses of an industrial distillation process using standard empirical modeling algorithms and data sequences of moderate length (e.g., $N = 1200$). Almost all of the resulting models were ultimately rejected due to unphysical instabilities, nonmonotonic step responses, incorrect signs for the steady-state gains, or dominant settling times that disagreed badly with known process behavior. Much better results were ultimately obtained when physically-motivated conditions like open-loop stability and agreement with the known signs of steady-state gains were imposed as explicit constraints in the parameter estimation process. This basic idea is illustrated in Fig. 8.10 to emphasize its relationship with cross-validation: here, a single dataset $\mathcal{D}$ is used for model identification, but structural and parametric model constraints are imposed on the basis of a description $\mathcal{B}$ of the *qualitative behavior* of the physical process $\mathcal{P}$, playing a role similar to that of the validation dataset $\mathcal{D}_V$ in cross-validation. This idea was discussed briefly in Chapter 1 (Sec. 1.1.4), where the extreme difference in the extrapolation behavior of polynomial and rational function fits to an exponential decay was demonstrated. Another application of this idea is the development of nonlinear empirical models that exactly match known steady-state behavior (Pottmann and Pearson, 1998), motivated by the fact that steady-state models are inherently simpler in structure and easier to develop than detailed dynamic models and are often developed for process design and scale-up.

Table 8.1: Sixteen model structures in class $\mathcal{C}$

Model	θ_1	θ_2	θ_3	θ_4	θ_5	Model class
1	0	0	×	0	0	pure delay
2	×	0	×	0	0	linear $AR(1)$
3	0	×	×	0	0	linear $AR(2)$
4	×	×	×	0	0	linear $AR(2)$
5	0	0	×	×	0	completely bilinear
6	×	0	×	×	0	
7	0	×	×	×	0	correct form
8	×	×	×	×	0	
9	0	0	×	0	×	completely bilinear
10	×	0	×	0	×	
11	0	×	×	0	×	
12	×	×	×	0	×	
13	0	0	×	×	×	completely bilinear
14	×	0	×	×	×	
15	0	×	×	×	×	
16	×	×	×	×	×	general form

8.2 Case study 1—bilinear model identification

To illustrate a number of the ideas discussed in Sec. 8.1, consider the problem of estimating the three constant coefficients in the diagonal bilinear model

$$v(k) = 0.8v(k-2) + u(k-1) - 0.2v(k-1)u(k-1). \qquad (8.16)$$

As is generally the case in practice, the correct model structure is *not* assumed to be known, and prediction models of the following form are considered:

$$\hat{y}(k) = \theta_1 y(k-1) + \theta_2 y(k-2) + \theta_3 u(k-1) \\ + \theta_4 y(k-1)u(k-1) + \theta_5 y(k-1)u(k-2). \qquad (8.17)$$

This working assumption provides a nice illustration of both the basic mechanics of cross-validation for structure selection and its limitations: parameters are estimated from input/output data and goodness-of-fit is compared for the 16 model structures shown in Table 8.1. These model structures are obtained by constraining some components of the parameter vector θ to be zero during the parameter estimation procedure; although it would be possible to consider 32 such constrained models, the 16 models with no linear input term (i.e., $\theta_3 = 0$) were omitted. Of the models considered, Model 1 necessarily exhibits poorest

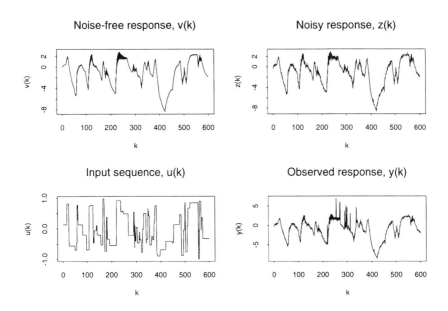

Figure 8.11: Input and response sequences

fit to the identification dataset because it is the most highly constrained and Model 16 exhibits the best fit because it is the least constrained.

The input sequence $\{u(k)\}$ on which these results are based belongs to the class of random step sequences described in detail in Sec. 8.4.3 and was generated as follows. First, $u(1)$ is drawn from the uniform distribution of values on the interval $[-1, 1]$; then, for each k, either the previous value is retained (i.e., $u(k) = u(k-1)$ with 90% probability), or a new value $u(k)$ is drawn from the uniform distribution on $[-1, 1]$. Following common practice, the observation noise is modeled as a zero-mean white Gaussian noise sequence $\{e(k)\}$; in the examples considered here, the standard deviation of this sequence is $\sigma = 0.2$, corresponding to about 10% of the root mean square (RMS) variation of the response $v(k)$. Outliers like those discussed in Sec. 8.1.2 are modeled by a sequence $\{o(k)\}$ taking the value 0 with 95% probability and taking the value $+5$ with 5% probability. To illustrate the mechanics of cross-validation, the observed data sequence is partitioned into 200 point subsequences as follows:

$$y(k) = \begin{cases} z(k) & 1 \leq k \leq 200 \\ z(k) + o(k) & 201 \leq k \leq 400 \\ z(k) & 401 \leq k \leq 600, \end{cases} \quad (8.18)$$

where $z(k) = v(k) + e(k)$ and the subsequences are denoted Segments 1, 2, and 3, respectively. A visual summary of the data sequences $\{u(k)\}$, $\{v(k)\}$, $\{z(k)\}$, and $\{y(k)\}$ is given in Fig. 8.11. In the following discussions, Segment 3 serves as a validation dataset and comparison of goodness-of-fit for Segments 1 and 3 illustrates the basic mechanics of cross-validation, whereas comparisons between

Table 8.2: Cross-validation results, Segment 1

Model	σ_{id}	Rank	σ_{val}	Rank
1	1.454	16	2.809	16
2	0.550	12	0.525	12
3	0.338	8	0.465	11
4	0.331	7	0.436	8
5	1.270	14	2.043	13
6	0.512	10	0.430	7
7	0.262	4	0.282	3
8	0.259	2	0.274	1
9	1.368	15	2.223	15
10	0.540	11	0.450	9
11	0.279	6	0.290	6
12	0.278	5	0.287	5
13	1.201	13	2.116	14
14	0.471	9	0.462	10
15	0.261	3	0.283	4
16	0.257	1	0.275	2

Segments 2 and 3 illustrate the influence of outliers on these results. In all cases, model parameters are estimated by minimizing the least squares criterion

$$\hat{\sigma} = \frac{1}{N-2} \sum_{k=3}^{N} [\hat{y}(k) - y(k)]^2. \quad (8.19)$$

Table 8.2 summarizes the results obtained from Segments 1 and 3 of the simulated input/output data, reflecting the effects of observation noise but free of outliers. The first column of this table specifies one of the 16 model structures defined in Table 8.1, the second column gives the optimum value of $\hat{\sigma}_{id}$ exhibited by the best fit model of this structure for the identification dataset (Segment 1), and the third column gives the relative ranking of these models on the basis of this $\hat{\sigma}_{id}$ value. As noted, Model 1 is consistently the worst of these structures (rank 16) and Model 16 is consistently the best (rank 1). The fourth and fifth columns of Table 8.2 give the corresponding values of $\hat{\sigma}_{val}$, the performance criterion computed from the validation dataset (Segment 3), and the relative ranking of these values. In terms of these rankings, note that the maximally unconstrained model (Model 1) moves from best to second-best, and Model 8 becomes the best model. Although this model structure is not correct (Model 7 is the correct structure), note that it includes the correct structure as a special

Table 8.3: Model parameter estimates, Segment 1

Model	θ_1	θ_2	θ_3	θ_4	θ_5
8	0.063	0.728	0.975	−0.216	0.000
16	0.081	0.704	0.984	−0.298	0.088
7	0.000	0.780	1.010	−0.221	0.000
15	0.000	0.776	1.021	−0.271	0.051
Exact	0.000	0.800	1.000	−0.200	0.000

case. Also, note that the correct model structure moves up in ranking from 4th to 3rd. The key points of this example are first, that cross-validation does lead to the selection of a less complex model (Model 8 vs. Model 16) and second, that it does not necessarily lead to the adoption of the *correct* model structure.

It is also instructive to examine the parameter estimates obtained for the four best models selected by cross-validation for this example. In decreasing order of performance, these models correspond to structures 8, 16, 7, and 15 from Table 8.1 and their estimated parameter values are listed in Table 8.3. Note that the correct model structure (Model 7) corresponds to a special case of all three of these other model structures, obtained by constraining one or more parameters to be zero. Constraining the model structure to be correct generally leads to the best parameter estimates, as seen here; in particular, the estimates for θ_2 and θ_3 obtained for Model 7 were the closest to the correct values from the four models compared here, and θ_4 is only slightly better estimated for Model 8. This observation is closely related to the results of Tulleken (1993) discussed in Sec. 8.1.3, which argue in favor of imposing explicit constraints during empirical model identification *when this is possible*. Clearly, there are many circumstances under which the "correct model structure" is either unknown or nonexistent, but when structural constraints can be imposed, they probably should be. Alternatively, the point has been made more forcefully (Lindskog and Ljung, 1997):

Don't estimate what you already know!

Table 8.4 presents the cross-validation results obtained from both the contaminated data sequence (Segment 2) and a cleaned version of this data sequence obtained by applying the Hampel filter described in Sec. 8.1.2 to the observed response sequence $\{y(k)\}$. Specifically, columns 2, 4, 6 and 8 of this table present the standard deviation estimates obtained from the identification dataset (columns 2 and 6, indicated by the subscript id) and the validation

Table 8.4: Cross-validation results, Segment 2

Model	σ_{id}^x	Rank	σ_{val}^x	Rank	σ_{id}^h	Rank	σ_{val}^h	Rank
1	1.861	16	2.782	16	1.564	16	2.773	16
2	1.410	12	1.027	12	0.592	12	0.510	12
3	1.342	8	0.894	11	0.380	8	0.443	10
4	1.259	4	0.628	6	0.370	7	0.411	7
5	1.758	14	2.125	14	1.399	14	2.024	13
6	1.369	10	0.774	8	0.542	10	0.413	8
7	1.322	7	0.702	7	0.304	4	0.284	3
8	1.241	2	0.462	2	0.297	2	0.273	2
9	1.777	15	2.163	15	1.471	15	2.169	15
10	1.398	11	0.883	10	0.584	11	0.450	11
11	1.312	6	0.635	5	0.313	6	0.295	6
12	1.246	3	0.478	3	0.312	5	0.292	5
13	1.755	13	2.087	13	1.369	13	2.095	14
14	1.364	9	0.781	9	0.454	9	0.430	9
15	1.312	5	0.634	4	0.304	3	0.284	4
16	1.241	1	0.458	1	0.295	1	0.271	1

dataset (columns 4 and 8, indicated by the subscript val). The values denoted with the superscript x (columns 2 and 4) were computed from the contaminated data sequence and those denoted with the superscript h (columns 6 and 8) were computed from the cleaned data sequence. As in Table 8.2, all 16 models are ranked according to these standard deviation estimates, and these rankings are given in the column to the right of these standard deviation estimates. In this example, the best model is consistently the most complex (Model 16) and the worst model is consistently the least complex (Model 1). The correct model structure (Model 7) is ranked 7th for both the identification and the validation datasets when computed from the outlier-contaminated data, but it is ranked 3rd by cross-validation when computed from the cleaned data sequence. These results illustrate two practically important points: first, that outliers can profoundly influence the results of model structure selection and second, that cross-validation does not always lead to the selection of a simpler model structure.

These conclusions are also evident from Table 8.5, which summarizes the estimated model parameters for the four best models according to the cross-validation procedure, analogous to Table 8.3. Here, results are presented for both the contaminated and the cleaned data, and two differences between these results are noteworthy. First, note that Models 16, 8, and 15 are ranked 1, 2, and 4 for both datasets, but the correct model structure (7) is replaced by a similar

Table 8.5: Model parameter estimates, Segment 2

Model	θ_1	θ_2	θ_3	θ_4	θ_5
16^x	0.313	0.403	0.931	−0.148	−0.027
8^x	0.317	0.400	0.930	−0.169	0.000
12^x	0.294	0.429	0.937	0.000	−0.150
15^x	0.000	0.610	1.180	−0.013	−0.218
16^h	0.147	0.668	0.892	−0.259	0.078
8^h	0.112	0.702	0.896	−0.188	0.000
7^h	0.000	0.800	0.958	−0.193	0.000
15^h	0.000	0.802	0.952	−0.163	−0.032
Exact	0.000	0.800	1.000	−0.200	0.000

model structure (12) when the results are based on the contaminated data. The difference between these model structures is that the term $\theta_4 y(k-1)u(k-1)$ appearing in Model 7 has been replaced by $\theta_5 y(k-1)u(k-2)$ in Model 12. Second, note the substantial differences in the parameter estimates obtained from the contaminated and cleaned data sequences, for the same model structures. For example, for Model 16, the spurious parameter θ_1 is twice as large when estimated from the contaminated data and the parameter θ_2 is underestimated three times as badly. The key point here is that the presence of outliers in a dataset generally has a significant adverse effect on both empirical model structure selection and model parameter estimation.

8.3 Model structure selction

It has been emphasized repeatedly throughout this book that the selection of a good model class $\mathcal{C}$ is in many respects the most important step in the empirical modeling process. To aid in making this selection, the following discussion summarizes some of the important relationships between structure and behavior for the model structure classes introduced in previous chapters. This discussion is divided into two parts: Sec. 8.3.1 considers the structural consequences of observable qualitative behavior and Sec. 8.3.2 considers the behavioral consequences of particular model structure choices. The first perspective provides a basis for identifying those model classes whose qualitative behavior matches

Table 8.6: Nine forms of nonlinear qualitative behavior

i	Abbreviation	Qualitative behavior class $\mathcal{B}_i$
1	HARM	Harmonic generation from sinusoidal inputs
2	SUB	Subharmonic generation from sinusoidal inputs
3	ASYM	Asymmetric responses to symmetric inputs
4	IDS	Input-dependent stability
5	SSM	Steady-state multiplicity
6	CHAOS	Chaotic responses to simple inputs
7	HOM	Nonlinear homogeneous behavior
8	PHOM	Nonlinear positive-homogeneous behavior
9	SL	Nonlinear static-linear behavior

that of the physical process and rejecting model classes whose members cannot exhibit the desired behavior. As a specific example, it has been noted repeatedly that nonlinear autoregressive terms are necessary for a discrete-time dynamic model to exhibit output multiplicity. Conversely, the second perspective is useful in making a preliminary assessment of the appropriateness of a proposed model structure before great effort is invested in model development.

8.3.1 Structural implications of behavior

Chapter 1 described six specific manifestations of nonlinearity, and Chapter 3 introduced three behaviorally-defined classes of nonlinear models, obtained by relaxing the superposition requirements that define the class of linear models. The following paragraphs consider each of these forms of nonlinear behavior in connection with preliminary model structure selection. To facilitate this and subsequent discussions, Table 8.6 defines a set of convenient abbreviations for these nine nonlinear phenomena. All of these phenomena are observable from input/output data and may be used to exclude certain model classes from further consideration.

As noted in Chapter 1, (super)harmonic generation (HARM) is one of the simplest manifestations of nonlinearity. Consequently, although harmonic generation provides a clear indication of system nonlinearity, it provides relatively little guidance in deciding between different nonlinear model structures. In particular, note that *all* of the nonlinear model classes considered in this book are capable of harmonic generation. Visually, this effect is seen clearly in the responses to sinusoidal input sequences: non-sinusoidal, periodic responses with the same period as the input imply superharmonic generation. Similarly, periodic responses whose period is an integer *submultiple* of the input period also imply harmonic generation, as in the case of the static nonlinearity $f(x) = x^2$. In general, the intensity of superharmonic generation decreases as the order of the harmonic increases, so harmonic components at frequencies $2f$ and $3f$ in response to an input sinusoid of frequency f can be expected to be more intense than higher harmonics, although exceptions exist. Probably the most important exception is the case of odd-symmetry static nonlinearities, $f(-x) = -f(x)$, that only generate harmonics of odd order.

Subharmonic generation (SUB) in response to periodic input sequences is a much rarer phenomenon, as discussed in Chapter 1. In particular, subharmonic generation corresponds to a *lengthening* of the period of oscillation relative to the period of the input sequence. In contrast to superharmonic generation, subharmonic generation requires the presence of a *dynamic* nonlinearity and this requirement excludes many of the model structures considered in this book. In particular, subharmonic generation requires nonlinear autoregressive terms in a discrete-time dynamic model and is therefore not possible in NMAX model classes like the finite Volterra models discussed in Chapter 5, nor in block-oriented models (e.g., Hammerstein and Wiener models) that combine linear dynamic subsystems with static nonlinearities, *unless feedback connections are allowed, as in the Lur'e model discussed in Chapter 4*. Further, note that bilinear terms of the form $u(k-i)y(k-j)$ are not sufficient to result in subharmonic generation except in the case of certain *unstable* responses that are approximately subharmonic in character. Conversely, polynomial NARMAX models involving quadratic and higher-order terms in $y(k-i)$ are capable of subharmonic generation, as are rational NARMAX models, additive NARMAX models, and linear multimodels involving output selection.

In contrast, asymmetric responses to symmetric input changes (ASYM) is a very common phenomenon, generally possible in all model classes *except* those satisfying symmetry restrictions like $\mathcal{M}[-u(k)] = -\mathcal{M}[u(k)]$. In particular, note that all homogeneous nonlinear models (HOM) satisfy this odd symmetry restriction, which may be viewed as a restricted form of homogeneity. More generally, asymmetric responses to symmetric input changes exclude symmetric model structures like the odd-order Volterra model class $O_{(N,M)}$, block-oriented models based on odd-symmetry static nonlinearities, or homomorphic systems based on odd-symmetry static nonlinearities. Conversely, this behavior is typical of generic Volterra and block-oriented models, bilinear models, polynomial or rational NARMAX models, or linear multimodels with arbitrary selection criteria. Finally, it is worth emphasizing that although asymmetric responses to symmetric input changes is a sufficient condition for nonlinearity, it is not a necessary condition. Consequently, informal nonlinearity tests like the one described in Chapter 1 for the crystallizer example (Eek, 1995) do not exclude the possibility of odd-symmetry nonlinearities. Further, note that such symmetry restrictions may be imposed by fundamental physical constraints.

Like subharmonic generation, input-dependent stability (IDS) also provides an indication of a "strong" nonlinearity. In particular, as in the case of subharmonic generation, input-dependent stability is not possible in nonlinear moving average (NMAX) models, thus excluding finite Volterra models as a special case.

In addition, this behavior is not possible for block-oriented structures like Hammerstein and Wiener models, although it is possible in feedback block-oriented structures like the Lur'e model. Further, this behavior appears to be generic for polynomial NARMAX models, including as a special case the class of bilinear models. Input-dependent stability may or may not be present for rational NARMAX models, additive NARMAX models, or linear multimodels with arbitrary selection criteria.

Steady-state multiplicity (SSM) provides an indication of the type of nonlinearity present in a discrete-time dynamic model. In particular, input multiplicity (one steady-state response y_s corresponding to more than one steady-state input u_s) requires the presence of nonlinear moving average terms. Consequently, this behavior is possible for NMAX model structures like Volterra models, for Hammerstein and Wiener models based on nonmonotonic static nonlinearities, and for input-selected linear multimodels. In contrast, input multiplicity is not possible for nonlinear autoregressive model structures like the Lur'e model, for Hammerstein or Wiener models based on monotonic nonlinearities, or for linear multimodels involving only output selection. Output multiplicity (one steady-state input u_s corresponds to more than one steady-state response y_s) requires the presence of nonlinear autoregressive terms. Consequently, this phenomenon is observable in NARX model structures like the Lur'e model, in output-selected linear multimodels, and in general polynomial NARMAX models. More complex steady-state phenomena like isolas that involve both input and output multiplicity are only possible in more complex model classes like the NARX* models, the polynomial and rational NARMAX models, and linear multimodels involving general selection. Conversely, note that bilinear models are incapable of either input our output multiplicity.

Chaotic responses to simple inputs (CHAOS) also provides evidence of a "strong" nonlinearity. In particular, it follows from the asymptotic constancy results presented in Chapter 4 for NMAX models that continuous NMAX models cannot exhibit chaotic impulse or step responses. Similarly, it follows from the preservation of periodicity that NMAX models cannot exhibit chaotic responses to periodic inputs. Also, bilinear models and input-selected linear multimodels cannot exhibit chaotic responses to simple inputs. Conversely, this behavior is possible for models involving nonlinear autoregressive terms and appears to be quite common in polynomial NARMAX models whose highest-order term is of even order in $y(k-i)$. In fact, chaotic step responses are seen in extremely simple models including quadratic NARX models and TARMAX models, both involving only first-order dynamics. As a corollary of this last observation, note that chaotic responses to simple inputs are possible in output-selected linear multimodels.

As an interesting contrast, note that homogeneity (HOM) is an extremely restrictive condition that immediately excludes many different model structures. In particular, nonlinear homogeneous behavior is not possible for polynomial NARMAX models, including as subsets the Volterra model class and the bilinear model class. More generally, nonlinear homogeneity is not possible for Hammerstein or Wiener models, nor for additive NARMAX models. As a consequence of this last observation, note that homogeneous NARX models are necessarily of at least second-order since first-order NARX models belong to the structurally additive NARMAX model class. Conversely, nonlinear homogeneous behavior is possible for rational NARMAX models and nonadditive NMAX, NARX, and NARX* models.

Interestingly, nonlinear positive-homogeneity (PHOM) is substantially less restrictive, although it is still a rather special form of nonlinear dynamic behavior. In particular, explicit constructions have been given in this book for positive-homogeneous Hammerstein and Wiener models, Uryson models, rational NARMAX models, and an entire class of structurally additive, positive-homogeneous linear multimodels (the TARMAX class). Conversely, like the homogeneous model class, the positive-homogeneous model class *excludes* all nonlinear polynomial NARMAX models, thereby excluding Volterra models and bilinear models as special cases of this result.

Finally, the class of static-linear (SL) nonlinear models appears to be very large, despite its extremely special character. In particular, note that the steady-state behavior of a static-linear model is completely characterized by a single steady-state gain K: $y_s = Ku_s$. Despite the apparent strength of this restriction, most of the structural model classes discussed in this book include non-trivial static-linear subsets. For example, explicit characterizations were given in previous chapters for static-linear members of the bilinear model family, the Volterra model family, and the TARMAX model family. Hence, in contrast to the more restrictive condition of homogeneity, static-linearity is possible in polynomial NARMAX models. In addition, note that the static-linear model class includes the nonlinear homogeneous model class as a proper subset. Conversely, because they involve only a single scalar nonlinearity, static-linear behavior is not possible for Hammerstein, Wiener, or Lur'e models.

8.3.2 Behavioral implications of structure

The results presented in the previous section were intended as a useful guide to preliminary model structure selection, providing a link between observed qualitative behavior and possible model structures. The following discussion presents the opposite view: given a candidate model structure, what can be said about its qualitative behavior, particularly in comparison to other model structures. Although some of this information is contained in the summaries just given, the following results should be useful in verifying that a particular model structure is reasonable before proceeding with a detailed model development.

Arguably, the simplest nonlinear discrete-time dynamic model is the Hammerstein model, consisting of a static nonlinearity followed in series by linear dynamics. As noted previously, there is no loss of generality in assuming the steady-state gain of this linear model is 1, implying that the static nonlinearity is simply the steady-state gain curve for the model. The qualitative character of the impulse and step and impulse response of this model is determined completely by that of the linear dynamic model since the static nonlinearity preserves the character of both of these inputs. Similarly, the static nonlinearity results in harmonic generation in response to sinusoidal inputs and, if the linear dynamic model exhibits lowpass characteristics (as is often the case in physical system models), the effect of this linear model is to attenuate the higher harmonics. Loosely speaking, this effect may be viewed as a "softening" of the effects of the static nonlinearity. Overall, the Hammerstein model may be viewed as a mildly nonlinear dynamic model capable of harmonic generation (HARM), asymmetric responses to symmetric input changes (ASYM), input multiplicity (SSM), and positive-homogeneity (PHOM). Conversely, Hammerstein models are not capable of either output multiplicity or the other five forms of nonlinearity discussed in Sec. 8.3.1 (SUB, IDS, CHAOS, HOM, and SL).

The Wiener model is of the same structural complexity as the Hammerstein model, consisting of the same two components connected in the reverse order. The behavioral implications of this order reversal are significant, however, and it appears that the Wiener model may be viewed as "slightly more nonlinear" than the Hammerstein model. As a specific example, consider the Wiener model formed from a lowpass linear dynamic model and the hard saturation nonlinearity $f(x) = \text{sign}\, x$. Sinusoidal inputs will be attenuated by the linear subsystem to a degree that depends on their frequency but unless they are completely rejected, the Wiener model output will be a unit-amplitude square wave of the same frequency as the input, shifted in phase by an amount that depends on the linear dynamics. The magnitude of the superharmonics generated by this Wiener model are therefore completely determined by the saturation nonlinearity. In contrast, the magnitude of the superharmonics generated by the dual Hammerstein model will depend the lowpass character of the linear dynamic model. Adopting the view that harmonic content represents a measure of nonlinearity, it follows that the Wiener model is inherently more nonlinear than the Hammerstein model. In addition, recall from discussion of the CSTR model given in Chapter 1 that the Wiener model is capable of exhibiting qualitatively asymmetric responses (i.e., oscillatory responses to positive steps and monotonic responses to negative steps), whereas the Hammerstein model is not. These observations are consistent with results based on numerical suitability measures (Menold, 1996; Menold et al., 1997a,b), which suggest that the Wiener model is inherently more flexible than its Hammerstein dual. This difference is somewhat subtle, however, since the Wiener and Hammerstein model are identical with respect to the nine behavioral criteria discussed in Sec. 8.3.1.

In both the Hammerstein and Wiener models, steady-state and dynamic behavior are very strongly connected; for example, neither of these models can exhibit nonlinear homogeneous or static-linear behavior. Although it does not appear to be as well known, the Uryson model introduced in Chapter 1 exhibits somewhat greater flexibility, consisting of several Hammerstein models connected in parallel. In particular, it was noted in Chapter 3 that the Uryson model can exhibit static-linearity. Also, like the Wiener model, the Uryson model is capable of exhibiting qualitatively asymmetric responses to symmetric input changes like those seen in the exothermic CSTR example discussed in Chapter 1. Consequently, it follows that the *dual Uryson model structure*, obtained by replacing the component Hammerstein models defining the Uryson structure with their dual Wiener models, should be more flexible than Hammerstein, Wiener, or Uryson models. Unfortunately, it must also be noted that dual-Uryson models do not generally exhibit finite-dimensional NARMAX representations. To see this point, consider a two-channel dual-Uryson model whose response is given by $y(k) = z_1(k) + z_2(k)$ where $z_i(k)$ is the response of the ith component Wiener model to the common input $u(k)$; in general, it is not possible to reconstruct both $z_1(k)$ and $z_2(k)$ from past values of $y(k)$ and $u(k)$. Conversely, any m channel dual-Uryson model may be represented as

$$y(k) = \sum_{r=1}^{m} g_r \left(\sum_{i=0}^{\infty} h_r(i) u(k-i) \right) \qquad (8.20)$$

where $h_r(i)$ is the impulse response of the rth linear model and $g_r(\cdot)$ is the corresponding static nonlinearity. Although this expression may be regarded as a NARMAX model (in fact, an NMAX model), it is an infinite-dimensional one; in general, no simpler NARMAX representation exists.

An important exception to this last observation is the class of dual-Uryson polynomial NMAX models. In fact, these models belong to the class of *projection-pursuit* models introduced in Chapter 4 and it was shown in Chapter 5 that they are equivalent to the family of finite Volterra models. Conversely, it was also noted in Chapter 5 that the number of parameters required to specify an arbitrary Volterra model rapidly becomes prohibitive, again arguing in favor of restricted special cases like polynomial Hammerstein, Wiener, Uryson, or dual-Uryson models. Alternatively, variations like the pruned Volterra models discussed in Chapter 5 or the "mixed Hammerstein-Wiener models" (*gemischtes Wiener-Hammerstein Modelle*) composed of Hammerstein *and* Wiener models in parallel (Kurth, 1995) also provide systematic approaches to restricting Volterra model complexity. It is important to note, however, that these block-oriented model structures are NMAX models and therefore cannot exhibit strongly nonlinear behavior like subharmonic generation, input-dependent stability, output multiplicity, or chaotic responses to simple inputs. Further, even the block-oriented extensions of these NMAX models based on arbitrary ARMAX linear models cannot exhibit these types of qualitative behavior unless feedback connections are considered.

The basic feedback block-oriented model structure is the Lur'e model discussed in Chapter 4, which is capable of a much wider range of behavior than the feedforward block-oriented models considered in the previous paragraphs. It is particularly interesting to compare the behavior of the Lur'e structure with that of the Hammerstein and Wiener structures, since all three combine linear dynamics with a static nonlinearity. The qualitative differences between Hammerstein and Lur'e models is illustrated in Table 8.7, which summarizes the ability of three model classes to exhibit each of the nine forms of nonlinear qualitative behavior discussed in Sec. 8.3.2: the Hammerstein model, the Lur'e model, and the structurally additive (NAARX) model class. Recall that both the Hammerstein and Lur'e model classes are proper subsets of the NAARX model class; consequently, any qualitative behavior that can be exhibited by

Table 8.7: Behavior of Hammerstein, Lur'e, and NAARX models

i	$\mathcal{B}_i$	Hamm.	Lur'e	NAARX
1	HARM	Yes	Yes	Yes
2	SUB	No	Yes	Yes
3	ASYM	Yes	Yes	Yes
4	IDS	No	Yes	Yes
5	SSM	Input	Output	Both
6	CHAOS	No	Yes	Yes
7	HOM	No	No	No
8	PHOM	Yes	Yes	Yes
9	SL	No	No	Yes

either the Hammerstein model or the Lur'e model can also be exhibited by the general NAARX model class. The general increase in behavioral flexibility in going from Hammerstein to Lur'e to general additive NARMAX models may be seen clearly by comparing the entries in this table.

To see that the Lur'e model can exhibit subharmonic generation, input-dependent stability and chaotic responses to simple inputs, consider the general first-order NARX model

$$y(k) = -b_1 f(y(k-1)) + b_1 u(k-1), \qquad (8.21)$$

corresponding to the Lur'e model obtained from the feedback connection of the static nonlinearity $f(\cdot)$ and the linear delay model $z(k) = b_1 u(k-1)$. Taking $b_1 = 1$ and

$$f(y) = \begin{cases} -2y & |y| \leq 2 \\ 0 & |y| > 2 \end{cases} \qquad (8.22)$$

leads to the threshold autoregressive model that Tong (1990) uses to illustrate the notion of subharmonics. Similarly, taking $f(y) = y(1-y)$ and $b_1 = -\alpha$ leads to the controlled logistic equation

$$y(k) = \alpha y(k-1)[1 - y(k-1)] - \alpha u(k-1), \qquad (8.23)$$

whose response to an impulse input of amplitude $\gamma = -y(1)/\alpha$ for $0 < y(1) < 1$ is given by the logistic equation discussed in Chapter 1; in particular, recall that for $\alpha = 4$, the solution to this equation is chaotic. Further, for any $\alpha > 0$, note that the response of this model to an impulse of positive amplitude or an impulse of amplitude more negative than $-1/\alpha$ leads to an unstable response.

Table 8.8: Behavior of NMAX, NARX, NARX*, and bilinear models

i	$\mathcal{B}_i$	NMAX	NARX	NARX*	Bilinear
1	HARM	Yes	Yes	Yes	Yes
2	SUB	No	Yes	Yes	(Unstable)
3	ASYM	Yes	Yes	Yes	Yes
4	IDS	No	Yes	Yes	Yes
5	SSM	Input	Output	Both	Neither
6	CHAOS	No	Yes	Yes	No
7	HOM	Yes	Yes	Yes	No
8	PHOM	Yes	Yes	Yes	No
9	SL	Yes	Yes	Yes	Yes

Conversely, it was shown in Chapter 6 that structurally additive models cannot exhibit nonlinear homogeneous behavior.

Intermediate in behavior between simple feedforward block-oriented structures like the Hammerstein model and more complex nonlinear autoregressive structures like the Lur'e model are the bilinear models discussed in Chapter 3. Table 8.8 compares the behavior of these models with that of the NMAX, NARX, and NARX* classes. With respect to the strong forms of nonlinearity (SUB, IDS, and CHAOS), the bilinear class lies between the NMAX class for which none of these behaviors are possible and the NARX class for which all are possible: although input-dependent stability is characteristic of bilinear models, subharmonic generation is only possible as a limiting behavior for unstable models, and chaotic responses to simple inputs is not possible. Also, note that bilinear models exhibit unique steady-states (i.e., neither input nor output multiplicity are possible), and they are incapable of either homogeneous or positive-homogeneous behavior, although static-linearity is possible.

The qualitative behavior of linear multimodels depends strongly on the selecion criteria on which they are based. This relationship is summarized in Table 8.9 where "IS" denotes input selection, "OS" denotes output selection and "GS" denotes general selection. More specific results are possible for the special case of TARMAX models introduced in Chapter 3 and these results are also summarized in Table 8.9, which gives necessary conditions on the nonlinear model coefficients c_i and d_i to exhibit each of the nine forms of nonlinear behavior considered here. In some cases, these conditions are also sufficient but in general they are not. For example, harmonic generation occurs if and only if at least one of these nonlinear model coefficients is nonzero, but the condition $c_i \neq 0$ for some i is not sufficient to guarantee either subharmonic generation or chaotic step responses.

Table 8.9: Behavior of linear multimodels

i	$\mathcal{B}_i$	IS	OS	GS	TARMAX Conditions
1	HARM	Yes	Yes	Yes	$c_i \neq 0$ or $d_i \neq 0$
2	SUB	No	Yes	Yes	$c_i \neq 0$
3	ASYM	Yes	Yes	Yes	$c_i \neq 0$ or $d_i \neq 0$
4	IDS	Yes	Yes	Yes	$c_i \neq 0$
5	SSM	Input	Output	Both	Input: $d_i \neq 0$, output: $c_i \neq 0$
6	CHAOS	No	Yes	Yes	$c_i \neq 0$
7	HOM	No	No	No	Linearity: $c_i = 0$, $d_i = 0$
8	PHOM	Yes	Yes	Yes	Always holds
9	SL	Yes	Yes	Yes	$\sum_{i=1}^{p} c_i = 0$ and $\sum_{i=0}^{q} d_i = 0$

Finally, note that "large" classes like polynomial NARMAX models, rational NARMAX models, structurally additive models, or projection-pursuit models can exhibit almost all of the types of nonlinear qualitative behavior considered in this book. Consequently, unless behavioral constraints are imposed either explicitly in the form of model parameter constraints or implicitly in the form of model subclass restrictions, empirically identified models are likely to exhibit complex behavior. In particular, note that the determination of steady-state loci, stability conditions, and conditions for subharmonic generation or chaotic responses are nontrivial for these models. Consequently, if such models are to be considered for a particular application, it is important to characterize them as completely as possible, at least through the use of extensive simulations, analysis of linearized model behavior in important operating regions, and any other characterizations that may be feasible.

8.4 Input sequence design

The following discussion considers the problem of specifying an effective input sequence for empirical model identification. In particular, this discussion considers three aspects of this problem:

1. The effectiveness criteria to be optimized
2. The practical constraints that must be satisfied
3. The design factors available for optimization.

A general discussion of the first two of these topics is given in Sec. 8.4.1 and Sec. 8.4.2 gives a brief overview of the third. Secs. 8.4.4 and 8.4.3 then consider these notions in connection with two specific classes of input sequences: random steps and the deterministic sine-power sequences.

8.4.1 Effectiveness criteria and practical constraints

In considering the effectiveness of an input sequence for empirical modelling, it is useful to distinguish between preliminary investigations in which several model classes are being considered and more focused investigations in which a single model class has been specified. In the first case, an effective input sequence is a good *screening input* and the most effective screening inputs are those that elicit visually obvious response differences between the model classes considered. To illustrate this idea, it is useful to consider the following four simple models:

1. Linear reference model:

$$y(k) = 0.8y(k-1) + u(k)$$

2. Superdiagonal bilinear model:

$$y(k) = 0.8y(k-1) + u(k) - 0.2y(k-1)u(k)$$

3. Nonadditive polynomial NARMAX model:

$$y(k) = 0.8y(k-1) + u(k) - 0.2y(k-1)y(k-2)u(k-1)$$

4. Linear multimodel (TARMAX model):

$$y(k) = 0.8|y(k-1)| + u(k).$$

In each case, the standard initial condition $y(k) = 0$ for $k \leq 0$ is assumed for transient inputs (e.g., steps and impulses), whereas for persistent inputs like sinusoids, the input sequence is assumed to extend to $k = -\infty$.

For impulses of amplitude $\alpha > 0$, all four of these models exhibit identical responses, decaying monotonically from the initial value $y(1) = \alpha$. Conversely, for $\alpha < 0$, Model 4 exhibits a qualitatively distinct response, corresponding to the oscillatory linear model $y(k) = -0.8y(k-1) + u(k)$. Hence, a pair of impulse responses—one positive and one negative—is sufficient to distinguish Model 4 from the other three, but the other three models exhibit identical responses to all impulses. For small amplitudes A, the step response does not clearly distinguish Models 1, 2, and 3 either, but it does for $|A|$ sufficiently large. This point is illustrated in Fig. 8.12, which shows the responses of these models to steps for $A = \pm 1$. Model 1 exhibits monotonic responses and odd symmetry [i.e., $u(k) \to -u(k)$ implies $y(k) \to -y(k)$], as it must due to its linearity. Conversely, Model 2 exhibits a pronounced asymmetry that is characteristic of bilinear models: the effective settling time is faster and the apparent steady-state gain is smaller for the positive step than for the negative step, and these differences are substantial. Model 3 also exhibits odd symmetry as may be seen by substituting $(-u(k), -y(k))$ into its defining equation, but is distinguishable from Model 1 because of the slight overshoot seen in its step responses; again, it is important to emphasize that these differences are only apparent if $|A|$ is sufficiently large. Conversely, if $|A|$ is large enough, both Models 2 and 3 exhibit

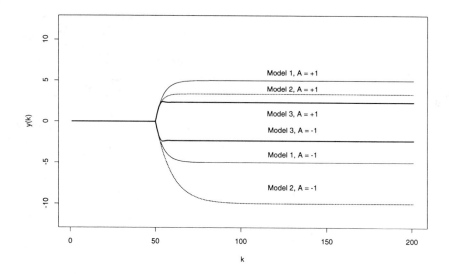

Figure 8.12: Step responses, Models 1, 2, and 3

unstable responses. At the other extreme, Model 4 is positive-homogeneous so its qualitative behavior can and does depend on the sign of A but not on $|A|$; in particular, this model exhibits a monotone response to positive steps and an oscillatory response to negative steps.

Sinusoidal responses for Models 1, 2, 3, and 4 are compared in Fig. 8.13, and similar conclusions hold as for the step responses: careful examination of the responses does indeed distinghish these four models, at least for sufficiently large input amplitudes. Model 1 exhibits a sinusoidal response, as it must by linearity, but the three nonlinear do not: Models 2 and 4 exhibit asymmetric responses, whereas Model 3 exhibits a symmetric response that is extremely nonsinusoidal. As with the step responses, Models 1, 2, and 3 are essentially indistinguishable on the basis of their responses to small amplitude sinusoids, and the responses of Models 2 and 3 become unstable for sufficiently large amplitude inputs. Again, since Model 4 is positive homogeneous, its response simply scales linearly with the input amplitude which therefore plays no role in the qualitative behavior.

The preceeding discussion also illustrates the difference between a good screening input that effectively distinguishes model classes and a good *identification input* that effectively distinguishes models *within* a class. For example, within the linear model class, the unit impulse response is—at least in principle—sufficient to distinguish all possible models, despite its inability to distinguish any of the four model classes considered in the previous example. Conversely, these model classes can be distinguished on the basis of their response to a single fixed-frequency sinusoid despite the fact that this input is *not* sufficient to characterize models within the linear model class.

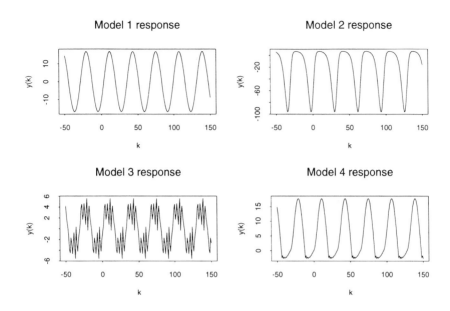

Figure 8.13: Sinusoidal responses, amplitude 4.5

Useful insights into identification effectiveness may be obtained by considering the criteria for good input sequences discussed in Chapter 7. Specifically, recall that a *root sequence* $\{r(k)\}$ for some model $\mathcal{M}$ is invariant under the action of that model; stated in simpler terms, if $\{r(k)\}$ is a root sequence for some model $\mathcal{M}$ then $\mathcal{M}[r(k)] = r(k)$. Further, it was noted that these sequences are *ineffective* as input sequences because if $\{y(k)\}$ is the response of the model $\mathcal{N}$ to the input sequence $\{r(k)\}$, it is also the response of the cascade $\mathcal{N} \circ \mathcal{M}$, the cascade $\mathcal{N} \circ \mathcal{M} \circ \mathcal{M}$, and so forth. Consequently, the input/output data pair $(r(k), y(k))$ will be unable to distinguish between these models. As a specific example, note that if $\{u(k)\}$ is any periodic sequence with period P, it is a root sequence for the trivial linear delay model $y(k) = u(k - P)$; this observation implies that effective input sequences should be aperiodic and excludes sinusoids as an immediate corollary.

The following example leads to a more general result; suppose $\{r(k)\}$ is a root sequence for the simple linear model

$$y(k) = ay(k-1) + bu(k). \tag{8.24}$$

It follows easily from the root condition $y(k) = u(k) = r(k)$ that

$$r(k) = \left(\frac{a}{1-b}\right)^k r(0) \tag{8.25}$$

for all $k \geq 0$, provided $b \neq 1$. Further, note that the cascade connection of the linear model (8.24) with itself is the second-order linear model

$$v(k) = 2av(k-1) - a^2 v(k-2) + b^2 u(k). \tag{8.26}$$

Hence, $\{r(k)\}$ is a root sequence for both of these linear models and cannot be used as an input sequence to distinguish either of them from the trivial linear model $y(k) = u(k)$. For linear model identification, this exclusion of root sequences reduces to the requirement that the input sequence be *persistently exciting*. Specifically, suppose $\{u(k)\}$ is a root sequence for the linear model

$$y(k) = \sum_{i=1}^{p} a_i y(k-i) + \sum_{i=0}^{q} b_i u(k-i). \tag{8.27}$$

This requirement implies $y(k) = u(k)$, from which it follows that

$$u(k) = [1-b_0]^{-1} \sum_{i=1}^{r} [a_i + b_i] u(k-i), \tag{8.28}$$

where $r = \max\{p, q\}$ and any undefined coefficients a_i or b_i appearing in this sum are taken as zero. Further, note that this recursion relation may be rearranged to the form

$$\sum_{i=0}^{r} \gamma_i u(k-i) = 0. \tag{8.29}$$

An input sequence $\{u(k)\}$ is said to be *persistently exciting of order* $r+1$ if no constants $\{\gamma_i\}$ exist such that Eq. (8.29) holds (Ljung, 1987, p. 362). Consequently, it follows that a sequence $\{u(k)\}$ is persistently exciting of order n if it is not a root sequence for any linear dynamic model of degree $\max\{p,q\} = n-1$. It is a standard result that persistence of excitation is a sufficient condition for a sequence $\{u(k)\}$ to be an effective input sequence for linear model identification. The practical consequences of this restriction are that effective input sequences for linear model identification cannot be too smooth.

It is also instructive to consider the following nonlinear example: suppose $\{u(k)\}$ is a binary sequence, assuming one of the two distinct values a or b for all k. It follows from direct substitution that any such binary sequence is invariant under the quadratic nonlinearity

$$g(x) = x + \gamma[ab - (a+b)x + x^2] \tag{8.30}$$

for any real γ. Hence, it follows that the response of any Hammerstein model constructed from this nonlinearity will exhibit exactly the same response to the input sequence $\{u(k)\}$ as the linear dynamic model on which this Hammerstein model is based. This example provides one of many illustrations of the unsuitability of binary sequences for nonlinear model identification noted in previous discussions. Extending this argument to m-level sequences for $m > 2$ with

The Art of Model Development 409

polynomials of order m leads to the result (Nowak and van Veen, 1994) that at least $m + 1$ distinct values are required in the identification of Volterra models involving nonlinearities of order m.

Taken together, the results just presented imply that effective input sequences for nonlinear model identification should be aperiodic, should involve transitions between enough different levels, and should not be too smooth. Conversely, practical limitations will generally restrict the class of input sequences that can be considered. Two of these limitations—sequence length and range of variation—are discussed in Sec. 8.4.2, but it is also important to note the following additional limitation. Despite the desirability of having many input transitions to elicit dynamic responses from the physical system, practical considerations often restrict this number. For example, in manufacturing processes it is frequently argued that rapidly changing inputs cause excessive wear on the control hardware (e.g., flow control valves). Conversely, two of the points illustrated in Sec. 8.5 are first, that *too few* input changes leads to highly variable parameter estimates and second, that the effective degree of nonlinearity of a physical system generally depends on both the range and dynamics of the input sequence. Therefore, while it is necessary to satisfy practical constraints, it is also necessary to perturb the physical system enough to exercise the nonlinear dynamics of interest.

8.4.2 Design parameters

In developing input sequences $\{u(k)\}$ that are both effective and practical, it is useful to consider the following four design characteristics (Pearson, 1998):

1. The sequence length N
2. The range of variation $[a, b]$
3. The distribution of $u(k)$ over this range
4. The *shape* or *frequency content* of $\{u(k)\}$.

The first two of these characteristics are fairly clear both in meaning and practical implications. As a general rule, the sequence length N should be made as long as possible to maximize the data available for model identification, but two important limitations should be borne in mind. First, the larger N is, the greater the probability that some unexpected (and generally undesirable) phenomena like outliers, unmeasurable external disturbances, or other anomalies will appear in the observed process response sequence $\{y(k)\}$. The second limitation is that data collection is often somewhat invasive or expensive, disrupting normal operation and requiring dedicated personnel, special equipment, etc. In practice, these considerations often limit the data sequence length N, sometimes rather severely. Similarly, for model identification effectiveness, it is generally desirable to make the range of variation of the input sequence as large as possible, particularly for nonlinear model identification since reasonable models cannot be obtained if the inherent nonlinearities of the physical system are not excited sufficiently by the input sequence. Conversely, the maximum allowable range of the input sequence is generally also limited by practical considerations;

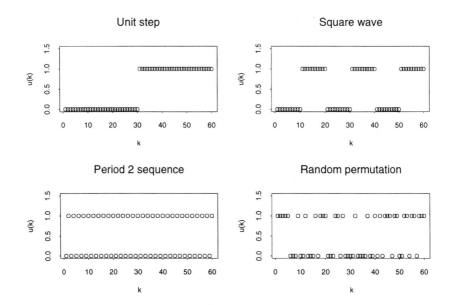

Figure 8.14: Four binary sequences

in manufacturing operations, for example, excessive changes in operating ranges can lead to the production of off-specification material or pose safety hazards. In contrast, one of the advantages of the indirect empirical modeling approach described in Chapter 1 is that practical restrictions on input sequence length and range are generally much less severe for simulations of fundamental models than for the physical process itself.

Generally, greater flexibility is possible with respect to the third and fourth input sequence characteristics listed here: distribution and shape. Further, as subsequent discussions illustrate, the influence of these design parameters can be quite profound. First, however, it is important to define these characteristics more precisely and distinguish clearly between them. Here, the term *distribution* is interpreted in the sense of traditional probability theory: it characterizes the relative likelihood that $u(k)$ assumes any particular value in the range $[a, b]$. In contrast, the *shape* of $\{u(k)\}$ determines its *smoothness* or *frequency content*. The distinction between these concepts is illustrated in Fig. 8.14, which shows four different sequences $\{u(k)\}$ all with the same distribution but each with a distinct shape.

More specifically, the four sequences shown in Fig. 8.14 all have the same binary distribution, equally likely to assume either of the two possible values $u(k) = 0$ and $u(k) = 1$. In addition, all of these sequences have the same range and the same length, so the *only* difference between these four sequences lies in their shape. In particular, the upper left plot shows the unit step sequence, equal to zero fof the first 30 samples and 1 for the last 30 samples. The upper right plot shows three cycles of a square wave with period 20 and a 50% duty

The Art of Model Development

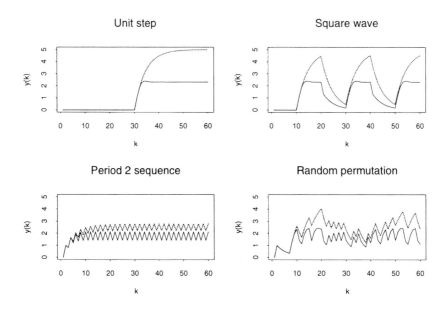

Figure 8.15: Responses of Models 1 and 3

cycle, and the lower left plot shows 30 cycles of the corresponding period 2 sequence, alternating at every time step between the two possible values 0 and 1. Finally, the lower right plot shows the results of *randomizing* this period 2 sequence: a random permutation is applied that takes the original 60 point into another, randomly re-ordered sequence. The upper two sequences may be regarded as "low frequency" input sequences since they exhibit fewer changes in value than the bottom two sequences do; analogously, the bottom two sequences may be regarded as "high frequency" input sequences.

Despite the general unsuitability of binary input sequences for nonlinear model identification, a comparison of the responses of Models 1 and 3 from Sec. 8.4.1 to these four sequences illustrates the influence of sequence shape on identification effectiveness. These models are both special cases of

$$y(k) = ay(k-1) + bu(k) + cy(k-1)y(k-2)u(k-1) \qquad (8.31)$$

with $a = 0.8$, $b = 1.0$ and either $c = 0$ (Model 1) or $c = -0.2$ (Model 3). Fig. 8.15 shows the responses for Models 1 (the upper curve in each plot) and 3 (the lower curve in each plot) to the input sequences shown in Fig. 8.14. Generally, it appears that the differences between these models are more evident from the low frequency responses shown in the upper two plots than from the high frequency responses shown in the lower two plots. This observation suggests that the low frequency input sequences would be better suited to distinguishing these models than the high frequency input sequences. The key point, however, is that the shape of the input sequence has a profound influence on its suitability, either as a screening input or as an identification input.

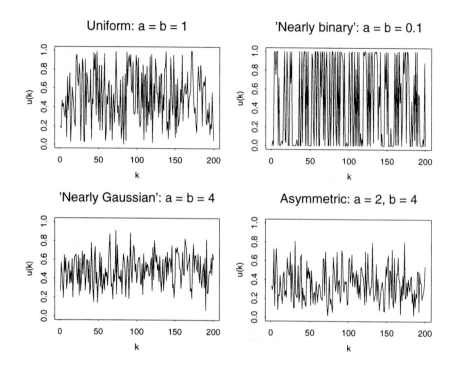

Figure 8.16: Four white noise sequences

To further illustrate the distinction between distribution and shape, Fig. 8.16 shows plots of four sequences, each with the same length ($N = 200$), the same range ($0 \leq u(k) \leq 1$ for all k), and the same shape (white noise), but different distributions. As discussed in Chapter 2, the term *white noise* refers to an independent, identically distributed sequence of random variables; a characteristic feature of these sequences is their irregularity, reflecting the fact that samples $u(k)$ and $u(j)$ are statistically independent if $k \neq j$. An extremely useful quantitative measure of the statistical dependence between samples $u(k)$ and $u(k+m)$ is the autocorrelation function, defined by

$$R_{uu}(m) = E\{u(k)u(k+m)\}. \tag{8.32}$$

For white noise sequences, it follows that

$$R_{uu}(m) = E\{u(k)\}E\{u(k+m)\} = \begin{cases} \bar{u}^2 + \sigma_u^2 & m = 0 \\ \bar{u}^2 & m \neq 0. \end{cases} \tag{8.33}$$

The term "white noise" has its origins in physics, and was coined in analogy with "white light," which is an incoherent mixture of all wavelengths or colors: if $\bar{u} = 0$, the power spectrum of a white noise sequence $\{u(k)\}$ is flat, implying energy is equally distributed at all frequencies. White noise is a theoretically attractive input sequence because it satisfies persistence of excitation conditions

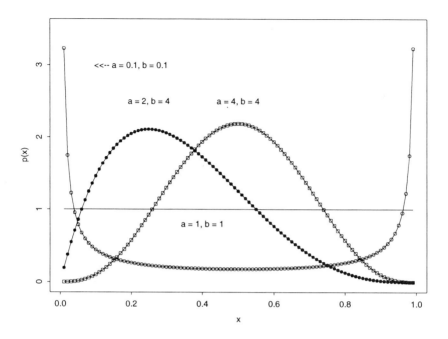

Figure 8.17: Four beta densities

like those discussed in Sec. 8.4.1 and because it can have significant advantages in developing parameter estimation algorithms, but it is often unpopular in practice for the reasons noted at the end of Sec. 8.4.1, providing one motivation for considering the input sequences introduced in Secs. 8.4.3 and 8.4.4.

Although the distributions of the four sequences shown in Fig. 8.16 are quite different, they all belong to the family of *beta distributions* (Johnson et al., 1995), whose probability density function is given by

$$p(x) = \frac{x^{a-1}(1-x)^{b-1}}{B(a,b)}. \tag{8.34}$$

Here, a and b are positive constants that determine the shape of the density function and $B(a, b)$ is a normalization constant (specifically, $B(x, y)$ is the beta function (Abramowitz and Stegun, 1972)). The four examples shown in Fig. 8.17 illustrate this range of behavior:

U Uniform: $a = b = 1.0$
B Nearly binary: $a = b = 0.1$
G Nearly Gaussian: $a = b = 4.0$
A Asymmetric: $a = 2, b = 4$,

and the white noise sequences shown in Fig. 8.16 represent i.i.d. sequences based on these four distributions. Probably the best-known member of this family is the uniform distribution, obtained by setting $a = b = 1$. More generally, the resulting density is symmetric if $a = b$, approaches a Gaussian limit as $a = b \to \infty$, and approaches a degenerate binary distribution supported on $x = 0$ and $x = 1$ as $a = b \to 0$. The key point illustrated by these four examples is that white noise need not be Gaussian, although these two terms are frequently heard together. More importantly, non-Gaussian input sequences can have significant advantages over Gaussian input sequences in some cases, a point demonstrated in Sec. 8.5. More immediately, however, it is important to illustrate the point that useful stochastic input sequences need not be white. Consequently, the following discussion introduces an important class of *dependent sequences* that appear to have certain practical advantages over i.i.d. sequences; the influence of this statistical dependence on parameter estimation and model structure selection is also considered in Sec. 8.5.

8.4.3 Random step sequences

Essentially, the distribution of a stochastic sequence $\{u(k)\}$ like the four white noise examples just considered defines the relative likelihood that $u(k)$ can assume any particular value in its admissible range. Conversely, the *dependence structure* of a stochastic sequence $\{u(k)\}$ defines the statistical relationship between samples $u(k)$ and $u(j)$ at different times $k \neq j$. An extremely useful example of a dependent input sequence is the family of *random step sequences*, defined as follows. The basic idea is to specify a certain fixed *switching probability* p and construct the sequence $\{u(k)\}$ so that it does not change value at every time step, but only every $K \sim 1/p$ steps, on average. A practical motivation for considering these sequences as identification inputs is that they require fewer changes in the value of a manipulated variable than i.i.d. sequences do, thus conforming better to the practical constraints discussed at the end of Sec. 8.4.1. More specifically, let $\{z(k)\}$ be an i.i.d. sequence defined for $k \geq 0$, with specified probability distribution; for $k = 0$, take $u(0) = z(0)$, and for $k > 0$, define $u(k)$ by:

$$u(k) = \begin{cases} z(k) & \text{with probability } p \\ u(k-1) & \text{with probability } 1-p. \end{cases} \quad (8.35)$$

Inherent in this definition is the assumption that the transitions occurring in this sequence are statistically independent: the transition probability at any time k is p, regardless of how many transitions may have occurred previously. Also, note that in the limit as $p \to 1$, the sequence $\{u(k)\}$ approaches the i.i.d. sequence $\{z(k)\}$.

Probably the most popular random step sequence in practice is that based on an i.i.d. uniform distribution (Chien and Ogunnaike, 1992; Hernandez and Arkun, 1993). One of the advantages of this choice—relative to a Gaussian alternative—is that the exact range of input values is precisely defined for the uniform distribution, but this advantage extends to the beta distributions more generally. Four examples of random step inputs are shown in Fig. 8.18, corresponding to a switching probability $p = 0.1$. These sequences were derived

The Art of Model Development

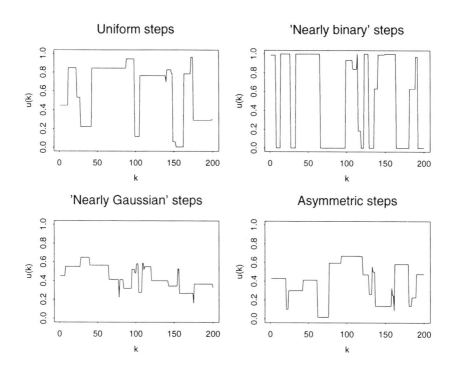

Figure 8.18: Four beta-distributed random step inputs

from the four i.i.d. beta-distributed input sequences described in Sec. 8.4.2, so the distribution of levels in these piecewise-constant sequences are uniform (U—upper left plot), nearly binary (B—upper right plot), nearly Gaussian (G—lower left plot), and asymmetric (A—lower right plot). All of these sequences are considered further in Sec. 8.5 where detailed comparisons of their identification effectiveness are presented, both in terms of the underlying distribution of values and the switching probability p.

For linear model identification, specific design criteria are available that ultimately specify the power spectrum of the input sequence (Ljung, 1987, p. 371), which is equivalent to a specification of the autocorrelation function $R_{uu}(m)$; in particular, these specifications do *not* fix the distribution of the input sequence. For nonlinear model identification, both the distribution and the autocorrelation function are important and one of the advantages of the random step sequences described here is that their autocorrelation has an extremely simple form. To see this point, first note that $E\{u(k)\} = E\{z(k)\} = \bar{z}$ and define $w(k) = z(k) - \bar{z}$, from which it follows that $\{w(k)\}$ is a zero-mean i.i.d. sequence with variance σ^2. Next, note that $u(k) = \bar{z} + v(k)$ where $\{v(k)\}$ is a zero-mean random step sequence whose autocorrelation function satisfies

$$R_{uu}(m) = \bar{z}^2 + R_{vv}(m). \qquad (8.36)$$

The advantage of this observation is that $R_{vv}(m)$ is easier to compute than

$R_{uu}(m)$. Specifically, note that for any $k \geq 0$, $v(k) = w(j)$ for some $j \leq k$ and $v(k+m) = w(i)$ for some $i \geq j$. If no transition occurs between time k and $k+m$, then $v(k) = v(k+m) = w(j)$. Thus, *conditional on this lack of transitions between time k and $k+m$*, the expectation $E\{v(k)v(k+m)\}$ has the value $E\{w^2(j)\} = \sigma^2$. Conversely, if one or more transitions occur between times k and $k+m$, it follows that $v(k) = w(j)$ and $v(k+m) = w(i)$ for some $i \neq j$; *conditional on this event*, the required expectation is

$$E\{v(k)v(k+m)\} = E\{w(i)w(j)\} = E\{w(i)\}E\{w(j)\} = 0, \quad (8.37)$$

since the sequence $\{w(k)\}$ is i.i.d. and zero mean. Hence, the autocorrelation function of $\{v(k)\}$ is given by

$$R_{vv}(m) = \text{Prob}\{\text{no transitions between } k \text{ and } k+m\} \cdot \sigma^2. \quad (8.38)$$

Define $P(k, k+m)$ as the probability that no transitions occur between times k and $k+m$, and note that $P(k, k+1) = 1 - p$ for all k. Further, because the transitions are assumed to be statistically independent, it follows that

$$P(k, k+m) = P(k, k+1) \cdot P(k+1, k+2) \cdots P(k+m-1, k+m)$$
$$= (1-p)^m. \quad (8.39)$$

The autocorrelation function for the random step sequences defined in Eq. (8.35) is therefore of the form

$$R_{uu}(m) = \bar{z}^2 + (1-p)^m \sigma^2. \quad (8.40)$$

Finally, the following three observations are useful. First, note that for beta-distributed random variables, the mean and variance are given by (Johnson et al., 1995, p. 217):

$$\bar{z} = \frac{a}{a+b} \qquad \sigma^2 = \frac{ab}{(a+b)^2(a+b+1)}. \quad (8.41)$$

Second, it is important to note that the random step sequences defined here are distributed on the range $[0, 1]$, but they may be re-scaled to any arbitrary range $[\mu_-, \mu_+]$ by defining

$$\mu(k) = \mu_- + [\mu_+ - \mu_-]u(k). \quad (8.42)$$

The mean and variance for this scaled sequence are

$$\bar{\mu} = \mu_- + [\mu_+ - \mu_-]\bar{z} \qquad \sigma_\mu^2 = [\mu_+ - \mu_-]^2 \sigma^2. \quad (8.43)$$

Third, it is interesting to note that the random step autocorrelation has the same form as that for a Gaussian $AR(1)$ autoregressive sequence with autoregressive coefficient $\phi = 1 - p$.

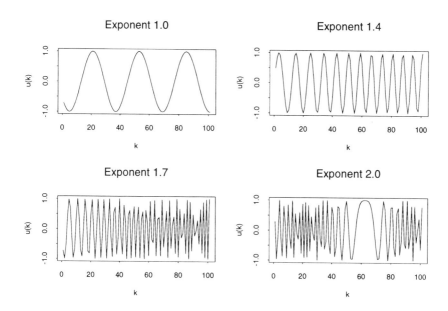

Figure 8.19: Four sine-power sequences, $\omega = \pi/16$

8.4.4 The sine-power sequences

The *sine-power sequences* constitute an interesting class of rather complex deterministic input sequences defined by the following deceptively simple equation:

$$u(k) = \alpha \sin\left[(\omega_0 k)^\gamma\right]. \tag{8.44}$$

Here, $\alpha > 0$ defines the amplitude of the sequence, which lies in the range $-\alpha \leq u(k) \leq \alpha$ for all k. The positive constant ω is a "frequency-like" parameter that determines the scaling of the discrete time-axis k, and the parameter γ lies between 1 and 2 and determines the shape of the sequence. This point may be seen in Figs. 8.19 and 8.20, which each show four plots of the sine-power sequence defined by Eq. (8.44) for $\alpha = 1$ and the same four values of γ (1.0, 1.4, 1.7, and 2.0). The difference between these two figures is the frequency parameter ω, which is $\pi/16$ for Fig. 8.19 and $\pi/3$ for Fig. 8.20. For $\omega = \pi/16$, the sinusoid obtained by setting $\gamma = 1$ (upper left plot in Fig. 8.19) corresponds to the standard sinusoidal input sequence introduced in Chapter 2, whereas for $\gamma = 1.4$ the resulting sequence exhibits a gradual increase in frequency. For $\gamma = 1.7$, the behavior of the sequence $\{u(k)\}$ is more complex, exhibiting burst-like phenomena near the end of the data record, and for $\gamma = 2.0$ the resulting sequence is highly irregular, containing segments exhibiting both slow and rapid variation.

For $\omega = \pi/3$, the behavior of the resulting sequences becomes even more irregular with increasing γ. In particular, note that for $\gamma = 1.4$ (upper right plot in Fig. 8.20) the resulting sequence appears quite similar to that obtained

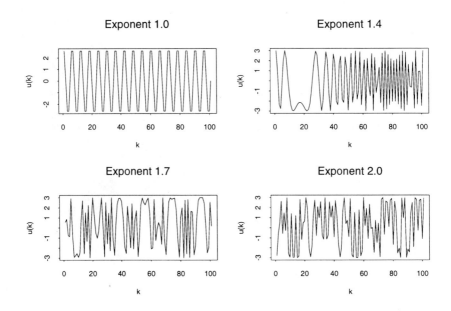

Figure 8.20: Four sine-power sequences, $\omega = \pi/3$

for $\gamma = 2.0$ and $\omega = \pi/16$; the greater irregularity of these sequences relative to those for $\omega = \pi/3$ is apparent in the bottom two plots in Fig. 8.20. In fact, the general behavior of these last two plots is reminiscnet of the irregularity of chaotic sequences. This observation should not be too surprising, since the most chaotic solution of the logistic equation may be expressed in a closed form that is quite similar to the defining equation for the sine-power sequence. Specifically, recall that the logistic equation is

$$y(k) = \alpha y(k-1)[1 - y(k-1)], \tag{8.45}$$

where $0 < y(0) < 1$ and $0 < \alpha \leq 4$. As discussed in Chapter 1, the sequence $\{y(k)\}$ is chaotic for sufficiently large α, becoming "most chaotic" for $\alpha = 4$. It may be verified by simple trigonometric identities that this particular chaotic sequence is given by

$$y(k) = \sin^2(2^k). \tag{8.46}$$

The inspiration for considering sine-power sequences as input sequences for model characterization was a problem proposed in *The American Mathematical Monthly* (MacKinnon, 1993), noting the exotic behavior of the sums:

$$S_\gamma(n) = \sum_{r=1}^{n} \sin(r^\gamma) \tag{8.47}$$

for $1 < \gamma < 2$. It is not difficult to show that these sums represent the response

The Art of Model Development

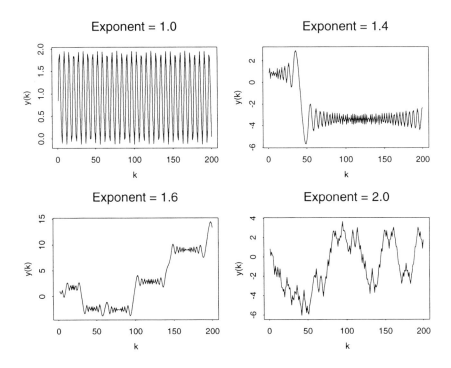

Figure 8.21: Sums of sine-power sequences

of the neutrally stable *unit root* model popular in economic time-series analysis:

$$y(k) = y(k-1) + u(k), \qquad (8.48)$$

to the sine-power input sequence defined in Eq. (8.44) for $\alpha = 1$ and $\omega = 1$. MacKinnon noted the transition behavior shown in Fig. 8.21 for these sums: they appear to oscillate about a constant value for some period of time and then switch suddenly, oscillating about some new constant value. He further notes that the duration of these oscillatory regions between transitions decreases with increasing γ. This point is illustrated in Fig. 8.21, which shows $S_\gamma(n)$ plotted vs. n for $\gamma = 1$, 1.4, 1.6, and 2.0. The case $\gamma = 1$ can be evaluated analytically (Gradshteyn and Ryzhik, 1965, 1.342) to obtain

$$S_1(n) = \sin\left(\frac{n+1}{2}\right) \sin\left(\frac{n}{2}\right) \csc\left(\frac{1}{2}\right). \qquad (8.49)$$

Manipulation of trigonometric identities further reduces this expression to

$$S_1(n) = \frac{1}{2}\cot\left(\frac{1}{2}\right) - \frac{\cos(n+1/2)}{2\sin(1/2)} \simeq 0.915 - 1.043\cos(n+1/2). \qquad (8.50)$$

Hence, it follows that $S_1(n)$ is an almost-periodic function, exhibiting persistent oscillation about the value 0.915. This behavior is clearly seen in the upper left plot in Fig. 8.21.

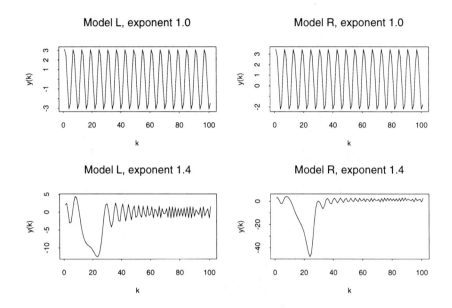

Figure 8.22: Sine-power responses, Models L vs. R

Viewing these sums as responses of the unit-root linear autoregressive model (8.48) to sine-power input sequences, it is clear that this linear model is being excited in a most unusual fashion. In particular, although the sine-power sequences are simple to generate and clearly determinisitc, for $\gamma > 1$ they appear to be able to elicit more information from the model than less complex deterministic sequences like impulses, steps, or sinusoids. In fact, these sequences may be useful screening inputs, as the example shown in Fig. 8.22 illustrates. The four plots in this figure compare the responses of the following two models:

L The linear model

$$y(k) = 0.8y(k-1) + u(k)$$

R Robinson's AR-Volterra model

$$y(k) = 0.8y(k-1) + u(k) + u(k-1) + u(k)u(k-1).$$

The left-hand plots show the responses obtained from Model L for sine-power input sequences with $\alpha = 3.0$, $\omega = \pi/3$, and $\gamma = 1.0$ (upper plot) or $\gamma = 1.4$ (lower plot). For comparison, the right-hand plots show the responses of Model R to the same input sequences. Close examination of the upper plots reveals a difference in amplitude, but no immediate evidence of the qualitative difference between these models. In contrast, the lower two plots exhibit visually pronounced differences both in detail and scale. These differences suggest that $\gamma = 1.4$ should be a much more effective input sequence for either distinguishing the model structures L and R or in identifying the parameters for model structure R than $\gamma = 1.0$.

The Art of Model Development

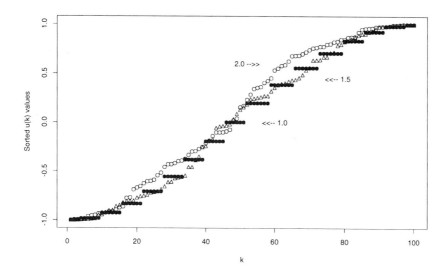

Figure 8.23: Distribution of values, $\gamma = 1.0, 1.5, 2.0$

In terms of the input sequence design parameters introduced in Sec. 8.4.2, the amplitude parameter α determines the range of the sine-power sequence and the parameters ω and γ determine the shape or frequency content of these sequences. Conversely, an interesting feature of these sequences is that they all have essentially the same *distribution* over the range $[-\alpha, \alpha]$. This point is illustrated in Fig. 8.23, which compares the distribution of 100 sample sequences obtained for $\gamma = 1.0$, 1.5 and 2.0; in all cases, $\alpha = 1$ and $\omega = \pi/16$. More specifically, Fig. 8.23 shows overlaid plots of the sequences obtained from $\{u(k)\}$ by rank-ordering them as

$$u_{(1)} \leq u_{(2)} \leq \cdots \leq u_{(100)}. \tag{8.51}$$

Because the sequence $\{u(k)\}$ for $\gamma = 1$ is periodic with period $\pi/16$, the corresponding rank-ordered sequence contains only 16 distinct values, indicated by the solid circles in Fig. 8.23. For $\gamma = 1.5$ and 2.0, the sequence $\{u(k)\}$ is no longer periodic, but the point of this plot is that the distributions of values for the sine-power sequences exhibits only small variation with γ. Similar conclusions hold for the frequency variable ω.

In contrast, the frequency content of the sine-power sequences varies greatly with both ω and γ. Since $\gamma = 1$ reduces this sequence to a sinusoid of frequency ω, variation of the frequency content of $\{u(k)\}$ with ω is to be expected, but the range of variation of the frequency content with γ is both less obvious and more significant, as shown in Figs. 8.24 and 8.25. The upper plots in Fig. 8.24 show 200 point sine-power sequences with $\alpha = 1$, $\omega = \pi/3$, and $\gamma = 1.0$ (left-hand plot) or $\gamma = 1.2$ (right-hand plot). The corresponding bottom plots show the

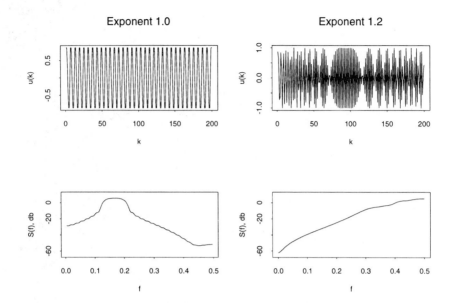

Figure 8.24: Sequences and spectra, $\gamma = 1.0, 1.2$

estimated power spectra obtained from these sequences using the same multi-window spectral estimation procedure (Riedel and Sidorenko, 1995). In these plots, the scaling is such that 0.5 corresponds $f = \pi$ and the broad peak around $f = 0.5/3 \simeq 0.167$ in the lower left plot represents the expected sinusoidal peak, broadened by the spectral estimation method used here. The point of this example is that the power spectrum shown in the lower right plot for $\gamma = 1.2$ was computed by the same procedure and is plotted on the same scale, so differences in these spectra reflect differences in the frequency content of the sine-power sequences for $\gamma = 1.0$ and $\gamma = 1.2$. In particular, note that the low-frequency peak present for $\gamma = 1.0$ is completely absent for $\gamma = 1.2$ and the power in the signal increases significantly witn increasing frequency.

The upper plots in Fig. 8.25 show the sine-power sequence with $\gamma = 2.0$ and the same α and ω parameter values as before (upper left plot), and a white noise sequence of the same duration, uniformly distributed over the same range. Visual comparison of these plots again emphasizes the irregularity of the sine-power sequence for $\gamma = 2.0$, particularly relative to the sequences shown in Fig. 8.24. Again, the lower two plots show the estimated power spectra for these sequences, obtained by the same method as before. The point of this comparison is to illustrate that both the sine-power sequence and the uniform white noise sequence exhibit essentially constant power spectra.

In summary, this discussion has introduced the class of sine-power sequences defined by Eq. (8.44) and characterized by three parameters. The parameter α defines the range of the resulting sequence $\{u(k)\}$ (i.e., $-\alpha \leq u(k) \leq \alpha$ for all k), and the parameters ω and γ define the shape of the sequence. In particular, the

The Art of Model Development

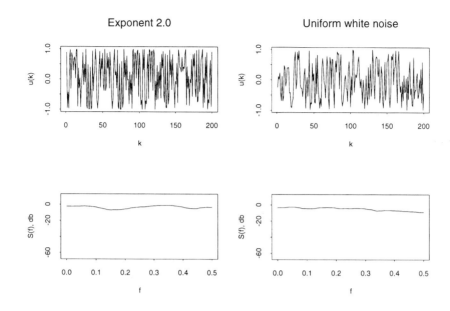

Figure 8.25: Uniform white noise vs. $\gamma = 2.0$

spectral concentration of the sequence (i.e., the frequency range over which the sequence exhibits significant power) increases from the infinitely narrow band sinusoidal sequence of frequency ω for $\gamma = 1$ to a sequence that is quite similar in character to white noise for $\gamma = 2$ and ω sufficiently large (e.g., $\omega = \pi/3 \simeq 1$ in the example considered here). In contrast, the distribution of these sequences is essentially independent of ω and γ except for $\gamma = 1$ and $\omega = \pi/m$ for small integers m, for which $u(k)$ assumes exactly m distinct values evenly distributed over the interval $[-\alpha, \alpha]$. Overall, for $\gamma > 1$, the sine-power sequences appear to be promising candidates for empirical modeling, particularly as screening inputs where even sinusoidal input sequences often yield useful evidence of nonlinearity.

8.5 Case study 2—structure and inputs

To illustrate many of the ideas discussed in the preceeding sections of this chapter, the following section presents a brief empirical model identification case study. The physical process considered in this study is a first-principles reactor model (Eaton and Rawlings, 1990) and, although this example is unrealistically simple, this simplicity permits the development of an exact discretization, just as in the case of continuous-time linear models. Unlike the linear case, however, the structure of this discretization is quite different from that of the original continuous-time model. The following sections present detailed descriptions of the Eaton-Rawlings reactor model (Sec. 8.5.1), its exact discretization (Sec. 8.5.2), analytic and empirical linear approximations of the reactor dynamics (Sec. 8.5.3), and 10 empirical nonlinear approximations of the reactor dynamics (Sec. 8.5.4).

8.5.1 The Eaton-Rawlings reactor model

The model considered here describes a second-order reaction occurring in an isothermal CSTR. Denoting the reactant concentration in the CSTR as $y(t)$, the dynamics of this reactor are described by the first-order, nonlinear ordinary differential equation

$$\frac{dy}{dt} = -hy^2 - \frac{yu}{V} + \frac{du}{V} \qquad (8.52)$$

where h is the kinetic rate constant for the reaction, V is the reactor volume, and d is the inlet concentration of the reactant. The maniuplated variable in this model is the inlet flow rate u, but the following normalization simplifies subsequent results:

$$\mu = \frac{u}{2hV}. \qquad (8.53)$$

In terms of this input variable, Eq. (8.52) becomes:

$$\frac{dy}{dt} = -h[y^2 + 2\mu y - 2d\mu]. \qquad (8.54)$$

Given a steady-state input $\mu(t) = \mu_s$, the corresponding steady-state output $y(t) = y_s$ is obtained by setting the time-derivative to zero in Eq. (8.54). This substitution leads to the quadratic equation

$$y_s^2 + 2\mu_s y_s - 2d\mu_s = 0 \qquad (8.55)$$

which has the two roots:

$$y_s = -\mu_s \pm \sqrt{\mu_s^2 + 2d\mu_s}. \qquad (8.56)$$

Because concentration must be a positive quantity, only the first of these roots is physically meaningful.

These constant model parameters are assumed to have the values considered previously (Eaton and Rawlings, 1990; Ogunnaike and Ray, 1994):

$$h = 1.50 \text{ liter/mole-hr}, \quad V = 10.51 \text{ liter}, \quad d = 3.5 \text{ mole/liter}.$$

These studies assumed that the flow rate u varies between 1.0 and 3.0 liters per hour and this operating range will be adopted as nominal or mid range in the discussion presented here. In addition, however, it will be useful to also consider a low range (0.5 to 1.0 liters per hour), a high range (3.0 to 5.0 liters per hour), and a wide range (0.5 to 5.0 liters per hour). The values of the scaled input variable $\mu = u/2hV$ for these four operating ranges are given in Table 8.10,

Table 8.10: Operating ranges and steady-state values

Range	Min μ	Max μ	μ_s	y_s
low	0.015	0.030	0.023	0.38
mid	0.030	0.090	0.060	0.59
high	0.090	0.150	0.120	0.80
wide	0.015	0.150	0.083	0.68

along with the associated steady-state values u_s and y_s defined as follows. For each of these four operating ranges, the steady-state input is defined as

$$\mu_s = \frac{\mu_- + \mu_+}{2}, \tag{8.57}$$

where μ_- is the lower limit of the operating range, and μ_+ is the upper limit of the operating range. The associated steady-state response values y_s are computed from Eq. (8.56).

The responses of the Eaton-Rawlings reactor model to uniformly distributed random steps with 10% switching probability are shown in Fig. 8.26 for each of these operating ranges. These responses demonstrate how much larger the range of variation of $y(t)$ is for the wide range input sequences than for the other three ranges. Consequently, the effects of the nonlinearity can be expected to be the most pronounced for this input range and subsequent discussions will be restricted to this case.

8.5.2 Exact discretization

In the case of continuous-time *linear* models, it is possible to derive an exact discretization if it is assumed that the control input can be changed only at regular sampling times t_k. That is, a discrete-time equation can be derived that exactly describes the evolution of the variable $y(t)$ from one sampling time to the next. This point was discussed in some detail in Chapter 1, where it was noted that the key to the development of this model is knowledge of the form of the exact solution to the continuous-time linear model equation. Although the Eaton-Rawlings reactor model is nonlinear, it is simple enough that its exact solution can be determined and, from this analytic solution, an exact discretization is possible just as in the linear case. More specifically, suppose the concentration $y(t_{k-1})$ is known at time $t_{k-1} = (k-1)T$ and that $\mu(t) = \mu(k-1)$ is constant for $(k-1)T \leq t < kT$. Eq. (8.54) may then be rearranged to yield

$$\frac{dy}{y^2 + 2\mu(k-1)y - 2d\mu(k-1)} = -hdt, \tag{8.58}$$

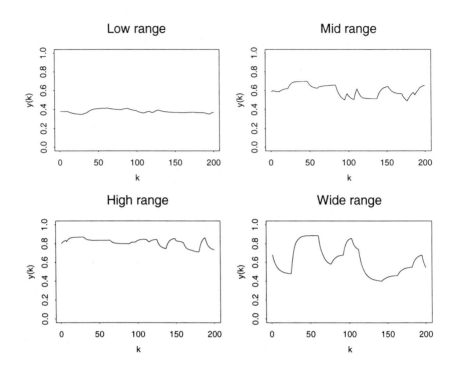

Figure 8.26: Typical identification output sequences

which can be integrated on the interval $[t_{k-1}, t_k]$ to obtain an explicit expression for $y(k) = y(t_k)$. The general form of this solution depends on the discriminant

$$\Delta = -8d\mu(k-1) - 4\mu^2(k-1). \qquad (8.59)$$

Because $\mu(k-1) \geq 0$ and $d \geq 0$, this quantity is negative, and integrating Eq. (8.58) yields the solution (Gradshteyn and Ryzhik, 1965, 2.172)

$$\frac{-1}{\phi(k-1)} \tanh^{-1}\left\{\frac{y(t) + \mu(k-1)}{\phi(k-1)}\right\} = -ht + C. \qquad (8.60)$$

Here, $\phi(k-1)$ is the quantity

$$\phi(k-1) = \sqrt{\mu^2(k-1) + 2d\mu(k-1)}, \qquad (8.61)$$

and C is a constant of integration to be determined by the initial condition $y(t) = y(k-1)$ at $t = t_{k-1}$. Solving for this constant of integration yields

$$C = h(k-1)T - \frac{1}{\phi(k-1)} \tanh^{-1}\left\{\frac{y(k-1) + \mu(k-1)}{\phi(k-1)}\right\}. \qquad (8.62)$$

The Art of Model Development

Substituting this result into Eq. (8.60) and rearranging slightly yields the following discrete-time model relating $y(k)$ to $y(k-1)$ and $\mu(k-1)$:

$$\tanh^{-1}\left\{\frac{y(k)+\mu(k-1)}{\phi(k-1)}\right\} = \tanh^{-1}\left\{\frac{y(k-1)+\mu(k-1)}{\phi(k-1)}\right\} + hT\phi(k-1). \tag{8.63}$$

This model may be simplified somewhat by taking the hyperbolic tangent of both sides and invoking the addition formula

$$\tanh(x+y) = \frac{\tanh x + \tanh y}{1 + \tanh x \tanh y}. \tag{8.64}$$

This simplification yields

$$y(k) = \{1 + \frac{y(k-1)+\mu(k-1)}{\phi(k-1)}\tanh[hT\phi(k-1)]\}^{-1}$$
$$\{y(k-1) + \mu(k-1) + \phi(k-1)\tanh[hT\phi(k-1)]\}$$
$$- \mu(k-1). \tag{8.65}$$

Further simplification of this expression is possible by defining the function

$$\tau(k-1) = \frac{\tanh[hT\phi(k-1)]}{\phi(k-1)}$$
$$= \frac{\tanh[hT\sqrt{\mu^2(k-1)+2d\mu(k-1)}]}{\sqrt{\mu^2(k-1)+2d\mu(k-1)}}. \tag{8.66}$$

Rearrangement of Eq. (8.65) in terms of this function yields the following final result:

$$y(k) = \frac{[1-\tau(k-1)\mu(k-1)]y(k-1) + 2d\tau(k-1)\mu(k-1)}{1+\tau(k-1)[y(k-1)+\mu(k-1)]}. \tag{8.67}$$

There are a number of important points to note concerning this discrete-time model. First, it is important to emphasize that no approximations were made here: if the input variable $\mu(t)$ is constant between sampling times t_{k-1} and t_k, Eq. (8.67) *exactly* corresponds to the solution of the original differential equation (8.52). Despite this exact correspondence, note the lack of structural similarity between these models. In particular, although the original continuous-time model involved the second-order nonlinear terms y^2 and uy, the discrete-time model may be viewed as a rational NARMAX model involving $y(k-1)$, $\mu(k-1)$, and the complicated static nonlinearity defining $\tau(k-1)$. One structural feature that *is* preserved between the continuous-time model (8.52) and the discrete-time model (8.67) is the dynamic order: both equations define first-order dynamic models. This point was made in Chapter 1 in discussing the exact discretization of state-space models, and the example presented here illustrates the observation made there that such order-preserving transformations tend to be extremely complicated, even when the starting model is a simple one.

8.5.3 Linear approximations

The traditional approach to developing an approximation of the Eaton-Rawlings CSTR model would be to first linearize the continuous-time model about its intended steady-state operating condition and then discretize this linear model. To provide a basis for comparison of subsequent results, the following paragraphs briefly discuss both this traditional derivation and empirical linear approximations of the reactor dynamics. In particular, these results provide a basis for comparing both goodness-of-fit and model parameter estimates for the nonlinear models considered in Sec. 8.5.4.

To derive the traditional analytic linearization, first assume steady-state values μ_s and y_s for the normalized inlet flow rate and reactor concentration, respectively. Then, define the deviation variables

$$z(t) = y(t) - y_s \qquad v(t) = \mu(t) - \mu_s. \tag{8.68}$$

Substituting these expressions into the original model (8.54) yields the following differential equation relating $z(t)$ and $v(t)$:

$$\frac{dz}{dt} = -2h(y_s + \mu_s)z + 2h(d + y_s)v - 2hvz - hz^2. \tag{8.69}$$

To obtain a linearized continuous-time approximation to the original nonlinear model equation, neglect the second-order terms vz and z^2 to obtain the approximate linear model:

$$\frac{dz}{dt} = -az + bu, \tag{8.70}$$

where the model coefficients are given by $a = 2h(y_s + \mu_s)$ and $b = 2h(d + y_s)$.

As discussed in Chapter 1, if $z(t)$ is sampled uniformly at times $t_k = kT$, it is possible to obtain a discrete-time equation that exactly describes the evolution of $z(t)$ at the sample times t_k. In particular, this equation is of the form

$$z(k) = \alpha z(k-1) + \beta v(k-1), \tag{8.71}$$

and the model coefficients α and β are determined by the coefficients a and b in Eq. (8.70). The simplest way to obtain the coefficient α is to consider the response of the models (8.70) and (8.71) to the same initial condition, say $z(0) = 1$. For the continuous-time model, this response is

$$z(t) = e^{-at}, \tag{8.72}$$

whereas for the discrete-time model, the response is

$$z(k) = \alpha^k. \tag{8.73}$$

Hence, setting $z(k) = z(t_k) = z(kT)$ yields the relation

$$\alpha = e^{-aT}. \tag{8.74}$$

Table 8.11: Affine model parameter values

Range	y_0	α	β
low	0.016	0.89	1.08
mid	0.038	0.82	1.13
high	0.058	0.76	1.12
wide	0.045	0.80	1.10

Given a, it is possible to determine b by requiring the continuous-time model (8.70) and the discrete-time model (8.71) to have the same steady-state gain. This requirement leads to the result

$$\beta = (1-\alpha)(b/a). \tag{8.75}$$

Since the empirical modeling results considered here are expressed in terms of the original variables $\mu(k)$ and $y(k)$ rather than the deviation variables $v(k)$ and $z(k)$, it is useful to reexpress Eq. (8.71) in terms of these variables. Expressing $z(k)$ and $v(k)$ in terms of $y(k)$, y_s, $\mu(k)$, and μ_s leads immediately to the following affine model:

$$y(k) = y_0 + \alpha y(k-1) + \beta \mu(k-1), \tag{8.76}$$

where the constant y_0 is given by

$$y_0 = (1-\alpha)y_s + \beta \mu_s. \tag{8.77}$$

In the present example, the parameters a and b in Eq. (8.70) depend on the steady-state operating values μ_s and y_s. Hence, the four operating ranges considered here lead to four different affine models of the form (8.76). The model coefficients y_0, α, and β defining these models are listed in Table 8.11. Note that these values all exhibit nonnegligible variation with the input range: β varies the least (on the order of $\pm 2\%$), followed by α (on the order of $\pm 8\%$), whereas y_0 varies the most (nearly $\pm 70\%$).

Before proceeding to the development and analysis of nonlinear empirical approximations to the Eaton-Rawlings reactor response, it is useful to briefly consider affine emprical approximations and compare them with the analytic linearization results just presented. Specifically, the exact discretization derived in Sec. 8.5.2 is used as a surrogate process: an input sequence $\{u(k)\}$ is generated, the corresponding output sequence $\{y(k)\}$ is computed, and affine models of the form (8.76) are identified empirically from this input-output data. The input sequences $\{u(k)\}$ belong to the family of random step sequences described in Sec. 8.4.3, based on the four specific beta distributions designated U, B, G, and A in Sec. 8.4.2. In all cases, the standard beta distribution is translated and scaled so that the standardized range $[0, 1]$ corresponds to the wide range defined in Table 8.10. For convenience, these sequences are designated "Uxyz" where x.yz represents the switching probability p (i.e., U010 represents a uniformly distributed random step sequence with switching probability $p = 0.10$). Once these choices are specified, 100 statistically independent sequences of this type are generated, each of length $N = 200$, and the reactor responses to these input sequences are then computed from the exact discretization described in Sec. 8.5.2, assuming a sampling time of $T = 0.1$ minutes. Finally, the parameters y_0, α, and β that minimize the least squares prediction error measure

$$J(y_0, \alpha, \beta) = \sum_{k=1}^{N} [\hat{y}(k) - y(k)]^2 \tag{8.78}$$

$$= \sum_{k=1}^{N} [y(k) - y_0 - \alpha y(k-1) - \beta u(k-1)]^2.$$

are determined for each of these 100 input-output sequence pairs. Comparing the identified affine model coefficients obtained for different input sequences yields some insight into the influence of both the distribution and the dependence structure of these sequences on the results.

Because it is of interest to summarize and compare several different identification scenarios, it is convenient to present these results in the form of box plots (Hoaglin et al., 1991) like those shown in Fig. 8.27. As the name implies, these plots consist of "boxes," each constructed from the following five numbers:

1. The minimum data value, v_1
2. The lower quartile, v_2
3. The median or middle value, v_3
4. The upper quartile, v_4
5. The maximum data value, v_5.

More specifically, these data values have the following interpretation:

1. 0% of the data values are less than v_1,
2. 25% of the data values are no larger than v_2,
3. 50% of the data values are no larger than v_3,
4. 75% of the data values are no larger than v_4,
5. 100% of the data values are no larger than v_5.

Given these numbers, a set of rectangular axes is constructed and horizontal lines are drawn at each value, v_1 through v_5. In the plots presented here, the median is indicated by a white line in the center of the plot, the upper and lower quartiles define the boundaries of a black box that contains this white line, and the extremes v_1 and v_5 are indicated by horizontal lines at the top and bottom of the plot, connected by vertical dashed lines to the black box in

The Art of Model Development

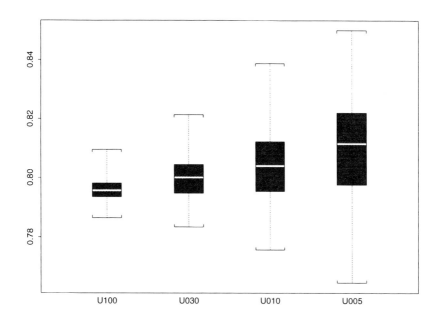

Figure 8.27: Estimated α parameter vs. switching probability

the center. Hence, in the results presented here, the white line at the center of each plot represents the typical estimate obtained from these simulations, and the black box surrounding this line defines the range of values seen in 50% of the simulations.

As a specific example, Fig. 8.27 shows the estimates obtained for the α parameter for four different input sequences. These results illustrate two general trends: first, the variability of the estimated parameter increases significantly as the switching probability p decreases and second, there is a significant increase in the median parameter estimate as p decreases. Analogous results are seen in the corresponding plots of the estimated β and y_0 parameters, but there is an important difference: the median α parameter estimates are in reasonable agreement with the linearized model value ($\alpha = 0.80$) listed in Table 8.11, whereas the estimated β and y_0 parameter values are not. In particular, all of the estimated β values are significantly smaller than the linearized value $\beta = 1.10$ and all of the estimated y_0 values are significantly larger than the linearized value $y_0 = 0.045$. These differences may be attributed to the fact that the empirical model seeks a best fit between the observed model response and the affine model prediction for a specific input sequence, whereas the analytic linearization approximates the response of the nonlinear model to infinitesimal deviations from a specified steady-state.

8.5.4 Nonlinear approximations

The following paragraphs present a detailed comparison of the results obtained by fitting the input/output behavior of the Eaton-Rawlings model to a nonlinear empirical model of the form:

$$y(k) = y_0 + \alpha y(k-1) + \beta u(k-1) + \gamma \nu(k-1). \tag{8.79}$$

In this model, y_0, α, β, and γ are unknown parameters, chosen to minimize the least square error criterion

$$J(y_0, \alpha, \beta, \gamma) = \sum_{k=3}^{N} [\hat{y}(k) - y(k)]^2 \tag{8.80}$$

$$= \sum_{k=3}^{N} [y(k) - y_0 - \alpha y(k-1) - \beta u(k-1) - \gamma \nu(k-1)]^2.$$

Note that if γ is constrained to be zero, this problem reduces to the affine empirical modeling problem considered in the previous section. Therefore, it follows that the optimum nonlinear model achieves at least as small a value of $J(y_0, \alpha, \beta, \gamma)$ as the optimum affine model does. *Conversely, it is important to note that this error measure is not the same as the prediction error criterion used to assess model adequacy in the following discussions.* In particular, the error measure $J(y_0, \alpha, \beta, \gamma)$ represents the *one-step prediction error* for the observation $y(k)$, given model parameters y_0, α, β, and γ, past input values $u(k-1)$ and possibly $u(k-2)$ (depending on the choice of $\nu(k-1)$—see below), *along with the past output observation $y(k-1)$ and possibly $y(k-2)$ (again, see below)*. In contrast, the *(total) prediction error* used to assess model performance in subsequent discussions summarizes the model's ability to predict $y(k)$ given the same model parameters *and past inputs alone*. The difference between this (total) prediction error and the one-step prediction error is important and is discussed further below.

The term $\nu(k-1)$ appearing in Eq. (8.79) defines the specific nonlinear model structure considered and is one of the ten nonlinear terms listed in Table 8.12. The rationale for selecting these particular models is as follows. First, note that all models are of comparable complexity in that they all involve exactly the same number of unknown parameters. Therefore, comparisons between these ten models represent comparisons of different *model structures* without having to correct for complexity. In particular, if Model i seems to out-perform Model j, this result *does not* arise from the fact that Model i has more degrees of freedom than Model j. Next, note that all of these models involve second-order nonlinearities of one of the following three forms:

$$\nu(k-1) = u^m(k-i)u^n(k-j),$$
$$\nu(k-1) = u^m(k-i)y^n(k-j),$$
$$\nu(k-1) = y^m(k-i)y^n(k-j),$$

Table 8.12: Ten nonlinear model terms $\nu(k-1)$

Model	Nonlinearity $\nu(k-1)$	Model structure
1	$u^2(k-1)$	Hammerstein
2	$u^2(k-2)$	Zhu and Seborg
3	$u(k-1)u(k-2)$	Robinson's AR-Volterra model
4	$u(k-1)y(k-1)$	bilinear, diagonal
5	$u(k-1)y(k-2)$	bilinear, superdiagonal
6	$u(k-2)y(k-1)$	bilinear, subdiagonal
7	$u(k-2)y(k-2)$	bilinear, diagonal, 2nd-order
8	$y^2(k-1)$	1st-order NARX (controlled logistic)
9	$y^2(k-2)$	2nd-order NAARX
10	$y(k-1)y(k-2)$	2nd-order NARX, non-additive

where $n + m = 2$, and both i and j are either 1 or 2. In fact, these ten terms represent all of the possible individual nonlinear terms in the NARMAX model:

$$y(k) = F(y(k-1), y(k-2); u(k-1), u(k-2)), \qquad (8.81)$$

where $F : R^4 \to R^1$ is restricted to be a second-order polynomial. Comparing goodness-of-fit measures for the ten models defined by the entries in Table 8.12 is therefore comparable to asking the question, In a stepwise regression procedure, which nonlinear term would be added first to the affine model developed in Sec. 8.5.3? One of the key points of this example is that the answer to such questions can depend strongly on the input sequences on which the results are based.

Fig. 8.28 presents four box-plot summaries of the relative RMS prediction errors obtained for the ten nonlinear models considered here, along with the affine reference model (Model 0) considered previously. As before, each of these box-plots represents a summary of 100 model identifications based on the exact discretization of the Eaton-Rawlings reactor model driven by 100 statistically independent input sequences, each of length $N = 200$. Model parameters are estimated by the method of ordinary least squares, and prediction errors are computed as the differences between the Eaton-Rawlings reactor responses and the empirical model responses to the input sequences alone. The four box-plots shown in Fig. 8.28 illustrate the dependence of this goodness-of-fit measure on the switching probability p on which the random step inputs are based; the input sequences considered here are uniformly distributed random steps with switching probabilities of $p = 1.00, 0.30, 0.10$, and 0.05.

Comparing all of these box plots, certain general trends are evident. First, note that the magnitude of the prediction errors generally increases with de-

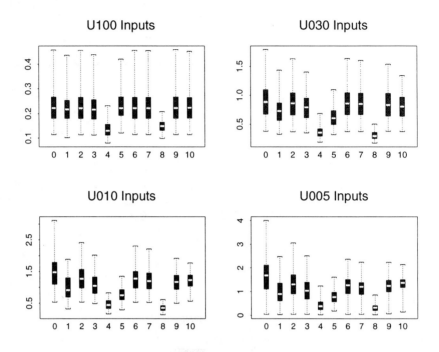

Figure 8.28: Influence of switching probability on RMS error

creasing p, a point that may be seen clearly by comparing either the median prediction errors or the worst-case prediction errors. Also, note that the differences in relative prediction error between the eleven models considered is a similarly strong function of p. For example, the difference between Models 5 and 6 is small for the U100 input sequences, and substantially greater for the U030 input sequences, and still larger for the U010 input sequences; the trend then reverses for the U005 input sequences, where the differences are again comparable to those seen for the U030 input sequences. Another, particularly important, observation is that Models 4 and 8 consistently exhibit the best performance, a point considered in some detail in subsequent discussions. Here, note that the U100 input sequences essentially divide the models into two groups: the "good" models (4 and 8) *vs.* the "poor" models (all of the others). As p decreases from 100% to 30% to 10%, differences between models become more pronounced (for example, compare the Hammerstein Model 1 with the modified Hammerstein Model 2). For $p = 5\%$, the differences seen between the eleven models appears comparable to that seen for $p = 10\%$. Finally, note that the variability in these goodness-of-fit results increases consistently as $p \to 0$. As a specific example, note that the maximum relative error is approximately 0.2% for Model 4 for the U100 input sequences, increases to $\sim 0.7\%$ for the U030 and U010 inputs, and is on the order of 1% for the U005 inputs. The corresponding numbers for Model 0 (the affine reference model) are $\sim 0.4\%$ for the U100 inputs, $\sim 1.5\%$ for

The Art of Model Development

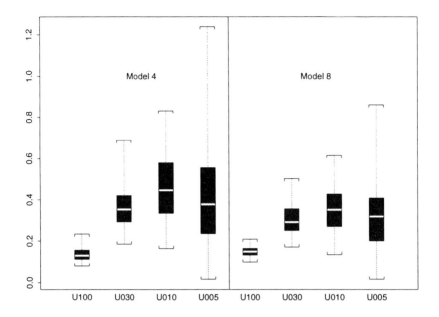

Figure 8.29: Comparison of Models 4 and 8, uniform steps

the U030 inputs, $\sim 3.0\%$ for the U010 inputs, and $\sim 4.0\%$ for the U005 inputs. This decrease in *consistency-of-fit* arises from the fact that it is the transitions in the input sequences that generate the responses from which the parameters are estimated. Hence, it is to be expected that more consistent parameter estimates and model fits will be obtained from sequences with more transitions (p large) than from sequences with fewer transitions (p small).

Because they are consistently the best models in terms of RMS prediction errors, Fig. 8.29 presents a detailed view of the box-plots shown in Fig. 8.28 for Models 4 and 8. The general decrease in consistency-of-fit with decreasing p just noted is even more clearly evident in these plots. A point that is not as clear from Fig. 8.28 is that the median RMS prediction error appears to exhibit a maximum value at $p \sim 0.10$ for both of these models. Since these models are the best of the eleven considered, it follows that among the uniformly distributed random step inputs considered here, the responses to the U010 sequences are the most difficult to model.

Part of the reason for this behavior may be seen in Fig. 8.30, which shows the influence of the switching probability p on the magnitude of the Eaton-Rawlings reactor response for uniformly distributed random step sequences. For $p = 1.00$, the input sequence changes values at every time step k, and the reactor dynamics are not fast enough to exhibit large responses to these input changes. Consequently, affine models can provide reasonable approximations

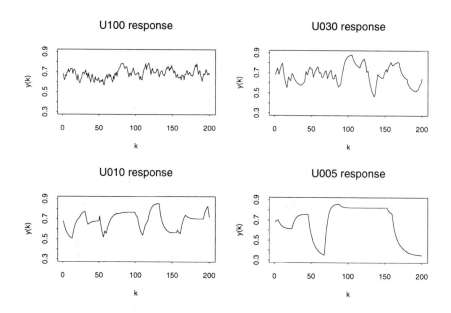

Figure 8.30: Influence of p on magnitude of response

to the process dynamics, a point that may be seen clearly in the parameter estimation results presented later in this discussion (refer to the discussion of Fig. 8.33). As the switching probability decreases, the input value remains constant longer and the reactor exhibits a larger magnitude response, implying the need for a more nonlinear model. Conversely, as p goes to zero, the input sequence exhibits few transitions between relatively long constant segments; in particular, for p small enough, the input sequence reduces to a single step or a constant with very high probability. Hence, an affine model with the correct steady-state gain and an appropriately chosen time constant should be able to match the response to such an input reasonably well, particularly for long sequences where the steady-state behavior is dominant. Consequently, it is to be expected that the nonlinearity of systems like the Eaton-Rawlings reactor model will be most pronounced at some intermediate input switching probability p. This point may be seen clearly in Fig. 8.29, where it appears that this maximum occurs for $p \sim 0.1$.

The input sequence distribution also has a significant influence on both the goodness-of-fit and the consistency-of-fit for the eleven models considered here. This point is illustrated in Fig. 8.31, which shows box-plots of the relative RMS prediction errors for Models 4 and 8 for the input sequences U010, B010, G010, and A010; the box-plots to the left of the vertical line describe Model 4 and those to the right of the vertical line describe Model 8. In both cases, considerable variation is seen in RMS prediction error with the input sequence distribution. In fact, the same general trend is seen for both Models 4 and 8: the relative prediction error increases in the order $G010 < A010 < U010 < B010$. Part of

The Art of Model Development

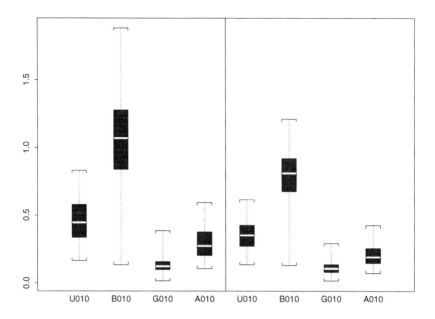

Figure 8.31: Comparison of Models 4 and 8, $p = 10\%$

this difference may be attributed to the differences in the variance of the different input sequences considered. In particular, recall that the fundamental criterion used to specify these input sequences was the total range of variation for the beta distribution that defined the probability of observing any possible value within that range. The distributional character of the resulting sequences— including their variance—is therefore defined by the shape parameters for the beta distribution. In fact, the standard deviations for the G, A, U, and B sequences may be computed from Eq. (8.41), yielding

$$\sigma_G \simeq 0.023, \quad \sigma_A \simeq 0.024, \quad \sigma_U \simeq 0.039, \quad \sigma_B \simeq 0.062.$$

Hence, it follows that the B and U input sequences "drive the process harder" than the G and A sequences, eliciting a more strongly nonlinear response. Conversely, the variability of the results obtained with these higher variance input sequences is also greater, particularly the B sequences.

Table 8.13 summarizes the relative model rankings obtained for the U010, B010, G010, and A010 input sequences on the basis of median RMS prediction error over the 100 simulations run for each input sequence class. As noted in the preceeding discussion, Models 4 and 8 consistently yield the best results for all of these input sequences, a point discussed further at the end of this section. At the other extreme, the affine reference model is generally the poorest one, exhibiting the highest relative RMS prediction error and generally covering a range of

Table 8.13: Model rankings *vs.* input distribution

Rank	U010	B010	G010	A010
1	8	8	8	8
2	4	4	4	4
3	5	5	1	1
4	1	10	3	5
5	3	9	2	3
6	9	7	7	7
7	7	6	6	2
8	10	3	9	6
9	2	1	0	9
10	6	2	5	0
11	0	0	10	10

large errors, relative to several of the other models. The only exception to this conclusion are, first, for the G010 input sequence, Models 5 and 10 actually exhibit larger median prediction errors than Model 0, and, second, for the A010 input sequence, Model 10 again exhibits a larger median prediction error than Model 0. This observation illustrates the point made earlier concerning the difference between the *one-step prediction error*—which is necessarily larger for the affine model, corresponding to a special case of Models 5 and 10 with $\gamma = 0$—and the *(total) prediction error*, used to compare models here. For good models, the differences between predicted and observed values of $y(k-i)$ should be small and these two error criteria should be approximately equal, but for the poorest approximations it is clear that these criteria can differ enough to yield unexpected rankings like the ones seen here.

It is the intermediate models that exhibit the greatest distributional dependence. For example, Model 10 ranks fourth best in terms of median prediction error for the B010 input but as was just noted, it exhibits even poorer performance than the affine reference model for the G010 and A010 inputs. Similarly, Model 5 ranks third best for the U010 and B010 inputs, next-worst for the G010 inputs, and fourth-best for the A010 inputs. It is also particularly interesting to compare Models 1 and 2: Model 1 is a quadratic Hammerstein model, whereas Model 2 is the modified Hammerstein model of Zhu and Seborg (1994). Although the example discussed in Chapter 5 showed little difference between these models, in *this* example there appears to be a significant difference. In particular, the Hammerstein model performs consistently better here than the modified Hammerstein model, although the extent of this difference varies significantly with the input sequence.

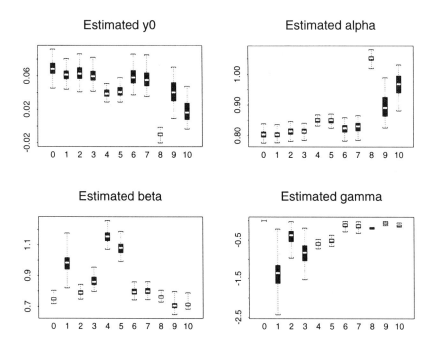

Figure 8.32: Summary of all parameter estimates, U010 inputs

Overall, the model rankings given in Table 8.13 provide useful guidance for model structure selection and practical insight into the influence of input sequence characteristics on this process. In this example, if enough model structures are considered to find good candidates (e.g., Models 4 and 8), the results are not strongly influenced by the choice of input sequence. Conversely, if practical considerations restrict consideration to intermediate models (e.g., Models 1, 2, and 3), the best choice *may* depend strongly on the input sequences used. Further, once a model structuree has been selected, the best fit parameters estimated for that structure appear to depend more strongly on the choice of input sequences, even for good models (e.g., Model 8 in this example). The remainder of this discussion is devoted to this point.

Fig. 8.32 summarizes the estimated y_0, α, β, and γ parameters obtained from the 100 simulations performed with U010 input sequences, indexed by model number. Recall that the constant term y_0 in the linearized model is $y_0 \simeq 0.045$ and note that Models 4 and 5 consistently exhibit y_0 estimates near this value, but Models 0, 1, 2, 3, 6, and 7 exhibit y_0 estimates that are both uniformly larger and more highly variable. Model 9 also exhibits a median value close to 0.045, but this estimate is also much more variable than that for Models 4 and 5; conversely, Model 8 exhibits the least variable parameter estimate, but at approximately −0.010 it is very far from the linearized parameter value. The results obtained for the affine model parameter α are qualitatively similar, but

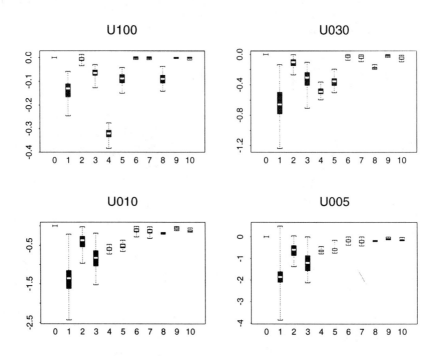

Figure 8.33: Estimated γ parameters, uniform steps

there are a few noteworthy differences. For example, the linearized parameter value is $\alpha \simeq 0.80$ and the estimates obtained for Models 0, 1, 2, and 3 are consistently close to this value. The estimates for Models 4 and 5 are the least variable, and again, the results obtained for Model 8 are noteworthy for their inconsistency with the other ten models: a value of $\alpha \simeq 1.05$ is obtained for Model 8, with variability consistent with that seen for Models 0, 1, 2, 3, 6, and 7. Generally greater dependence on model structure is seen in the β parameter. For the linearized model, this parameter has the value $\beta \simeq 1.10$ but only Models 4 and 5 yield estimates near this value; all of the other estimates are significantly smaller, including the affine Model 0. The most variable estimates are obtained for Model 1 and the least variable estimates are obtained for Models 0 and 8.

These last two conclusions extend to the estimated γ parameters: the most variable estimates are obtained for Model 1 and the least variable for Model 8. Fig. 8.33 gives an overall summary of the estimates of the γ parameter obtained from each of 100 identifications using the U100, U030, U010, and U005 input sequences. Note that since $\gamma = 0$ corresponds to the affine limit of all of the models considered here, the magnitude of the estimated γ parameter may be viewed as an approximate measure of the nonlinearity of the response of the Eaton-Rawlings reactor model to the specified input sequence. Comparing the U100 results with those obtained for the U030 input sequences, it is clear that the estimated γ parameters are consistently larger for the U030

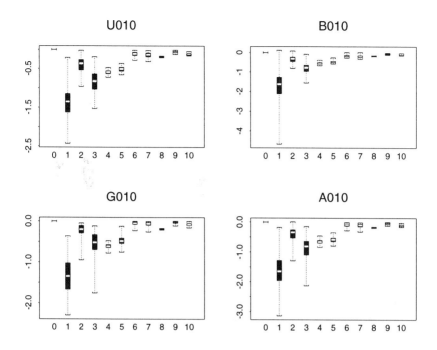

Figure 8.34: Estimated γ parameter, $p = 10\%$

sequences, suggesting that these responses are "more nonlinear," in agreement with the relative RMS error results discussed previously. It is particularly interesting to note the behavior of the Hammerstein model (Model 1): both the median estimated γ parameter and the range of parameter estimates increase as $p \to 0$. Conversely, for the modified Hammerstein model (Model 2), note that the estimated γ parameter is significantly smaller than for the unmodified Hammerstein model, consistent with the observation that Hammerstein model offers greater improvement in RMS error over the affine model than the modified Hammerstein model does, for all of the input sequences considered here.

The dependence of the estimated γ parameter on the input sequence distribution is illustrated in Fig. 8.34, which presents box-plots of the estimates obtained for all eleven models using the U010, B010, G010, and A010 input sequences. Here, the range of variation is smaller than it is with the switching probability p, but variations are significant nonetheless. It is interesting to note that Models 1 and 3 generally exhibit the largest median parameter values and that they generally also rank among the better models in terms of relative RMS error performance for all but one input sequence, not far behind the consistently superior Models 4 and 8. The sole exception is the B010 input sequence, for which Models 1 and 3 are among the worst, probably because Model 1 (the Hammerstein model) cannot be identified from binary input sequences (Pearson et al., 1996) and Model 3 is structurally quite similar to Model 1.

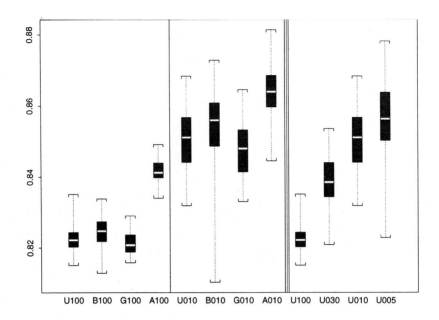

Figure 8.35: Estimated α parameters, Model 4

Fig. 8.35 presents a box-plot summary of the α parameter estimates for Model 4, illustrating their dependence on both the distribution (U, B, G, or A) and the shape (i.e., the switching probability p) of the input sequences. The leftmost four box-plots compare the estimates obtained for $p = 1.00$ and the most pronounced difference seen in these plots is between the asymmetrically distributed A100 sequence and the three symmetrically distributed inputs. The middle four box-plots summarize the results obtained for $p = 0.10$ and these plots show both a significant increase in the magnitude of the estimated α parameter and an increase in variability with decreasing p, for all distributions. The rightmost four box-plots illustrate this trend more clearly, comparing the results obtained with uniformly distributed random step sequences for $p = 1.00$, 0.30, 0.10, and 0.05. Analogous results are obtained for the parameters y_0, β, and γ for Model 4. In fact, roughly the same behavior is seen in all of the estimated parameters, for all of the model structures considered here, except the α and γ parameters for Model 8, which merit closer examination.

The estimated α parameters for Model 8 are shown in Fig. 8.36, presented in the same format as Fig. 8.35. Here, note that the parameter estimates obtained for $p = 1.00$ are significantly *more* variable than those obtained for $p = 0.10$, in marked contrast to the results just described for Model 4. Also, note that all of the α parameter estimates obtained for $p = 0.10$ (i.e., the middle four box-plots) violate the linear stability limit $|\alpha| < 1$ for the affine model, whereas the param-

The Art of Model Development 443

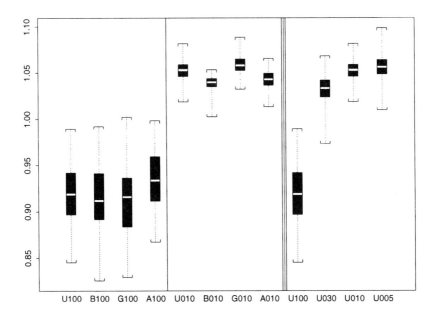

Figure 8.36: Estimated α parameters, Model 8

eters estimated for $p = 1.00$ are reasonably consistent with the (stable) affine model parameter estimates. Nevertheless, Model 8 is stable over the operating range considered here due to the influence of the nonlinear term $\gamma y^2 (k-1)$ for γ small and negative. Comparing the results for $p = 1.00, 0.30, 0.10$, and 0.05 (the right-most four plots) suggests that the variability in these parameter estimates achieves a minimum for $p \sim 0.1$. Analogous behavior is seen in the estimated γ parameter values, but the general behavior of the β and y_0 parameters is like that of the α parameter of Model 4 just described.

Overall, the results presented here suggest that Model 8 is the best of the ten nonlinear model structures considered, largely independent of the input sequence $\{u(k)\}$. In contrast, the relative rankings of many of the other models—for example, Models 1 and 5—depend strongly on the input sequence parameters, particularly the switching probability p. Also, the estimated parameter values are more consistent for Model 8 than for any of the other models considered. Finally, the dependence of the parameter variability for Model 8 on the switching probability is *opposite* in character of that observed for all of the other models. Taken together, these results suggest Model 8 is extremely interesting and possibly different in character from the other ten candidates considered here, thus raising the obvious question, Why?

To answer this question, consider the *Euler discretization* of the continuous-time Eaton-Rawlings model obtained by replacing the derivative dy/dt with the finite difference $[y(k) - y(k-1)]/T$ and replacing $y(t)$ on the right-hand side of the equation with $y(k-1)$:

$$\frac{y(k) - y(k-1)}{T} = -h[y^2(k-1) + 2\mu(k-1)y(k-1) - 2d\mu(k-1)]. \quad (8.82)$$

This expression may be rearranged to obtain

$$\begin{aligned} y(k) &= y(k-1) - hTy^2(k-1) - 2hT\mu(k-1)y(k-1) + 2dhT\mu(k-1) \\ &= y(k-1) - 0.150y^2(k-1) - 0.300\mu(k-1)y(k-1) + 0.525\mu(k-1). \end{aligned} \quad (8.83)$$

Suppose only one of the nonlinear terms in this equation is to be retained: since $y_s \sim 10\mu_s$ for the input sequence range considered here, the quadratic term in Eq. (8.83) is approximately five times larger than the bilinear term, so the quadratic term would be retained and the bilinear would be dropped. In fact, this approach leads to the structure of Model 8 and the corresponding parameter values are actually quite close to those obtained by dropping the bilinear term. Conversely, dropping the quadratic term and retaining the bilinear one leads to the structure of Model 4, although the best fit parameters obtained empirically for Model 4 are *not* particularly close to those given in Eq. (8.83).

Fig. 8.37 compares the prediction errors for three models in response to the same U010 input sequence: the Euler discretization defined by Eq. (8.83), Model 4, and Model 8. The solid line corresponds to the Euler discretization, which exhibits an RMS prediction error of approximately 0.511%. The solid triangles correspond to the prediction errors for Model 4, obtained by a least squares fit to the U010 input sequence considered here and its response from the exact Eaton-Rawlings reactor discretization. Here, the RMS prediction error is 0.159%, more than three times smaller than that of the Euler discretization. Similarly, the open circles represent the prediction errors for Model 8, obtained by the same procedure and having an RMS prediction error of 0.144%. For comparison, the RMS prediction error for the best fit Hammerstein model (Model 1) is 0.355%, more than twice that of Models 4 and 8 but still substantially better than the Euler discretization. Also, note that the RMS prediction errors associated with the analytic linearization described in Sec. 8.5.3 is 4.070%, an order of magnitude larger than that of the Euler discretization.

Overall, this case study may be viewed as a "toy" in many respects, not least of which is the extreme simplicity of the model on which it is based. Tukey (1987) notes that the term toy is "usually used for someone else's specification, and often implies a belief that the specification referred to is too special to be really helpful." Certainly, this criticism is justifiable here—the simplicity of the Eaton-Rawlings reactor model stands in marked contrast to the complexity of mechanistic models discussed at length in Chapter 1. Conversely, Tukey also makes the following observation:

> Notice that *starting* with a toy may be good judgment, and that even stopping *temporarily with it may be* necessary, *but that stopping with a toy* on a continuing basis is *not* acceptable.

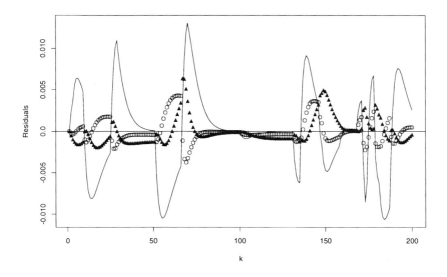

Figure 8.37: Prediction errors: Euler, Model 4, and Model 8

Here, the Eaton-Rawlings toy problem has been considered because it provides a nice illustration of a number of extremely important practical points. First and foremost, note that by considering a large enough collection of nonlinear empirical models (ten, in this example), one or two very good model structures were identified (Models 4 and 8). Further, these models were ultimately shown to have a natural connection to the original physical system (i.e., they approximated the Euler discretization of the continuous-time model) and the best of these models (Model 8) exhibited a much weaker dependence on the identification input sequence than the other models did. Conversely, it is important to note that the parameter estimates obtained exhibited a significant input-dependence, even for these two best models. Also, it is important to note that even the poorer model structures gave consistently better results than the Euler discretization and *substantially* better results than the analytic linearization.

8.6 Summary: the nature of empirical modeling

This chapter began with a simplified four-step description of the empirical modeling process and it is useful to revisit that procedure, comparing the problem of linear model identification with that of nonlinear model identification. In the linear case, the first step in this procedure—specification of a model structure—essentially reduces to one of specifying a linear model *representation*. For example, linear model predictive control applications have traditionally been based on truncated impulse response models (or equivalently, truncated step response models). In contrast, the design of optimal linear regulators generally requires a linear state-space model (Kwakernaak and Sivan, 1972). In any case, the choice of linear model structure is typically dictated by the intended application and strongly influences the choice of computational procedures used subsequently. The second step in the linear model development process—specification of an effective input sequence—has been the subject of substantial research effort and many results are available. For example, the book edited by Godfrey (1993) consists of 14 chapters and is primarily devoted to pseudorandom binary sequences for linear model identification, although other input sequences are also discussed, including sums of sinusoids, periodic signals based on linear chirps, and various stochastic sequences. The nature of the parameter estimation problem—the third step in this procedure—is largely determined by the initial model choice, which dictates both the parameters to be estimated and the algorithms available for estimating them. Finally, the fourth step in linear model development— validation of the result—may be approached from various perspectives, but an extremely important point is that goodness-of-fit is *not* an adequate criterion on which to base this validation. This point is illustrated clearly by the results of Tulleken (1993) who applied standard linear model identification procedures to input/output data from an industrial distillation process, obtaining many different models and rejecting almost all of them because their general qualitative behavior did not match that of the physical process. This problem was ultimately overcome through *constrained* linear model identification that guaranteed stability and other physically important behavioral attributes.

More generally, Johansen (1996) notes that, without some such physically-motivated constraints, empirical modeling is an ill-posed problem in the sense that the results need not be unique and they need not depend continuously on the available data. This last observation argues strongly against excessive reliance on data-dependent procedures for model structure selection, particularly in the case of nonlinear model development where the range of *potential* qualitative behavior is vast and largely unexplored. One of the principal aims of this book has been to provide some general insights into this range of behavior, focusing on a few nonlinear model classes that have been discussed fairly extensively in the modeling and control literature. It is hoped that these insights will prove useful in the first step of nonlinear model development, providing a basis for the selection of nonlinear model structures that are compatible with the dominant qualitative behavior of the physical system of interest.

Once a model structure has been tentatively chosen, it is important to generate input/output data using input sequences that are effective in either preliminary model structure comparisons or in subsequent parameter estimation. As discussed in Sec. 8.4, four basic design variables for input sequences are:

1. Sequence length
2. Sequence range
3. Sequence distribution
4. Sequence shape or frequency content.

Of these design variables, the last two are generally the least limited by practical constraints; general guidelines are to avoid sequences like sinusoids that are too regular and other choices like pseudorandom binary sequences that exhibit only a small number of discrete amplitude levels. Better alternatives are the random step sequences described in Sec. 8.4.3 or the determinstic sine-power sequences described in Sec. 8.4.4.

As in the case of linear models, the initial structure choices made in nonlinear model development essentially determine the range of possible computational procedures that can be used in fitting the model to available input/output data. For many popular model structures, this range is extremely broad and can be expected to increase in the forseeable future, particularly as ideas from robust statistics for reducing sensitivity to outliers (Dodge, 1987; Hampel, 1985; Huber, 1981; Rousseeuw and Leroy, 1987) find their way into the dynamic modeling literature (Dai and Sinha, 1989; Martin and Thomson, 1982; Pearson, 1999; Poljak and Tsypkin, 1980). In fact, the question of which data features should be regarded as anomalous and which ones should be regarded as essential is a central issue in nonlinear dynamic model development. As a specific example, Le et al. (1996) propose a non-Gaussian time-series model that can be used to predict series containing flat streches, bursts, and outliers but they also note that bilinear models (driven by white Gaussian noise) are capable of generating bursts and that threshold models (again, driven by white Gaussian noise) can generate flat stretches. The essential question in developing these models is, What is to be modeled? In the context of modeling for control applications, this question frequently reduces to one of whether the controller should respond to a particular data feature: if yes, the model should predict it accurately enough that the controller performs satisfactorily; if no, the prediction model should approximate the "true" response that has been obscured by the anomalous behavior precisely so the controller does *not* respond to this data feature. Clearly, answers to questions like this will influence both model structure selection and the choice of parameter estimation algorithm.

Once a tentative model has been developed, it is vital to assess its reasonableness, and one extremely helpful approach is the comparison of *several different model structures*. In particular, since linear (or affine) models have such a strong historical precedent, it is generally a good idea to compare the performance of any nonlinear dynamic model with that of a linear reference model. This linear model should be chosen either to optimize goodness-of-fit with respect to the same data used for nonlinear model identification or on the basis of historical acceptance (e.g., the Taylor-series linearization of a simplified mechanistic model). In addition, if "obvious" nonlinear model structures exist (e.g., Euler discretizations of simplified mechanistic models like the Eaton-Rawlings reactor model), these structures also provide reasonable candidates for consideration, although it is worth emphasizing that these model structures *may or may not* yield reasonable fits to observed input/output data. For example, recall from Chapter 1 (Sec. 1.1.3) the difference in qualitative behavior between the continuous-time Velhurst equation and its approximate discretization for certain choices of sampling times. Similarly, if a particular model structure offers significant application advantages like the modified Hammerstein model of

Zhu and Seborg (1994), it may be reasonable to include this structure in the collection just to see how well it approximates the process dynamics.

Another extremely useful observation concerning model validation is the following: exact optimization of one particular performance criterion often leads to unsatisfactory performance with respect to other important criteria (Tukey, 1987). As a simple example, Tukey notes that the designers of old-fashioned glass milk bottles had to consider multiple design criteria, including the amount of glass required for a bottle of specified volume, the ability to pack well in refrigerators, ease of cleaning and the ability to handle them without significant breakage. Consequently, although spheres minimize surface area requirements for given volume, the practical advantages of geometrically suboptimal (i.e., nonspherical) milk bottles generally outweighed the disadvantage of their additional glass requirements. To avoid excessive reliance on a single criterion, Tukey advocates the use of multiple objective approaches that seek acceptable performance with respect to several useful criteria rather than optimal performance with respect to one specific criterion, however important. In the context of the modeling problem considered in this book, it is reasonable to consider the following criteria in judging the performance of an empirical model:

1. Compatibility with the intended application
2. Goodness-of-fit with respect to identification data
3. Goodness-of-fit with respect to validation data
4. Goodness-of-fit compared to simpler models
5. Apparent reasonableness of the steady-state locus
6. Stability of linearized approximations
7. Qualitative nature of the extrapolation behavior
8. Outlier-sensitivity of model predictions
9. Outlier-sensitivity of model parameters
10. Compatibility with known qualitative behavior

Here, Criterion 1 refers to *structural compatibility* of the model with a particular model-based control strategy that is under consideration. Criteria 2, 3, and 4 may be based on any numerical goodness-of-fit measure, but if a particular measure has strong historical acceptance for a given application (e.g., mean square prediction error), this measure should at least be included among those considered. Criteria 5 and 6 may require a certain additional computational effort, but they are also likely to uncover unphysical model behavior if it is present, as in the results described by Tulleken (1993). Criterion 7 is intended to avoid the behavior illustrated in Chapter 1 (Sec. 1.1.4): the identified model fits the available data extremely well, but fails catastrphically outside this range. The question of what constitutes "extrapolation" is somewhat subtle, however, as extrapolated input sequences can include those covering a wider range of values, having markedly different distributions (e.g., uniformly distributed random steps vs. PRBS sequences with comparable switching patterns), or having significantly different frequency contents. Criteria 8 and 9 measure the robustness of the identified model; one possible implementation of this idea would be to apply a data cleaning procedure like the Hampel filter described in Sec. 8.1.2 to the input and output data, re-estimate the model parameters and compare the results, both with respect to model parameters and model predictions. Large differences might indicate the presence of outliers in the data and should be investigated; in particular, those data points modified by the data cleaning procedure should be examined from an application context to determine whether there is some *physical* basis for declaring them outliers. Finally, Criterion 10 refers to qualitative behavior like the presence or absence of inverse responses, monotonic or oscillatory step responses, input-dependent apparent time-constants, and other observable phenomena. Note that such an evaluation requires a careful analysis of both the model and the physical system under consideration since the qualitative behavior of both of these systems may be far from obvious. As a specific example, Kumar and Daoutidis (1999) develop a detailed differential-algebraic equation model for an ethylene glycol reactive distillation column, demonstrating that this model exhibits both a five-fold output multiplicity and a change from minimum-phase to nonminimum-phase behavior at high product purity; these conclusions were based on a careful analysis of a system of 79 ordinary differential equations and 54 algebraic relations.

Bibliography

Abramowitz, M. and Stegun, I. (1972). *Handbook of Mathematical Functions*. Dover.

Aczel, J. and Dhombres, J. (1989). *Functional Equations in Several Variables*. Cambridge University Press.

Adamek, J., Herrlich, H., and Strecker, G. (1990). *Abstract and Concrete Categories*. Wiley.

Adams, G. and Goodwin, G. (1995). Parameter estimation for periodic ARMA models. *J. Time Series Anal.*, 16:127–145.

Agarwal, R. (1992). *Difference Equations and Inequalities*. Marcel Dekker.

Arbel, A., Rinard, I., Shinnar, R., and Sapre, A. (1995). Dynamics and control of fluidized catalytic crackers. 2. multiple steady states and instabilities. *Ind. Eng. Chem. Res.*, 34:3014–3026.

Arce, G. and Gallagher, N. (1988). Stochastic analysis for the recursive median filter process. *IEEE Trans. Information Theory*, 34:669–679.

Arfkin, G. (1970). *Mathematical Methods for Physicists*. Academic Press.

Astola, J., Heinonen, P., and Neuvo, Y. (1987). On root structures of median and median-type filters. *IEEE Trans. Signal Proc.*, 35:1199–1201.

Astola, J. and Kuosmanen, P. (1997). *Fundamentals of Nonlinear Digital Filtering*. CRC Press.

Baheti, R., Mohler, R., and Spang, H. (1980). Second-order correlation method for bilinear system identification. *IEEE Trans. Automatic Control*, 25:1141–1146.

Bai, E.-W. (1998). An optimal two-stage identification algorithm for Hammerstein-Wiener nonlinear systems. *Automatica*, 34:333–338.

Banerjee, A., Arkun, Y., Ogunnaike, B., and Pearson, R. (1997). Estimation of nonlinear systems using linear multiple models. *AIChE J.*, 43:1204–1226.

Bangham, J. (1993). Properties of a series of nested median filters, namely the data sieve. *IEEE Trans. Signal Proc.*, 41:31–42.

Bartee, J. and Georgakis, C. (1994). Bilinear identification of nonlinear processes. In *Proc. ADCHEM'94*, pages 47–52, Kyoto, Japan.

Basser, P. and Grodzinsky, A. (1994). Electrostatic interactions between polyelectrolytes within an ultracentrifuge. In *Proc. 16th Annual Int'l. Conf. IEEE Engineering in Medicine and Biology Society*, volume 2, pages 750–751.

Bekiaris, N., Meski, G., Radu, C., and Morari, M. (1993). Multiple steady states in homogeneous azeotropic distillation. *Ind. Eng. Chem. Res.*, 32:2023–2038.

Bemporad, A. and Morari, M. (1999). Control of systems integrating logic, dynamics, and constraints. *Automatica*, 35:407–427.

Benallou, A., Seborg, D., and Mellichamp, D. (1986). Dynamic compartmental models for separation processes. *AIChE J.*, 32:1067–1078.

Bentarzi, M. and Hallin, M. (1994). On the invertibility of periodic moving-average models. *J. Time Series Anal.*, 15:263–268.

Bequette, B. (1991). Nonlinear control of chemical processes: A review. *Ind. Eng. Chem. Res*, 30:1391–1413.

Beran, J. (1994). *Statistics for Long-Memory Processes*. Chapman and Hall.

Berman, A. and Plemmons, R. (1979). *Nonnegative Matrices in the Mathematical Sciences*. Academic Press.

Billings, S. (1980). Identification of nonlinear systems—a survey. *IEE Proc., Part D.*, 127:272–285.

Billings, S. and Fakhouri, S. (1982). Identification of systems containing linear dynamic and static nonlinear elements. *Automatica*, 18:15–26.

Billings, S. and Voon, W. (1983). Structure detection and model validity tests in the identification of nonlinear systems. *IEE Proc., Part D*, 130:193–199.

Billings, S. and Voon, W. (1986a). Correlation based model validity tests for nonlinear models. *Int. J. Control*, 44:235–244.

Billings, S. and Voon, W. (1986b). A prediction-error and stepwise-regression estimation algorithm for nonlinear systems. *Int. J. Control*, 44:803–822.

Bird, R., Steward, W., and Lightfoot, E. (1960). *Transport Phenomena*. Wiley.

Blondel, V. and Tsitsiklis, J. (1999). Complexity of stability and controllability of elementary hybrid systems. *Automatica*, 35:479–489.

Bloomfield, P., Hurd, H., and Lund, R. (1994). Periodic correlation in stratospheric ozone data. *J. Time Series Anal.*, 15:127–150.

Bloomfield, P. and Steiger, W. (1983). *Least Absolute Deviations*. Birkhauser.

Blyth, T. (1986). *Categories*. Longman.

Bockris, J. and Reddy, A. (1970). *Modern Electrochemistry*, volume 1. Plenum.

Box, G., Jenkins, G., and Reinsel, G. (1994). *Time Series Analysis*. Prentice-Hall, 3rd edition.

Boyd, S. and Chua, L. (1985). Fading memory and the problem of approximating nonlinear operators with Volterra series. *IEEE Trans. Circuits Systems*, 32:1150–1161.

Brandt, J. (1987). Invariant signals for median-filters. *Utilitas Math.*, 31:93–105.

Brillinger, D. (1977). The identification of a particular nonlinear time series system. *Biometrika*, 64:509–515.

Brockwell, P. and Davis, R. (1991). *Time Series: Theory and Methods*. Springer-Verlag.

Bruni, C., DiPillo, G., and Koch, G. (1974). Bilinear systems: An appealing class of nearly linear systems in theory and applications. *IEEE Trans. Automatic Control*, 19:334–348.

Chen, H. (1995). Modeling and identification of parallel nonlinear systems: Structural classification and parameter estimation methods. *Proc. IEEE*, 83:39–66.

Chen, R., Liu, J., and Tsay, R. (1995). Additivity tests for nonlinear autoregression. *Biometrika*, 82:369–383.

Chen, R. and Tsay, R. (1993). Nonlinear additive ARX models. *J. Amer. Statist. Assoc.*, 88:955–967.

Chen, S., Billings, S., and Grant, P. (1992). Recursive hybrid algorithm for non-linear system identification using radial basis function networks. *Int. J. Control*, 55:1051–1070.

Chien, I.-L. and Ogunnaike, B. (1992). Modeling and control of high-purity distillation columns. In *Preprints,1992 Annual AIChE Meeting*, Miami.

Cho, Y. and Powers, E. (1994). Quadratic system identification using higher order spectra of i.i.d. signals. *IEEE Trans. Signal Proc.*, 42:1268–1271.

Corduneanu, C. (1991). *Integral Equations and Applications*. Cambridge University Press, New York.

Coxson, P. and Bischoff, K. (1986). Cluster analysis and lumpability of large linear systems. In *Proc. American Control Conf.*, pages 1187–1191, Boston.

Coxson, P. and Shapiro, H. (1987). Positive input reachability and controllability of positive systems. *Linear Algebra Appl.*, 94:35–53.

Curtain, R. and Pritchard, A. (1977). *Functional Analysis in Modern Applied Mathematics*. Academic Press.

Curtain, R. and Pritchard, A. (1978). *Infinite Dimensional Linear Systems Theory*. Springer-Verlag.

Cybenko, G. (1989). Approximation by superpositions of a sigmoidal function. *Math. Control Signals Systems*, 2:303–314.

Dai, H. and Sinha, N. (1989). Robust recursive least-squares method with modified weights for bilinear system identification. *IEE Proc., Part D*, 136:122–126.

Davies, L. and Gather, U. (1993). The identification of multiple outliers. *J. Amer. Statist. Assoc.*, 88:782–801.

Davis, H. (1962). *Introduction to Nonlinear Differential and Integral Equations*. Dover.

Dodge, Y. (1987). An introduction to statistical data analysis: L_1-norm based. In Dodge, Y., editor, *Statistical Data Analysis Based on The L_1-Norm and Related Methods*, pages 1–21. North-Holland.

Doherty, M. and Ottino, J. (1988). Chaos in deterministic systems: Strange attractors, turbulence, and applications in chemical engineering. *Chem. Eng. Sci.*, 43:139–183.

Draper, N. and Smith, H. (1998). *Applied Regression Analysis*. Wiley.

Duchesne, B., Fischer, C., Gray, C., and Jeffrey, K. (1991). Chaos in the motion of an inverted pendulum: An undergraduate laboratory experiment. *Amer. J. Phys.*, 59:987–992.

Eaton, J. and Rawlings, J. (1990). Feedback control of chemical processes using on-line optimization techniques. *Comput. Chem. Eng.*, 14:469–479.

Eek, R. (1995). *Control and Dynamic Modelling of Industrial Suspension Crystallizers*. PhD thesis, Delft University of Technology.

Efron, B. and Tibshirani, R. (1993). *An Introduction to the Bootstrap*. Chapman and Hall.

Elaydi, S. (1996). *An Introduction to Difference Equations*. Springer-Verlag.

Eskinat, E., Johnson, S., and Luyben, W. (1991). Use of Hammerstein models in identification of nonlinear systems. *AIChE J.*, 37:255–268.

Eskinat, E., Johnson, S., and Luyben, W. (1993). Use of auxiliary information in system identification. *Ind. Eng. Chem. Res.*, 32:1981–1992.

Fan, J. and Gijbels, I. (1996). *Local Polynomial Modelling and Its Applications*. Chapman and Hall.

Flaig, A., Arce, G., and Barner, K. (1998). Affine order-statistic filters: "medianization" of linear FIR filters. *IEEE Trans. Signal Proc.*, 46:2101–2112.

Foss, B., Johansen, T., and Sorensen, A. (1995). Nonlinear predictive control using local models - applied to a batch fermentation process. *Control Eng. Practice*, 3:389–396.

Fulton, W. (1969). *Algebraic Curves: An Introduction to Algebraic Geometry*. Benjamin.

Gaffuri, P., Faravelli, T., Ranzi, E., Cernansky, N., Miller, D., A.d'Anna, and Cjajolo, A. (1997). Comprehensive kinetic model for the low-temperature oxidation of hydrocarbons. *AIChE J.*, 43:1278–1286.

Gallagher, N. and Wise, G. (1981). A theoretical analysis of the properties of median filters. *IEEE Trans. Acoustics, Speech, Signal Proc.*, 29:1136–1141.

Georgakis, C. (1986). On the use of extensive variables in process dynamics and control. *Chem. Eng. Sci.*, 41:1471–1484.

Godfrey, K., editor (1993). *Perturbation Signals for System Identification*. Prentice-Hall.

Gradshteyn, I. and Ryzhik, I. (1965). *Table of Integrals, Series, and Products*. Academic Press.

Granger, C. and Anderson, A. (1978). *An Introduction to Bilinear Time Series Models*. Vandenhoeck and Ruprecht, Gottingen.

Gravador, E., Thylwe, K.-E., and Hökback, A. (1995). Stability transitions of certain exact periodic responses in undamped Helmholtz and Duffing oscillators. *J. Sound Vibration*, 182:209–220.

Greblicki, W. (1992). Nonparametric identification of Wiener systems. *IEEE Trans. Information Theory*, 38:1487–1493.

Greblicki, W. and Pawlak, M. (1989). Nonparametric identification of Hammerstein systems. *IEEE Trans. Information Theory*, 35:409–418.

Greblicki, W. and Pawlak, M. (1991). Nonparametric identification of nonlinear block-oriented systems. In *Preprints, 9th IFAC/IFORS Symposium*, pages 720–724, Budapest, Hungary.

Gross, F., Baumann, E., Geser, A., Rippin, D., and Lang, L. (1998). Modelling, simulation and controllability analysis of an industrial heat-integrated distillation process. *Comput. Chem. Eng.*, 22:223–237.

Guckenheimer, J. and Holmes, P. (1983). *Nonlinear Oscillations, Dynamical Systems, and Bifurcations of Vector Fields*. Springer-Verlag.

Güttinger, T., Dorn, C., and Morari, M. (1997). Experimental study of multiple steady states in homogeneous azeotropic distillation. *Ind. Eng. Chem. Res.*, 36:794–802.

Guzzella, L. and Amstutz, A. (1998). Control of diesel engines. *IEEE Control Systems Magazine, October*, pages 53–71.

Haggan, V. and Ozaki, T. (1980). Amplitude-dependent exponential AR model fitting for non-linear random vibrations. In Anderson, O., editor, *Time Series*, pages 57–71. North Holland.

Halanay, A. and Ionescu, V. (1994). *Time-Varying Discrete Linear Systems*. Birkhauser, Boston.

Hamer, J., Akramov, T., and Ray, W. (1981). The dynamic behavior of continuous polymerization reactors—ii: Nonisothermal solution homopolymerization and copolymerization in a CSTR. *Chem. Eng. Sci.*, 36:1897–1913.

Hamming, R. (1983). *Digital Filters*. Prentice-Hall, 2nd edition.

Hampel, F. (1985). The breakdown points of the mean combined with some rejection rules. *Technometrics*, 27:95–107.

Hardle, W. (1990). *Applied Nonparametric Regression*. Cambridge University Press.

Hastie, T. and Tibshirani, R. (1990). *Generalized Additive Models*. Chapman and Hall.

Hernandez, E. and Arkun, Y. (1993). Control of nonlinear systems using polynomial ARMA models. *AIChE J.*, 39:446–460.

Hlavacek, V. and van Rompay, P. (1981). On the birth and death of isolas. *Chem. Eng. Sci.*, 36:1730–1731.

Hoaglin, D., Mosteller, F., and Tukey, J. (1991). *Fundamentals of Exploratory Analysis of Variance*. Wiley.

Horn, R. and Johnson, C. (1985). *Matrix analysis*. Cambridge University Press.

Horton, R., Bequette, B., and Edgar, T. (1991). Improvements in dynamic compartmental modeling for distillation. *Comput. Chem. Eng.*, 15:197–201.

Huber, P. (1981). *Robust Statistics*. Wiley.

Hudson, J. and Mankin, J. (1981). Chaos in the Belousov-Zhabotinskii reaction. *J. Chem. Phys.*, 74:6171–6177.

Jacobsen, E. and Skogestad, S. (1991). Multiple steady states in ideal two-product distillation. *AIChE J.*, 37:499–511.

Johansen, T. (1996). Identification of non-linear systems using empirical data and prior knowledge—an optimization approach. *Automatica*, 32:337–356.

Johansen, T. and Foss, B. (1993). Constructing NARMAX models using ARMAX models. *Int. J. Control*, 58:1125–1153.

Johansen, T. and Foss, B. (1995). Identification of non-linear system structure and parameters using regime decomposition. *Automatica*, 31:321–326.

Johansen, T. and Foss, B. (1997). Operating regime based process modeling and identification. *Comput. Chem. Eng.*, 21:159–176.

Johnson, A. and Newman, J. (1971). Desalting by means of porous carbon electrodes. *J. Electrochem. Soc.*, 118:510–517.

Johnson, N., Kotz, S., and Balakrishnan, N. (1994). *Continuous Univariate Distributions*, volume 1. Wiley.

Johnson, N., Kotz, S., and Balakrishnan, N. (1995). *Continuous Univariate Distributions*, volume 2. Wiley.

Jones, D. (1978). Nonlinear autoregressive processes. *Proc. Roy. Soc., Ser. A*, 360:71–95.

Kantor, J. (1987). An overview of nonlinear geometrical methods for process control. In Prett, D. and Morari, M., editors, *Shell Process Control Workshop*. Butterworths.

Kayihan, F. (1998). Kappa number profile control for continuous digesters. In Berber, R. and Kravaris, C., editors, *Nonlinear Model Based Process Control*, pages 805–829. Kluwer.

Kazantzis, N. and Kravaris, C. (1997). System-theoretic properties of sampled-data representations of nonlinear systems obtained via Taylor-Lie series. *Int. J. Control*, 67:997–1020.

Keshner, M. (1982). $1/f$ noise. *Proc. IEEE*, 70:212–218.

Kirnbauer, T. (1991). *Nichtlineare prädiktive Regelung unter Verwendung von autoregressiven Volterra-Reihen zur Modellierung der Regelstrecke*. PhD thesis, Technischen Universität Wien.

Kirnbauer, T. and Jorgi, H. (1992). Nonlinear predictive control using Volterra series models. In *IFAC-NOLCOS 2 Proc.*, pages 558–562.

Klambauer, G. (1975). *Mathematical Analysis*. Marcel Dekker.

Koh, T. and Powers, E. (1985). Second-order Volterra filtering and its applications to nonlinear system identification. *IEEE Trans. Acoustics, Speech, Signal Proc.*, 33:1445–1455.

Kreyszig, E. (1978). *Introductory Functional Analysis with Applications*. Wiley.

Krischner, K., Lubke, M., Eiswirth, M., Wolf, W., Hudson, J., and Ertl, G. (1993). A hierarchy of transitions to mixed mode oscillations in an electrochemical system. *Physica D*, 62:123–133.

Kumar, A. and Daoutidis, P. (1999). Modeling, analysis and control of ethylene glycol reactive distillation column. *AIChE J.*, 45:51–68.

Kurth, J. (1995). *Identifikation nichtlinearer Systeme mit komprimierten Volterra-Reihen*. PhD thesis, RWTH Aachen.

Kwakernaak, H. and Sivan, R. (1972). *Linear Optimal Control Systems*. Wiley.

Le, N., Martin, R., and Raftery, A. (1996). Modeling flat stretches, bursts, and outliers in time series using mixture transition distribution models. *J. Amer. Statist. Assoc.*, 91:1504–1515.

Lee, J. (1998). Modeling and identification for nonlinear model predictive control: Requirements, current status and future research. In Allgöwer, F. and Zheng, A., editors, *International Symposium on Nonlinear Model Predictive Control: Assessment and Future Directions*, pages 91–107, Ascona, Switzerland.

Lee, J. and Mathews, V. (1994). A stability condition for certain bilinear systems. *IEEE Trans. Signal Proc.*, 42:1871–1873.

Leontaritis, I. and Billings, S. (1987). Experimental design and identifiability for nonlinear systems. *Int. J. Systems Sci.*, 18:189–202.

Leontartis, I. and Billings, S. (1985). Input-output parametric models for nonlinear systems. *Int. J. Control*, 41:303–344.

Lindskog, P. and Ljung, L. (1997). Ensuring certain physical properties in black box models by applying fuzzy techniques. In *Proc. SYSID'97*, Kitakyushu.

Ljung, L. (1987). *System Identification: Theory for the User*. Prentice-Hall.

Lundstrom, P. and Skogestad, S. (1995). Opportunities and difficulties with 5×5 distillation control. *J. Process Control*, 5:249–261.

MacKinnon, N. (1993). Problem 10319. *Amer. Math. Monthly*, 100:590.

MacLane, S. (1998). *Categories for the Working Mathematician*. Springer-Verlag, 2nd edition.

Mallows, C. (1980). Some theory of nonlinear smoothers. *Ann. Statist.*, 8:695–715.

Mandler, J. (1998). Modeling for control, analysis and design in complex industrial separation and liquefaction processes. In *Preprints, 5th IFAC Symposium Dynamics Control Process Systems*, pages 405–413, Corfu.

Martin, R. and Thomson, D. (1982). Robust-resistant spectrum estimation. *Proc. IEEE*, 70:1097–1114.

McLeod, A. (1994). Diagnostic checking of periodic autoregressive models with applications. *J. Time Series Anal.*, 15:221–233.

Meadows, E. and Rawlings, J. (1997). Model predictive control. In Henson, M. and Seborg, D., editors, *Nonlinear Process Control*, chapter 5, pages 233–310. Prentice Hall.

Mejdell, T. and Skogestad, S. (1991). Estimation of distillation compositions from multiple temperature measurements using partial-least-squares regression. *Ind. Eng. Chem. Res.*, 30:2543–2554.

Melcher, J. (1981). *Continuum Electromechanics*. M.I.T. Press.

Menold, P. (1996). Suitability measures for model structure identification. Diplomarbeit, Universität Stuttgart, Institut für Systemdynamik und Regelungstechnik.

Menold, P., Allgöwer, F., and Pearson, R. (1997a). Nonlinear structure identification of chemical processes. In *Proc. PSE'97-ESCAPE-7*, Trondheim, Norway.

Menold, P., Allgöwer, F., and Pearson, R. (1997b). On simple representations of distillation dynamics. In *Proc. ECCE-1*, Florence, Italy.

Michelsen, F. and Foss, B. (1996). A comprehensive mechanistic model of a continuous Kamyr digester. *Appl. Math. Modelling*, 20:523–533.

Mohler, R. (1991). *Nonlinear Systems: Volume II, Applications to Bilinear Control*. Prentice-Hall.

Morari, M. and Zafiriou, E. (1989). *Robust Process Control*. Prentice-Hall.

Morse, A. (1999). Introduction to the special issue on hybrid systems. *Automatica*, 35(3).

Morse, P. and Feschbach, H. (1953). *Methods of Theoretical Physics*, volume 2. McGraw-Hill.
Munkres, J. (1975). *Topology: A First Course.* Prentice-Hall.
Murray-Smith, R. and Johansen, T., editors (1997). *Multiple Model Approaches to Modelling and Control.* Taylor and Francis.
Nahas, E., Henson, M., and Seborg, D. (1992). Nonlinear internal model control strategy for neural network models. *Comput. Chem. Eng.*, 16:1039–1057.
Narendra, K. and Gallman, P. (1966). An iterative method for the identification of nonlinear systems using a Hammerstein model. *IEEE Trans. Automatic Control*, 11:546–550.
Narendra, K. and Parthasarathy, K. (1990). Identification and control of dynamical systems using neural networks. *IEEE Trans. Neural Networks*, 1:4–26.
Nayfeh, A. and Mook, D. (1979). *Nonlinear Oscillations.* Wiley, New York.
Nicholls, D. and Quinn, B. (1982). *Random Coefficient Autoregressive Models.* Springer-Verlag.
Nikias, C. and Petropulu, A. (1993). *Higher-order Spectra Analysis.* Prentice-Hall.
Nowak, R. and van Veen, B. (1994). Random and pseudorandom inputs for Volterra filter identification. *IEEE Trans. Signal Proc.*, 42:2124–2135.
Ogunnaike, B. and Ray, W. (1994). *Process Dynamics, Modeling, and Control.* Oxford.
Oldham, K. and Spanier, J. (1974). *The Fractional Calculus.* Academic Press.
Oppenheim, A. and Schafer, R. (1975). *Digital Signal Processing.* Prentice-Hall.
Pajunen, G. (1984). *Application of a model reference adaptive technique to the identification and control of Wiener type nonlinear processes.* PhD thesis, Helsinki University of Technology.
Pajunen, G. (1992). Adaptive control of Wiener type nonlinear systems. *Automatica*, 28:781–785.
Papoulis, A. (1965). *Probability, Random Variables, and Stochastic Processes.* McGraw-Hill.
Pawlak, M., Pearson, R., Ogunnaike, B., and Doyle, F. (1994). Nonparametric identification of a class of nonlinear time series systems. Presented at Bernoulli Soc. World Congress, Chapel Hill, North Carolina, USA.
Pearson, R. (1995). Nonlinear input/output modelling. *J. Process Control*, 5:197–211.
Pearson, R. (1998). Input sequences for nonlinear modeling. In Berber, R. and Kravaris, C., editors, *Nonlinear Model Based Process Control*, pages 599–621. Kluwer.
Pearson, R. (1999). Data cleaning for dynamic modeling and control. Accepted for 1999 European Control Conference, Karlsruhe, Germany,.
Pearson, R. and Ogunnaike, B. (1997). Nonlinear process identification. In Henson, M. and Seborg, D., editors, *Nonlinear Process Control*, chapter 2, pages 11–110. Prentice-Hall.
Pearson, R., Ogunnaike, B., and Doyle, F. (1992). Identification of discrete convolution models for nonlinear processes. In *Preprints, Annual AIChE Meeting*, Miami, Florida.
Pearson, R., Ogunnaike, B., and Doyle, F. (1996). Identification of structurally constrained second-order Volterra models. *IEEE Trans. Signal Proc.*, 44:2837–2846.
Pearson, R. and Pottmann, M. (1999). Gray-box identification of block-oriented nonlinear models. Submitted for publication.
Podlubny, I. (1999). Fractional-order systems and $PI^\lambda D^\mu$-controllers. *IEEE Trans. Automatic Control*, 44:208–214.
Poljak, B. and Tsypkin, J. (1980). Robust identification. *Automatica*, 16:53–63.

Pottmann, M. and Pearson, R. (1998). Block-oriented NARMAX models with output multiplicities. *AIChE J.*, 44:131–140.
Pottmann, M. and Seborg, D. (1992). Identification of non-linear processes using reciprocal multiquadric functions. *J. Proc. Control*, 2:189–203.
Priestley, M. (1981). *Spectral Analysis and Time Series*. Academic Press.
Priestley, M. (1988). *Non-linear and Non-stationary Time Series Analysis*. Academic Press.
Qin, S. and Badgwell, T. (1998). An overview of nonlinear model predictive control applications. In Allgöwer, F. and Zheng, A., editors, *International Symposium on Nonlinear Model Predictive Control: Assessment and Future Directions*, pages 128–145, Ascona, Switzerland.
Rao, T. S. (1981). On the theory of bilinear time series models. *J. Roy. Statist. Soc., Ser. B*, 43:224–255.
Rao, T. S. and Gabr, M. (1984). *An Introduction to Bispectral Analysis and Bilinear Time Series Models*. Springer-Verlag.
Ray, W. and Jensen, K. (1980). Bifurcation phenomena in stirred tanks and catalytic reactors. In Holmes, P., editor, *New Approaches to Nonlinear Problems in Dynamics*, pages 235–255. SIAM.
Richards, J. (1983). *Analysis of Periodically Time-Varying Systems*. Springer-Verlag.
Riedel, K. and Sidorenko, A. (1995). Minimum bias multiple taper spectral estimation. *IEEE Trans. Signal Proc.*, 43:188–195.
Robinson, P. (1977). The estimation of a nonlinear moving average model. *Stochastic Proc. Appl.*, 5:81–90.
Rousseeuw, P. and Leroy, A. (1987). *Robust Regression and Outlier Detection*. Wiley.
Rumelhart, D., McClelland, J., and the PDP Research Group (1986). *Parallel Distributed Processing*, volume 1. M.I.T. Press.
Russo, L. and Becquette, B. (1995). Impact of process design on the multiplicity behavior of a jacketed exothermic CSTR. *AIChE J.*, 41:135–147.
Saaty, T. (1981). *Modern Nonlinear Equations*. Dover.
Saaty, T. and Bram, J. (1964). *Nonlinear Mathematics*. Dover.
Saravanan, N., Duyar, A., Guo, T., and Merrill, W. (1993). Modelling of the space shuttle main engine using feed-forward neural networks. In *Proc. American Control Conf.*, pages 2897–2899.
Schmidt, A., Clinch, A., and Ray, W. (1984). The dynamic beahvior of continuous polymerization reactors — iii: An experimental study of multiple steady states in solution polymerization. *Chem. Eng. Sci.*, 39:419–432.
Scott, A. (1970). *Active and Nonlinear Wave Propagation in Electronics*. Wiley.
Seborg, D. and Henson, M. (1997). Introduction. In Henson, M. and Seborg, D., editors, *Nonlinear Process Control*, chapter 1, pages 1–9. Prentice Hall.
Segal, B. and Outerbridge, J. (1982). A nonlinear model of semicircular canal primary afferents in bullfrog. *J. Neurophysiology*, 47:563–578.
Shilov, G. (1974). *Elementary Functional Analysis*. M.I.T. Press.
Simonoff, J. (1996). *Smoothing Methods in Statistics*. Springer-Verlag.
Sriniwas, G., Arkun, Y., Chien, I.-L., and Ogunnaike, B. (1995). Identification and control of a high-purity distillation column: A case study. *J. Process Control*, 5:149–162.
Steen, L. and Seebach, J. (1978). *Counterexamples in Topology*. Dover.
Stoica, P. (1981). On the convergence of an iterative algorithm used for Hammerstein system identification. *IEEE Trans. Automatic Control*, 26:967–969.
Stoyanov, J. (1987). *Counterexamples in Probability*. Wiley.

Stromberg, K., Toivonen, H., Haggblom, K., and Waller, K. (1995). Multivariable nonlinear and adaptive control of a distillation column. *AIChE J.*, 41:195–199.
Su, H.-T. and McAvoy, T. (1993). Integration of multilayer perceptron networks and linear dynamic models: A Hammerstein modeling approach. *Ind. Eng. Chem. Res.*, 32:1927–1936.
Teymour, F. and Ray, W. (1989). The dynamic behavior of continuous solution polymerization reactors—iv: Dynamic stability and bifurcation analysis of an experimental reactor. *Chem. Eng. Sci.*, 44:1967–1982.
Teymour, F. and Ray, W. (1992). The dynamic behavior of continuous solution polymerization reactors—vi: Complex dynamics in full-scale reactors. *Chem. Eng. Sci.*, 47:4133–4140.
Tong, H. (1990). *Non-linear Time Series*. Oxford.
Tseng, C.-H. and Powers, E. (1995). Identification of cubic systems using higher order i.i.d. signals. *IEEE Trans. Signal Proc.*, 43:1733–1735.
Tukey, T. (1987). Configural polysampling. *SIAM Review*, 29:1–20.
Tulleken, H. (1993). Grey-box modelling and identification using physical knowledge and Bayesian techniques. *Automatica*, 29:285–308.
Uppal, A., Ray, W., and Poore, A. (1974). On the dynamic behavior of continuous stirred tank reactors. *Chem. Eng. Sci.*, 29:967–985.
van Lint, J. and Wilson, R. (1992). *A Course in Combinatorics*. Cambridge University Press.
Weischedel, K. and McAvoy, T. (1987). Feasibility of decoupling in conventionally controlled distillation columns. *Ind. Eng. Chem. Fund.*, 19:379–384.
Wendt, P. (1990). Nonrecursive and recursive stack filters and their filtering behavior. *IEEE Trans. Acoustics, Speech, Signal Proc.*, 38:2099–2106.
Wheeden, R. and Zygmund, A. (1977). *Measure and Integral*. Marcel Dekker.
Wigren, T. (1993). Recursive prediction error identification using the nonlinear Wiener model. *Automatica*, 29:1011–1025.
Wigren, T. (1994). Convergence analysis of recursive identification algorithms based on the nonlinear Wiener model. *IEEE Trans. Automatic Control*, 39:2191–2206.
Willems, J. (1991). Paradigms and puzzles in the theory of dynamical systems. *IEEE Trans. Automatic Control*, 36:259–294.
Wimp, J. (1984). *Computation with Recurrence Relations*. Pitman.
Yang, J. and Honjo, S. (1996). Modeling the near-freezing dichothermal layer in the Sea of Okhotsk and its interannual variations. *J. Geophysical Research*, 101(C7):16,421–16,433.
Zhang, S. and Ray, W. (1997). Modeling of imperfect mixing and its effects on polymer properties. *AIChE J.*, 43:1265–1277.
Zhu, Q. and Billings, S. (1993). Parameter estimation for stochastic nonlinear rational models. *Int. J. Control*, 57:309–333.
Zhu, Q. and Billings, S. (1994). Identification of polynomial and rational NARMAX models. In *Proc. SYSID'94*, pages 259–264, Copenhagen, Denmark.
Zhu, X. and Seborg, D. (1994). Nonlinear predictive control based on Hammerstein models. *Control Theory Appl.*, 11:564–575.

Index

affine model
  category, 333
  common Hammerstein-Wiener, 172
  definition, 247
  initial conditions, 254
  motivation, 247
Akaike Information Criterion (AIC), 386
AR-Volterra model, 148, 182, 184, 230,
    234, 420, 432
ARIMA model, 78
ARMA model
  category, 322
  definition, 61
  vs. homomorphic system, 358
ARMAX model, 17, 62
asymmetric responses (ASYM)
  CSTR dynamics, 43
  definition, 32
asymptotic constancy
  category, 329
  definition, 149
  homomorphic systems, 346
  linear conditions, 340
  linear vector space, 315
  NMAX models, 149
autoregressive model
  linear $AR(p)$, 61
  nonlinear, (see NARX model)
  random coefficient, 105

beta distribution, 413
BIBO stability
  category, 328
  definition, 67
  linear conditions, 339
  NAARX models, 178
  NARX models, 159
  NARX* models, 166
  neural NARMAX models, 200
  NMAX models, 149
  slow decay models, 76
  TARMAX models, 279
bilinear model
  burst phenomena, 95, 135
  completely bilinear, 96, 104, 238
  CSTR dynamics, 47
  definition, 95
  diagonal, 96
  distillation column, 40
  impulse response, 97
  NMAX, NARX comparison, 403
  non-homogeneity, 137
  stability conditions, 108
  state-space, 94
  static-linearity, 110, 133
  steady-state response, 109
  step response, 98
  subdiagonal, 96
  superdiagonal, 96, 238
  unstable subharmonics, 101, 107
  Volterra representation, 236
block-oriented model
  continuous- vs. discrete-time, 358
  definition, 25, 224
  Volterra equivalence, 226
box plots, 430
Brownian motion, 77, 102
burst phenomena
  bilinear models, 102, 135
  vs. outliers, 447

cascade connection
  composition of morphisms, 310
  of $MA(q)$ models, 311
  of NARX models, 152
  of NMAX models, 151
  of Volterra models, 312
  polynomial NARMAX, 313
category
  connected, 331

definition, 303
discrete, 309, 356
opposite, 348
thin, 330
category, specific
  ARMA, 322
  AR, 321
  ASYMP, 329
  Add$_\infty$, 367
  Aff, **Aff**^{ss}, 333
  BIBO, 328
  DTDM, 314
  Gauss, 334
  Hom$_\phi$, 324, 344
  H, 327
  INARX, 351
  LN$_\infty$, 325
  LN$_\infty^0$, 360
  L, 326
  MA, 321
  MONO, 329
  Median, 335
  NARX, 323
  NMAX, 323
  NState, 324
  PH, 327
  SL, 327
  Sand$_\psi$, 361
  State, 322
  TMAX$_\infty$, **TARX**$_\infty$, 367
  Volt, 324, 342
Cauchy's equation, 90
Cauchy's power equation, 345
chaotic responses (CHAOS)
  H^0-ARMAX models, 120
  impossible for NMAX models, 150
  in continuous-time models, 29
  in discrete-time models, 30
  in physical systems, 29
  Lur'e models, 402
  NARX models, 159
  TARMAX models, 283
class (vs. set), 304
continuous stirred tank reactor (CSTR)
  bilinear model, 47
  definition, 9
  Eaton-Rawlings model, 13
  Hammerstein model, 45
  linear multimodel, 49
  Lur'e model, 49

NARMAX model, 48
nonuniform mixing, 10
output multiplicity, 43
qualitative asymmetry, 43
TARMAX model, 129, 293
Uryson model, 50
Wiener model, 45
cross-validation, 387, 392

data pretreatment
  Hammerstein strategy, 385
  Hampel filter, 380
  linear FIR filters, 383
data sieve, 337
distillation column
  bilinear model, 40, 291
  Hammerstein model, 36, 292
  NARMAX model, 41
  qualitative dynamics, 35
  Wiener model, 38
Duffing equation, 27

Eaton-Rawlings reactor model
  affine empirical models, 429
  description, 424
  Euler discretization, 443
  exact discretization, 13, 425
  linearization, 428
  nonlinear empirical models, 432
EXPAR model
  definition, 174
  modified, 175
  output multiplicity, 162
exponential stability, 67, 316
extensive variable control, 40

fading memory systems, 241
finite escape time, 85, 191
finite impulse response (FIR)
  linear, 64, 383
  nonlinear, (see NMAX model)
fractional ARIMA model, 79
fractional Brownian motion, 78
fractional-order PID controller, 74
functor
  definition, contravariant, 348
  definition, covariant, 347
  inclusion functor, 347
  isomorphism, 353
  linearization functor, 349

model inversion functor, 351
realization functor, 354
uniform sampling functor, 356
fundamental model
  complexity of, 8
  definition, 7
  reduction, 11
  working assumptions, 10

generally selected multimodel
  definition, 253
  isola example, 272
  Wiener model, 275
gray-box model, 18

Hadamard product, 308
Hammerstein model
  comparison with Lur'e, 402
  CSTR dynamics, 45
  definition, 25
  distillation column, 36
  Eaton-Rawlings reactor, 432
  finite $H_{(N,M)}$ class, 219
  limitations, 38
  linear multimodel, 274
  structural additivity, 170
Hammerstein strategy, 385
Hammerstein vs. Wiener
  affine common class, 172
  common linearization, 350
  CSTR dynamics, 45
  distillation column, 39
  linear multimodels, 274
  NARX vs. NARX* models, 173
  sinusoidal responses, 26
  steady-state agreement, 26
Hampel filter, 380
harmonic, (see superharmonic)
homogeneity (HOM)
  category, 327
  definition, 32, 90
  H^0-ARMAX models, 117
  homogeneous functions, 114
  homogeneous models, 111
  impossible for bilinear models, 137
  median filter, 147
  vs. homomorphic systems, 359
  vs. polynomial NARMAX, 300
  vs. Volterra, 300
homogeneous function of order ν, 115

homomorphic system
  asymptotic constancy, 346
  category, 324, 344
  definition, 115
  homogeneous, 116, 359
  limitations, 116, 120
  linear system isomorphism, 358
  positive-homogeneity, 346
  static-linearity, 344
  vs. NARX models, 165
hybrid systems, 292

impulse response
  and initial objects, 331
  bilinear models, 97
  H^0-ARMAX models, 118
  homomorphic systems, 117
  input-selected multimodel, 257
  linear models, 64
  matching vs. linearization, 350
  median filter, 146
infinite-dimensional model
  BIBO stability, 77
  diffusion phenomena, 72
  fractional Brownian motion, 78
  linear, 8
  nonexponential decay, 74
  nonrational transfer functions, 74
  Volterra model, 211, 233
initial conditions
  affine models, 254
  autoregressive simulations, 72
  standard, 63
initial object, 330
input multiplicity
  definition, 33
  generally selected examples, 269
  impossible for NARX models, 162
  input-selected examples, 260
  TARMAX models, 282
  vs. output-selected models, 264
input sequences
  and initial objects, 330
  design criteria, 404
  distribution vs. shape, 409, 421
  persistently exciting, 408
  random steps, 391, 414
  screening inputs, 370, 405, 419
  sine-power sequences, 417
  standard four, 68

Index

input-dependent stability (IDS)
  bilinear models, 31
  definition, 30
  hydrocarbon oxidation, 30
  Lur'e models, 402
  NARX models, 158, 204
  subtle NARMAX example, 186
input-selected multimodel
  definition, 253
  Hammerstein model, 274
  impulse vs. step responses, 257
  relation to NMAX models, 255
  steady-state, 256
inverse response
  definition, 58
  in reduced models, 12
isola
  definition, 35
  linear multimodel, 272
  NARX* example, 165
isomorphism
  between **AR** and **MA**, 354
  between $\mathbf{C}^W$ and $\mathbf{D}^T$, 357
  between $\mathbf{Hom}_\phi$ and **ARMA**, 358
  functor, 353
  morphism, 335

LARNMAX model
  AR-Volterra models, 182
  definition, 148
  PPOD structure, 230
linear model
  categories, 320, 326
  definitions, 89
linear multimodel
  continuous-time example, 244
  CSTR dynamics, 49
  definition, 49, 248
  generally selected, (see generally selected multimodel)
  H^0-ARMAX example, 119
  input-selected, (see input-selected multimodel)
  Johansen-Foss, 248
  modified Johansen-Foss, 250
  NARX example, 153
  output-selected, (see output-selected multimodel)
  qualitative behavior, 403
  Tong's TARSO class, 251

linear vector space, 315
Lipschitz condition, 160
Lur'e model
  chaotic responses, 402
  comparison with Hammerstein, 402
  CSTR dynamics, 49
  definition, 19
  gray-box modeling, 19
  input-dependent stability, 402
  structural additivity, 169
  subharmonic generation, 402

Mallows' data smoothers, 136
median, 380
median absolute deviation (MAD), 381
median filter
  category, 335
  definition, 136
  extensions, 337, 361
  qualitative behavior, 145
  root sequences, 146, 335
morphism
  composition law, 303
  definition, 303
  isomorphism, 335
moving average
  linear $MA(q)$, 61
  nonlinear, (see NMAX model)

NAARX model, (see structurally additive model)
NARLMAX model, 162
NARMAX model
  additive, (see structurally additive model)
  classes of, 141
  definition, 18, 141
  polynomial, (see polynomial NARMAX model)
  rational, (see rational NARMAX model)
NARX model
  BIBO stability, 159
  category, 323
  definition, 152
  instability condition, 158
  inverse model, 152, 351
NARX* model
  BIBO stability, 166
  definition, 153
  vs. NARX model, 163

neural network
  approximations, 201
  feedforward network, 198
  NARMAX model, 199
NMAX model
  category, 323
  definition, 143
  discontinuous, 209
  qualitative behavior, 148
NMPC, 6, 230
noise, $1/f$, 77
noise, white, (see white noise)
nonexponential decay
  linear models, 74
  rational NARMAX models, 193
nonlinear function
  additive, 90
  asymmetric, 25
  continuous, 149
  even, 25
  homogeneous, 90
  invertible, 148
  Lipschitz, 160
  odd, 25
  positive-homogeneous, 123
  radially unbounded, 159
  rational, 186
  sector-bounded, 161
  static nonlinearity, 24
nonlinearity criterion
  Billings and Voon, 218
  Eek's graphical method, 33

object, 303
Ockham's razor, 386
outliers
  direct empirical modeling, 21
  Mallows' data smoothers, 136
  median filter response, 146
  real data examples, 376
  sensitivity of least squares, 373, 376, 393
  vs. burst phenomena, 447
output multiplicity
  CSTR dynamics, 43
  definition, 33
  generally selected examples, 269
  impossible for NMAX models, 148
  NARX models, 204
  output-selected examples, 266

TARMAX models, 282
  vs. input-selected models, 256
output-selected multimodel
  definition, 253
  relation to NARX models, 264
  steady-state, 264

parallel connection
  of additive NMAX models, 179
  of NMAX models, 151
  of Volterra models, 212
PARMA model, (see time-varying linear models)
periodic component removal, 383
Pexider equation
  homomorphic $NARX(1)$ model, 165
  Wiener vs. Hammerstein, 171
PHADD model, 288
phenomenological model, 10
polynomial NARMAX model
  CSTR dynamics, 48
  definition, 180
  distillation column, 41
  subtle instability, 186
positive systems, 287, 342
positive-homogeneity (PHOM)
  category, 327
  homomorphic systems, 346
  of p-norms, 122
  PH^0-ARMAX models, 125
  PHADD models, 288
  PHOM functions, 122
  PHOM models, 121
  Pythagorean models, 122
  TARMAX models, 50, 127, 276
  Uryson models, 52
  vs. polynomial NARMAX, 300
  vs. Volterra, 300
power set, 328
PPOD model
  AR-Volterra model, 230
  definition, 230
  modified Hammerstein model, 231
  Volterra representation, 232
projection-pursuit model
  definition, 197
  finite $P^r_{(N,M)}$ class, 223
  finite $V_{(N,M)}$ equivalence, 223
pseudorandom binary sequence (PRBS), 373, 375, 448

Pythagorean model, 122

qualitative behavior
 bilinear vs. NMAX vs. NARX, 403
 explicit model constraints, 388
 Hammerstein vs. Lur'e, 402
 linear multimodels, 403
 summary of types, 396

radial basis functions, 202
rational NARMAX model
 definition, 188
 structurally additive, 192
 Zhu and Billings example, 189
rectification, 24, 102, 216
RMS-to-DC converter, 122
root sequences
 as ineffective inputs, 407
 median filter, 146, 335
 vs. initial objects, 336

sandwich model, 225, 361
semi-integral equation, 74
state-space model
 bilinear, 31
 category **NState**, 324
 category **State**, 322
 control-affine, 32
 discrete, 12, 13
 fermentation reactor, 245
 linear, 5, 7, 66
 nonlinear, 7, 9
 pendulum, 30
static-linear model
 bilinear models, 110, 133
 category, 327
 definition, 130
 homomorphic systems, 344
 relation to homogeneity, 131
 TARMAX models, 282
 Uryson model, 132
 Volterra models, 343
steady-state
 bilinear models, 109
 gain curve, 25
 gain zero, 59
 input-selected multimodels, 256
 matching, 12, 19
 multiplicity (SSM), 33
 NAARX models, 179

NARX models, 162
NMAX models, 148
output-selected multimodels, 264
qualitative mismatch, 19
structurally additive model
 BIBO stability, 178
 definition, 168
 factorization result, 177
 Hammerstein models, 170
 Lur'e models, 169
 NMAX-Uryson equivalence, 179
 positive-homogeneous, 288
 rational NARMAX, 192
 subset of projection-pursuit, 197
 TARMAX models, 128, 276
 Wiener exclusion, 170
subcategory
 definition, 319
 IO subcategory, 329
 joint subcategory, 337
subharmonic
 bilinear instability, 107
 generation (SUB), 26
 generation conditions, 24, 26
 Lur'e models, 402
 NARX models, 158
 NMAX impossibility, 150
 pendulum example, 28
 sea-ice dynamics, 28
superharmonic
 definition, 23
 generation (HARM), 22
superposition condition, 89

TARMAX model
 chaotic responses, 283
 CSTR dynamics, 50
 definition, 127, 276
 multimodel representation, 277, 280
 qualitative behavior summary, 403
 static-linearity, 282
terminal object, 330
threshold autoregressive model, 153
time-varying linear models
 cyclostationarity, 87
 first-order, 81, 85
 Hill's equation, 86
 Meissner equation, 87
 PARMA models, 87

unit root model, 77, 419
Uryson model
   additive NMAX models, 179
   CSTR dynamics, 50
   definition, 50
   finite $D_{(N,M)}$ equivalence, 221
   finite $U^r_{(N,M)}$ class, 220
   TARMAX representation, 285

validity function, 246
Volterra model
   block-oriented equivalence, 226
   category, 324, 342
   diagonal $D_{(N,M)}$ class, 219
   examples, 144
   finite $V_{(N,M)}$ class, 211
   infinite-dimensional, 233
   odd-symmetry, 212, 217
   ordered representation, 227
   projection-pursuit equivalence, 223
   pruned, 228

static-linearity, 343
symmetric representation, 227

white noise
   beta distributed, 413
   Gaussian, 62, 77, 101
   interpretation, 411
   vs. sine-power sequences, 422
Wiener model
   CSTR dynamics, 45
   definition, 26
   distillation column, 38
   finite $W_{(N,M)}$ class, 221
   invertible nonlinearity, 172
   linear multimodel, 275
   non-additivity, 170
   non-NARMAX example, 173

zero dynamics, 167, 317
zero object, 330